THEORETICAL AND APPLIED ASPECTS OF EYE MOVEMENT RESEARCH

ADVANCES
IN
PSYCHOLOGY
22

Editors

G. E. STELMACH

P. A. VROON

NORTH-HOLLAND
AMSTERDAM · NEW YORK · OXFORD

THEORETICAL AND APPLIED ASPECTS OF EYE MOVEMENT RESEARCH

Selected/Edited Proceedings of
The Second European Conference on
Eye Movements, Nottingham, England
19-23 September, 1983

Edited by

Alastair G. GALE

Division of Radiology
Queen's Medical Centre
Nottingham
England

and

Frank JOHNSON

Oxford Medical Systems
Abingdon
Oxford
England

1984

NORTH-HOLLAND
AMSTERDAM · NEW YORK · OXFORD

ISBN: 0 444 87557 3

Publishers:
ELSEVIER SCIENCE PUBLISHERS B.V.
P.O. Box 1991
1000 BZ Amsterdam
The Netherlands

Sole distributors for the U.S.A. and Canada:
ELSEVIER SCIENCE PUBLISHING COMPANY, INC.
52 Vanderbilt Avenue
New York, N.Y. 10017
U.S.A.

Library of Congress Cataloging in Publication Data

European Conference on Eye Movements (2nd : 1983 : Not-
    tingham, Nottinghamshire)
    Theoretical and applied aspects of eye movement
research.

    (Advances in psychology ; 22)
    Bibliography: p.
    Includes indexes.
    1. Eye—Movements—Congresses. I. Gale, Alastair G.
II. Johnson, Frank. III. Title. IV. Series: Advances
in psychology (Amsterdam, Netherlands) ; 22.
QP477.5.E9 1983   612'.846   84-10240
ISBN 0-444-87557-3

PRINTED IN THE NETHERLANDS

PREFACE

When Galileo protested 'but does it move' he may well have been describing that much smaller globe, the eye. For this motion, so necessary for our vision, is a fascination to which the work here attests. The study of eye movements encompasses researchers from different scientific disciplines and progresses in relatively disparate directions ranging from fundamental research to the problems encountered in various visual tasks. Thus while on the one hand state of the art technology allows such niceties as display modification during a saccade, in other situations far less complex arrangements can also contribute to our understanding of visual behaviour. With such diversity in mind we set out to organise a Conference which would bring together representatives from these different areas in the belief that we could all learn something new from one another.

In September 1983 the Second European Conference on Eye Movements was held at the Queen's Medical Centre, Nottingham, England. The Conference was the second meeting of the European Group for Eye Movement Research and was the successor to the meeting in Bern in 1981. It was run under the auspices of the Biological Engineering Society in association with the Applied Vision Association. Originally envisaged as a small European Conference the response was such that it rapidly grew into an international meeting which still met our original aim of promoting the wider exchange of information concerning eye movement research in all its diverse fields. Some 160 delegates attended and were presented with 66 papers over the three day period. This volume presents the edited proceedings of that meeting and its title reflects the wide range of topics addressed by the authors. The chapters are gathered together in sections adhering to the conference structure. Four of these were collections of individually submitted papers and five were organised symposia, namely; properties of the saccadic eye movement system, medical image perception, visual search, reading and peripheral vision and neurophysiology of eye movements. Each section has a short introduction by the appropriate chairperson.

The organisation of the Conference and subsequent production of this volume could not have succeeded without the dedicated help of a few individuals to whom we would like to express our gratitude: Mrs. Elaine Wilkin and Mrs. Evelyn Pawley have tirelessly performed the miriad tasks so necessary to the smooth running of an international meeting and in the preparation of these proceedings. Mr. Keith Copeland, Mrs. Margot Gale and Mukesh Patel assisted greatly with the Conference arrangements. Professor Brian Worthington allowed us to overrun his department with the paperwork. Professor Rudolf Groner, the originator of the European Group for Eye Movement Research, offered much helpful advice. Dr. John Findlay, Dr. Geoffrey Underwood, Dr. Laurence Harris and Mr. Murray Sinclair acted as symposia organisers and Dr. Ted Megaw, Mr. Keith Copeland and Dr. Deborah Levy kindly agreed to be chairpersons for the other sessions. Dr. Kees Michielsen of Elsevier Science Publishers B.V. (North-Holland) has been most helpful with the preparation of this work. Finally special appreciation is accorded to Professor E. Llewellyn Thomas who was our guest at the Conference and whose address concluded the meeting.

Alastair Gale
Frank Johnson
Nottingham
March, 1984

CONTENTS

Section 3
READING AND PERIPHERAL VISION

Section 4
EXAMINING DISPLAYS

Section 9
THEORETICAL ASPECTS

Report: THE RECORDING EYE

Section 1

EYE MOVEMENT RECORDING

Theoretical and Applied Aspects of Eye Movement Research
A.G. Gale and F. Johnson (Editors)
© Elsevier Science Publishers B.V. (North-Holland), 1984

EYE MOVEMENT RECORDING

Frank Johnson

Oxford Medical Systems,
Abingdon, England.

This section contains contributions on three aspects of instrumentation.
Two papers (Frietman and Frecker) describe eye movement recording techniques. Two papers address the problems of analysis (Widdel and Kliegl), and
the paper by McConkie is concerned with using eye movement recording as a
basis for making stimulus manipulations.

At a conference where NAC revealed their latest "Eye Mark" system which included 3 solid-state head-mounted cameras, it may be wondered whether
worthwhile instrumentation development is maintained by individual research
workers. These papers indicate that valuable work is being done to improve
the hardware.

The Eye-sistant described by Frietman develops the processing of the signals
from reflection to provide an accuracy of 6 minutes of arc. No drift is
reported and the bandwidth extends to 1.5 kHz.

The system is used as a communication aid and has easy extension to function
as a pupillometer. Head movement was coped with by measuring the position
of further light sources positioned on the frames of the device. This
could resolve head movements to 1.5 minutes of arc over 150 mm movement
range.

Frecker and colleagues describe a system which is the result of ten years
development along many avenues at the University of Toronto. The system
utilizes the first Purkinje image formed when infra-red light is reflected
from the eye. The position of the image is recorded in two dimensions by
linear detector arrays, processed using techniques analogous to those used
in gamma cameras. This system records eye movements with a noise level of
30 arc seconds over the dynamic range of 30 arc seconds to 36 degrees.
Applications of this device are to psychopharmacology and contingent display control is feasible.

The volume of data generated by the detectors is recognised by the papers
by Widdel and Kliegl. Widdel first addresses the problem of definition of
a fixation. This has been rarely considered in previous work and yet is
fundamental to any work on analysis of eye movements. He recommends appropriate window sizes for visual search experiments.

The modular programmes of Kliegl form a valuable contribution towards a
generalised approach to eye movement analysis. Routines which many laboratories have written are included in a cohesive suit of programmes and
Kliegl was willing to make them available to other research workers.

The use of contingent stimulus control was mentioned by Frecker and is
fully considered by McConkie and colleagues. His primary concern was in
the choice of a display which could react within a sufficiently short time.
The eye movement recording system clearly has to have a comparable real-time
response. Performance using a system with a throughput of 4 msec. was

described.

This section on instrumentation confirms that there is more work to be done and demonstrates the contributions being made by individuals. Despite growing commercial interest there remains scope for development by research workers.

Theoretical and Applied Aspects of Eye Movement Research
A.G. Gale and F. Johnson (Editors)
© Elsevier Science Publishers B.V. (North-Holland), 1984

5

THE DETECTION OF EYEBALL MOVEMENTS WITH THE EYE-SISTANT

Edward E.E. Frietman

Dept.of Applied Physics, Delft University of Technology, Lorentzweg 1,2628 CJ

Maarten M. Joon, Gijs K. Steenvoorden

Institute of Applied Physics, Stieltjesweg 1, 2600 AD, Delft, Holland

The EYE-SISTANT[1], a portable instrument developed for the detection of eyeball movements detects horizontal and vertical eyeball movements separately on a non-contact basis. The final result, expressed in analogue form, is proportional to the amount of reflected Infra Red (IR) energy coming from the iris, pupil and sclera. The IR sources and light-sensitive elements are mounted on oculist spectacles, making the device also suitable for persons wearing spectacles. The EYE-SISTANT has already been applied in fundamental research and in the domain of communication aids for the disabled.

INTRODUCTION

The EYE-SISTANT consists of two separate sensor assemblies, suitable for the detection of horizontal and vertical eyeball movements, mounted on a pair of spectacle frames (see Fig. 1). The electronics perform the amplification, analysis and processing of the signals derived from the sensor assemblies (see Fig. 2). Both eyes are simultaneously lit by the energy from two different Infra Red Light Emitting Diode (IRLED) parts. The IR sources used are modulated with a 5 kHz square wave, in order to make the detection of the eyeball movements independent of noise and disturbances from surrounding light sources. Horizontal eyeball movements are sensed by one pair of silicon Photo Transistors (PTR's), which measure directly the difference in reflectivity of the iris-to-sclera boundary, called the limbus. Fig. 1a indicates the detection areas of the horizontal part. Both the PTR's and the IR sources are mounted in front of the eyes so as to

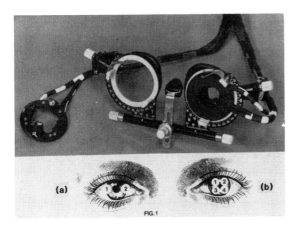

Fig. 1. Photograph of the detection areas, as seen by the PTR's, and the sensor assemblies mounted in a Universal Measuring frame.

minimise obstruction of the field of view, while maintaining the capability
to accurately monitor the position of the eyeball. In this way a field of
view of about 18° to the left and right is assured. Vertical eyeball movements are sensed by two pairs of silicon PTR's, which measure the difference
in reflectivity of the pupil-and-iris boundaries. Fig. 1b indicates the detection areas of the vertical part. In this case the field of view will be
8° (upwards) and -10° (downwards). The reason for the asymmetry is that the
upper eyelid usually covers the upper part of the eye itself, which results
in a smaller deflection area upwards. The final result in the form of an
analogue voltage is obtained by the technique of synchronous detection.
Safety precautions are taken to prevent any part of the eyes from being damaged by the IR energy. The IRLED source is adjusted in such a way that the
energy produced is less than the maximum given by the American National
Standard Institute[2] (ANSI 1978 and 1980). Precautions are also taken to make
the instrument fail-safe. The standards of the IEC 601/1 guarantee the
electrical safety.

EYEBALL MOVEMENT DETECTION.

Basic Principles:
In both horizontal and vertical cases the recording of eyeball movements
(Dutch patent NO.0.A.78.01616) is based on the detection of differences in
the reflection of diffused circular surface areas located on the eyes, as
can be seen in Fig. 1a and 1b. In the case of the detection of the horizontal eyeball movement the input to the synchronous detector (SD) consists of
merely the combined result of two PTR's (see Fig. 2). For the vertical eye-

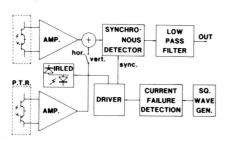

ball movement detection, however, the
SD is provided with a combination of
signals coming from four PTR's (see
Fig. 2). This means that the EYE-SIS-
TANT must consist of at least two separately functioning analysing parts,
which will provide signals proportional to the horizontal and vertical
eyeball movements. The SD contains a
suppressed carrier multiplier. The
output signal is then fed into a low
pass filter (LPF), whose function is
to remove higher-order harmonics and
their derivitives.

Fig. 2. Block diagram of the
EYE-SISTANT

Both IRLED sections, driven in a chopped mode, illuminate the eyes homogeneously with monochromatic light with a wavelength of 930 nm, so that no
pupil diameter variations will occur. The amount of energy produced is
7 mW/cm^2. The temperature rise of the cornea due to the irradiation has been
measured by means of thermography. In an unfavourable situation it will rise
0.4° C at the most. The values lie within the physiological limits of the
normal variations in cornea temperature. A current failure part will switch
off the current through the IRLED in case of a malfunction.

Horizontal eyeball movement detection:
A shift of the eyeball in horizontal direction causes an alteration of the
reflected energy due to the change of the reflection coefficients of the
iris-to-sclera boundary (see Fig. 1). This change, characterized by S_n, is
detected by the two PTR's and presents a different output signal e_h.

$$S_n = r_i A_{ni} + r_s A_{ns} \qquad\qquad n=1,2 \qquad\qquad (1)$$

where r_i and r_s represent the reflection coefficients of iris and sclera and A_i and A_s their corresponding detection areas.

$$e_h = \alpha_1 (S_1 - S_2) \qquad\qquad\qquad (2)$$

with α_1 as a constant to match the dimensions. The indices 1 and 2 refer to the circular areas 1 and 2 in Fig. 1a. The position of the two PTR's with respect to the eye is such that no virtual image of the IRLED is allowed to be detected. A most demanding condition required for a linear relation between the eyeball movements and output signal is to avoid direct radiation from the IRLED's into the PTR's. Wrongly positioned components will cause a severe distortion.

To decrease the influence of disturbances, such as 50 and 100 Hz from surrounding light sources, the technique of synchronous detection is used to restore information. The input of the SD consists of a Pulse-Amplitude-Modulated (PAM) signal, carrying amplitude information proportional to the position of the eyeball with respect to the two detection areas (see Fig. 1a), together with the information containing the disturbances. The output voltage e_h of the SD is at every moment proportional to the angular roll β_h of the eyeball. The contribution of the disturbances to the output voltage e_h is negligible. The spurious spectral components coming from the harmonics of the chopped signal and the contribution of the disturbances are removed by an LPF with a cutoff frequency of 1,5 kHz and a slope of 48 dB/octave. Filtering can be done without affecting the spectrum of the original signal because of the fact that both spectra fall in different regions (see also the section on spectral analysis).

Vertical eyeball movement detection:
The vertical eyeball movements, on the other hand, are measured by four PTR's by collecting the diffused IR energy from the iris-pupil border (see also Fig. 2). The output voltage e_v is the result of summing two preamplifier sections, as shown in Fig. 2. A uniform expression for S_n and e_v is:

$$S_n = r_{np} A_{np} + r_{ni} A_{ni} + r_{ns} A_{ns} \qquad n=1,2,3,4 \qquad (3)$$

where r_p represents the reflection coefficient of the pupil and A_p his corresponding detection area.

$$e_v = \alpha_2 \{(S_1 + S_2) - (S_3 + S_4)\} \qquad\qquad (4)$$

with α_2 as a constant to match the dimensions. The indices 1, 2, 3 and 4 refer to the circular surface areas, as can be seen in Fig. 1b. The relation between the output voltage e_v and the vertical angle displacement β_v (see Fig. 3) is linear within limits depending on pupil diameter and eyeball position.

Spectral Analysis:
Fig. 3 is a schematic representation of the analysing involved in the processing of the PAM signal. $e_h(t)$ and $e_v(t)$ are bandwidth-limited causal signals. There is a linear relationship between any output signal $e(t)$, available at the output of the amplifier (see Fig. 2) and an angular displacement $\beta(t)$ of the eyeball, which can be expressed as follows:

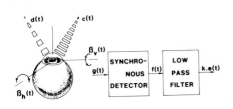

Fig. 3. Schematic representation of
the analysing and processing.

$$\bar{\beta}(t) = \bar{\beta}_{hor.}(t) + \bar{\beta}_{vert.}(t) \qquad (5)$$

$\bar{\beta}_{hor.}(t)$ and $\bar{\beta}_{vert.}(t)$ represent a
vector.

$$|\bar{e}(t)| = \alpha.r.|\bar{\beta}(t)| \qquad (6)$$

where r represents a factor proportional to a certain reflection coefficient. So the signal e(t) comes
into existence through the multiplication of a certain eyeball position $\beta(t)$ and the amount of reflected energy, which comes from a
pulsating IRLED {c(t)} and/or from surrounding intermittent disturbances
{d(t)}. The sum of the signal c(t) and d(t), with corresponding frequencies
f_c and f_d, multiplicated by e(t) results in a signal g(t). The equivalent
process in the frequency domain is the convolution of the spectrum E(f) with
the sum {C(f) + D(f)} which yields G(f).

$$g(t) = \{e(t).\{c(t) + d(t)\}\} \xrightarrow{F} G(f) = \{E(f) \bullet \{C(f) + D(f)\}\} \qquad (7)$$

$$G(f) = 2\pi AE \sum_{\substack{n=-\infty \\ n\neq0 \ ; \ |f-nf_c| \leq f_e}} SINC(n.\frac{\pi}{2}).E(f-nf_c) + 2\pi BE \sum_{\substack{m=-\infty \\ m\neq0 \ ; \ |f-mf_d| \leq f_e}} SINC(m.\frac{\pi}{2}).E(f-mf_d) \qquad (8)$$

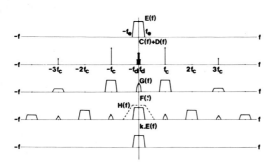

Fig. 4. Graphical presentation of the processing of the signals in the frequency domain.

A,B and E represent respectively the amplitude of a
pulsating signal, an intermittent disturbance and a
bandwidth-limited eyeball
position signal. Multiply
ing g(t) by a derivitive
(see Fig. 2.: "SYNC signal")
of the pulsating signal c(t)
of appropriate dimensions
leads to a signal f(t) in
which both a sample of the
original signal and the
disturbances are worked up.
The convolution of the
spectrum G(f) and C(f) is
performed in the frequency

domain in order to yield F(f).

$$f(t)=g(t).c(t) \xrightarrow{F} F(f)=G(f)\bullet C(f)$$

$$F(f) = 4\pi^2A^2E \sum_{\substack{n=-\infty \\ n\neq0 \ ; \ n\neq k}}^{\infty} SINC\{(k-n).\frac{\pi}{2}\}.SINC(n.\frac{\pi}{2}).E(f-kf_c) +$$

$$+ 4\pi^2ABE \sum_{\substack{n=-\infty \\ n\neq0}}^{\infty} \sum_{\substack{m=-\infty \\ m\neq0}}^{\infty} SINC(n.\frac{\pi}{2}).SINC(m.\frac{\pi}{2}).E\{f-(nf_c + mf_d)\} \qquad (9)$$

F(f) shows the complete complex spectrum under the following conditions

$$|f-nf_c| \leq f_e \text{ and } |f-mf_d| \leq f_e$$

The convolution of f(t) with an impulse response h(t) of a low pass filter (LPF) results in a filtered version "k.e(t)".

Results:

Drift-free detection with an accuracy of 6 minutes of arc is obtained during the detection of the horizontal and vertical eyeball movements. The ranges obtained are as stated before. The bandwidth of 1,5 kHz is sufficient to detect eyeball movements in a broad range of interest.

APPLICATIONS:

EYEBALL-POSITION-CONTROLLED COMMUNICATION SYSTEM: EPCOS[4,5]

The EPCOS consists of three main parts: the EYE-SISTANT, a one-line 80-character display, a function matrix and a microprocessor system based upon the Motorola MC 6809 and its matching peripherals such as EPROM and RAM memories, two parallel I/O interfaces, two 10-bit Analogue-to-Digital (ADC) convertors and two serial I/O interfaces which support the communication with a video-screen and keyboard. The ADC converts the analogue signals representing a momentary position of both eyeballs, in horizontal and vertical directions, into a digital word.

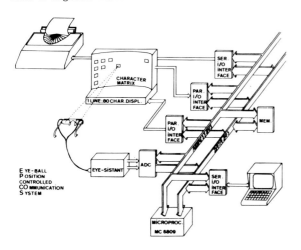

The character matrix consists of six rows, on each of which a maximum of fifteen signal bulbs can be mounted. The panel in which the matrix is integrated is bent in a radius of one metre, so that the relative distance between two adjacent bulbs is $2°$ of an arc, assuming the distance from the head to the panel is one metre. The alphanumeric and functional symbols are etched in the lenses of the bulbs. The most frequently used characters are clustered together in the centre. The one-line 80-character display unit will reproduce the text as it was composed out of the character matrix. The text can be written down at a typewriter which is permanently connected to the matrix.

Fig. 5. Block diagram of the EPCOS.

Use of the EPCOS:

The EPCOS is designed for the category of totally disabled persons. The eyeballs only need to be positioned with reference to the centre of the matrix because of the fact that head movements are absent.

As a start a test procedure has to be performed, during which the deviation of the eyeballs with respect to each character of the matrix must be calibrated. The acquired information is stored in a memory page, which will

function as a look-up table during the actual communication itself. After
the test procedure the system will function as a normal typewriter. Words
are selected by the eyes by fixating, one by one, the desired character.
During a certain access time the EYE-SISTANT signals digitised through the
ADC's are compared with the data from the look-up table, prior to the display at the one-line 80-character display. Errors can be corrected at any
time throughout the whole selection procedure. The text is printed at the
typewriter.

Concluding Remarks: After a certain evaluation time, in which it was used as
a communication aid, the EPCOS is going to be available for the category of
totally disabled persons.

THE EYE-SISTANT AS A PUPILLOMETER.

The universal design of the EYE-SISTANT enables the user to implement very
easily a new function by adding electronic components to the system (see Fig.
6). As an example a design of a pupillometer is given. The basic signals required to determine the momentary
pupil diameter are derived from the
vertical eyeball movement detection
assembly. Involuntary pupil reactions will not occur because the
eyes are homogeneously lit by an IR
source with a wavelength of 930 nm.
At the beginning a quick visual
alignment of the position of the
iris with respect to the vertical
assembly with the aid of a ceratos-
cope (Disc of Placido) enables the
user to position the phototransistors accurately. A combination of
the four signals from the PTR's in
a summation amplifier will provide

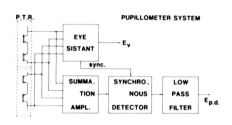

Fig. 6. Block diagram of the
         pupillometer

an output signal $e_{p.d.}$ proportional to the pupil diameter variations.

$$e_{p.d.} = \alpha_3 \sum_{n=1}^{4} S_n = \alpha_3 \sum_{n=1}^{4} (r_{np}A_{np} + r_{ni}A_{ni} + r_{ns}A_{ns}) \qquad (10)$$

With α_3 as a constant to match the dimensions. S_n represents the proportional
influences of the surface areas 1 through 4 situated around the pupil. r_i,
r_s and r_p represent the physical reflection properties of iris, sclera and
pupil with A_i, A_s and A_p as the corresponding surface areas (see also Fig.1).
Combining the two signals e_v and $e_{p.d.}$ enables the pupil diameter variations
to be quickly processed. The detection of the vertical eyeball movements can
serve as a compensation signal in the final result of the pupillometer system.

Concluding Remarks: A possible application stems from an investigation into
the influence of the disease "multiple sclerosis" on the speed and/or amount
of the pupil reactions.

HEAD MOVEMENT DETECTION: AN ADDITIONAL USE OF THE EYE-SISTANT.

A preliminary investigation has been made into, how the results derived from
the EYE-SISTANT and incidentally occurring head movements could be combined.

The accuracy with which eyeball movements can be detected depends largely on the amount of fixation of the head with respect to the field of view in which any visual stimuli is offered. The head movements occupy roughly a frequency range of DC– 25 Hz.

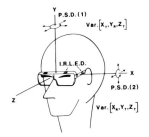

First the detection of the displacement (X_t, Y_t, Z_t) and rotation (X_R, Y_R) of the head with respect to an imaginary X,Y and Z-axis has to be performed. The Universal Measuring frame used in the EYE-SISTANT must at all times occupy a stable position within the above-mentioned axis system. In this particular project only a contactless optical detection method was permitted by the future users.

Fig. 7. Schematic representation of the system parameters

In the final version four IRLED's – one pair transmitting its energy upwards and the other sideways – were mounted on the Measuring frame (see Fig. 7). The two pairs of IRLED's are positioned with respect to each other so that a mutual coherence will always exist. Any movement of the head will result in a change in directions of the IRLED's. Combining this result with that of the EYE-SISTANT determines the field of view in a three-dimensional space. Only in this way is a fixed relationship between the results of eyeball and head movements guaranteed.

For the analysis of a displacement, only the detection of the energy from one IRLED is sufficient. In the case of a rotation, however, the processing of the information of at least two IRLED's is necessary.

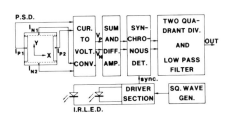

Fig. 8. Block diagram of the two-dimensional headmovement detection

As light-sensitive sensors two two-dimensional Position Sensitive photo Detectors (PSD's) are used as a measuring device (see Figs. 7 & 8). Both PSD's are equipped with a lens system to transform the projection of a beam of light into a signal proportional to the position of the spot at the surface. Additional electronics see to the analysis and processing of the signals.

The dual-axis PSD, as was constructed by Noorlag et al. consists of a reversed biased square P^+-N junction, with two extended lateral contacts at opposite sides of the P-layer and two at the remaining opposite sides of the N-layer. The photocurrent will only be divided into the electrical currents I_n and I_p between the lateral contacts. The magnitude of the currents, available at the x- and y-contacts, is proportional to the position of the light spot at the surface of the PSD with respect to the distance to the individual lateral contacts (see Fig. 8). After signal processing the position of the light spot expressed in the parameters s and y, is written as follows:

$$x = \{(I_{p2}-I_{p1})/(I_{p2}+I_{p1})\} \; ; \; y = \{(I_{n2}-I_{n1})/(I_{n2}+I_{n1})\} \; ; \; I_{p2}+I_{p1}=I_{n2}+I_{n1} \quad (11)$$

Background illumination, determined by a current I_d, smoothly spread out over the entire PSD surface, will lead to a proportional increase in the current across the lateral contacts.

$$x=\{(I_{p2}+I_d)-(I_{p1}+I_d)\}/(I_{p2}+I_d+I_{p1}+I_d) =\{(I_{p2}-I_{p1})/(I_{p2}+I_{p1}+2I_d)\} \quad (12)$$

A similar expression results for the parameter y. In general, the transformation of a light spot into an output current I_n or I_p must be considered to be linear.

Similar to the EYE-SISTANT the IRLED's (see Fig. 8) used are modulated by a 5 kHz square wave. The current available from the PSD is transformed into a voltage via a current-to-voltage convertor. To fulfill the requirements of the light distribution centre of gravity both the sum and difference of the voltage V_n and V_p must be generated. Both signals are recognised as PAM-signals: the envelope of the signal contains the information of displacement and rotation, while the magnitude indicates the amount of detected energy. Involuntary influences from surrounding disturbances in the final result are prevented by using the technique of synchronous detection. The harmonic components are removed by a low pass filter. Two identical sections are necessary for the three-dimensional analysis. (Only one section is drawn). To achieve the proposed relationship the results of both the EYE-SISTANT and head movement detection must be processed in an acquisition system.

Results:
A distance of 1 m with respect to the IRLED part was maintained during the test procedure. A lens with a focal distance of 3 cm was available. The sensitivity in the case of displacement, determined by the voltage variations, was related to the shift in the IRLED's over a range of 100 mm: 0.14 V/mm. The accompanying signal-to-noise ratio was 50 dB. The resolution defined by the discrimination of the signal in noise was 1.5 minutes of arc. The deviation from the theoretical lapse was 1.5%. The resolution in the case of rotation, over a total range of -90° to 90° was $2^{\circ}17'$. The corresponding discrimination number was $1^{\circ}30'$. The total field of motion is bounded in both the x- and y-directions to a resolution of 150 mm due to the physical dimensions of the PSD (6×6 mm^2) and the focal distance of the lens.

ACKNOWLEDGEMENTS.
The author would like to thank the Institute of Applied Physics for the encouraging assistance, and specially Wim Klumper and Bram Hardenbol for helpful comments on the topic of head momements, Susan Massotty for correcting my English, Joyce van Middelkoop for all the diagrams   and last but not least Marjan Mulder for all the typing.

REFERENCES:
(1) Frietman E.E.E., Joon M.M., The EYE-SISTANT: a two-dimensional eye-movement detection system, Technical Report nr: 805-201 and 109-262, Delft, Institute of Applied Physics, TNO/TH, 1980.
(2) American National Standard: For the safe use of lasers, ANSI, Z136.1-1980, American National Standards Institute,INC,1430 Broadway, New York.
(3) Papoulis A., The Fourier integral, Mc Graw Hill, New York.
(4) Kate J.H. ten, Frietman E.E.E., Stoel F.J.M.L., Willems W., Eye-controlled Communication Aids, Med.Progr.Technol.8,1-21, (1980).
(5) Frietman E.E.E. et al., Eyeball-Position-controlled COmmunication Systems: EPCOS (to be published).
(6) Noorlag D.J.W., Middelhoek S., Two-dimensional position-sensitive photodetector with high linearity made with standard i.c.-technology, Solid State and Electron Devices, May 1979, vol 3, no. 3.

Theoretical and Applied Aspects of Eye Movement Research
A.G. Gale and F. Johnson (Editors)
© Elsevier Science Publishers B.V. (North-Holland), 1984

HIGH-PRECISION REAL-TIME MEASUREMENT OF EYE POSITION
USING THE FIRST PURKINJE IMAGE

R.C.Frecker*   M.Eizenman**   P.E.Hallett***

Institute of Biomedical Engineering, and also, Departments of
*Pharmacology, **Ophthalmology, and ***Physiology,
University of Toronto, Toronto, Canada M5S 1A4

ABSTRACT

Using a pulsed IR source, and novel signal processing, the centre
of gravity of the 1st Purkinje image on an array of discrete
phototransistors is computed in real time. An overall system
noise of better than 30 arc seconds of ocular rotation is
achieved with a dynamic range of from 30 arc seconds to 30 arc
degrees.   Linearity is better than 2%, and a new eye position is
generated every msec with a velocity resolution of 2 degrees/sec.
Details of the residual movements of fixation are clearly
resolved. Subjects may be set up for measurement in less than 60
seconds.

INTRODUCTION

A variety of useful non-contacting eye position monitors have been
developed by researchers at the University of Toronto. For example, the
corneal reflex has been recorded cinematographically, superimposed on the
scene [1]; the position of the corneal reflex has been recorded on a linear
array of photodiodes with alternating electrical polarity [2,3]; the
positions of the left and right edges of the pupil have been electrically
subtracted to yield an eye position signal insensitive to change in pupil
size [4,5]; and, the direction and magnitude of ocular rotation can
theoretically be obtained from pupillary shape independent of translation
[6].   None of these methods allows the speedy set up of the subject and
rapid analysis of data which are necessary for clinical studies; and none
provides the high precision of contact lens methods that is appropriate for
pharmacological and physiological research. Accordingly, the authors have
developed a complete eye-monitoring facility to further their research into
non-invasive clinical tests [7,8], the   psychopharmacology of   drug
dependence [9-14], and visual physiology [15-17]. This has involved the
application of novel signal processing methods normally found in gamma-ray
cameras [18-21].   Requirements for precision and real-time data analysis
led to the development of a system which can also be used for the
contingent control of image projection, and for objective visual field
mapping (perimetry).

OPTICAL CONFIGURATION

The optical system of the 2-dimensional version of the eye tracker is shown below (Figure 1). The basic operation of both the 1- and 2-dimensional systems is identical, the difference being that in the latter, optical information from the cornea is split into two orthogonal channels and processed in parallel.

OPTICAL LAYOUT

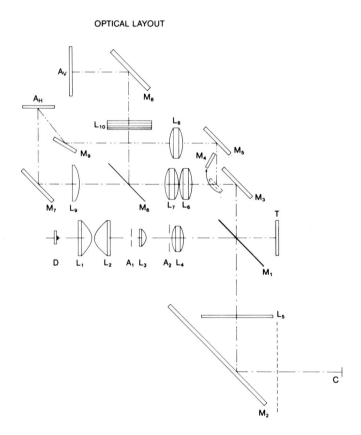

Figure 1

The optical layout of the 2-dimensional eye tracker

The light source is a high-output (40 mW/sr) gallium arsenide diode (D). The system of lenses L1 to L4, two apertures A1 and A2, a pellicle splitter (M-1), the Fresnel lens (L-5), and dichroic mirror (M-2), create a circular collimated beam, 10 mm in diameter at the cornea (C). This beam is reflected by mirror M-2, collected by lens L-5, transmitted through splitter M-1, reflected from front-surfaced mirror M-3 (past M-4, which is shown out of position), magnified by the doublet L-6/L-7, and split into two images by splitter M-6 to cylindrical lenses L-9 and L-10. The two images formed here are orthogonal to one another and are passed by mirrors M-7 and M-8 to linear arrays AH (horizontal) and AV (vertical). The optical path from the mirror M-4, when it is interposed between L-6 and M-3, is through L-8 and, by way of M-9, to array AH. The latter configuration is used during the initial subject alignment, the mirror M-4 being controlled from the outside.

OPTICAL SAFETY CONSIDERATIONS

The intensity of the light incident on the cornea in the continuous mode of illumination is 100 uW/cm^2. Peak spectral intensity is at 930 nm and the spectral bandwidth (full width half maximum) is 45 nm. In the pulsed mode of operation which is used in the 2-dimensional system, the peak intensity is increased proportionally to the shortening of the 'on-time' of the source diode, which is typically 200 usec every msec. The American National Standards Institute (ANSI) standard Z136:1 was met with respect to ocular radiation to ensure subject safety during prolonged periods of exposure. In the pulse mode, this system is operated at 20% of the permitted maximum. The distribution of light intensity on the phototransistor detection array, and the positioning of the optical system with respect to the subject's line of sight, are controlled automatically by three high-torgue DC servo motors -- using optical feedback from both the entire array and selected clusters of sensor elements. At the end of the automatic optical alignment procedure a pre-defined light intensity distribution, which is approximately Gaussian, irradiates the central portion of the array.

GENERATION OF POSITION INFORMATION

In the sensor plane(s), image intensity is converted to electrical currents by two densely packed 20-element phototransistor arrays. Typical outputs from array elements are in the range of 500 nA, and the current-to-voltge conversions are made by 2-stage transimpedence amplifiers (40 analog channels). Since amplifier input equivalent noise and expected phototransistor currents at the input stage might be of the same order of magnitude, special attention was given to minimizing equivalent noise currents at the input stages of the amplifier circuits.

The amplified output of each phototransistor is multiplied by a time-varying function, and all the outputs are summed to form another function, the zero-crossing of which is proportional to the instantaneous position of the eye. This technique is similar to that used in delay line gamma cameras to locate a scintillation event using an array of photomultipliers [19]. The weighting function is derived by an optimization procedure in which the variance of the estimated eye position is minimized, while the slope of the transfer function relating the computed eye position to the true eye position is held constant. More details are provided elsewhere [18-21].

The output of the system is generated as a 15-bit digital word (1 bit = 4 arc sec of ocular rotation) and sent to a DEC PDP 11/34 minicomputer on 16 driven digital lines for signal analysis. The output is also made available for monitoring purposes as a 0-15 volt analog signal. The noise floor of the signal processing unit is equivalent to an ocular rotation of some 6 arc sec, while the overall system noise is less than 30 arc sec. A new eye position is made available 1,000 times each second, and a velocity resolution of 2°/sec is achieved, over a velocity bandwidth of 125 Hz. The dynamic range of 30 arc sec to 36 degrees of ocular rotation spans a range of more than 3 decades. System linearity (maximum deviation from the best fitting straight line relating observed to true position) is 1.5%, expressed as a percentage of the total range.

METHOD OF SIGNAL ANALYSIS

Straight forward techniques were employed to identify eye movements, to reject artifacts, and to calculate the various eye movement parameters. A procedure based on velocity criteria was utilized to determine the beginnings and ends of both saccades and microsaccades. In each case the position information was low-pass filtered and differentiated. The low-pass filtering was done by fitting the position data with a parabolic polynomial. The polynomial is calculated to fit five points in the position sequence using the least squares criteria. For a 1 kHz sampling rate, this is equivalent to a low-pass filter with a 3-db point at 125 Hz.

In searching fixation records for microsaccades, absolute eye velocities greater than 5°/sec (twice the maximum velocity due to system noise, eye tremor, and eye drift) are recognized. At this point, a backwards (in time) search is implemented to identify two successive velocities of less than 2.5^{o}/sec. The first point of this pair is chosen to be the beginning of the microsaccade. The end of a microsaccade is found using a similar procedure. Peak velocity is simply defined as the highest velocity achieved between the beginning and end of an identified movement. Duration and magnitude are easily calculated with reference to the available postion/time information. The characteristics of microsaccadic overshoots are also calculated in this manner. Initial recognition of a primary macrosaccade occurs at velocities greater than 50°/sec, while that of a secondary saccade is 20°/sec. For both primary and secondary saccades, algorithms similar to those used for microsaccadic characterization and parameter extraction are employed.

Waveform analysis can be implemented at the end of the collection of a given batch of eye position information (e.g., after 25, or after 100 stimuli are presented); or in virtual real time, within the time frame of successive stimulus presentations if it is desired to ajudicate a particular response and derive subsequent stimulus properties from this analysis. Further, contingent display control can be achieved by algorithms which deal with the stream of real time position information which is available msec by msec. This can be quite useful in circumstances where the new target position is to be a function of the anticipated temporal and spatial course of a given eye movement.

TYPICAL RECORDING OF HORIZONTAL EYE MOVEMENTS

Figure 2-A/B shows the output signals for a typical microssaccadic sequence. The horizontal axes represent time plotted in msec, while the vertical axes are eye position, in minutes of arc (upper panel), and eye velocity for the same ocular excursion, in degrees of arc/sec (lower panel).

In Figure 2-A, after a slow drift of less than 10 arc min, a microsaccadic excursion of some 20 arc min (labelled 'm') is observed away from the primary position. The primary position is shown as 0°), and 'plus' excursions are nasal, while 'minus' excursions are temporal in direction. An oppositely directed microsaccade of similar amplitude is also labelled 'm'. Microsaccadic reversals are labelled 'r', and 3 are seen in this particular record. In Figure 2-B the corresponding velocity traces are shown, and in the cases of the larger microsaccadic components these approach 30°/sec.

Comparable traces for a larger saccadic sequence are seen on the right hand side of Figure 2. In Figure 2-C the solid trace represents stimulus position and shows two transitions – one of approximately 11.5^o, an a second of approximately 3°. The saccadic responses to these stimuli are seen in the digitized trace, the 1 msec intervals being clearly visualized during the higher velocity portions of the movements themselves. Two primary saccades ('p'), and reversals ('r'), as well as one secondary saccade ('s') are clearly seen. The corresponding velocity traces are shown in Figure 2-D.

These records are typical of those routinely obtained with the eye tracker. The set-up time (once a bite plate has been fashioned) is less than 60 seconds, and subsequent sessions may use the same bite plate without problems. While subject restraint is required to achieve this resolution, a helmet-mounted tracker using the same principles is now under development. Automated image acquisition renders the apparatus well suited to routine laboratory and clinical application.

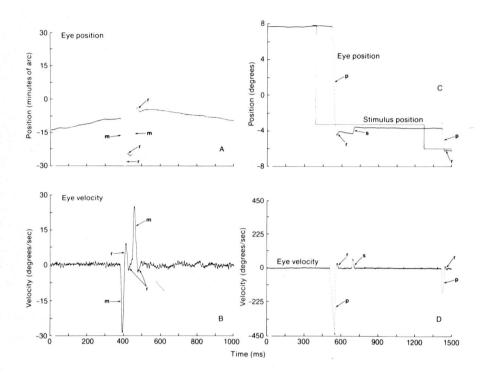

Figure 2

Position and velocity traces for horizontal saccades

ACKNOWLEDGEMENTS

Special appreciation is expressed to Tony Jares for his valued technical
contributions. The authors are also indebted to the A.E. MacDonald
Foundation for Ophthalmology for support of M.E.; to the Natural Sciences
and Engineering Research Council of Canada for grant A-3902 to R.C.F.; and
to the Medical Research Council of Canada for grant MA-7673 to P.E.H.

REFERENCES

[1] Llewellyn-Thomas, E. (1968). Movements of the eye. Scientific American, August, 1968:88-97.

[2] Gentles, W. (1969). Effects of Chlordiazepoxide and Diazepam on the Oculomotor System. M.A.Sc. Thesis, University of Toronto.

[3] Frecker, R.C. (1973). Effects on Human Saccadic Eye Movements of Diazepam, Pentobarbital, and Dextroamphetamine. Ph.D. Thesis, University of Toronto.

[4] Lightstone, A.D. (1973). Visual Stimuli For Saccadic and Smooth Pursuit Movements. Ph.D. Thesis, University of Toronto.

[5] Hallett, P.E., Lightstone, A.D. (1976). Saccadic eye movements towards stimuli triggered by prior saccades. Vision Research 16:99-106.

[6] Bechai, N.R.L., Hallett, P.E. (1977). Measurement of the rotation of a disc from its elliptical projection, with an application to eye movements. J. Opt. Soc. Am., 67:1336-1339.

[7] Hallett, P.E. (1978). Primary and secondary saccades to goals defined by instructions. Vision Research 18:1279-1296.

[8] Guitton, D., Buchtel, H.A., Douglas, R.M. (1982). Disturbances of voluntary saccadic eye movement mechanisms following discrete unilateral frontal lobe removals. in: G. Lennerstrand, et al (Eds), Functional Basis of Ocular Motility Disorders. Oxford Pergamon Press, Oxford.

[9] Gentles, W., Llewellyn-Thomas, E. (1971). Effects of benzodiazepines upon saccadic eye movements in man. Clin. Pharm. & Therapeut. 12:563-574.

[10] Frecker, R.C., Llewellyn-Thomas E. (1975). Saccadic eye movement velocity an indicator of dose level of diazepam, pentobarbital, and D-amphetamine in humans. J. Pharmacol. Clin. 2(1):36-40.

[11] Frecker, R.C. (1976). Ergonomic considerations in quantifying psychoactive drug effects. Proc. 9th Annual Scientific Meeting, Human Factors Association of Canada, Bracebridge, Canada. pp 27-28.

[12] Rothenberg, S.J., Selkoe, D. (1981). Specific oculomotor deficit after diazepam. Psychopharmacology, 74:232-236.

[13] Rothenberg, S.J., Selkoe, D. (1981). Specific oculomotor deficit
after diazepam - II: smooth pursuit eye movements. Psychopharmacology,
74:237-240.

[14] Wilkinson, I.M.S., Kine, R., Purnell, M. (1974). Alcohol and eye
movement. Brain, 97:785-792.

[15] O'Beirne, H., Llewellyn-Thomas, E. (1967). Curvature in the saccadic
movement. AMA Arch. Ophthalmology, 77:105-109.

[16] Hallett, P.E. (1971). Physiology of Vision. In: A. Sorsby (Ed).
Modern Ophthalmology, VOl 1, pp 203-330.

[17] Hallett, P.E., Adams, B.D. (1980). The predictability of saccadic
latency in a novel, voluntary oculomotor task. Vision Research,
20:329-339.

[18] Eizenman, M. (1983) Precise Non-Contacting Eye-Movement Monitoring
System. Ph.D. Thesis, University of Toronto.

[19] Ueda, K., Kawaguchi, F., Takami, K., Ishimatsu, K. (1978).
Investigation of contribution functions in a scintillation camera. J.
Nucl. Med. 19:825-835.

[20] Eizenman, M., Frecker, R.C., Joy, M.L.G., Hallett, P.E. (1980). A
new mathematical approach to the high-resolution measurement of eye
movements. Proc. 8th Canadian Medical and Biological Engineering
Conference, Hamilton, Canada. pp 72-73.

[21] Eizenman, M. Frecker, R.C., Hallett, P.E. (in press). Precise
non-contacting measurement of eye movements using the cornea reflex.
Accepted for publication by Vision Research, July 1983.

Theoretical and Applied Aspects of Eye Movement Research
A.G. Gale and F. Johnson (Editors)
© Elsevier Science Publishers B.V. (North-Holland), 1984

OPERATIONAL PROBLEMS IN ANALYSING EYE MOVEMENTS

Heino Widdel

Forschungsinstitut für Anthropotechnik
Wachtberg-Werthhoven
FRG

Eye movement measurement with cornea-pupil reflection technique requires an operational definition of a fixation to be
able to analyse eye movements by data reduction strategies.
The present paper illustrates some effects of variable fixation definitions on the calculation of eye movement characteristics conducting an experiment with simple visual scanning tasks. The results of eye movement evaluation are dependent on the defined size of a fixation. They are varying in
alternation with stimulus aspects provoking different densities of fixation locations as well as with individual eye
movement behavior.

INTRODUCTION

In the last years a considerable development in eye movement monitoring
technology could be ascertained corresponding with an increase in eye
movement literature. In comparison with the amount of eye movement
measurement data the number of publications dealing with methodological
issues as, e.g., reliability and validity, turns out to be rather low. A
general concept was presented by McConkie (1980) suggesting how to develop
a strategy of giving information about confidence and quality criteria of
experimental results using eye movement records as data. Previous investigations about the reliability of eye movement measurement have been mainly
concerned with the confidence of hardware aspects of specific recording
systems, e.g., Schroiff (1983) and Boecker and Schwerdt (1981).

Research interest here was focused, instead, on the analytical algorithms
used in processing eye movement data. Measurement systems operating with
cornea-pupil reflections record electro-optical signals which are correlates of eye movements. Parameters of eye movements as, e.g., fixations,
can exclusively be the results of reduction strategies and algorithms used
by specific computer programs. This implies that fixations are not an
evident part of eye movements but must be defined operationally as a
fundamental aspect of the data reduction concept. This consideration leads
to the questions "What is a fixation?" and "How is a fixation defined?".
Generally, definitions of fixations or other eye movement parameters are
not described in the research literature.

Only a few papers have been found which give fixation definitions, e.g.,
Buurman et al. (1981). They define eye movements as fixations remaining
within a square with horizontal and vertical sides of 1.3 deg of visual
angle for at least 100 msec. Other publications prefer a duration of
200 msec as the minimum duration and a square of 2 deg (e.g., Salthouse

et al. (1981), Moffitt (1980), and Widdel and Kaster (1981)). McConkie (1980) suggested that when reporting and describing eye movement data, the identification characteristics of a fixation or a saccade used by the algorithms of data reduction, generally should be specified.

In our own investigations difficulties occurred in discriminating two or more expected fixations during a visual search process. The hypothesis arose that variations in operational definitions of a fixation may influence eye movement measurement results. This would limit the comparability and generalization of research findings. In order to test these considerations an experimental investigation was conducted. The experimental approach used was to have subjects scan visual patterns with a fixed number of dots. The subjects would be required to look at each dot, the number of which differs in each pattern. The fixations were calculated with reduction algorithms using different operational definitions. With this approach the effect of various fixation definitions on measurement results should be determined.

APPARATUS

The eye movement recording system used for the following experimental investigation was a Honeywell Oculometer. Its function is based on the principles of the pupil-cornea reflection method, also called point of regard measurement, as described in detail by Young and Sheena (1975). A schematic illustration of this recording system is given in figure 1. The eye is illuminated with infrared radiation from a single light source reflected by a mirror on its way to the eye. Some of this light reaching the eye is reflected by the cornea. Another small amount of this light enters the eye and is reflected by the retina back through the pupil. The different parts of reflected radiation are monitored with an infrared sensitive television camera. It provides an enlarged image of the eye with a bright pupil and a brighter small image of the corneal reflection, superimposed on the pupil. When the eye rotates around its center, the corneal reflection moves differentially with respect to the pupil, because the cornea and the rest of the eye have different radii of curvature. Thus, shifts of the corneal reflection relative to the pupil correspond to shifts in the direction of regard.The optical signal returning from the eye is fed to an electro-optical tracker for conversion to an electrical signal, which is generated in the circuitry of the electro-optical tracker and processed by a mini-computer. This computer using the programmed processing procedure determines the line of sight or the eye position relative to the display by calculating the x- and y-coordinates in correspondence with the difference of the corneal reflection and the calculated center of the pupil. The system operates with a sampling rate of 50 Hz, i.e., every 20 msec voltages are given which correspond to a pair of coordinates. The experimenter selected the time periods from which data were stored on tape.

During the eye movement measurement procedure, subjects are seated and their heads are immobilized by placing the chin on a chinrest and pressing the forehead against a headband. Small lateral head movements parallel to the display plane are allowed by the system but not rotational movements. At the beginning of a session an automatic 3-point-calibration has to be performed which is supposed to obtain a horizontal and vertical accuracy of better than + or -.5 deg. Calibration is necessary, because the digital output for the eye position varies with individual differences in the shape of the eye.

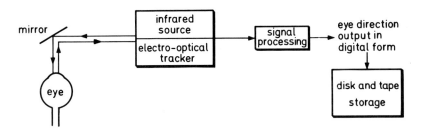

Figure 1
Schematic representation of eye movement measurement setup

During the investigation the subject sat in front of a screen with an eye
to screen distance of 84 cm. A slide projector was placed out of sight
behind the white translucent screen for the purpose of projecting stimulus
pattern to the subject.

The stimulus material projected on the screen consisted of a set of four
different simple dot patterns. Each pattern was identical in size but
differed in the density, respectively in the number of the black dots. The
size of each pattern was 51 cm or 33.8 deg in horizontal and 35 cm or 23.5
deg in the vertical. They included 12, 24, 35, and 48 black dots which
were arranged in a matrix structure, and each dot had a size of 0.5 cm or
0.34 deg in diameter. The detailed dimensions of the four patterns are
presented in table 1.

Table 1
Dimensions of the dot patterns representing the stimulus material

No. of dot pattern	number of dots horizontal	vertical	distances in degrees between single dots horizontal	vertical
1	4	3	11.3°	11.8°
2	6	4	6.8°	7.8°
3	7	5	5.6°	5.9°
4	8	6	4.8°	4.7°

EXPERIMENT

Eight subjects participated in this experiment. All had normal visual
acuity without glasses and their age centered around 30 years.

After the 3-point-calibration, but before the experimental presentation, a
linearization procedure was performed. This involved the projection of a
matrix of 5 x 7 dots onto the screen, and each dot had to be fixated
separately, i.e., eye movement data were collected only when subjects
looked at a dot. The linearization procedure serves the purpose of generating correction values, if data were distorted, which mainly occurs in the
peripheral field of view. Moreover, these data were used to calculate the
individual size of fixations in the static case of measurement.

Then the four stimulus patterns were presented in the order as illustrated
in table 1. When a dot pattern appeared, subjects had to scan it by
successively fixating each dot in a fixed scanpath. The duration time of
looking at a dot should be at least 0.5 sec before the subjects look at
the following dot. Eye movement data were collected during the entire
procedure of scanning the stimulus pattern. After a short pause, the next
pattern was presented. In the first session, subjects had to practice with
this scanning procedure to get used to the rhythm of eye movement jumps
from one dot to the next. With the main experimental pattern sets, subjects had to accomplish two identical sessions separated by a time period
of two days. When all experimental sessions were finished, eye movement
data were available for evaluation from eight subjects, each with two sets
of four patterns, or 64 recorded patterns.

DATA REDUCTION

Eye movement data of the experiment have been stored on tape and were
evaluated in a second step with a reduction program. First, the linearization procedure was initiated to correct possible distortions between view
direction of subjects and corresponding oculometer data. In principle, the
differences of adjacent oculometer coordinates of two fixations are related to the differences of the two corresponding dot coordinates of the
display. The visual angles of these differences, changed in oculometer
units, are calculated for all display areas. Since a dot matrix of 5 x 7
has been fixated, 35 areas of the display were available for which the
correction factors could be calculated. The detailed procedure is described in Kaster (1981) which allows elimination of distortions for each
single eye movement data point. It seemed to be more economic, when reflecting on the present problem, to correct the defined window for analysing fixations than to correct each single oculometer datum of eye movements. For example, the 1.5 deg-window covers 9 x- and 5 y-units (see tab.
2) in the center of the display, but 12 x- and 7 y-units in the left upper
area of the display. Both x- and y-differences are equivalent to 1.5 deg.

A window represents the operationally defined fixation concerning the
spatial size. A square shape was used with five different values of .5, 1,
1.5, 2, and 3 deg for each side of the window. The degrees were converted
in oculometer units depending on the linearization results. The minimum
duration of a fixation was fixed to 200 msec which is equivalent to 10
oculometer coordinates. When analysing the recorded eye movements the
window was moved over the coordinates until a sample of at least 10

coordinates was found. Window movement was continued until less than 10 coordinates were included. The center of the whole sample covered by the moving window was calculated. Subsequently, this center and the center of the used window were brought to coincidence. The eye movement units now being located in the window were identified as a fixation.

The practical operation works as shown in table 2. The difference of 9 x-units is equivalent to 1.5 deg determining the size of the window. The moving window covered 14 x-coordinates ranging from 81 to 91 which indicates a difference of 10 x-units exeeding 1.5 deg. The mean of the sample of coordinates is 87, so that the fixation-window of 1.5 deg includes the x-units from 86 to 91. This fixation has a duration of 260 msec. The y-coordinates did not change more than 1.5 deg, i.e., the fixation was identified while scanning in the horizontal direction.

Occasionally, data was missed because of blinks or technical recording failures as indicated by the PEN-signal (see table 2). In most cases, an interpolation of the reliable data could fill the data gaps.

Table 2
Identification of a fixation with a 1.5°-window during horizontal
eye movements

	X	Y	PEN	
	61	133	1	
	68	133	1	
	81	133	1	
	86	133	1	
	87	134	1	
	86	134	1	
moving window	85	134	1	
	86	135	1	
	86	134	1	
	86	134	1	
identified fixation	87	135	1	
	87	135	0	— lack of
	89	134	0	records
	88	136	0	
	91	136	1	
	91	136	1	
	104	137	1	
	109	137	1	

equivalent to 1.5° is a delta-X of 9 units
equivalent to 1.5° is a delta-Y of 5 units

RESULTS

A general survey of the measured eye movement data shows that between 2 % - 14 % of the individual data had to be eliminated because of technical failures, blinks or other specific eye problems. First, the empirical estimation of the fixation size extracted from the linearization pattern, when each dot was fixated separately, was conducted. The fixation size of

individuals ranged from .4 deg to .6 deg and indicates a high homogeneity
of fixation size in a static measurement case. This relationship changed
when the eye movements were analyzed in dynamic scanning situations, as
will be discussed later.

The reliability of fixation frequency as a retest measure including the
data of the first and the second session for four window sizes is characterized by high correlations (table 3). They indicate that the individual
number of fixations retain the same relative frequency rank among subjects
when eye movements were recorded at two different times using the same
stimulus material. The high stability of individual eye movement characteristics may be due, in part at least, to the low number of subjects.

Table 3
Reliability of fixation measurement

	window—size of fixation			
	1^O	1.5^O	2^O	3°
retest-measurement	.89	.92	.92	.96

Table 4
Frequencies of excessive fixations depending on fixation definition and
stimulus pattern

dot matrix of stimulus pattern	number of dots	expected number of fixations for 16 presentations	window—size of fixation				
			0.5^O	1^O	1.5^O	2^O	3°
3 x 4	12	192	—	93	70	61	49
4 x 6	24	384	—	111	94	93	55
5 x 7	35	560	—	224	189	129	17
6 x 8	48	768	—	262	201	133	—

Table 4 shows fixation frequencies, which exceed the expected frequencies,
i.e., the number of dots which subjects were to look at. The values are
summations of the excessive fixations from eight subjects who have scanned
each stimulus pattern twice. These fixation frequencies are calculated

with five different fixation definitions. The vacant cells represent invalid estimations of fixations because in three or more cases observed fixation frequencies are lower than expected. When using a .5 deg-window to collect ten or more eye movement coordinates (200 msec), which establish a fixation, the attempt often failed to find as many fixations as expected. Furthermore, unrealistic saccade durations were found. On the other hand, analysing the scanning of the 6 x 8 dot matrix with a 3 deg-window turns out to be insufficient on account of a lack of differentiation between expected adjacent fixations.

A view on the interpretable values indicates some tendencies not based on statistical significance. The dot patterns of the 3 x 4 and 4 x 6 matrices provoked an eye movement structure from which the expected fixation frequencies could be reproduced better than from the 5 x 7 and 6 x 8 matrices. This difference is emphasized in distinct clearness with the use of window-sizes of 1 deg and 1.5 deg as against window-sizes of 2 deg and 3 deg. Generally, the results show that the 3 deg-definition of a fixation seems to verify the best reproduction of the expected fixation patterns, when the fixation distances are larger than approximately 5 deg (see table 1). When the distances become smaller a window-size of 2 deg turns out to be the preferable analysing definition.

Table 5
Individual frequencies of excessive fixations depending
on fixation definition

window-size of fixation	no. of subject							
	1	2	3	4	5	6	7	8
0.5°	53	-	157	124	-	-	100	-
1°	41	99	141	91	54	90	134	40
1.5°	38	77	102	80	64	56	98	39
2°	30	70	61	58	35	24	88	50
3°	-	36	-	13	-	-	26	13

The expected number of fixations of each subject is **238** when presenting the four stimulus patterns two times

The results in table 5 represent the summations of excessive fixations combining the data of all stimulus patterns having been scanned twice. They are presented for each subject and each window-size. It is discern-

ible that the insufficient data for the .5 deg-window originated with four
subjects, viz., no. 2, 5, 6, and 8 and for the 3 deg-window with four
subjects, viz., no. 1, 3, 5, and 6.

Aside from these data deficiencies, the best examples of the expected
fixation frequencies are realized with the data of subjects no. 1, 2, 5,
6, and 8. Furthermore, there are only small intraindividual differences
with regard to window-sizes. In contrary to these homogeneous findings,
subjects no. 3, 4, and 7 show differences between small and large window-
sizes. Generally, the lowest deviations from the expected fixation frequencies can be realized with a window-size of 2 deg.

DISCUSSION

The measurement of dynamic eye movements using a recording system based on
the cornea-pupil reflection method requires useful data reduction strategies. These can be realized with appropriate computer software. For the
analysis of fixations an operational definition of this eye movement
parameter must be postulated. In the present investigation the size of a
fixation, which is a relevant definition aspect, was systematically varied
and the consequences upon ascertained findings were studied. A dependency
on stimulus characteristics provoking different densities of fixations and
a dependency on individual eye movement behavior were pointed out. These
results make evident that an optimal definition of general validity does
not exist. But they confirm the necessity for researchers in this area to
give information about fixation definition when reporting eye movement
data.

Some moderate recommendations can be suggested on the basis of these
findings. When analysing dynamic eye movements in a search task with
closely structured elements it may be advisable to use a 2 deg-window.
This holds true when visual material of high density is to be looked at,
e.g., fixations at alternatives of a computer display menu or scanning a
radar monitor with very dense symbology positioning. On the other hand,
one, perhaps, should use a 3 deg-window for lower densities of visual
elements, e.g., investigating fixation patterns during the scanning process of some displays in a vehicle cockpit or looking at traffic signs
when driving a car.

The problem becomes more subtle when individual differences are examined.
It seems to be advantageous to use 2 deg- or 3 deg-fixation definitions
depending on visual scanning tasks. The use of smaller windows yielded
distinct interindividual differences. Considering the homogeneous results
of static fixation sizes, these differences have to be attributed to
movement related functions, e.g., over- or undershooting saccades. A hypothesis is that the correcting saccades may have small distances which can
be analysed with small window-sizes but they are lost using larger window-
sizes.

Further investigations should be focused on additional reduction strategies, e.g., on operational definitions of saccades including direction and
speed of eye movements. An interactive balancing of both definitions
during an iterative process may lead to more valid results.

Acknowledgement
The author wishs to thank Dipl.-Ing. Karl-Reinhard Kimmel for his help in
handling the technical equipment, Dipl.-Ing. Jürgen Kaster and Franz
Molitor for their fruitful comments and the program development, and
Dipl.-Tech.-Psych. Frank Pitrella for contributing to the clarity of this
paper.

REFERENCES

[1] Boecker, F. and Schwerdt, P., Die Zuverlässigkeit von Messungen mit
    dem Blickaufzeichnungsgerät NAC Eye-Mark-Recorder 4, Zeitschrift für
    experimentelle und angewandte Psychologie 28 (1981) 353-373.

[2] Buurman, R.; Roersma, T. and Gerrissen, J.F., Eye movements and the
    perceptual span in reading, Reading Research Quarterly 16 (1981) 227-
    235.

[3] Kaster, J., Graphic system supports for eye movement analysis, SID
    Digest of Technical Papers (1981) 54-55.

[4] McConkie, G.W., Evaluating and reporting data quality in eye movement
    research, Technical Report No. 193, University of Illinois (1980).

[5] Moffit, K., Evaluation of the fixation duration in visual search,
    Perception and Psychophysics 27 (1980) 370-372.

[6] Salthouse, T.A.; Ellis, C.L.; Dienier, D.C. and Somberg, B.L.,
    Stimulus processing during eye fixations, Journal of Experimental
    Psychology: Human Perception and Performance 7 (1981) 611-623.

[7] Schroiff, H.-W., On the reliability of eye-movement data, in: Lüer,
    G.(ed.), Bericht über den 33. Kongress der Deutschen Gesellschaft für
    Psychologie (Hogrefe, Göttingen, 1983) 157-160.

[8] Widdel, H., A method of measuring the visual lobe area, in:
    Groner, R., Menz, C., Fisher, D.F., and Monty, R.A. (eds.), Eye
    movements and psychological functions (Lawrence Erlbaum, Hillsdale,
    N.J., 1983 in press).

[9] Widdel, H. and Kaster, J., Eye movement measurement in the assessment
    and training of visual performance, in: Moraal, J. and Kraiss, K.-F.
    (eds.), Manned Systems Design. Methods, Equipment, and Applications
    (Plenum Press, New York, 1981) 251-270.

[10] Young, L.R. and Sheena, D., Survey of eye movement recording methods,
    Behavior Research Methods and Instrumentation 7 (1975) 379-429.

Theoretical and Applied Aspects of Eye Movement Research
A.G. Gale and F. Johnson (Editors)
© Elsevier Science Publishers B.V. (North-Holland), 1984

EMAN: A MODULAR AND ITERATIVE EYE-MOVEMENT ANALYSIS PROGRAM

Reinhold M. Kliegl

Max Planck Institute for Human Development
and Education
Berlin, West Germany

EMAN is an eye-movement analysis program that consists of
four modules. The first module rescales eye positions
to coordinates of the display. The second and third modules reduce data to a fixation format and identify areas
of bad measurement by means of iterative passes over the
data. In the fourth module iterative algorithms are employed for the identification of line numbers and for
achieving congruence between fixations and display.

INTRODUCTION

Videobased systems for monitoring eye movements - such as the Applied Science Models or Demel's DEBIC 80 - generate information about eye position
and pupil diameter at 60 or 50 Hz. The program EMAN was written for the
analysis of this eye-position information. EMAN first rescales machine
output to coordinates of the display as described in Kliegl and Olson
(1981). The rescaled values are then processed by three other program modules that are reported in this paper. First, reduction of valid samples
to position clusters and reduction of low-data-quality samples to blink
clusters is accomplished in the Module REDUCE. The second module, CLEANUP,
performs additional checks on the data-quality of position clusters and removes saccadic position clusters from the data base. Finally, the output
of CLEANUP, a sequence of fixations, is matched to the text display by Module TEXT. This module permits interactive changes of parameters to achieve
congruence between fixations and lines of the text display. Of course, the
last module is only meaningful for the analysis of eye movements that were
collected during reading.

The program was parameterized on the basis of the following three assumptions: One, the machine resolution or the display size allow reliable identification of fixations at the level of letters. Two, with a 60 Hz sampling rate, only clusters encompassing at least six samples (i.e. lasting
longer than 100 ms) will be called a fixation. Three, clusters with fewer
samples may occur between fixations and indicate the time necessary to
shift the eye. Since it takes up to six samples to complete a return sweep
across a line of text of about 20 degrees of visual angle, the time between
two fixations within a line should not exceed five samples.

In programming EMAN it was not assumed that data obtained during a certain
trial are of equally good quality. To the contrary, each program module
was motivated by a different set of problems all or some of which may apply
during a given trial. In the description of the modules, emphasis will be
placed on the problems they address and how solutions were implemented. In

general, conservative solutions are advocated since eye-movement researchers are usually blessed with enough data to compensate for segments of low-
quality data.

MODULE REDUCE

Prior to reduction there must be a criterion to distinguish clearly invalid
from potentially valid measurements. In videobased systems invalid measurement most obviously occurs when the subject blinks or when the identification of pupil and corneal reflexion fails. Both conditions lead to a
zero or greatly reduced pupil diameter. To determine a criterion for valid
measurement the program computes mean and standard deviation of non-zero
pupil diameters. Based on either a multiple of standard deviations or a
certain percentage of the mean, a lower boundary for pupil diameter is set.
After determining the criterion for potentially valid measurements, the
data are processed by the algorithm displayed in Figure 1. If the pupil
diameter is below the criterion, the sample is called a BLINK. Successive
blinks are collapsed into a BLINK CLUSTER. A sequence of valid samples is
combined to a POSITION CLUSTER if their coordinates are sufficiently similar. Note, however, that some clusters may contain only one sample.

```
.
.

BLINK?

        YES:    WAS LAST SAMPLE (CLUSTER) BLINK?

                YES:  UPDATE BLINK CLUSTER

                NO:   FINISH POSITION CLUSTER
                      START NEW BLINK CLUSTER


        NO:     WAS LAST SAMPLE (CLUSTER) BLINK?

                YES:  FINISH BLINK CLUSTER
                      START NEW POSITION CLUSTER

                NO:   IS PRESENT SAMPLE (CLUSTER) WITHIN WINDOW?
                      YES:  UPDATE POSITION CLUSTER
                      NO:   FINISH POSITION CLUSTER
                            START NEW POSITION CLUSTER
.
.
```

FIGURE 1. Reduction algorithm in Module REDUCE

If the pupil diameter is below the criterion, the top block of the algorithm is executed. If the last sample was a blink as well, the present
sample is simply added to the blink cluster. If the last sample was valid,
then the previous position cluster is closed and a blink cluster is started.

If a sample passes the first validity test, it will enter the second block
of the algorithm. There, if the last sample was a blink, the blink cluster

is closed and, since positions during blinks are not valid, its x- and y-values are set to zero. The present sample then becomes the seed for a new cluster of valid positions. If the last sample was valid as well, then, if the coordinates of the present sample are within a window around coordinates based on the running mean of a cluster of previous samples, the present sample is added to this cluster. Otherwise, the previous cluster is finished and the present sample becomes the seed for a new cluster. The algorithm that decides whether a sample is within a window was suggested by Mason (1976). A detailed description of the present adaptation can be found in Kliegl and Olson (1981). Since the positions were rescaled to display coordinates and since the resolution of the eye-tracker under the viewing conditions in the laboratory were plus or minus one character, the minimal distance between clusters was set at one.

New position clusters are formed if a single sample falls outside the window. The cluster resulting from this seed, however, may be within the window of the preceding position cluster. To ensure that successive clusters are at least the minimum distance apart from each other, the clusters themselves are processed by the reduction algorithm. Successive clusters are collapsed into one, if they are closer than the minimum distance. The algorithm is applied iteratively until no further reduction in the number of clusters occurs. Thus, the reduction iteration will stop, once all neighbouring position clusters have a minimum distance from each other.

Experience has shown that occasionally two clusters with virtually identical positions were separated by a single cluster containing mostly one or two samples. Since it is impossible that an eye movement can leave and come back to a certain position without an intervening fixation, it is assumed that these samples reflect spurious measurement and that the two clusters are actually one. To identify and remove outliers of this kind a second iterative algorithm compares triplets of neighbouring clusters. The outlier is counted for the duration, but its position information is ignored for the computation of cluster coordinates. The iteration stops once no further reduction in number of clusters occurs.

At the end of this two-stage iterative reduction algorithm the data are available as a sequence of clusters. For position clusters we know how many samples they encompass, their average horizontal and vertical positions and their average pupil diameters. Blink clusters have zero values in position and pupil diameter vectors. So far no fixation-saccade distinction was made and no sample was removed from the data base. The next module, CLEANUP, serves this purpose.

MODULE CLEANUP

The criterion for distinguishing valid from invalid measurements alone is not sufficient. There are other conditions that yield doubtful data quality, but are not necessarily reflected in loss of, or sufficiently reduced values of pupil diameters. For example, prior to a blink or loss of delimiters, the position information is already contaminated by the ensuing loss of measurement. Analogously, low data quality may be observed after an unstable measurement situation. To ensure data quality around blink clusters, the algorithm shown in Figure 2 is used. The data base is searched for blink clusters. Once a blink cluster is encountered, the program engages in a backward search until two consecutive position clusters with similar vertical position are found. Position clusters too discrep-

ant from the previous ones are recoded as blink clusters. A similar algorithm is used for a forward search after a blink cluster.

      :
      :

IS CLUSTER N A BLINK CLUSTER?

    NO: N: = N + 1

    YES: NI: = N
         FLAG IS "FALSE"
         UNTIL FLAG IS TRUE: ITERATE

ITERATE: HAVE POSITION CLUSTERS (NI - 1) AND (NI - 2) SIMILAR
          VERTICAL POSITION?

    YES: FLAG IS "TRUE"
         N = N + 1

    NO: RECODE POSITION CLUSTER (NI - 1) AS BLINK CLUSTER
        NI = NI - 1

END OF ITERATE
      :
      :

FIGURE 2. Ensuring stable measurement around blink clusters

In this case, only vertical information is used since unstable measurement
mainly occurs at the end or beginning of lines. At these locations horizontal positions are of course maximally discrepant. Since unstable measurement usually results in a big variation of vertical positions, the similarity between two consecutive position clusters can be set fairly liberal,
e.g., at two or three lines in a reading situation.

The next step towards a clean sequence of fixations is to remove blink and
saccadic position clusters (i.e. clusters with fewer than six samples) from
the data base. While deleting these clusters, the program records how many
samples are removed between two fixations. This permits the identification
of areas of low data quality that never resulted in an actual loss of delimiters. Usually a sequence of position clusters with few samples per
cluster is indicative of a condition in which the time between two fixations might well exceed estimates of durations of saccadic movements. As
mentioned previously, the maximum time between fixations within a line
should be shorter than the time for a linesweep. If the time is longer
than five samples, it may be inferred that the two fixations delimit an
area of bad measurement.

To retain information about areas of blink clusters and about areas of bad
measurement, the fixations delimiting these conditions are flagged in the
vector coding the pupil diameter. The fixation preceding the bad area contains the number of samples that were lost; the fixation following the bad
area codes whether it was a blink cluster or unstable measurement that
caused the loss. This information is quite important in future analyses

since it prevents wrong inferences about information processing between the two fixations. For example, we most certainly must not conclude that this area was skipped with a large saccade.

MODULE TEXT

For eye movements collected during reading, the described data processing was often not sufficient to achieve congruence between fixations and the appropriate lines of the text paragraph. Reasons for the failure could have been a temporarily invalid calibration table, greater machine noise associated with the vertical position, or a subject's fixation habits. In general, even the presence of only a small systematic drift may cause a great similarity in vertical positions of fixations at opposite ends of successive lines and may lead to misplaced fixations.

The TEXT module was written to remove systematic drift in vertical position within a line (see Figure 3). The program compares successive pairs of fixations. If their vertical and horizontal positions are within a certain range they are considered to be from the same line. In this case, if their vertical positions are not close to identical, the vertical position of the first fixation is replaced with the average of both. An uninterupted sequence of fixations judged to be within the same line will be called a cluster of fixations within a line (FWL). The program iteratively processes the data base until all fixations within FWL clusters have virtually identical positions.

```
    :
    :
FLAG IS "TRUE"

WHILE FLAG IS "TRUE" DO:

        FLAG IS "FALSE"

        FOR N = 1 TO (NUMBER OF FIX -1) DO:

                ARE NI AND (N + 1) FIXATION ON THE SAME LINE?

                        YES: ARE VERTICAL POSITIONS IDENTICAL?

                                YES: N = N + 1

                                NO:  VERT. POS. OF FIX. N = AVERAGE OF VERT. POS.
                                     OF FIX. N AND FIX. (N + 1)
                                     FLAG = "TRUE"
                                     N = N + 1

                        NO:  N = N + 1

        END OF FOR LOOP

END OF WHILE LOOP
    :
    :
```

FIGURE 3. Determining vertical position for fixations within a line

Once the iteration stops it is likely that more FWL clusters were identified than the text contains lines. If two fixations within a line were
separated by a large saccade or if the subject jumped back to re-read part
of the text, sequences of fixations within one line will yield separate
FWL clusters. To arrive at the appropriate set, FWL clusters indicating
the same lines need to be collapsed.

Under default parameterization the program assumes a fixed minimum distance between "true" lines and combines FWL clusters that are within this
margin. Ideally, the number of lines identified by the program should
correspond to the number of lines the subject read. An immediate evaluation of the quality of the computations is possible by inspecting th plot
of fixations and text which the program generates on the terminal.
this plot it is immediately obvious if the program found too many o   o
few lines. If the plot is bad, the user has the option of telling he
program the number of lines it should find. This will invoke another iterative procedure. At the beginning, the minimum distance between lines is
zero, but will be slowly increased with every iteration until the distance
has been identified that generates the desired number of lines. Again,
a terminal plot may be used for evaluative purposes. If the plot is still
not adequate, the user has the option of changing parameters that determine
whether fixation pairs should be allocated to a FWL cluster. A change in
these parameters will result in a different set of FWL clusters and, consequently, in different line values. If the program fails again, one
should probably delete parts of the text for which congruence could not be
achieved - or improve on the algorithms proposed in this paper.

TECHNICAL INFORMATION

EMAN was written in VAX/VMS Fortran 77 and is implemented on DEC VAX 11/780
machines. Input to the program are x,y-coordinates and pupil diameters
collected with a fixed sampling rate. The program reads all the records
of a trial into three vectors. Two additional vectors store the number of
samples in a cluster and the record number of the first sample in the cluster. At the beginning, the first vector is a unit vector; the second vector simply contains the record numbers. These five vectors form a workspace matrix common to all modules. All computations are carried out on
workspace elements with results replacing original values. In Module RE-
SCALE x and y coordinates in machine units are replaced with x and y coordinates of display dimensions. Modules REDUCE and CLEANUP effect elements of all vectors. Whenever samples are combined in a cluster, the number of samples and their average position and pupil diameter are stored in
corresponding elements of the first sample of the cluster. Then, the samples following the cluster are shifted up the appropriate number of elements. Thus, with every iteration the length of the workspace vectors is
reduced. It is possible to obtain output after every iteration to monitor
the reduction and "cleaning" process. After data were processed in TEXT,
the workspace vectors contain number of samples, horizontal display position, line number, pupil diameter (or a flag indicating loss or unstable
measurement following/preceding the fixation), and the record number of the
first sample of this fixation. The latter permits immediate identification
of the fixation in the raw data. The output is suitable for input to programs FIXPLT and MATCH described in Kliegl (1981). Substitution of the
TEXT module with algorithms designed for analyses of other than paragraph-
reading data is possible. Modules available are for analyses of tracking-
a-light task (Pavlidis (1981), Olson, Kliegl, & Davidson (in press)) and

the analyses of single-word reading data. Despite the large number of iterative procedures the analysis is quite fast. For example, an analysis of eye movements collected during reading of a 20-line long text paragraph (6538 records, reduced to 354 fixations) was accomplished within one minute. The program is available upon request.

REFERENCES

Kliegl, R., Automated and interactive analysis of eye fixation data in reading, Behavior Research Methods and Instrumentation 13 (1981) 115-120.

Kliegl, R. and Olson, R. K., Reduction and calibration of eye monitor data, Behavior Research Methods and Instrumentation 13 (1981) 107-111.

Mason, R. L., Digital computer estimation of eye fixation, Behavior Research Methods and Instrumentation 8 (1976) 185-188.

Olson, R. K., Kliegl, R. and Davidson, B. J., Dyslexic and normal reader's eye movements, Journal of Experimental Psychology: Human Perception and Performance (in press).

Pavlidis, G. Th., Do eye movements hold the key to dyslexia?, Neuropsychologia 19 (1981) 57-64.

Theoretical and Applied Aspects of Eye Movement Research
A.G. Gale and F. Johnson (Editors)
© Elsevier Science Publishers B.V. (North-Holland), 1984

INSTRUMENTATION CONSIDERATIONS IN RESEARCH INVOLVING
EYE-MOVEMENT CONTINGENT STIMULUS CONTROL

George W. McConkie, Gary S. Wolverton, & David Zola

Center for the Study of Reading
University of Illinois at Urbana-Champaign
Champaign, Illinois
U. S. A.

In the study of perception during reading the use of eye
movement contingent control of the stimulus display has
proved to be a useful research technique. With such a
system it is possible to experimentally manipulate, in
real time, the characteristics of the stimulus display
that is present on selected fixations as reading is in
progress, and to observe the effects of the manipulation
on the eye movement pattern. This technique can also be
used in the study of other on-going, visually based
tasks. This paper provides examples of how the technique has been used, and describes a number of the
instrumentation concerns which are important to consider
when setting up a system to do this type of research.

INTRODUCTION

There are three possible reasons for monitoring eye movements in psychological research. First, the eye movement records can serve as a source of
data. Sometimes aspects of these data, such as the durations of fixations
or the lengths or directions or velocity patterns of the eye movements, are
the dependent variables in the study. At other times, the data are used to
determine which trials should be excluded from analysis. Second, real time
information about the eye movements can be used as a basis for making on-
line experimental manipulations in the stimulus. For example, a tone can
sound or a change can be made in the visual array when the eyes fixate a
certain location. Third, real time information about the eye movements can
be used as a basis for other forms of measurement. EMG recording can be
enabled only when the eyes fixate a certain location, or EEG data can be
selected and averaged based on the location and/or movements of the eyes.

The focus of the present paper will be on the second of these alternatives:
using eye movement monitoring as a basis for making stimulus manipulations.
For a number of years we have been studying the nature of the perceptual and
eye movement control processes taking place as people read. This required
the development of a computer-based system that was capable of monitoring
people's eye movements as they read, and, on the basis of that information,
making real time changes in the text from which they were reading (McConkie,
Zola, Wolverton, & Burns, 1978). Using this system, we are able to allow
subjects to read text displayed on a cathode-ray tube (CRT), and as they are
reading, to manipulate the stimulus pattern that is present in the region
where they look for a particular fixation, or at a particular time during
the fixation. We then examine the effect which these display manipulations

produce on the eye movement patterns as a way of learning about the
processes being studied.

These techniques have been extremely fruitful in investigating aspects of
the perceptual processes as they occur during reading (McConkie, 1983;
Rayner, 1983). They have also been used to study perception in simpler
tasks (Bridgeman, Hendry, & Stark, 1975; Irwin, Yantis, & Jonides, 1983;
Levy-Schoen & Rigaut-Renard, 1981). We believe that they could be equally
useful in the study of perception in other visually-based tasks, such as
visual search and picture perception. Anticipating that other researchers
are likely to attempt to develop research of this sort, the purpose of the
present paper is to describe some of the considerations that must be taken
into account in selecting equipment for this type of research. While it
will deal primarily with eye movement contingent control of visual displays,
many of the points made will also be relevant to on-line control of other
forms of stimulus manipulation or data collection.

In order to illustrate some of the perceptual issues which can be investigated using these techniques, we will briefly describe four examples from
our own research.

1. We were interested in whether or not it is necessary for the eyes to be
   centered at the exact location to which they are sent on a saccade in
   order for processing to proceed normally. This was investigated by
   causing certain fixations to be "misplaced" slightly as people were
   reading. During certain saccades, the text was shifted two character
   positions left or right on the CRT, so that when the eyes stopped for
   the next fixation they were centered at a location different from where
   they normally would have been (McConkie, Zola, & Wolverton, 1980; see
   also O'Regan, 1981).

2. In order to determine whether visual information is acquired and used
   from specific visual regions during fixations in reading, the letters
   in these regions were replaced with other letters on certain fixations.
   Thus, use of this information would produce processing difficulties,
   discernible in the eye movement pattern (Underwood & McConkie, 1983).

3. We investigated whether particular aspects of the visual stimulus pattern present on successive fixations are brought together into a single
   mental representation. Certain characteristics of the text pattern
   were changed during occasional saccades, such as the spacing between
   words or the forms of the letters. If the system attempts to integrate
   these aspects of the visual array across successive fixations some
   degree of difficulty should be encountered, again being reflected in
   the eye movement pattern (McConkie & Zola, 1979; Rayner, McConkie, &
   Zola, 1980).

4. We studied the time characteristics of the perception and eye movement
   systems by producing changes in the display at specific times following
   the onset of a fixation and observing the effect which this had on the
   shapes of the distributions of fixation times (McConkie, Underwood,
   Zola, & Wolverton, 1983; Wolverton, 1979).

These types of studies make strong demands both on the eye movement monitoring equipment and on the equipment used to produce and manipulate the
stimulus. Decisions in the choice of equipment to purchase or develop are

crucial; wrong choices can greatly limit the research which can be done and introduce undetected artifacts into the data. Computer programming must be tight and well controlled. However, the nature of the constraints on the equipment and programming depends on the characteristics of the research to be carried out: these constraints are very strong for some types of studies and less strong for others. In order to describe the concerns that must be taken into consideration we will discuss two cases in which eye movement contingent control of a CRT display is required, one in which the display must be changed during the period of a saccadic eye movement, and a second in which it is necessary to make a change at some time after the beginning of a fixation.

CHANGING THE DISPLAY DURING A SACCADE

In order to change the stimulus display during a saccade, there are four things that must be accomplished prior to the beginning of the following fixation: 1) detecting the onset of the saccade, 2) determining that this is the saccade on which a change is desired, 3) changing the computer's display instructions so they can create the new image, and 4) actually having the new image present on the CRT. How much time there is to complete all this depends on the length of the saccade on which the change is desired. In reading, the shortest saccades can be completed in less than 20 msec, for instance. If the task is to look from one specified point to another which is some distance away, this time can be 90 msec or more, depending on the distance of the two points.

Detecting the onset of the saccade. How early in the saccade the eye movement monitoring system can detect that the eyes have begun to move depends on several factors. First is the speed of throughput of the eye movement monitor. Given that the monitor indicates that the eyes are at a given position, how much time has passed since they were actually in that position? With the scleral reflection approach to eye movement monitoring this throughput time can be a matter of no more than three or four msec or even less, unless longer delays have been introduced into the circuitry in order to filter out noise. With some forms of filtering, this delay can be over 20 msec. Therefore, in some cases a saccade could be completed before the eye movement monitor indicates that it has begun. With equipment using television technology to photograph the eye, and then processing the digitized image to identify the eye position, the eyes can begin to move during the 16 msec period required to complete one scan of the eye. Whether the output shows the eyes beginning to move depends on when during the scan the movement began, and whether the critical information in the image used to track the eyes lies toward the top or bottom of the frame. Thus, it would be quite possible to miss the beginning of the movement in one frame, and only detect it in the next, so that a short movement may be finished.

The second factor influencing how early an eye movement can be detected is the rate at which the eye position is sampled. How much time elapses between taking successive samples of the eyes' location? At least that much time can pass after the eyes begin to move before that movement is detected. In our own work we sample the eye position every msec, and compute how much movement has occurred in the last 4 msec. If this movement is above threshold, we then require a certain number of additional samples to be above threshold, as well. Depending on other factors, it may be possible in this way to reliably detect the onset of a saccade within a relatively few msec of the time that the eye movement signal begins to show the movement.

The third factor is the amount of noise in the eye movement signal. The
noise level essentially defines a region of indeterminacy around the eye's
position. The eye movement signal must move outside this region before an
eye movement can be reliably detected. If the noise level is high, then the
eyes must move further before the movement can be reliably detected. This
increases the likelihood that an eye movement sample will fail to indicate
that the eye has begun to move, thus delaying the time until the movement is
detected.

These three factors are additive. That is, the total amount of delay which
can occur in detecting the onset of a saccade is at least the total of the
maximum delay possible from each of these three sources. This total delay
can be sizeable. A combination of slow throughput, slow sampling rate, and
high noise level can result in total delays of 50 msec or more. In this
case, all but long saccades would be completed before their initiation was
detected. The combination of fast throughput, high sampling rate and low
noise level can permit saccades to be reliably detected within less than 10
msec after their initiation.

Determining whether a change should occur on this saccade. The amount of
time required for this stage varies widely, depending on the requirements of
the study. For example, in the simple case, display changes are called for
on every saccade, every leftward saccade, or on the 5th, 10th and 15th saccades made. In a more complex case, a change is made only if the prior fixation is in a certain region of the display.

The most difficult cases are those where the display change is made contingent on aspects of characteristics of the saccade itself. For instance,
a change may be desired only if the saccade will be of at least a certain
length, or will take the eyes to a certain location. These decisions
require waiting during the saccade to see if it reaches a certain velocity
or passes a certain boundary, or until sufficient information is available
to permit an accurate prediction of the location of the following fixation.
Obviously, these latter types of decisions require eye movement data
obtained at high sampling rates and with low noise, and they leave only the
latter part of the saccade time available for making stimulus changes prior
to the beginning of the following fixation.

Making changes in the display instructions. The time required to make
changes in the display instructions can also be quite variable from study to
study. This time consists of the total time required to calculate the
necessary changes, and, if necessary, to transmit those changes (or a copy
of the changed list of display commands) to the display device. In the simplest case, alternative images have been previously prepared and are present
within the computer's high-speed memory. The CPU, or a special display processor having access to high-speed memory, controls the display device
directly. In this case, when a change in the display is required, it is
simply a matter of taking display commands from a different region of
memory. Many display changes can be accomplished in this simple manner. On
the other hand, if there are many possible forms which the next display can
take, contingent upon eye position, then each alternative must be computed
when it is needed. This computation time varies with the complexity of the
displays involved. Also, once the display commands have been modified, if
they must be transmitted to the memory of the display device, there will be
an additional delay, the amount of which depends on the speed of transmission of information between the devices involved.

Realizing the image on the display device. The most common electronic
displays for psychological research involve illuminating the image on the
screen a point at a time. This is done either as a complete raster scan or
in a sequence more optimally related to the characteristics of the display
itself, using a point-plotting device. In either case, the process takes
time. The amount of time for raster displays is usually either 16 or 32
msec; for point-plotting equipment the time depends on the complexity of the
image and the efficiency with which it was coded. However, this does not
mean that the image can be realized on the screen within this period of
time after it is called for. In many instances, it is not possible to
begin "painting" a new image on the screen until the beginning of a refresh
cycle. That means that with a 16 msec refresh cycle, if a new image were
called for just after the beginning of a cycle, it would be necessary to
complete that cycle and then to display the new image on the next cycle.
Thus, a total of 31 msec could elapse before the initial display of the next
image were complete. With a 32 msec refresh rate, that time could be as
much as 63 msec. Thus, substantial delays can result at this point.

Two approaches can be taken to reducing this source of delay. The first is
to use simple images with a point-plotting device so that the image can be
refreshed rapidly. The second approach is to have the facility to begin
displaying a new image in mid-cycle. With a point-plotting device it is
possible to break the cycle at any point and begin displaying an alternative
image. Thus the change can be completed within the refresh cycle time.
Also, with some graphics equipment it is possible to point the controller to
a new region of high-speed memory at the end of any horizontal scan during
the refresh process. It is therefore possible to display the new image
within the period of a single refresh cycle after it is requested.

Finally, some recent raster scan graphics controllers have a degree of flexibility in the refresh rate, which makes it possible to attain refresh
cycles requiring less than the normal 16 msec period. Further development
of this equipment could greatly facilitate eye movement contingent display
control where complex displays are required.

Summary. When making a change in the visual display during the period of a
saccadic eye movement, there are a number of steps which must occur within a
time ranging from 20 to 90 msec, depending on the length of the saccades
involved in the study. During this period of time it is necessary to detect
the onset of the saccade, determine whether a display change should occur
during this saccade, make the necessary modifications in the display
instructions within the computer and perhaps transmit these to the display
device, and actually realize the new image on the screen. In order to
accomplish this, it is necessary to have equipment with the required characteristics and to program it with a concern for minimizing delays.

Desirable system characteristics are the following:

1. Eye movement monitoring equipment with fast throughput, low noise, and
    which yields new information with high frequency.

2. A program which samples the eye position with a high frequency, which
    can detect saccade onset and make the decision about whether to ini-
    tiate a display change as early as the study permits, and which minim-
    izes the amount of computation involved in changing the image.

3. A display device capable of being rapidly refreshed and of initiating
the presentation of a new image part way through the refresh cycle.

How many non-optimal characteristics can be tolerated in the equipment
depends, of course, on the demands of a given study.

CHANGING THE DISPLAY DURING A FIXATION

When it is necessary to make an experimental manipulation at a certain time
after the beginning of a fixation, many of the concerns described in the
prior section again apply. This is particularly true if the change must
occur relatively early in the fixation, such as 20 to 50 msec after it
begins. Again, one is faced with the problems of carrying out all of the
steps which are required in the time which is available. If the change is
not required until 100 msec or more after the onset of the fixation, more
time is available for the steps required.

However, there is another concern which arises when dealing with changes
during a fixation. When manipulating the stimulus during saccades, there is
less concern about just when the change takes place. It has been our
experience that the visual system is quite insensitive to display changes
made while the eyes are moving. Blanking out the display is detected, but
replacing text strings with other strings is not. Thus, it is not critical
exactly when during the saccade the change is made. However, when making
changes during the fixation, the timing of the change often becomes the
point of the research. In order to make the change at the specific time
desired, it is necessary to accurately identify the beginning of the
fixation, since that is the base for timing. There are two problems which
arise here, one having to do with the equipment being used, and the other
with characteristics of the eye movements themselves.

With respect to the equipment used for monitoring the eye movements, all of
the concerns described earlier related to detecting of the onset of a saccade apply here in detecting the onset of the fixation. The same sources of
delays are present. Furthermore, there can be considerable variability in
how soon after the beginning of a fixation it is detected. For instance,
with a 16 msec sampling rate there is an inherent 16 msec variability in
when the fixation is said to begin. With the noise in a relatively clean
signal this variability can become 32 msec. With a noisy signal, the variability increases still further. Variability in determining the point from
which timing should begin results in variability in the time that elapses
from the actual beginning of the fixation until the display change takes
place. Furthermore, as described earlier, characteristics of the refresh
process can add variability to the amount of time that elapses from the time
a display change is called for until it is actually displayed. How much
variance in timing can be tolerated depends, of course, on the nature of the
experiment. If relatively precise control is needed, equipment for the
research must be selected with care.

The preceding discussion was based on the assumption that an accurate,
noise-free signal sampled at a high rate would clearly indicate when a fixation begins. However, this is not the case. As the eyes decelerate during
a saccade there is typically a period of overshoot, with the signal coming
to a peak and then moving back the other direction and gradually stabiliz-
ing. This probably represents a settling time of the eyes, during which
they center themselves in the socket, and perhaps regain their shape after

responding to the forces of the ocular muscles. This overshoot is exaggerated in equipment which monitors reflections from the lens as well as from the eye's surface, suggesting that the torque applied to the eye may induce distortions in the internal parts of the eyes, which must also return to normal at the end of the eye movement. The problem, of course, is what to identify as the beginning of the fixation; whether this should be the peak of the overshoot at which time the forward component of the movement is completed, whether it should be at the end of the settling period, or whether it should be at some other time. There is no clear answer at the present time and this adds variance both within and across studies in this area. It is our contention that the beginning of the fixation should be identified with the point at which visual information of the type needed for the task being used can first be acquired from the display. We will shortly be conducting research to attempt to identify where this point occurs.

One final comment should be made with regard to the amount of variability involved in identifying the beginnings of saccades and fixations. In many eye movement studies the primary data of interest are the durations of fixations. The duration of a fixation is, of course, simply the time from the end of one saccade until the beginning of the next. The error variance in fixation durations, then, is a sum of the error variance in identifying each of these defining events, since it is reasonable to assume that the two sources of variance are uncorrelated. Thus, while the previous discussion has been concerned primarily with the problems which this sort of variance produces for employing eye movement contingent stimulus control, in fact many of the same concerns arise even in the case in which no display changes are required, but where accurate eye movement data are desired. Those factors that contribute to accurate identification of saccade and fixation beginnings also contribute to accuracy in fixation duration data.

Summary. Making display changes at precise times following the onset of a fixation requires the ability to reliably detect when the fixation begins. This requires equipment which has fast throughput, a low noise level, and which can be sampled at a high rate. Even then, noise free eye movement data would not indicate a clear point at which the fixation begins. This adds another source of variance. Research aimed at identifying at what point in the fixation perception of visual detail is possible may help resolve this problem. Finally, the eye movement monitoring issues which have been discussed are not only of importance in controlling the stimuli contingent upon eye movements. They are also of concern if the desire is to obtain an accurate measure of fixation times.

ADDITIONAL CONCERNS

There are two additional equipment concerns that should be mentioned. First, the necessity of making fast display changes requires that the display image fade quickly. Thus, for CRT displays a fast-decay phosphor is required. Second, rapid sampling of the eye position generates a great deal of data, requiring large amounts of storage space. It may seem reasonable to bypass this requirement by doing on-line reduction of the data as they are being collected. We believe this to be unwise. Having a complete data record is useful for three purposes. If the time at which display changes take place is recorded in the data, for instance by setting a bit pattern in a data word collected at the time a display change is called for, it is possible to verify that the system was operating properly. In this type of research, there is no other way to be sure that this is the case. Also,

more accurate data reduction programs can be developed when reduction is
off-line. There are not the time limitations, and the program can move forward or backward along the data stream to find saccade beginnings and endings. Finally, there are frequently irregularities in the eye movement
data, probably resulting from blinking and squinting. These can lead to
strange patterns in the reduced data, and examining the raw data can indicate whether or not these data are usable.

AN EXAMPLE

In our eye movement contingent display control system, we use the SRI Dual-
Purkinje Image Eyetracker (Cornsweet & Crane, 1973). It is claimed to have
a throughput of about 4 msec. The noise level places a band of indeterminacy around the mean signal value equivalent to less than 2 min of arc of
eye movement. Thus, an eye movement can be reliably detected by the time
the eyes have moved 5 min of arc or less. We sample the eye position every
msec, checking the distance moved over the preceding 4 msec. The peak of
the overshoot at the end of the saccade is detected by a change in direction
of the values being obtained over a 4 msec period, and the end of the
overshoot period is detected by finding less than 4 min of arc of movement
over a 4 msec period. Furthermore, we have found that we can predict the
location of the next fixation, usually within 40 min of arc, once we have
identified the point of peak velocity within a saccade.

The eye movement signal is sampled by a PDP-11/40 computer, which, in addition to the CPU, has a display processor which has access to high-speed
memory. The display processor controls a point-plotting CRT. With this, we
can present a single line of text with a 3-msec refresh rate. Thus, the
entire line can be changed within 3 msec.

Most of our studies involve a relatively few alternate lines which can be
displayed, contingent upon the eye's location and state. Thus, the alternative lines are typically stored in high-speed memory. Display changes simply involve the CPU directing the display processor to a new region of
high-speed memory where an alternative display list resides. Thus, the
change requires minimal time, with no transmission time required.

In our most recent study, we were able to detect the onset of a saccade,
predict the location of the following fixation, and change the line of text
if the fixation was going to be on a particular word, within the period of
all but the shortest saccades. In this way, it was possible to study the
value of obtaining peripheral visual information from a word on its later
identification.

The ability to exert eye movement contingent control over visual and other
stimuli provides a powerful research technique which permits detailed investigation of perception and eye movement control as people are engaged in
on-going tasks. This technique has been used in the study of reading and in
some simpler tasks. It should now be extended to the investigation of
visual search, picture perception, and other visually-based tasks.

REFERENCES

Bridgeman, B., Hendry, D., & Stark, L. Failure to detect displacement of the visual world during saccadic eye movements. Vision Research, 1975, 15, 719-722.

Cornsweet, T. N., & Crane, H. D. Accurate two-dimensional eye tracker using first and fourth Purkinje images. Journal of the Optical Society of America, 1973, 63, 6-13.

Irwin, D. E., Yantis, S., & Jonides, J. Evidence against visual integration across saccadic eye movements. Perception & Psychophysics, 1983, 34, 49-57.

Levy-Schoen, A., & Rigaut-Renard, C. Pre-perception ou activation motrice au cours du T. R. oculomoteur? In J. Requin (Ed.), Anticipation et comportement. Paris: Centre National de la Recherche Scientifique, 1981.

McConkie, G. W. Eye movements and perception during reading. In K. Rayner (Ed.), Eye movements in reading: Perceptual and language processes. New York: Academic Press, 1983.

McConkie, G. W., Underwood, N. R., Zola, D., & Wolverton, G. S. Some temporal characteristics of processing during in reading. Unpublished manuscript, University of Illinois, 1983.

McConkie, G. W., & Zola, D. Is visual information integrated across successive fixations in reading? Perception and Psychophysics, 1979, 25, 221-224.

McConkie, G. W., Zola, D., & Wolverton, G. S. How precise is eye guidance? Paper presented at the annual meeting of the American Educational Research Association, Boston, MA., April, 1980.

McConkie, G. W., Zola, D., Wolverton, G. S., & Burns, D. D. Eye movement contingent display control in studying reading. Behavior Research Methods and Instrumentation, 1978, 10, 154-166.

O'Regan, K. The "convenient viewing position" hypothesis. In D. F. Fisher, R. A. Monty, & J. W. Senders (Eds.), Eye movements: Cognition and visual perception. Hillsdale, NJ: Erlbaum, 1981.

Rayner, K. The perceptual span and eye movement control during reading. In K. Rayner (Ed.), Eye movements in reading: Perceptual and language processes. New York: Academic Press, 1983.

Rayner, K., McConkie, G. W., & Zola, D. Integrating information across eye movements. Cognitive Psychology, 1980, 12, 206-226.

Underwood, N. R., & McConkie, G. W. Perceptual span for letter distinctions during reading (Tech. Rep. No. 272). Urbana: 1983.

Wolverton, G. S. The acquisition of visual information during fixations and saccades in reading. Paper presented at the annual meeting of the American Educational Research Association, San Francisco, Calif., 1979.

Section 2

**PROPERTIES OF THE
SACCADIC EYE MOVEMENT SYSTEM**

Theoretical and Applied Aspects of Eye Movement Research
A.G. Gale and F. Johnson (Editors)
© Elsevier Science Publishers B.V. (North-Holland), 1984

PROPERTIES OF THE SACCADIC EYE MOVEMENT SYSTEM

INTRODUCTION

John M. Findlay,
Department of Psychology, University of Durham,
Durham, England.

The study of saccadic eye movements is currently experiencing a period of enormous fertility, yielding much of interest both for physiologists and for psychologists. The reasons for this are not hard to find. In contrast to most of the motor activity that the body produces that of the eye is subject to very close constraints (three rotational degrees of freedom only) and is also achieved with a constant mechanical load. This renders the oculomotor system amenable to analysis by precise quantitative modelling, and at the level of neurophysiology it is probably true to say that our understanding of the neural networks and channels leading this motor output (see for example Leigh and Zee, 1983), is at least as good as that of the visual sensory pathways. Even more awe-inspiring is the fact that in many cases it is possible to link these two research areas and, in the case of simple oculomotor reflexes, to trace the whole sensorimotor links and see in detail the ingenious and beautiful systems by means of which the brain achieves its goals.

In the case of the saccadic system, this also forms a starting point for the investigation of certain aspects of voluntary activity. Saccadic eye movements in almost all cases represent a co-ordination of both sensory and voluntary factors. The situation does occur in which an observer makes a totally reflex saccade to a visual stimulus which 'catches the eye'. This is, however, a relatively rare response. Equally rare is the hypothetical case of an entirely 'voluntary' saccade where the observer moves her eyes into empty space simply for the fun of it rather than for the purpose of looking at something. Thus examination of the properties of human eye movements with carefully controlled tasks and instructions can be expected to lead to a precise understanding of the way in which sensory and voluntary factors interact, and this must surely form a productive research area for many years to come.

The papers in the Symposium formed a representative and informative sample of this activity. Deubel, Wolf and Hauske present an elegant model of the visual control of saccades. Crucial to the model is the concept of spatial and temporal filtering of the visual input. The existence of such filtering is indicated by several behavioural studies on saccades, and is also a concept widely used in current work on visual psychophysics (De Valois and De Valois, 1980). The paper by De Bie and Van Den Brink also addresses a fundamental issue in eye movements; the miniature eye movements that occur during fixation. They point out that all previous studies of this work have been hampered by looking only at spontaneous activity. By introducing controlled, small, stimulus movements and offsets they demonstrate that two processes are involved and also produce a quantitative model. Technical advances often underlie scientific ones, and their study was made possible by the use of a relatively new technique for recording; the scleral search coil method. Use of this method enabled Findlay and Harris to examine the paradigm of double-step tracking without the restriction, imposed in most previous studies, to movements along the horizontal axis.

They demonstrate that direction and amplitude do not appear to be separately
programmed, and also present further evidence for a goal seeking stage in
saccadic programming, characteristic of many recent models.

It is gradually being realised that the conventional text book description
of the saccade as a stereotyped reflex-like response is, at best, an
oversimplification. Several papers make this point. Wolf, Deubel and
Hauske confirm and extend previous findings which show that if it is arranged that saccades consistently fail to achieve the intended target (by the
experimenter cunningly moving the target before the eye gets there), then
an adaptive resetting of the 'gain' of the system occurs, recalibrating the
relationship between target eccentricity and saccade amplitude. Several
papers make the point that the trajectory of saccades is much more variable
than often supposed. The careful analysis by Van Gisbergen, Van Opstal and
Ottes shows this by analysing a skewness parameter in both normal and
diazepam influenced movements. Voluntary influences can also affect the
trajectory. Crawford establishes, using feedback training procedures,
that subjects can reduce the peak velocity of their saccades. Similar
reductions of peak velocity have been reported in several studies to result
from 'fatigue', a concept which is generally ill-defined and particularly
obscure in relation to visual tasks. Sen and Megaw show clearly that prolonged task performance can be achieved without any consistent peak velocity
reduction. Modifiability of saccade trajectories is again implicated in the
work of Zangemeister and Huefner on saccades during active head movements.
As well as the expected modification of trajectory that occurs as a result
of the interaction of the VOR with the saccadic movement, their analysis
of head-eye trajectories suggests that more than one centrally programmed
saccade profile may be occurring.

Moving increasingly to papers concerned with the involvement of voluntary
factors in saccades, several contributions show that such processes are not
too elusive to be attacked by laboratory methods. The ability to produce
a tradeoff between the accuracy of a movement and its speed of execution
has long been recognised as a fundamental characteristic of motor systems,
whether operating in a ballistic or a guided mode. Surprisingly, little
attempt has been made to investigate this in relation to the control of eye
movements. Kapoula presents data to show that such a tradeoff is indeed
characteristic of the saccadic system. When subjects were asked to perform
a task which involved acquisition of a target whose characteristics could
only be discriminated with precise fixation, then the variability of the
resulting movements was reduced. The effect is quite small, but consistent
enough to offer a positive answer to the question. All the papers discussed
so far restrict their attention to one single individual saccadic movement.
Rather different factors become important when sequences of movements are
concerned. One attractive hypothesis is that the size of saccades in read-
ing-like tasks is determined by the area in the periphery of vision from
which information can be extracted on the prior fixation. This accounts for
data in many situations, but Levy-Schoen, O'Regan, Jacobs and Coeffe show
that when a character sequence is scanned with constraints designed to eliminate all other influences, the 'visibility span' concept just described
does not fare very well in predicting saccade sizes. While many workers in
the traditions relating to saccadic eye movements are loth to use the concept of 'attention', current psychological work in other areas is gradually
refining this notion and it may be that the time is not far off when it can
be defined with sufficient precision to take its place in oculomotor models.
Vaughan reviews some of this work and explores the finding that saccades to

recently stimulated locations in space show significantly longer latencies than those to control non stimulated locations.

References

De Valois, R.L. and De Valois K.   Spatial Vision; Annual Review of Psychol. 31 (1980) 309-341.

Leigh, R.J. and Zee, D.S.   The Neurology of Eye Movements, (F.A. Davis, Philadelphia, 1983).

Theoretical and Applied Aspects of Eye Movement Research
A.G. Gale and F. Johnson (Editors)
© Elsevier Science Publishers B.V. (North-Holland), 1984

THE EVALUATION OF THE OCULOMOTOR ERROR SIGNAL

H. Deubel, W. Wolf and G. Hauske

Lehrstuhl für Nachrichtentechnik
Technische Universität München
Arcisstr. 21, 8000 München 2
FRG

A spatial-temporal model of saccadic control is proposed which predicts saccadic responses to complex spatial and temporal target configurations and is close to underlying physiological structures. The model consists of an afferent preprocessing stage organized in parallel channels and a Spatial-Temporal Translator which evaluates the oculomotor error signal by determining the center of gravity of the visual input signals. It is demonstrated that spatial preprocessing forms an important part in oculomotor control.

INTRODUCTION

A large number of investigations have been performed to study basic features of saccadic programming using single luminous dots on homogeneous background as targets. They led to current models of saccadic control which predict the response of the oculomotor system to step patterns of increasing complexity /1,2/. Basic to these models is that the visual input is (hypothetically) reduced to the "retinal error signal" (i.e. the angle between foveal gaze and target position) which serves as the model input. At least outside our laboratories, however, the visual environment consists of complex spatial structures offering a variety of potential targets for the eye. For the selection of one target among many alternatives pattern recognition processes are necessary. Then, the oculomotor error signal has to be evaluated by determining the coordinate of the visual structure which has been "labeled" as target. With the step outside our labs a conceptual shift occurs in the philosophy of models of oculomotor control: Preprocessing of spatial visual information becomes an essential component of oculomotor behaviour. In conformity with a suggestion of Robinson /1/ we want to introduce a new concept of saccadic control which embodies the spatial nature of the real input signal by considering space and time equally. Further, the involvement of complex pattern recognition in the generation of the error signal is demonstrated.

A SPATIAL-TEMPORAL MODEL OF SACCADIC CONTROL: THEORETICAL CONSIDERATIONS

An essential feature of the visual system is its capability of coding visual information in parallel channels which are retinotopically organized at lower cortical levels. As to sensory-motor processing, it is well-known from neurophysiological studies that retina and visual cortex project onto the superior colliculus (SC) forming a retinotopic sensory map. Deeper layers of the SC form a motor map with movement fields which correspond to the sensory map. Local stimulation of these neurons leads to a saccade within only 20 msec which is directed to the corresponding retinal field /3/. This makes clear that close to saccade onset target position is coded in a retinotopic map by the site of the firing of an individual cell. At the next stage of the oculomotor branch, the brainstem, saccade

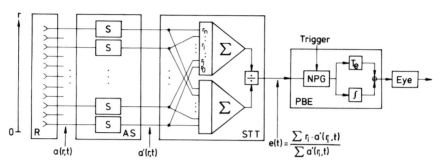

Fig 1: Spatial-temporal model of saccadic control.

size is coded in the discharge rate of the burst neurons /4/. Hence an essential question in saccade generation concerns the translation of spatially distributed information into the scalar "oculomotor error signal" specifying the amplitude of the desired saccade. For this mechanism which Robinson /1/ coined the "Spatial-Temporal Translator" (STT) Sparks and Mays /5/ postulate a neural circuit which performs an integration of the input excitation weighted by the retinal eccentricity of the stimulus. This concept however predicts that simultaneous stimulation of two retinal sites leads to a larger saccade than stimulation of one site alone. Further, saccades are largely independent of luminance and size of a single peripheral stimulus. Therefore, an additional, normalizing operation is demanded. These considerations lead to a "spatial" model which is shown in Fig. 1.

The visual information $a(r,t)$ in space (r) and time (t) given by the luminance distribution on the retina (R) is processed by parallel channels of the afferent preprocessing stage (AS). The input/output relationship of a single channel is described by the operator S. The preprocessed signals $a'(r,t)$ are fed into the STT which performs a spatial integration of the input signals weighted by their eccentricities r_i, and a normalization to the mean input intensity. The scalar oculomotor error signal $e(t)$ is input to the premotor brainstem elements (PBE, adopted from Robinson /1/). Triggered by an independent decision mechanism the neural pulse generator (NPG) forms the appropriate innervation to the eye muscles. The model suggests the following two hypotheses which we investigated experimentally:

1. THE EFFECTIVE SACCADIC GOAL IS THE CENTER OF GRAVITY OF THE TARGET CONFIGURATION

Weighted spatial integration and a normalization to the mean input intensity define a system which determines the coordinate of the center of gravity of the input signal. Consequently, the model predicts the eye to land at the center of gravity of the target configuration. Experimental evidence for this assumption came from recent investigations of Findlay /6/. In his experiments subjects had to saccade to targets in the periphery which consisted of two equiluminant squares with large spatial separation. The finding was that the saccades land at an intermediate position between the targets which in fact can be roughly described by the term "center of gravity" of the global target configuration. Our first experiment from which preliminary data were published recently /7/ confirms

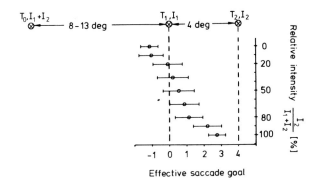

Fig. 2: Mean eye position after primary saccade (abszissa) plotted as a function of relative target intensity (ordinate). Bars indicate ± 1 SD.

and extends Findlay's basic findings demonstrating that the concept of the center of gravity is also valid for various target intensities.

The subjects fixated a bright LED (T_0) on a screen (see Fig. 2). After a random delay, two peripheral targets (T_1, T_2) with different intensities (I_1, I_2) were presented for 100 msec. Their intensities were varied from trial to trial keeping their sum constant ($I_1+I_2=60 \text{ cd/m}^2$). Since the subjects generally did not recognize that two targets were simultaneously present, they were simply told to "follow the target as fast as possible". Triggered with the primary saccade, either T_1 or T_2 reappeared and served as starting point for the next trial (Detailed information about stimulus generation, eye movement recording and data analysis are given elsewhere /8/).

The experimental results in Fig. 2 show the mean eye position at the end of the primary saccades relative to the targets as a function of relative target intensity. If only a single target is present (0% and 100% condition) the saccades fall short of the respective target due to the usual undershoot of about 10% of target angle. For differing intensities, the data show a linear relationship of the saccadic amplitudes to relative target intensity. This means that following the prediction by the proposed concept of the STT a spatial average is formed in which eccentricity and intensity are equal factors.

2. AFFERENT INFORMATION PROCESSING DETERMINES THE SACCADIC AMPLITUDE TRANSITION FUNCTIONS.

Experiments in which a target is displaced before the saccade reveal a gradual transition of the effective saccade goal from one target position to the next /2,9/. These amplitude transition functions demonstrate temporally integrative properties of saccadic control which are incorporated in the model of Becker and Jürgens /2/ by an input-controlled integration of the retinal error signal. In terms of our spatial-temporal model (Fig. 1) step-like displacements of a target mean that different retinal sites are subsequentially stimulated. This is exemplified in Fig. 3a showing the input excitation a(r,t) for the case of an impulse-like target displacement in a spatial-temporal representation. Assume

H. Deubel et al.

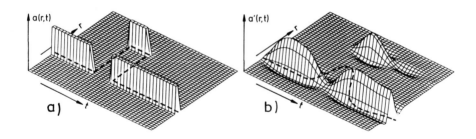

Fig. 3: a) Spatial-temporal representation of the retinal input
signal a(r,t) for an impulse-like target displacement. b) Spatial-
temporal excitation a'(r,t) after filtering by the preprocessing stage.
The dashed line gives the time course of the oculomotor error signal.

now that the effect of afferent preprocessing can be described by filtering the signal in space and time /10/. The output signal a (r,t) is
what the STT "sees". The dashed line in Fig. 3b then illustrates the time
course of the center of gravity of the filtered signal as calculated by
the STT. From the analysis of the model in Fig. 1 it follows that this
curve is equal to the response of a single channel S to flash-like stimulation. This means that the transfer characteristic of the afferent preprocessing stage is reflected in the saccadic amplitude transition functions. Therefore, it should be of interest to compare impulse responses
of the sensory pathways derived from perceptual experiments /11,12/
with amplitude transition functions to impulse-like target displacements.

The se amplitude transition functions were determined with the experimental paradigm given in Fig. 4. We studied correction saccades provoked
by an artificially induced refixation error which proved to be programmed
on basis of visual reafference /8/. The primary saccade evoked by the
first target step triggered a second target displacement of 2 deg. After
a delay varied randomly between 0 and 200 msec a further, impulse-like
target displacement with duration D and positive or negative amplitude
A was performed.

Fig. 5 a-c display the effect of the impulse-like target displacement
on the amplitudes of the correction saccades as a function of the temporal separation ITS of correction saccade onset from test impulse onset.
The data reveal that the amplitudes of the saccades are systematically
increased or decreased if they occur later than 70 msec after impulse
onset. The effect is slightly reversed for saccades later than 180 msec
after impulse onset which means that the test stimulus can make these
saccades fall short of both targets. The similarity of the curves with
normalized impulse responses of the afferent visual system determined
in perceptual experiments is demonstrated by Fig. 5d (redrawn from
/11/ and /12/) . It is therefore suggested that common visual pathways
are involved in perception and sensory-motor processing.

The data demonstrate that the results from two basically different
types of experiments can be comprised in a model which incorporates the
spatial nature of the input signal as well as the spatial-temporal pro-

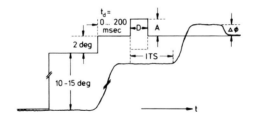

Fig. 4: Definition of stimulus and response parameters.

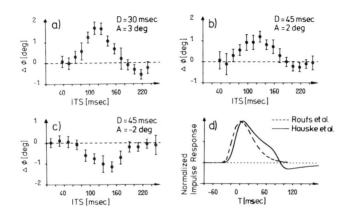

Fig. 5: a-c) Mean increase Δ∅ of correction saccade amplitudes as a function of ITS for different values of the parameters D and A. Vertical bars indicate twice the 99% significance level for the difference to the mean amplitude without an impulse.
d) Normalized impulse responses of the sensory visual system derived from perceptual experiments.

cessing in the afferent system. An essential element of this concept is the STT which serves to reduce spatially coded information to the scalar oculomotor error signal. According to our model, the STT evaluates the center of gravity of the input by means of weighted integration. Encouraging neurophysiological evidence for the validity of the center of gravity concept of the STT comes from an observation of Robinson /3/ that simultaneous stimulation of two sites in the SC leads to a saccade which is in fact determined by a weighted average of the saccade vectors that result when each site is stimulated by itself. Weighting factor is the relative intensity of the two stimulation currents. These data imply that the averaging occurs at or below the level of the SC.

On the other hand, some of Findlay's data /6/ reveal deviations from the predicted center of gravity showing that the effective saccadic goal tends to be closer to that stimulus which is nearer to the fovea. Further it was shown that the effect is, at least partially, under voluntary control /13/. These data may reflect a refined weighting of the visual stimuli according to their behavioural significance. Clearly, our model is only valid for fast, reflexive saccades to simple targets. Flexibility of the saccadic behaviour, on the other hand, can be implemented in the concept by increasing the complexity of afferent preprocessing. The following experiment suggests that complex pattern recognition processes are in fact involved in saccadic control.

PROGRAMMING OF CORRECTION SACCADES IN COMPLEX ENVIRONMENT

In order to evaluate the postsaccadic oculomotor error signal from the visual reafference the visual structure which was selected as target has to be relocalized after the primary saccade. This seems to be an easy task if the target structure can be clearly distinguished from the background. But what happens if nonfigural structures which we cannot immediately reidentify are used as targets for the saccade? In the following experiment we investigated whether perceptual reidentification of the target is a necessary condition for the appropriate oculomotor response.

Our subjects had to perform horizontal goal-directed saccades on extended, electronically generated pseudo-noise patterns of vertical bars which changed from trial to trial. Patterns and experimental sequence are illustrated in Fig. 6. The subject fixated on a given fixation line (I). After a random delay, the saccadic target was defined by a dark/bright inversion of a small (0.75 deg) area, 5 deg in the periphery (II). During the primary saccade a refixation error was artificially induced by displacing the whole scene by 0.5 to 1 deg (Type 1). In order to test whether background or, alternatively, target structure is primarily used to determine the postsaccadic error, a second experiment (Type 2) was performed in which a target area of 2 deg was displaced intrasaccadicly while the other parts of the scene remained stationary (III). Since target and background had similar structure it was not possible to recognize the target after the primary saccade nor to perceive its intrasaccadic displacement. Hence, this experimental paradigm allows separation of visual information processing in the saccadic system from perception.

Fig. 7 represents the distributions of the final eye position after the correction saccades in the two experiments. Parameter is the size d of the target displacement. The data clearly show that, for both types of experiments, the distributions are shifted according to the size of target displacement. This means that involuntary correction saccades (with short latencies of 160-200 msec) occurred which accurately eliminated the postsaccadic refixation error.

The data demonstrate the existence of an evaluation process which is based on an interaction between pre- and postsaccadic visual information and is independent from perception. We conclude that defining a target as saccadic goal, visual information about the target structure must be stored in a spatially organized buffer and, after the saccade, compared with the actual foveal reafference in order to determine the oculomotor error signal. These findings are in agreement with other investigations /14,15/ giving evidence for an interaction between pre-and postsaccadic visual information.

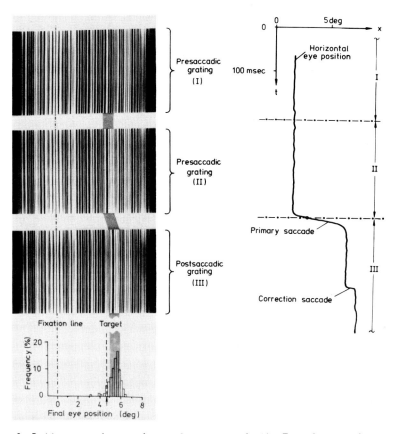

Fig. 6: Patterns and experimental sequence in the Type 2 experiment.

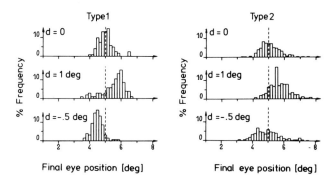

Fig. 7: Final eye position after correction saccades. The vertical dashed line gives the original target position before displacement.

A considerable gap exists between the simple model derived for the programming of saccades to isolated targets and the more complex information processing in the sensory-motor system as demonstrated in our last experiment. It mainly concerns our knowledge about the selective "gating" functions of the preprocessing stage. Investigating goal-directed eye movements to increasingly complex spatial patterns, however, should allow to cast some light on the underlying mechanisms.

REFERENCES

/1/ Robinson, D.A., Models of the saccadic eye movement control system, Kybernetik 14 (1973) 71-83.
/2/ Becker, W. and Jürgens, R., An analysis of the saccadic system by means of double step stimuli, Vision Res. 19 (1979) 967-983.
/3/ Robinson, D.A., Eye movements evoked by collicular stimulation in the alert monkey, Vision Res. 12, (1972) 1795-1808.
/4/ Robinson, D.A., Oculomotor unit behaviour in the monkey, J. Neurophysiol. 33 (1970) 393-404.
/5/ Sparks, D.L. and Mays, L.E., The role of the monkey superior colliculus in the control of saccadic eye movements: A current perspective, in: Progress in oculomotor research, A.F. Fuchs and W. Becker (eds.), Elsevier North Holland, Amsterdam and New York, (1981) 137-144.
/6/ Findlay, J.M., Global processing for saccadic eye movements, Vision Res. 22 (1982) 1033-1045.
/7/ Deubel, H. and Wolf, W., Secondary saccades induced by altering visual feedback, Invest. Ophthal. Vis. Sci. 22, No. 3, (1982), Suppl., 26.
/8/ Deubel, H., Wolf, W. and Hauske, G., Corrective saccades: Effect of shifting the saccade goal. Vision Res. 22, (1982) 353-364.
/9/ Findlay, J.M., Amplitude transition functions for small saccades, same Vol.
/10/ Marko, H., The z-model - a proposal for spatial and temporal modeling of visual threshold perception, Biol. Cybern. 39, 111-123 (1981).
/11/ Roufs, J.A.J. and Blommaert, F.J.J., Temporal impulse and step responses of the human eye obtained psychophysically by means of a drift-correcting pertubation technique, Vision Res. 21, (1981) 1203-1221.
/12/ Hauske, G. Lupp, U. and Wolf, W., Comparison of two methods to analyse temporal properties of the visual system, Invest. Ophtal. Vis. Sci., 22, No. 3 (1982) Suppl., 253.
/13/ Findlay, J.M., Local and global influences on saccadic eye movements. A. Hein and M. Jeannerod (eds.) Eye Movements: Cognition and Visual Perception (1981) Lawrence Erlbaum, Hillsdale N.J.
/14/ Wolf, W., Hauske, G. and Lupp, U., Interaction of pre-and postsaccadic patterns having the same coordinates in space, Vision Res. 20 (1979) 117-125.
/15/ Breitmeyer, B.G., Kropfl, W. and Julesz, B., The existence and role of retinotopic and spatiotopic forms of visual persistence, Acta Psychologica 52 (1982) 175-196.

Acknowledgement: This investigation was supported as a project of the "Sonderforschungsbereich 50" by the Deutsche Forschungsgemeinschaft. We wish to thank Dipl.-Ing. C. Zetzsche for his engagement in performing the third experiment in this study and G. Kürzinger for preparing the photographs.

Theoretical and Applied Aspects of Eye Movement Research
A.G. Gale and F. Johnson (Editors)
© Elsevier Science Publishers B.V. (North-Holland), 1984

SMALL STIMULUS MOVEMENTS ARE NECESSARY
FOR THE STUDY OF FIXATIONAL EYE MOVEMENTS

J. de Bie and G. van den Brink

Department of Applied Physics
Delft University of Technology
P.O.Box 5046, 2600 GA Delft
The Netherlands

A correlation technique has been applied to prove the
control function of both drift and saccades during fixation. It has been shown that behaviour of the system
while using small step stimuli did not differ from that
during fixation. Using step reactions and (partly) stabilized images, it has been proved that micro-saccades
are triggered by a retinal error threshold mechanism,
which is isotropic. A model for the slow fixation control system has been tested, consisting of an integrator, with feedback from fixation error and velocity.

INTRODUCTION

Much research in the past concerned the question whether steady fixation was
controlled by micro-saccades or by slow movements. The early view was that
the abrupt, easily detectable saccades had a control function, and that the
slow movements were just what they looked like: noise and drift (see (1) for
a review). Nachmias (2), however, found that "drift" also contained corrective information. This view was supported by St.Cyr and Fender (3), who
showed that both saccades and drift controlled fixation. Later, it was found
that micro-saccades could be suppressed, without deterioration of fixation
accuracy (4, 5, 6). Drift control was obvious, but saccades were now suspected to be voluntary, or a bite-board artefact, and not necessary for fixation.

METHOD

Our experiments were all carried out with the method of the scleral search
coil in a magnetic field (8), with the coil embedded in a silicon contact
ring, as developed by Collewijn et al. (9). The accuracy of the system, as
far as could be measured with an after image method, was better than $1'^1$.
Four subjects, including the authors, participated in the experiments. Except
for refractive corrections, all had normal vision. They looked at the stimulus monocularly, the other eye being closed with adhesive tape. Extensive experiments were done with one subject, but the other subjects showed very similar results in control experiments. The stimulus was a bright spot generated
under computer control on a television monitor at 4.5 m. distance, in a dark
room. The luminance was 2 log units above absolute foveal threshold. The
stimulus luminance has been temporally modulated in all experiments with
(partially) stabilized images (modulation depth 50%, frequency 2 Hz) in order
to avoid fading of the image. The eye movements were sampled at 100 Hz.

FIXATION CONTROL

Our own experiments support the view that both saccades (if not suppressed)

and slow movements control fixation. We developed a correlation method to
show this more quantitatively. We correlated the drift velocity at any moment with the actual distance from the mean fixation point. This distance
was measured some time earlier. The result is shown in Fig. 4 (crosses), using a delay time of 200 ms. It clearly proves slow control, except for a
small drift to the left for this subject. The mean drift velocity is always

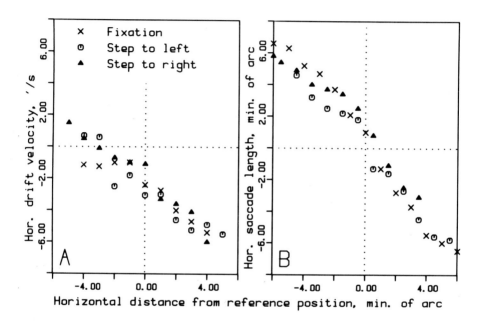

Fig. 1.: Horizontal drift velocity (A) and saccade length (B) as a function
of the distance from the reference position, measured 200 ms earlier, fixation and 2.5' stimulus step.

in the direction of the mean fixation point. During this experiment the subjects also made one or two micro-saccades per second. These saccades also
controlled fixation: in Fig. 1B (crosses) the mean saccade size has been
plotted as a function of the actual distance from the mean fixation point at
the beginning of the saccade. It is clear from this that the micro-saccades
have a recentering function too.

It is evident from our and from other experiments that drift and saccades
control fixation. However, no models exist for these systems, in sharp contrast to a variety of models for smooth pursuit and larger saccades. The reason for this might be, that the visual input is stationary during fixation.
All eye movements are a result of internal noise, which is filtered by the
control systems. It is impossible to separate noise and control in this case.
For this reason we applied small stimulus movements. The first question that
needed to be answered was, whether the eye's reaction on its own noise during
fixation of a stationary target was the same as its reaction on stimulus
movements. First we want to state that this is the fixation condition we normally meet during everyday life: body movements, head movements and stimulus

movements always move the image on the retina. Secondly, it would not be very probable if there would be a difference in reaction on stimulus movements that are even smaller than normal fixation eye movements. The following experiment also indicates no difference between the two conditions. The stimulus made small horizontal jumps (2.5'), with randomly chosen time of occurrence and direction. The resulting eye movements during the first second after the stimulus jump have been analysed in the same way as we did for normal fixation. The results can be seen in Fig. 1A and B (circles and triangles). There is no difference: the slow movements and the saccades behave similarly with small stimulus jumps, as they do with fixation errors caused by the eye movements itself. This result gives us the possibility to analyse the reaction on small stimulus movements more extensively, and to apply the results in a model of the fixation control systems.

MICRO-SACCADES

The first problem concerning micro-saccades is their triggering. Although micro-saccades can be suppressed, normally a fixating subject makes one or two saccades per second. It is an interesting point whether these saccades are caused by the slow movements itself or not. In order to answer this question, the image was stabilized. In this situation it often occurred that the subject made a cascade of saccades, 200-500 ms apart, often in combination with drift, all in the same direction. This behaviour indicates that the image had not been stabilized precisely at the normal fixation point. When, however, the image was stabilized only during periods of slow movements, and not during saccades, the saccade-rate dropped to 10-20% of the normal value, measured under the same conditions! It is important that in

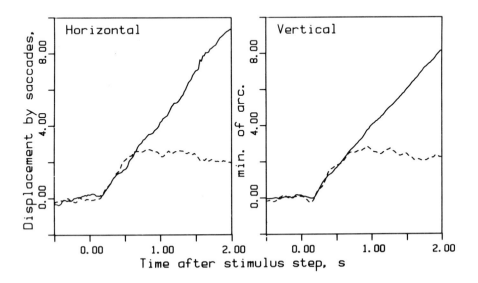

Fig. 2: Saccadic reactions on a stimulus step (3.7', oblique). Continuous curves: complete image stabilization, dashed lines: stabilization only during slow movements, mean curves, N=350

this condition no non-stabilized cues are present: the saccades re-appear as soon as a non-stabilized image is seen, even when it is some degrees in the periphery. These facts indicate that the major part of the saccades during fixation is "meant to be corrective", and is triggered in some way by the fixation error, using a threshold mechanism. It also proves that most saccades during fixation are caused by "drift". This is in contradiction with the assumptions, that micro-saccades are (also) triggered by the time elapsed since the last saccade (2), or that micro-saccades occur as a nervous tic (7), or as a means to bring the fixation point to another part of the fovea.

Is it the fixation error itself that triggers the saccade, or is the error extensively filtered, as is often assumed in other biological reaction mechanisms (10, 11)? In trying to solve this problem, we again applied small step stimuli. Experiments described in the literature concerning large steps (above some degrees of arc) show saccades to appear about 200 ms after the step. With smaller steps this is no longer the case. The mean reaction time is larger, and the variability is also much larger (12).

In Fig. 2 the mean saccadic reaction on a 3.7' stimulus jump in an oblique direction is shown (dashed line). In this experiment the image had been stabilized between saccades, which means that only saccades could be used to reduce the fixation error. It is clear that they do this between 150 and 700 ms after the stimulus jump. In this period there was no difference with what we found when the image was completely stabilized (continuous line), which indicates that usually only one saccade took place in that period. There was no directional asymmetry, the saccades behaved exactly the same in all directions. Note that the very early reactions (150 ms) already have a component in the right direction. This supports the view that the actual size of the saccade is subject to modification after its triggering (13).

The question arises why a reaction saccade in one case occurs after 200 ms, and in another after 700 ms. Therefore we divided the reactions in four different classes, based on the reaction time (Fig. 3). In this experiment we used normal conditions, without stabilization. It is clear that early ones

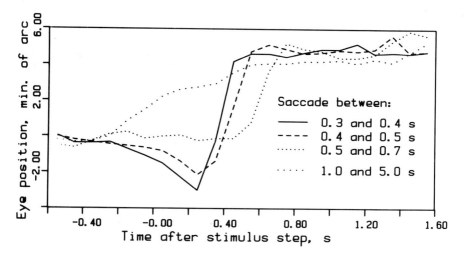

Fig. 3.: Different types of mean eye movements; classification was based on the moment of the first saccade after the stimulus step (5'), N=30.

take place when this error is small. There can be at least two explanations for this effect. It is possible that the fixation error is integrated in some way, and that the resulting filtered error is the input for a threshold mechanism. This would also explain why short stimulus pulses do not always result in a saccade. However, an alternative explanation may be found in the noise in the determination of the fixation error by the control system. The chance that a saccade occurs, due to an inaccurate fixation error detection, increases when the real fixation error approaches the threshold, thus causing a shorter mean reaction time. This simple fact alone could cause the effect shown in Fig. 3. Until now, we were not able to discriminate between these alternative hypotheses, because the properties of the noise in the fixation error detection are not known.

SLOW CONTROL

The nature of the slow control system is entirely different: it is continuous, while the saccades are abrupt events. Three facts make the analysis of the system very difficult:
1. Normally, the slow movements are interrupted once or twice per second by a saccade.
2. Visual feedback after a delay (150 ms) makes a description with techniques from system theories more complex.
3. The noise of the system makes it impossible to draw any conclusions on the basis of individual records.
The first two problems could be solved by using a stabilized image: the feedback loop is now open, and very few saccades occur in this situation. Averaging was used to make the results more accurate.
Fig. 4 (continuous lines) shows the "drift" reaction when a small step was

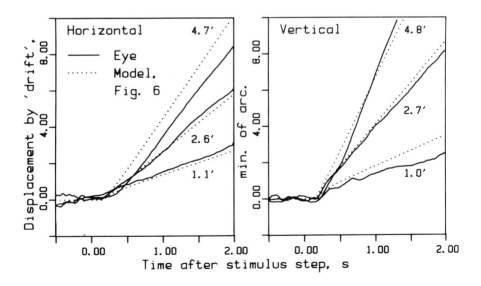

Fig. 4.: Slow eye movement reactions on different stimulus steps, with complete image stabilization. Mean curves, N=350.

applied on top of a stabilized image. The result with a very small step (1 to 5') is a constant velocity, which strongly suggests a simple integrating system. Note that the gain of the system, which means the velocity per unit fixation error, is different for the horizontal and vertical systems.
We repeated the same experiment with the image not stabilized during the periods between micro-saccades. In order to make the influence of saccades as small as possible, we stabilized the image only during the saccades. This procedure is only justified if saccades and slow movements are independent. In order to check this assumption, we also selected from our material all records without a saccade in the first second after the stimulus jump (30% of the records), and compared the mean with the average of the other reactions. No difference was found, which confirms the theory that micro-saccades and slow movements are the outputs of completely independent s stems. Our method seems therefore justified, and the results, presented as fixation errors, can be seen in Fig. 5. The continuous lines represent the means of

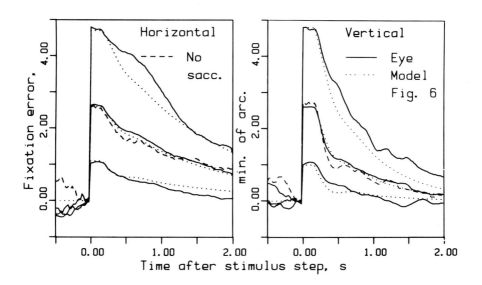

Fig. 5.: Slow eye movement reactions on different stimulus steps; image stabilization only during saccades. Mean curves, N=350-700. Dashed lines: mean of records without a saccade in the first second after the step (N=100).

all records, while the dashed lines are the means of the no-saccade records. The influence of the visual feedback is clear, beginning at 300 ms (twice the reaction time).

The results cannot be described by a simple model with an integrator (Fig. 4) and feedback of the fixation error, a model proposed by Robinson ((14), p. 533, Fig. 10). With this model the influence of the feedback would not have been very abrupt after 300 ms. "Smooth pursuit" models use the error velocity as input signal. These models are not suited for a position control system, because a constant position error (with zero error velocity) would not produce any reaction. When, however, both inputs are used, it is possible to develop a model that is able to predict our results within the experimental

error (Fig. 6). It consists of an integrator with feedback from both the
fixation error and the error velocity, and is in fact a modification of
Robinson's model, without a threshold for the velocity.

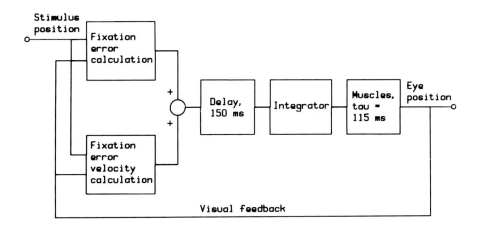

Fig. 6.: Model of the slow control system during fixation.

The gain of the integrator in the model is assumed to be constant, and calculated from the best fit in the stabilized image reactions (Fig. 4, dotted
lines). As can be seen, this linear behaviour is a simplification.
The fixation error velocity is very high for a short time during the stimulus
step. It cannot be expected that the velocity detection mechanism is able to
determine this correctly. Therefore, a maximum velocity has been assumed,
based upon the best fit. No other assumption concerning the velocity detection mechanism has been made. It's gain has also been calculated using the
best fit. A reaction time of 150 ms has been used and the eye muscles have
been represented by a first order system (τ=115 ms). Table 1 gives a summary
of the parameters used. Our conclusion from the fit in Fig. 5 (dotted lines)
is, that this simple model resembles the results quite well, and it even may
be a basic model for slow control as well as smooth pursuit.

	Hor.	Vert.	( )
Gain of the integrator	1.2	1.8	1/s
Gain of the velocity feedback	0.7	0.35	s
Maximal fixation error velocity	25	70	'/s

Table 1.: Parameters used in the model (Fig. 6).

CONCLUSIONS

From our experiments it can be concluded that both saccades and slow control
help to maintain accurate fixation, confirming the view of Nachmias (2),
St.Cyr and Fender (3) and others. Also, these systems seem to be independent,
and reacting on eye movement induced fixation errors in the same way as on
stimulus movements.
The micro-saccades are triggered when the fixation error, possibly in a fil-

tered form, exceeds a small threshold (3-5'). During normal fixation, 80-90% of the saccades are caused by the slow movements (or "drift"). We found indications that the size of a micro-saccade can be altered after it is triggered. It also seems that the saccadic system behaves the same in horizontal and vertical directions, while the slow control system is not isotropic. This slow system can be accurately described by an integrator with visual feedback of the fixation error as well as the error velocity. It may be possible to combine this model with the properties of velocity detectors and the smooth pursuit, in order to find a basic model for slow control as well as smooth pursuit.

Our final conclusion is, that small stimulus movements, together with (partial) image stabilization, proved to be very useful in the study of the fixation control systems.

FOOTNOTE:

[1]. Angular movements of the eye are given in minutes of arc in this paper (symbol ', 1' = 2.9. 10-4 radian).

REFERENCES:

(1)   Ditchburn R.W., Eye movements and visual perception (Clarendon Press, Oxford, 1973), 310-337.
(2)   Nachmias J., Two-dimensional motion of the retinal image during monocular fixation, J. Optical Society of America 49 (1959) 901-908.
(3)   St. Cyr G.J. and Fender D.H., The interplay of drifts and flicks in binocular fixation, Vision Res. 9 (1969) 245-265.
(4)   Steinman R.M., Cunitz R.J., Timberlake G.T. and Herman M., Voluntary control of microsaccades during maintained monocular fixation, Science 155 (1967) 1577-1579.
(5)   Haddad G.M. and Steinman R.M., The smallest voluntary saccade: implications for fixation, Vision Res. 13 (1973) 1075-1086.
(6)   Kowler E. and Steinman R.M., The role of small saccades in counting, Vision Res. 17 (1977) 141-146.
(7)   Kowler E. and Steinman R.M., Small saccades serve no useful purpose, Vision Res. 20 (1980) 273-276.
(8)   Robinson D.A., A method of measuring eye movements using a scleral search coil in a magnetic field, IEEE Trans. Bio-Med. Electr. BME-10 (1963) 137-145.
(9)   Collewijn H., van der Mark F. and Jansen T.C., Precise recording of human eye movements, Vision Res. 15 (1975) 447-450.
(10)   Hildreth J.D., Bloch's law and a Poisson counting model for simple reaction time to light, Perception and Psychophysics 26 (1977) 153-162.
(11)   Pacut A., Mathematical modelling of reaction latency: The structure of the models and its motivation, Acta Neurobiol. Exp. 40 (1980) 199-215.
(12)   Wyman D. and Steinman R.M., Latency characteristics of small saccades, Vision Res. 13 (1973) 2173-2175.
(13)   Hou R.L. and Fender D.H., Processing of direction and magnitude by the saccadic eye-movement system, Vision Res. 19 (1979) 1421-1426.
(14)   Robinson D.A., Models of oculomotor neural organization, in: Bach-y-Rita P., Collins C.C. and Hyde J.E. (eds.), The control of eye movements (Academic Press, New York, 1971).

Theoretical and Applied Aspects of Eye Movement Research
A.G. Gale and F. Johnson (Editors)
© Elsevier Science Publishers B.V. (North-Holland), 1984

SMALL SACCADES TO DOUBLE-STEPPED TARGETS
MOVING IN TWO DIMENSIONS

John M. Findlay and Laurence R. Harris

Department of Psychology, University of Durham,
Durham DH1 3LE, England

We studied saccadic eye tracking, concentrating particularly on
the situation where a target makes a jump movement in the period whilst a
saccade is being prepared to a previous jump. Such a perturbation affects
the saccadic system in several ways. The spatial characteristics of the
movement, amplitude or direction equally, can be modified on the basis of
new information arriving up to 80 msec before the initiation of the
saccade. The perturbation rarely produces any substantial effects on the
trajectory of the movement itself, but the occasional exceptions reveal the
presence of a goal seeking feedback mechanism underlying saccadic
production.

INTRODUCTION

Making a saccadic eye movement towards a visual target represents a
sophisticated example of neural information processing but one which is
open to precise quantitative modelling. Recently it has been shown that,
even after a saccade has been initiated, its course is by no means fixed
and can be influenced by external constraints.

A particularly useful experimental technique for demonstrating this is
'two-step tracking'. If a subject is asked to follow a target which moves
in a step jump, he will do this by means of a saccade. It may be now
arranged that a second, perturbing, step occurs during the preparatory
period of this movement. Becker and Jurgens (1979) showed that the effect
of the second step depends very systematically on the temporal interval
between its occurrence and the onset of the saccade elicited by the first
step. From this, they developed a model for the human saccadic system
which was based on extensive experimental data, but with saccades which
were rather larger than those normally made by the saccadic system (Bahill,
Adler and Stark, 1975). For this reason we felt it worthwhile to replicate
their studies with smaller target movements. Also they studied exclusively
horizontal movements; thus movement direction was not separable from movement amplitude. We study here target jumps in two dimensions.

We concentrate particularly on two questions. The first concerns whether
saccade direction is programmed independently of saccade amplitude. The
second concerns the details of the trajectory of the saccade. The
traditional 'ballistic' view of saccades has given way recently to a viewpoint which holds that saccades are produced by a goal directed process,
controlled not directly by visual information, but by some internal
representation of the desired end position (Mays and Sparks, 1981, have
produced the clearest evidence for this). We thus looked to see whether

the appearance of a perturbing step during the course of preparation for
a saccade could affect its trajectory.

EXPERIMENT 1

The first experiment closely followed the methodology of Becker and Jurgens
(1979). Their paper, however, concerns only saccades over 15 degs in
amplitude. Our aim was to see how closely their results could be replicated with saccades of smaller sizes.

METHODS

Subjects were seated faced a CRT screen (Tektronix 608; P31 phosphor) at a
distance of 65 cm. Targets were squares with sides of 4 min arc made of
9 points in a 3 x 3 matrix. Subjects were instructed to follow jumps of
the target as accurately and as quickly as possible. The target was
presented in the central position for a time which varied randomly from
trial to trial between 1 and 5 seconds and then moved either in a single
jump, or with two jumps. It remained at its final destination for 1 sec
before moving back to the centre for the start of the next trial. The
initial movement was either 2 or 4 degrees either rightwards or leftwards.
In an attempt to avoid anticipation and familiarity effects, a large variety
of second jump combinations were included. For the 2 degree initial jump
the target either remained in the 2 degree position (control single step),
moved a further 3 degrees in the same direction, or else moved in the
opposite direction by 2 degrees (return to the central position), 3, 4 or
6 degrees. For the 4 degree initial step, the possibilities were no
further movement, or reverse movement of 2, 4, 6 or 8 degrees. There were
thus 22 possible stimulus movements of which only two involved a second
jump in the same direction as the initial movement. Interstep intervals
(ISI) were 50, 100, 150 and 200 msecs. Each stimulus type occurred four
times in the course of a block, once with each ISI. Two subjects
(laboratory workers) participated in 8 blocks. Horizontal eye movements
were recorded with an infra-red device (Findlay, 1974). An LSI Alpha
computer presented the stimuli and recorded the eye movements. The
computer later identified and measured the initial saccades. Trials containing blink artefacts or anticipatory saccades were removed from the
analysis.

RESULTS

Amplitude transition functions (plots of saccade end point against the
interval between the second step and the saccade onset) were plotted
for the initial saccades obtained with each stimulus configuration Four
typical plots are shown in Figure 1 (a-d) for two subjects. There is the
expected variation of first saccade amplitude dependent on its temporal
relation to the second step. When the saccade starts less than 80 msecs
after the second step the first saccade goes to the position reached by
the target in the first step (dotted line P1). For large intervals,
saccades go to the second target position (dotted line P2). When the first
and second target positions are on the same side of the original fixation
(Figure 1a and b) there is a transition region in which saccades of intermediate amplitude occur. Note the decreased frequency of saccades in this
region shown as a histogram below (Figure 1e). If the two target positions
are on opposite sides of the original fixation point (Figure 1c-d), this
reduction becomes even more pronounced and hardly any saccades occur in the
transition region.

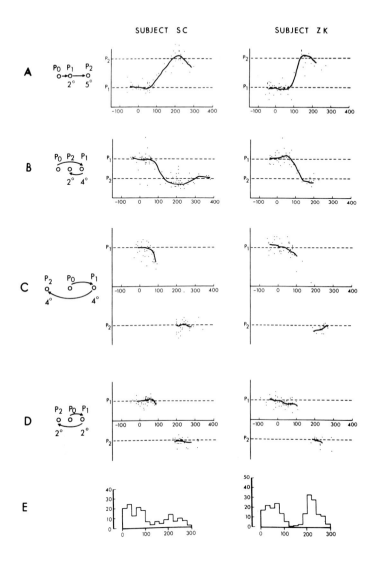

Figure 1. (a-d) Examples of amplitude transition functions for
subjects tracking various double step stimuli. Dotted
lines show median amplitudes to control single steps.
(e) Histogram showing decreased frequency of saccades
in the transition region in cases (a) and (b).

Qualitatively, this is the same pattern that Becker and Jurgens found with
large saccades (Becker and Jurgens, 1979). However an important difference
emerges when the quantitative details are considered. For all double jumps
studied the transition region commences 70 msecs after the second jump
(+/-10 msecs). The transition region seems to be much more tightly time-
locked to the interval after the second step than it was for the larger
jump sizes used by Becker and Jurgens (1979).

EXPERIMENT 2

In our second experiment we sought to distinguish changes in direction from
changes in amplitude. We also hoped that by introducing a change in the
required vector of the saccade we might reveal modification of a saccade
actually during the course of its trajectory.

METHOD

In this experiment the target jumped from the central position to one of
eight stations eight degrees away. The stations were arranged at the
points of the compass with North corresponding either to straight up or, in
some blocks, rotated by 22 degs (Figure 2a). Having jumped to one of these
points, the target could either remain stationary or jump to either of the
adjacent stations. The interval between the two target jumps was 50, 80,
110 or 140 msecs. The target was a square (sides 10 min arc) made up of a
matrix of 3 x 3 points displayed on a Hewlett Packard 1321 screen (P31
phosphor).

Eye movements were recorded by a scleral search coil (Collewijn et al.
1975). Our system had a resolution of 0.1 deg in both the horizontal and
vertical dimensions. Eye movements were recorded and analysed as before.
The programme measured the amplitude and trajectory of the initial saccade.
The trajectory was measured in terms of two vectors, firstly that between
the eye position 10 msec before the start of the saccade and that 60 msec
later (overall trajectory); secondly that between the eye position 10 msec
before the start of the saccade and 30 msec later (initial trajectory).
The difference between the initial trajectory and the overall trajectory
gave a measure of the curvature of the saccade.

Four subjects (postgraduates and laboratory workers) were tested, each
carrying out four blocks (384 saccades). The target was left on at its
final destination for a period equal to the time that it had been on at
the centre before the trial began. The screen was then blanked for one
second before the next trial.

RESULTS

No major differences were found between the various initial step directions,
nor between clockwise and anti-clockwise second jumps. All the double step
trials have therefore been pooled. Figure 2 shows amplitude (Figure 2b)
and direction (Figure 2c) transition functions respectively

A common pattern of results, with minor interesting variations, emerged in
all four subjects. As in the unidimensional case, saccades were of
appropriate amplitude and direction to fixate the first target position
when the interval between the second step and the start of the saccade was
less than 60-80 msec. If the interval was longer than 160 msecs, saccades

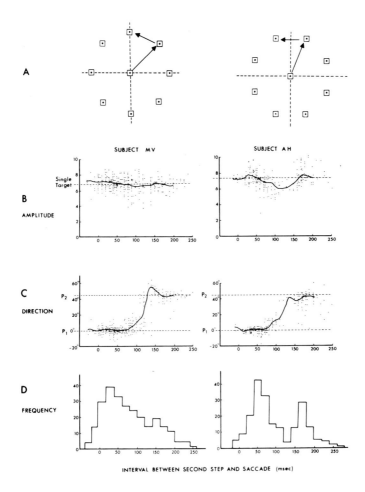

Figure 2.   (a) Illustration of stimuli used in Experiment 2.
            (b) Examples of transition functions for saccade amplitude.
            (c) Examples of transition functions for saccade direction.
            (d) Histograms showing distribution of saccade occurrence
                with respect to the second step.

were (generally) directed to the second target position. In between, there
was a direction transition region in which saccades were directed along
intermediate vectors (Figure 2c). Saccades in the transition region showed
some tendency to be hypometric. Once again, a reduced frequency of
saccades occurred in this region (Figure 2d). One subject (MV) produced a
number of saccades which were directed beyond the second target, rather
than to intermediate positions. These occur systematically at the end of
the transition period, giving an overshoot in her direction transition

function. The other two subjects produced data similar to those of AH.

Figure 3 is a plot of the change in the direction vector (curvature: initial -overall) trajectory as a function of the saccade onset latency after the second jump.

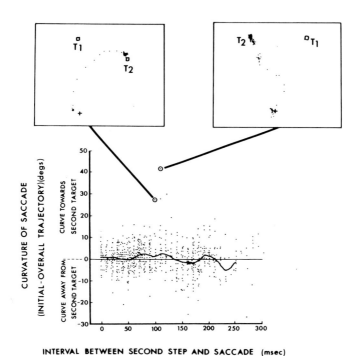

INTERVAL BETWEEN SECOND STEP AND SACCADE (msec)

Figure 3.   Saccade curvature, plotted in relation to the interval
             between the second step and the saccade. The insets
             show two instances of unusual, grossly curved, movements,
             together with their identifications on the plot.

The sign of the curvature measure has been inverted for anticlockwise second steps so that the points are plotted with positive values indicating curvature towards the second target and vice versa. In the transition region there is a small, but definite, trend for saccades to be slightly curved, on average, towards the second target by an angle of some 2 degrees. This is, however, well within the normal range of variability of saccade trajectories. Occasionally though, saccades in the transition region can be found where the trajectory shows a very pronounced swerve towards the second target. Two examples are shown in the insets to Figure 3. In one of these the horizontal component reverses direction during the course of the movement. It must be emphasized however that this behaviour occurs very rarely.

DISCUSSION

Eye movements elicited by a target that jumps twice in quick succession reveal some interesting properties of the saccade control system. If the initial saccade precedes the second step, or follows the second step by an interval of less than 80 (+/-10 msec), its characteristics are those of a normal saccade to the first target movement. Likewise, if the first saccade does not occur before 180 (+/-20 msec) then a saccade of normal characteristics appropriate to a step to the second target position is observed. Between these times, a transition region is found in which the second target can influence the saccade intended for the first.

The perturbing effects are manifest on frequency, amplitude and direction. Firstly the new visual information renders the release of a saccade less likely: a trough appears in the saccade occurrence histograms (Figure 1e and 2d). This suggests that the new visual information exercises an inhibitory effect at some point in the saccadic generation process. The strength of this inhibition varies with the configuration; it is strongest and longest lasting when the second jump takes the target to the opposite hemifield in such a way as to demand a complete backtracking of the saccade (Figure 1c, d). Interestingly, this intense inhibition was not found in the case when a target jumped across the vertical meridian but did not pass across the central point.

The saccades which occur in the transition region possess several unusual properties. In general they are directed to a position in between the target's positions. Furthermore, this end-point position shows a transition as a function of the latency of the initial eye movement after the second step. The amplitude transition functions shown in Experiment 1 are quite similar to those observed by Becker and Jurgens (1979) although they are for much smaller saccades. In Experiment 2 we show that analogous direction transition functions can occur.

MODIFICATION OF TRAJECTORIES (MOTOR MODIFICATION)

In general we find that saccade trajectories are only minimally modified by movement of their target. Usually a saccade to an intermediate position or the second target position simply replaces the one being prepared to go to the first location. This would suggest that saccades are in fact pre-programmed. But very occasionally the new visual information DOES modify the saccade in mid-flight so that the gaze swings over to the second target position (Figure 3 inserts). These two apparently conflicting observations: saccades being preprogrammed and saccades being modifiable are, perhaps, compatible. If saccades are produced by means of a goal seeking feedback loop, it would seem highly desirable to have a 'gate' which could be closed to prevent perturbation of a saccade in course of execution. This would be particularly important in normal visual environments with many potential sources of stimulation.

We conclude that the target for a saccade is normally selected well before the eye moves and some inhibitory gate is effective in removing competing goals. The gate may very occasionally fail to operate (perhaps whilst open-ing or closing) giving rise to grossly abnormal saccade trajectories. It may, indeed, always be slightly leaky thus creating very small curves in the trajectories of our saccades (the 'pull' of the new target: Figure 3) and 'noise' in the saccades of everyday life when there are always many competing goals.

MODIFICATION OF THE GOAL (SENSORY MODIFICATION)

Saccades that occur in the transition region do not generally have a modified trajectory. They appear aimed, from the start, towards a goal that is
not actually present: one between the actual target locations. Becker and
Jurgens (1979) suggest that this represents a temporal integration of the
sensory information. Our experiments show that this integration can
operate across two dimensional space.

Perturbations which effect a change in saccade direction appear to be made
in precisely the same way as those which effect a change in saccade amplitude only. We did not replicate the finding of Hou and Fender (1979) of
saccades with direction appropriate to one target position but amplitude
reduced. This led them to conclude that saccade direction and saccade
amplitude are programmed separately. On the contrary our data suggest
that saccade direction and saccade amplitude are not separately evaluated,
but rather that in some way, perhaps along the lines suggested by McIlwain
(1976), a direct translation can occur between the activation in a retinotopic spatial map, and the appropriate eye movement.

REFERENCES

Behill, A. T., Adler, D. and Stark, L., Most naturally occurring saccades
    have an amplitude of 15 degrees or less, Invest. Ophthal. 14 (1975)
    468-469.

Becker, W. and Jurgens, R., An analysis of the saccadic system by means of
    double step stimuli, Vision Research 19 (1979) 967-983.

Collewijn, H., Van der Mark, F. and Jansen, T. C., Precise recordings of
    human eye movements, Vision Research 15 (1975) 447-450.

Findlay, J. M., A simple apparatus for recording microsaccades during
    visual fixation, Quart. J. Exp. Psychol. 26 (1974) 167-170.

Hou, R. L. and Fender, D. H., Processing of direction and magnitude by the
    saccadic eye-movement system, Vision Research 19 (1979) 1421-1426.

Mays, L. E. and Sparks, D. L., Saccades are spatially, not retinotopically,
    coded, Science 211 (1980) 1163-1165.

McIlwain, J. T., Large receptive fields and spatial transformations in the
    visual system, in: Porter, R. (ed.), International Review of Physiology,
    Volume 10, Neurophysiology II (1976).

Theoretical and Applied Aspects of Eye Movement Research
A.G. Gale and F. Johnson (Editors)
© Elsevier Science Publishers B.V. (North-Holland), 1984

PROPERTIES OF PARAMETRIC ADJUSTMENT IN THE SACCADIC SYSTEM

W. Wolf[1], H. Deubel[2] and G. Hauske[2]
[1] Bundeswehr University Munich, ET/WE1,
8012 Neubiberg, F.R.G.
[2] Technical University of Munich,
Lehrstuhl für Nachrichtentechnik,
Arcisstr. 21, 8000 München 2, F.R.G.

Properties of the "parametric adjustment effect" (PAE) were
studied using single dot targets which moved in single or double
steps. When the first target step of randomly chosen 10-15 deg
occurred in a predetermined direction saccade onset triggered
a second intrasaccadic target displacement of 4 deg either into
the same or into the opposite direction of the first step.
The experimental data show that the size of the primary saccades
adapts to this specific stimulus configuration by changing the
gain of the system. The adaptation is directionally selective
and asymmetrical with different time constants for gain increase
and decrease. A model describing the PAE is presented.

INTRODUCTION

The oculomotor system represents a highly complex machinery which
fascinated scientists of all centuries. One prominent property of the
system is its ability to perform jerky goal-directed eye movements
reaching high velocities up to 700 deg/sec. The control of these so-called
saccades is assumed to be performed by a particular branch of the oculomotor system, the so-called "saccadic controller".

The high velocity of saccades implies that the controller operates
in a preprogrammed open loop manner during the movement. An open
loop system, however, suffers from a higher sensitivity to fluctuations
of its parameters than a closed loop system. Introducing his concept
of "self-repair", Robinson /1/ mentions that there are many sources
for these fluctuations, e.g. growth, aging, fatiguing and disease
of nerves as well as eye muscles under which the accuracy of saccadic
eye movements is well preserved. Therefore we have to assume a mechanism
which maintains proper calibration of the system by readjusting it
as soon as untolerable dysmetria of the movement is detected. Such
an adaptive behaviour was originally found by McLaughlin /2/ and coined
"parametric adjustment effect" (PAE). He demonstrated that a persistent
mismatch between intended and actual eye movement induced by an intrasaccadic shift of the saccade goal leads to a gradual adaptive change
of saccadic amplitudes.

In a number of psychophysical studies /3,4,5,6/ McLaughlin's basic
findings were confirmed and extended, but three fundamental questions
still have not been solved completely, namely: Is the adjustment effect-
i) really parametric?
ii) directionally selective?
iii) asymmetrical, which means different characteristics for gain
     increase and decrease?
We have performed a series of experiments with the aim to cast some
light upon these problems and to describe the PAE by a functional model.

METHODS

A horizontally concave stimulus board subtending 45 deg in the horizontal and 18.5 deg in the vertical plane was centered in front of the
subject's head which was fixed by a biteboard and a forehead holder.
Viewing distance was 1.50m and vision was binocular. On the horizontal
meridian of the stimulus board which was illuminated by a projector at
a photopic luminance of 21 cd/m^2 40 red light emitting diodes (LEDs)
were placed in equidistant steps of 1 deg. The LEDs subtended a visual
angle of 0.2 deg and were mounted behind a translucent white foil which
made inactive LEDs invisible. The effective luminance of an activated
LED was set to 90 cd/m^2.

Eye movements were measured by bitemporal EOG-technique. The amplified
EOG signal was digitized at a sampling rate of 1kHz and stored on disc
for later evaluation. The resolution of the eye movement recording was
about 0.5 deg.

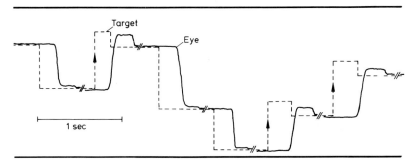

Fig. 1: Example of an experimental sequence in Exp. II. The dashed
trace indicates the target position (within the possible range of
39 deg); the solid trace shows the eye position.

The computer-controlled timing of the experimental procedure is
shown in Fig. 1. The actually illuminated LED represents the target on
which the subject had to fixate. After the subject indicated his readi-
ness by pressing a push-button the target stepped away from the fovea
with a random delay. The amplitude of the first target step was randomly
varied between 10 and 15 deg, and the direction was randomly chosen to
the right or to the left. The three types of experiments in this study
(Exp. I-III) differed in the modification of the target position which
occurred during the initial saccades. In Exp. I which served as control
experiment to determine basic parameters of the oculomotor reaction the
stimulus was left unchanged; Exp. II and III were designed to induce PAE
by shifting the saccade goal by 4 deg into the opposite direction (Exp. II)
or into the same direction (Exp. III) of the saccade. In order to
test the directional selectivity of the PAE the second target step
in a certain session occurred only when the first target step had occurred in a predetermined direction, varied from session to session.
The final target location served as the starting point for the next
trial thus eliminating return saccades. The subjects (3 female students
naive with respect to the aim of the study) had to follow the target
as fast as possible. They were additionally asked to press a push-button

if they had detected the second intrasaccadic target displacement. Each experimental session consisted of about 220 trials and lasted 30 minutes. Between the experimental sessions a pause of several days was allocated to avoid possible effects of long-term adaption.

The evaluation of stimulus and eye movement data was done by digital computer extracting the amplitudes r_1, r_2 and the latencies l_1, l_2 of the primary and the secondary saccades together with the target step sizes a_1 and a_2 for each trial. The parameters mentioned are depicted in Fig. 2. Amplitude calibration necessary to compensate for fluctuations of the EOG was done by assuming that $(r_1 + r_2)$ meets the target step size $(a_1 + a_2)$. If no correction saccade was detected, r_2 was set to 0.

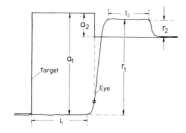

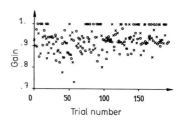

Trial number

Fig. 2: Definition of stimulus and response parameters

Fig. 3: Saccadic gain for single saccades in Exp. II (circles: saccades to the left; crosses: saccades to the right).

Under the assumption of a linear relationship between the angles of target step and saccade the following equation between the basic variables exists:

$$r_1 = G \cdot a_1 + B \qquad (1)$$

where G defines the gain of the saccadic system and B the response bias.

RESULTS

The basic setting of the system gain (Exp.I)

Exp. I was performed without manipulation of the saccade goal to reveal the basic setting of the system parameters. Fig. 3 shows the saccadic gain as a function of consecutive trial numbers in a single experiment. No difference exists between saccades to the left (crosses) and those to the right (circles). The data points at the 1.0 level represent saccades without a secondary saccade. Their exact gain may scatter between 0.97 and 1.03 (see Methods). The parameters G and B of equation (1) were fitted to the experimental data by regression analysis which reveals no significant system bias B. Further, the gain for both directions proved to be about 0.9 as previously reported /7/.

Directional selectivity of the PAE

To induce parametric adjustment, the saccade goal was shifted into the opposite direction of the first step by a second intrasaccadic target displacement which leads to a large saccadic overshoot (Exp. II). Fig.4 shows the gain of single saccades for consecutive trials in experiments

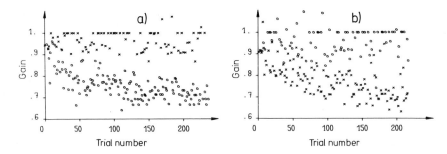

Fig.4: Saccadic gain in Exp.II. Intrasaccadic target displacements occurred
a) with leftward saccades (circles), b) with rightward saccades (crosses).

with the second step occurring only when saccades were performed to the
right (Fig. 4a) or to the left (Fig. 4b). Rightward saccades are indicated by crosses, leftward by circles. The data demonstrate that the
gain is rapidly reduced for saccades into that direction which was
predetermined for the target modification. The time course of the
adaptation varies between subjects, but the initial steep decrease
happens within the first 50 trials.

Asymmetry of the PAE
In Exp. III the second target step occurred in the same direction
as the first step, inducing an increase of the gain. A comparison of
the data from Exp. II and III is given in Fig. 5 presenting only the
data with intrasaccadic second target step. The figures display the
mean saccadic gain as a function of trial number determined in Exp.II
(crosses) and Exp.III (circles) for three subjects. Each curve
represents the data of 3 experimental sessions. All subjects show
a clear and fast decrease of their gain in Exp.II as already demonstrated in Fig. 4. Gain increases elicited in Exp.III, however,
occur considerably slowly in comparison to the decrease.

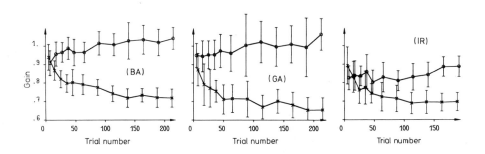

Fig. 5: Saccadic gain (mean + SD) in Exp.II (crosses) and Exp.III (circles)
for three subjects. Only the data of trials with intrasaccadic target displacement are shown.

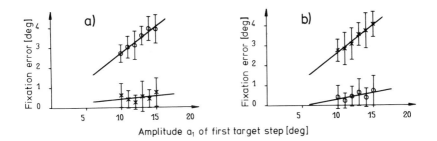

Fig. 6: Absolute fixation error (a_1-r_1) as a function of the size of the first target step a_1. (a,b: Data replotted from Fig. 4 a,b). The data points show the mean $\pm$ 1 SD.

The adjustment effect for different saccadic amplitudes

Figs. 3-5 do not provide information about the influence of the size of the first target step which was varied between 10 and 15 deg throughout all experiments. Therefore, we replotted the data from Figs.4 a,b to display the absolute fixation error of the primary saccades as a function of the step size (Fig.6a,b). For the calculation of the mean values the data from the first 50 trials were discarded in order to exclude the transient part of the adaptation process. The regression lines in these diagrams demonstrate that the fixation error is proportional to the target step in all cases. Therefore, the saccadic gain G as defined in equation (1) can be assumed to be equal for saccades of all amplitudes.

DISCUSSION

The data show that the "parametric adjustment effect" can be easily provoked by our experimental procedure. Before discussing the three basic topics outlined in the introduction the following two findings should be noted. The rates for the detection of the intrasaccadic target shifts by the subjects were usually smaller than 20%, which excludes a conscious strategy as source of the PAE. Further, the intrasaccadic target displacements induced correction saccades with latencies down to 90-100 msec which are significantly shorter than those found in Exp. I and in other related studies /7,8/. These latency values are close to the latencies of the "express saccades" found by Fischer and Boch in the monkey /9/. They can be interpreted as an adaptive behaviour of saccade timing.

The PAE is really parametric

An important question refers to the complexity of the adaptive controller. On the basis of the data from previous studies /2,3,4,5,6/ it cannot be excluded that the saccadic system is able to specifically adapt to the conditioning stimulus configuration with an adaptive change of saccadic amplitudes for a certain target eccentricity leaving saccades of other eccentricities unaffected. In a more sophisticated scheme amplitudes of saccades might be related to the actual values of innervation by means

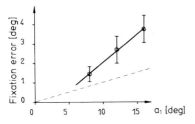

Fig. 7: Absolute fixation error.
Data points show the mean $\pm$ 1 SD.
The dashed line indicates the
fixation error for the normal
gain of 0.9.

of a "look-up table". In this case the adaptive controller should be able
to adapt to specific cases of the conditioning stimulus.

The term "parametric adjustment", however, implies that a parametric
relationship between target step size and resulting saccadic amplitude
exists. In the special case of a linear relationship a modification
of the gain should affect saccades according to their amplitudes. The
data in Fig. 6 a,b clearly support such a parametric model with an
adaptive system gain. Further evidence for this fact was achieved by
a pilot experiment using the procedure of Exp. II with the following
modifications: a_1 was chosen from only 3 different values (8, 12 and
16 deg) and the conditioning second target step occurred only with the
12 deg target step. Fig. 7 shows the resulting absolute fixation errors.
The fact that the amplitudes of the saccades are proportionally reduced
for all three target step sizes strongly argues for the parametric
nature of the effect.

The PAE is directionally selective

Directional selectivity of the PAE is confirmed by the data plotted in
Fig. 4 which is in conformity with earlier results. Its existence seems
plausible, because a mechanism which has to "repair" peripheral changes
of the oculomotor system with its different extraocular muscles needs
some independence in the adjustment mechanisms associated with
different directions.

The PAE is asymmetrical

In Fig. 5 the asymmetry of the PAE is shown by the significantly different time constants for gain increase and decrease. This finding is
in good agreement with other investigations /6/ and might be a consequence of different significance for over- and undershooting as discussed by Henson /5/. He concluded that a crucial constraint to the
saccadic behaviour is not to overshoot the target. Accordingly, large
overshoots occurring in the initial phase of Exp. II should lead to a
rapid reduction of the saccadic gain whereas undershoots can be easily
handled by fast correction saccades. The slow increase of the gain can
also be explained by the impending danger of instability of the controller because a gain >1 for both directions would result in (damped)
oscillations.

A model for the parametric adjustment effect

A functional model of the adaptive controller is given in Fig. 8. The
"oculomotor error signal" e which is fed into the decision stage in
order to trigger the saccade represents the intended saccade size. It
is multiplied by the actual system gain and disturbed by the unbiased

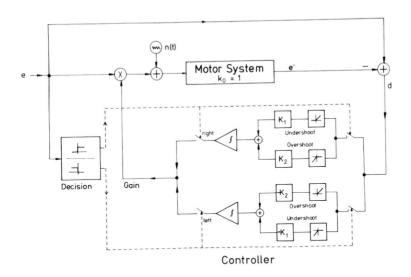

Controller

Fig. 8: A functional model for the PAE.

noise n(t) which leads to the scattering of saccadic amplitudes. The actual output e'of the motor system is subtracted from the intended value e
resulting in the refixation error signal d which is input to the adaptive controller. Because of the directional selectivity and the asymmetry
of the PAE the controller is split   into different parallel branches
which are alternatively selected by the decision elements. The weighted
refixation errors $k_1 \cdot d$ and $k_2 \cdot d$, resp., are averaged over time by
integrating units which preserve the actual system gain. Therefore, the
parameters k_1 and k_2 determine the time constants for the gain
increase and decrease, resp. It is important to note that the model needs

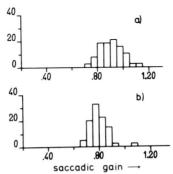

saccadic gain $\longrightarrow$

Fig. 9: Histograms of saccadic gain. Each histogram is determined
from the gain values of 40 consecutive trials. a) Trials 1-40.
b) Trials 121-160.

no reference value to obtain the normal saccadic gain of 0.9. Because
of its nonlinear character the gain of the model is provided by the
difference (k_1-k_2) and the power N of the unbiased noise function $n(t)$.
In fact the model yields a gain of about 0.9 for the values k_1, k_2 and N
estimated from the experimental data.

As demonstrated in /10/, it is obvious that the generation of the oculomotor error signal e is the salient point in such models. Therefore,
we have tried to provoke the PAE under a more complex situation with
pseudo-noise patterns as saccadic stimuli as we used in /10/. Fig. 9
shows histograms of gain values calculated from the 160 individual saccades in a pilot experiment of this kind. Each histogram represents the
data of 40 consecutive trials, Fig. 9a for the first population in the
experiment, Fig. 9b that for the last population. The clear decrease
of the gain indicates that the PAE was elicited and suggests that the
PAE is not restricted to simple isolated stimuli.

REFERENCES

/1/ Robinson, D.A., How the oculomotor system repairs itself. Invest.
    Ophthalmol. 14:6 (1975) 413-415.
/2/ McLaughlin, S.C., Parametric adjustment in saccadic eye movements,
    Percept. Psychophys. 2 (1967) 359-362.
/3/ Vossius, G., Adaptive control of saccadic eye movements, Bibl.
    Ophthal. 82 (1972) 244-250.
/4/ Weisfeld, G.E., Parametric adjustment to a shifting target alter-
    nating with saccades to a stationary reference point. Psychon.
    Sci. 28 (1972) 72-74.
/5/ Henson, D.B., Corrective saccades: Effect of altering visual
    feedback. Vision Res. 18 (1979) 63-67.
/6/ Miller, J.M., Anstis, T. and Templeton, W.B., Saccadic plasticity:
    parametric adaptive control by retinal feedback. J. Exp. Psychol.:
    Human Perception and Performance 7 (1981) 356-366.
/7/ Deubel, H., Wolf, W. and Hauske, G., Corrective saccades: Effect
    of shifting the saccade goal. Vision Res. 22 (1982) 353-364.
/8/ Becker, W., The control of eye movements in the saccadic system.
    Bibl. Ophthal. 82 (1972) 233-243.
/9/ Fischer,B. and Boch, R., Saccadic eye movements after extremely
    short reaction times in the monkey. Brain Res. 260 (1983) 21-26.
/10/ Deubel, H., Wolf, W. and Hauske, G., The evaluation of the oculo-
    motor error signal. This volume.

Acknowledgement - This work was supported as a project of the SFB 50
by the Deutsche Forschungsgemeinschaft.

PARAMETRIZATION OF SACCADIC VELOCITY PROFILES IN MAN

J.A.M. van Gisbergen, J. van Opstal and F.P. Ottes
Department of Medical Physics and Biophysics
University of Nijmegen
Nijmegen
The Netherlands

In order to characterize their shape objectively, saccadic
velocity profiles were fit with a so-called gamma function
whose parameter γ is a direct measure for the degree of asymmetry (skewness). It was found that there is a tight relation
between saccade duration and skewness which, remarkably,
holds for normal fast and abnormally slow saccades: longer
lasting saccades have relatively shorter acceleration periods
but need more time to come to a halt. Mathematical expressions
summarizing relations among the saccadic parameters amplitude,
duration peak-velocity and skewness are presented.

INTRODUCTION

As a means to characterize properties of the saccadic system, peak velocity/
duration/amplitude plots ('main sequence'; Bahill et al., 1975) have become
widely used. Although obviously a very useful tool, this characterization,
based on just three data points in the velocity profile, cannot provide a
complete description of saccade dynamics. For example, it has been noticed
(Hyde, 1959; Baloh et al., 1975) that velocity profiles of large saccades
are skewed whereas in small saccades the accelerating and the decelerating
parts are about equal. This phenomenon, as far as we know, has never been
studied systematically. A second feature of the saccadic system, which has
received little attention, is the occurrence of slow saccades. In studies
attempting curve fitting of main-sequence plots, these 'atypical' saccades
are sometimes weeded out. This would not be necessary if mathematical
expressions interrelating the various saccade parameters could be found,
which are also valid for slow saccades. In this report we propose a procedure to quantify differences in the shape, and especially the skewness, of
velocity profiles. We think that this objective method of characterizing
shapes of velocity profiles can be a useful tool to quantify abnormalities
caused by pathology, drugs or fatigue which cannot be fully portrayed by the
usual main-sequence plots. Our results may also be of interest for model
simulations.

Our data confirm earlier suggestions in the literature that saccades with
larger *amplitudes* (A) have more skewed velocity profiles, but we could establish that the skewness parameter (S) has a tighter relation with saccade
duration (D). Recently it was discovered in the cat that the product of
duration and peak velocity is linearly related to A (Evinger et al., 1981).
We have found that a similar relation holds in the human. This makes it possible, by also incorporating the S-D relation, to express interrelations
among saccade parameters which appear to remain valid in different behavi-

oural states.

METHODS

Saccades (4 subjects) were elicited by a spot of light (0.4 deg; 5 cd/m^2) which jumped, after a randomly varying period of 0.5 to 1.1 sec, from the primary position to a peripheral location on a screen at 57 cm. Background intensity was 1.2 cd/m^2. Horizontal and vertical movements of the left eye were measured with an improved version of the double magnetic induction method (Reulen and Bakker, 1982). In our version of this method (Bour et al., 1983) the disturbance in an alternating magnetic field, created by a thin metal suction ring (thickness: 0.3 mm) on the eye, is measured in a contact-free manner with a nearby detection coil. Raw eye position signals were low-pass filtered (-3 dB at 150 Hz) and sampled at a rate of 500 Hz in each channel. The nonlinearity inherent in the double magnetic induction method was corrected in the computer using the raw eye position signals for 85 different fixation positions. Occasionally data where this linearization procedure could not be used with confidence had to be rejected. By positioning the detection coil eccentrically relative to the suction ring on the eye, a measuring range of 0-35 deg was available in one quadrant of the visual field. Resolution was 15 min of arc, or better, in the range 0-25 deg. Track velocity of the eye was computed from the linearized position signals by differentiation (central-difference algorithm) followed by smoothing with a symmetrical digital filter (-3 dB at 72 Hz).

RESULTS

Saccadic velocity profiles were fit with a mathematical function in order to characterize their shape. Because large saccades have pronounced skewness whereas small saccades are nearly symmetrical, the fit function should have the potential to mimick these shapes. The present results were obtained using the density function of the gamma distribution (gamma function for short):

$$v(t) = c(t/\beta)^{\gamma-1} \exp(-t/\beta) \qquad \beta > 0 \;\; ; \;\; \gamma \geq 0 \qquad (1)$$

where $\{v(t), t \geq 0\}$ is the saccade velocity profile; c and β are scaling constants for velocity and duration respectively, and γ is a form parameter determining the degree of asymmetry. Small γ values imply asymmetrical $v(t)$ profiles. Curve fitting was done using the least squares error criterion. The iteration procedure started from initial parameter estimates based on computation of the first three central moments (Abramowitz and Stegun, 1972). When these estimates were changed, the iteration program still converged on the same solution. The fit, judged from the correlation coefficient, was good (r $\geq$ 0.90; see Fig. 1).

Since our interest is mainly in the *shape* of saccadic velocity profiles, the parameters c and β, which reflect differences in velocity and duration, will not be considered here. The form parameter γ has a clear relation with A; for small saccades we found larger γ values than in large-amplitude saccades (Fig. 1). Since the goodness of fit was quite satisfactory, (see Fig. 1) these differences in γ must reflect changes with A which cannot be accounted for by any combination of amplitude and time scaling and, thus, do not trivially reflect the fact that in larger saccades the eye moves faster and longer than in smaller saccades. This can be appreciated from Fig. 2A,C where velocity curves of horizontal saccades with various amplitudes are

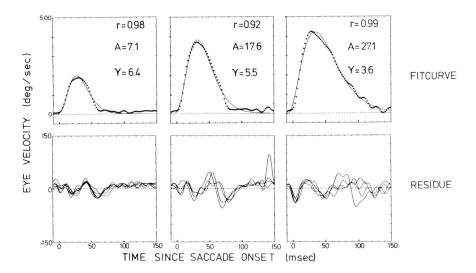

Figure 1
Examples of track velocity profiles of horizontal saccades
(sampled every 2 msec) with their best-fit curves (continuous line). Subject FPO. Error in fit (residue) for a number
of saccades of the same size is shown below (note difference
in scale). Positivity in residue means that data points lie
above fitcurve. The residue is only slightly greater than the
noise level but errors tend to be systematic. Abbreviations:
r, correlation coefficient; A, amplitude in deg; γ, see text.

displayed before and after time- and velocity axis normalization. As can be
seen, the velocity profile is almost symmetrical in the small saccade but
clearly skewed in the largest saccade.

Skewness, defined as the normalized third central moment for a statistical
distribution, can be derived directly from γ using the relation

$$S = 2/\sqrt{\gamma} \qquad \text{(Abramowitz and Stegun, 1972)} \qquad (2)$$

When a plot is made of S versus A for normal saccades a clear linear relation is found (Fig. 3A). This plot confirms earlier suggestions in the literature (Introduction) that S increases with A. Contrary to what would be
expected, however, the plot in Fig. 3A also assigns some positive skewness
to small saccades. It is very probable that this is due to a systematic
failure of the gamma function to fit the tail of saccade velocity profiles
(Fig. 1). While the fit curve has an exponential decay and takes an in-
finitely long time to reach zero velocity, the eye seems to stop more ab-
ruptly. As appears from computations of the third central moment directly
from the measured data, the S values derived from γ are systematically too
high by a nearly constant amount of 0.5. This alternative method of
characterizing the shape of the velocity profile has its own problems: for
example, it is necessary to define a point where the saccade has stopped
and to substitute the noisy fluctuations occurring after this point artificially by zero's to get meaningful results. Also, the values obtained

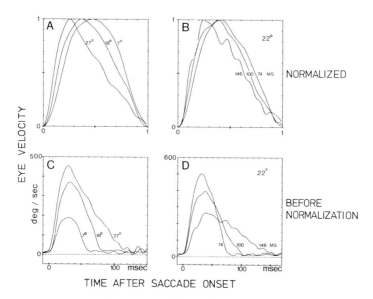

Figure 2

Differences in shapes of saccade velocity profiles. A: horizontal saccades with different *amplitudes* plotted on normalized
axes. Note increase in skewness with amplitude. Subject FPO. B:
normalized plot to show that saccades with longer *durations* are
more skewed. These normalized profiles are from 22 deg horizontal saccades obtained in the alert state (duration 74 msec) or
after a diazepam injection (durations 100 and 146 msec). Subject
JVG. C,D: Same saccades as shown in A and B but displayed with
normal axes.

show considerably more scatter. This difference is probably due to the fact
that the fit procedure is based on the shape of the entire velocity profile
and therefore more immune to small noisy variations near saccade offset.
Therefore, in what follows, γ-derived S values will be used unless indicated
otherwise.

By comparing plots of S versus A and D of normal fast saccades it can be noticed that both relations are about equally tight (Fig. 3A,B). This is not
too surprising since A and D in such saccades are linearly related. To explore whether, perhaps, one of these relations is more fundamental, we have
also compared both plots in conditions where the normal A-D relation is disrupted. This was achieved in one subject, by an injection of diazepam; further relevant data were obtained from a fatigued subject near the end of a
long session. Interestingly, the S-A relation deteriorated while the same
S-D relation remained valid under these conditions (Fig. 3C,D). This means
that when small saccades of a given size have different maximum velocities
(V_m), and accordingly different durations, their velocity profiles cannot be
matched by any combination of amplitude and/or time scaling (see Fig. 2B,D).
Since the S-D relation is linear, at least when based on γ-fit results, it
can be put in de form of a simple equation (a and b constants; $0.70 \leq r \leq 0.97$):

$$S = a D + b \qquad\qquad (3)$$

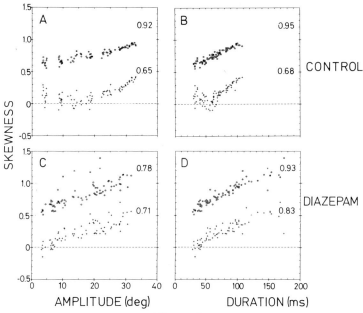

Figure 3
Relation between skewness and saccade amplitude (A,C) and between skewness and saccade duration (B,D) before and after an
intravenous injection of 7 mg diazepam (Valium) which caused
abnormally slow saccades. Subject: JVG. Dots: derived from gamma
fit. Crosses: third central moment. Numbers denote correlation
coefficients. Note that S-D relation is stronger (gamma fit results: P ≤ 0.005) than S-A relation when D-A relation is disturbed by diazepam. The third central moment data, which show
more scatter, indicate the same trend but are significant only
at the P = 0.10 level. In the other subject, where slow saccades
were caused by fatigue, the S-D relation was significantly (P ≤
0.01) more tight than the S-A relation, independent of the method
of computing skewness.

Following up on the work of Evinger et al. (1981) we have confirmed, in the
human, that plots of the product V_m. D versus A yield a strong linear relationship (r ≥ 0.98) which, if a small intercept is ignored, can be represented as

$$V_m D = c A \qquad (4)$$

Linear regression yields a mean value of 1.64 for constant c (4 subjects;
see Table 1) which is rather close to the value of 1.9 found in the cat by
Evinger et al. (1981). When subjects made many slow saccades due to diazepam or fatigue, equation (4) remained valid (r = 0.97) but c became slightly higher (4 - 13%). By recombining the relations (3) and (4) into a single
equation:

$$V_m = \frac{a \ c}{S - b} A \qquad (5)$$

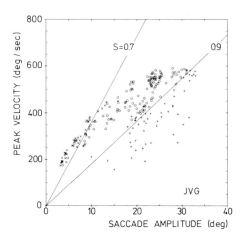

Figure 4
To show that, according to equation (5), the V_m/A plane can be
subdivided into an array of sectors, fanning out from the origin, where saccades have similar skewness. Skewness and saccade
duration (see equation (3)) increase in the clockwise direction;
actual datapoints (crosses: $S < 0.7$; circles: $0.7 \leq S \leq 0.9$;
plus signs: $S > 0.9$) were obtained from a set of experiments
which yielded both normal and abnormally slow horizontal saccades (diazepam experiment) in subject JVG.

the interrelations among the various saccadic parameters (V_m, A and S) can
be succinctly summarized. Table 1 gives a survey of the values found for the
various constants in our four subjects (rightward horizontal saccades):

Table 1.

subject	a	b	c
JVG	4.54	0.49	1.65
MJVG	4.27	0.55	1.59
HG	4.40	0.56	1.57
FPO	3.06	0.69	1.75
mean	4.07	0.57	1.64

What equation (5) entails can perhaps best be appreciated from a graph like
in Fig. 4 where the V_m/A plane has been subdivided into a few sectors by
lines of equal skewness, whose location can be computed if the three constants in equation (5) are known (see Table 1). Two points are worth noting:
1) The graphical expression of equation (5) in Fig. 4 incorporates the fact
   that, for normal main-sequence saccades, S increases with A. Since the
   V_m/A relation of normal saccades in man is nonlinear, the data points
   from large saccades lie in a sector which is closer to the horizontal
   axis than is the case for small saccades.
2) The skewness sectors also give a graphical illustration of our finding
   that, for a given amplitude, V_m and S are inversely related (i.e., slower
   saccades are more skewed; see also Fig. 2B,D).
As can be seen, the actual S values of normal saccades and abnormal slow sac-

cades (Fig. 4) conform at least qualitatively with the 'predictions' from equation (5). The mild violations which occur probably reflect mainly the scatter in relation (3). Yet, we feel that Fig. 4 must be interpreted with caution. First, as explained above, the precise value for intercept b in the S-D relation remains to be determined: the values in Table 1 are systematically too high. Second, there is some uncertainty as to whether the linear S-D relation, which emerges from the gamma fit results, is really valid for small (short duration) saccades. The S values computed directly from the eye velocity data (third central moment; Fig. 3), despite the considerable amount of scatter, hint that equation (3) and by implication also equation (5) are perhaps over simplified. Therefore we regard equation (5) as an interesting working hypothesis, which requires further investigation, rather than as an established fact.

DISCUSSION

The main finding of the present work is that there appears to be a relation between the duration of a saccade and the shape of its velocity profile. The skewness of large 'alert' saccades has been noticed before (Hyde, 1959; Baloh et al., 1975) but the linear relation between S and D, valid for different behavioural states, is new. Skewness of saccade profiles seems to be very pronounced in the cat, an animal with rather slow saccades (Evinger et al., 1981). It remains to be investigated whether the relations expressed in equation (5) have perhaps a more general meaning and allow a better understanding of these interesting inter-species differences.

At present it is not clear why saccades which last equally long, but whose amplitudes may be quite different, should have the same shape of velocity profile. If the oculomotor plant can be assumed to behave as a linear system, differences in the time course of saccade velocity profiles must reflect differences in the neural control signal. The saturation in V_m for large saccades probably reflects saturation of the high-frequency saccadic burst in motoneurons. Likewise, the increase of saccade duration with A is generally thought to reflect an increase in the duration of the motoneuron burst. Reasoning along the same lines, the differences in skewness of saccade velocity profiles may reflect shape differences in the structure of the neural control signal generated for saccades of different amplitudes. Model studies, currently in progress in our laboratory, are needed to establish these hypothetical shape differences more precisely. An alternative to the idea that the peculiarities in saccade velocity profiles described in this paper reflect properties of the neural control signals would be to assume that nonlinearities in the plant play a role. While this possibility cannot be excluded, our finding that the *shape* of saccade velocity profiles clearly depends on the level of alertness strengthens the belief that the time structure of input signals must be a major determining factor. A more definite answer to questions such as these must, of course, come from single-unit recording studies in awake animals.

Finally, the possibility that our skewness findings reflect some kind of artifact related to the method of measuring eye movements should be considered. To rule this out, the study should be repeated using a different method. It should be added, however, that we have no reason to suspect our method since neither the main-sequence plots nor the velocity profiles proper show anything abnormal. Furthermore, it is not clear how the result that the shape of saccade velocity profiles depends on the level of alertness could be due to such an artifact.

In conclusion, we think that skewness is an interesting characteristic of saccades which deserves further attention. The gamma function gives a very reasonable first-order approximation of the saccadic velocity profiles. It is quite possible, of course, that other functions, with more parameters, may provide a better fit. An intriguing problem, of a more fundamental nature, is what property of the saccadic pulse generator (or oculomotor plant?) causes abnormally slow saccades to have the same skewed velocity profiles as large, high-velocity saccades with the same duration.

ACKNOWLEDGEMENTS

This study was supported by the Netherlands Organization for the Advancement of Pure Research (ZWO). We thank Niek van den Berg, Jan Bruijns, Lo Bour and Jos Eggermont for useful discussions and technical help.

REFERENCES

[1] Abramowitz, M. and Stegun, J.A., Handbook of mathematical functions, p. 930 (Dover Publications, New York, 1972).

[2] Bahill, T.A., Clark, M.R. and Stark, L., The main sequence, a tool for studying human eye movements, Math. Biosci. 24 (1975) 191-204.

[3] Baloh, R.W., Sills, A.W., Kumley, W.E. and Honrubia, V., Quantitative measurement of saccade amplitude, duration, and velocity, Neurology 25 (1975) 1065-1070.

[4] Bour, L.J., Van Gisbergen, J.A.M., Bruijns, J. and Ottes, F.P., The double magnetic induction method for measuring eye movement: results in monkey and man. Submitted for publication in IEEE Trans. Bio-Med. Eng.

[5] Evinger, C., Kaneko, C.R.S. and Fuchs, A.F., Oblique saccadic eye movements of the cat, Exp. Brain Res. 41 (1981) 370-379.

[6] Hyde, J.E., Some characteristics of voluntary human ocular movements in the horizontal plane, Am. J. Opthalmol. 48 (1959) 85-94.

[7] Reulen, J.P.H. and Bakker, L., The measurement of eye movement using double magnetic induction, IEEE Trans. Bio-Med. Eng., BME29 (1982) 740-744.

Theoretical and Applied Aspects of Eye Movement Research
A.G. Gale and F. Johnson (Editors)
© Elsevier Science Publishers B.V. (North-Holland), 1984

THE MODIFICATION OF SACCADIC TRAJECTORIES

Trevor J. Crawford

Department of Psychology, University of Durham,
Durham DH1 3LE, England.

This study reports the outcome of a series of tests using
two forms of feedback in attempts to obtain modifications
of human saccadic trajectories. Modifications occur more
readily when a velocity reduction, rather than an increase,
is signalled. Saccadic peak velocities were reduced for
large saccades with continuous feedback but not with trial
by trial feedback. Some implications for models of the
saccadic system are discussed.

INTRODUCTION

Early descriptions of saccadic eye movements developed the view that
saccades were largely stereotyped, preprogrammed movements. This conclusion
was advanced from at least two lines of research. Westheimer (1954) reported that saccades were produced 'ballistically' to the movement of visual
target, even when the timing of the displacement meant that the target had
returned to the line of the visual axis before the start of the eye movement.
Studies on saccade durations and peak velocities also seemed to support the
view that saccades were produced in a stereotyped manner. Robinson (1964)
demonstrated that the duration of the saccade was dependent on the amplitude
of the movement. Yarbus (1967) claimed that this duration could not be
controlled voluntarily when subjects were directed to increase or decrease
the speed of the their saccades.

Subsequent research*(e.g. Becker and Jurgens, 1981) has undermined the original Westheimer (1954) results by showing that a saccade can in fact be
modified by visual information received up to 80 mscs before the saccade is
released and adjustments in the saccade can occur during the saccade flight
(Van Gisbergen, Ottes and Eggermont, 1982). However, the Yarbus experiment
has been taken by some workers as conclusive evidence that saccade velocities cannot be modified by human subjects. It also provided an important
foundation for formulations in which the saccade was seen as a totally preprogrammed eye movement (e.g. Fuch 1971, 1976; Shaknovich, 1977). There are
a number of problems with the experiment, however, which must cast some
doubt on the validity of deductions based on this result.

The absence of several important details of Yarbus' report means that it is
difficult to be certain about the precise nature of the conditions used.
From the data presented, it appears that only between 8 and 10 trials were
used in each condition. His subjects therefore had little time to become
faimiliar with their tasks. Also it appears that a mixed rather than a
blocked trials design was adopted, which would decrease the possibility for

subjects to develop a consistent strategy for the tasks. Yarbus also apparently only investigated saccades within an amplitude range of 6 to 8 degrees. It may be that there are important operational differences in the control of small and large saccadic eye movements (Frost and Poppel, 1976; Posner, 1978). His apparatus for measuring eye movements, which only allowed saccade durations to be measured with a temporal resolution of 10 msecs, would have left room for appreciable errors in the estimation of the durations of these small saccades, which usually only take 30-40 msecs. Finally, other measures of the saccadic trajectory such as the peak velocity were not discussed in Yarbus' brief account.

Recent research has re-opened the idea that subjects may be able to influence directly the course of an individual saccadic eye movement. Viviani and Berthoz (1977) showed that the saccade trajectory could be slowed when subjects were asked to detect a key word during a large oblique saccade across a text display containing the word. Van Gisbergen, Ottes and Eggermont (1982) showed, in a double target step paradigm, that saccadic velocities can be modified in mid-flight. Also Jergens et al (1981) found that in a population of saccades to a given target location there will be a range of peak velocity and duration values but the saccades will still arrive accurately at the target. It should be noted, however, that the results reported by Viviani and Berthoz (1977) and Van Gisbergen et al. (1982) were obtained for saccades of approximately 30 and 40 degrees which is well beyond the normal amplitude range for naturally occurring saccades (Bahill et al., 1975).

Such observations suggested that the modification of saccades was certainly possible. Yet there has been no clear demonstration that the duration and peak velocity of horizontal saccades can be systematically modulated in the absence of general manipulations of the nervous system (e.g. with drugs; Carpenter, 1977).

It was predicted that it should be possible, under certain conditions, for examples as with highly informative feedback, to induce adjustments in the parameters of the saccadic trajectory, in particular, the duration and peak velocity. However, if saccades are the product of preprogrammed, stereotyped processes the modification of a saccade trajectory should not be possible.

EXPERIMENT 1

Three post-graduate students participated in this experiment. The subjects were all naive concerning the detailed characteristics of eye movements. All subjects had emmetropic vision.

This experiment used a method of trial by trial feedback to examine the modification of saccade speeds in two saccade motor tasks. In both tasks the subjects had to move their eyes to a 5 degree target jump. Two biofeedback (BF) conditions were used: A 'BF-slow' task where subjects were required to reduce the speed of their saccades and a 'BF-fast' condition where subjects were required to increase the speed of their saccadic movements. The subjects were informed that the purpose of these conditions was to determine whether they were able to adjust the speed of saccadic eye movements, and that to help them achieve this visual and auditory feedback on the peak velocity of their eye would be provided in the form of a computerised DAC unit measureof the saccade peak velocity. This was presented on a P-31 phosphor oscilloscope approximately 1 second after the eye movement for the duration of that trial. In addition, an auditory tone whose frequency was related to the saccade peak velocity was also given. Each subject also took part in a control session, with no feedback, in which they were instructed to track as accurately as possible the movement of a target step which occurred on that trial.

Two further subjects took part in a condition in which they were given practice in the BF-slow paradigm for ten days. The details of the training procedure were identical to the BF-slow condition described above. On day 1 baseline control data on saccade characteristics were obtained. Subjects were practiced for 10 trials only (to reduce any possible effects of fatigue) on each day for ten consecutive days. Eye movements were sampled at 500 Hz. The data presented in this paper were taken from the initial saccades on a given trial

TABLE 1

		% Change in Peak Velocities and Durations		Correlations		
		BF-FAST	BF-SLOW	CONTROL	BF-FAST	BF-SLOW
S_1	PV	5.73	22.00	A/PV .9936	.9925	.9305
	D	1.73	10.49	A/D .8272	.2294	.3870
S_2	PV	8.65	0.44	A/PV .9603	.8775	.9576
	D	33.40	52.34	A/D .8204	.4808	.6415
S_3	PV	4.69	2.80	A/PV .9320	.8451	.8632
	D	0.12	14.60	A/D .9116	.4465	.5768

Modifications in peak velocity (PV) and duration (D) from the control conditions.

A/PV - Saccade Amplitude/Peak Velocity correlations
A/D - Saccade Amplitude/Duration correlations.

RESULTS

Saccade Peak Velocity and Duration

The analysis of individual records revealed that subjects were often using a specific strategy in their efforts to reduce saccade speeds in the BF-slow task. On a large proportion of trials in this condition subjects produced a regular sequence of hypometric saccades (or 'multiple saccades') which ultimately took the eye to the target position. Therefore, as saccade peak velocities and durations are a function of saccade amplitudes they were treated by using the amplitude of the saccade in the calculation of any differential effects produced in the feedback conditions as a function of the control values by calculating the duration and velocity of saccades of equivalent amplitudes from the linear regression functions obtained in the control condition. The difference between feedback performance and the control values was expressed in percentage terms. A positive value indicated an increase and a negative value a decrease in the feedback performance. Table 1 shows that the peak velocities are similar in each condition. In contrast to the results for the peak velocities, Table 1 also illustrates that saccade durations are consistently increased in the BF-slow condition relative to both the control and the BF-fast condition.

One subject appears to have longer durations in the CF-fast condition than
in the control condition. This subject, however, has even longer durations
in the BF-slow task.

Correlations

Table 1 shows that the amplitude/peak velocity correlations are largely
maintained in each condition. Therefore any adjustments in peak velocities
are assumed to be the outcome of contingent modifications of saccade amplitudes. In the control conditions correlations of approximately 0.9 and 0.8
are found for the amplitude/duration relationships. These correlations are
clearly reduced in the BF-fast and BF-slow conditions showing that in the
feedback conditions there was an increased variability in this relationship.
This indicates, firstly, that the increased durations in the BF-slow condition cannot have been the result of simply increasing the saccade amplitudes. Secondly, this result shows that a weakening in the saccade
duration/amplitude relationship can occur indpendently of any overall
changes in the baseline level of saccade duration as manifested in the
BF-fast conditions.

Extended Training Condition

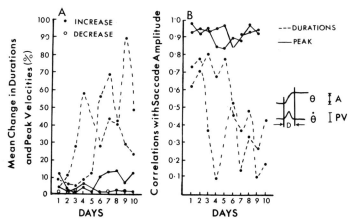

Figure 1.   (A)   The training effects on saccade peak velocities
                   and durations.

             (B)   Modifications of saccadic velocity and duration
                   correlations.

             A - Amplitude; PV - Peak Velocity; D - Duration

Peak Velocity

Figure 1a shows the changes in peak velocities relative to the initial
velocities measured on day 1. No significant trend for peak velocities
were observed over the training sessions. The mean changes in peak
velocities across the entire training period were 1.14% for S1 and 7.93%

for S2. The correlations for saccade amplitudes and peak velocities were
also unchanged. There were uninterrupted reliable amplitude/velocity
correlations on every day (Fig. 1b).

Saccade Duration

Figure 1a also illustrates the changes in saccade durations. The level of
saccade durations increases from initial values of 8%(+) and 3.6%(-) on day
1 to final values of 49%(+) and 21%(+) on day 10. Figure 1b shows that the
lengthening of saccade durations was largely isomorphic with a fall in the
amplitude/duration correlations, confirming that in contrast to peak
velocities neither the within nor the between session modulation of saccade
durations can be adequately accounted for solely in terms of adjustments in
the saccade amplitudes.

Discussion

The extension of saccade durations has been previously observed in a patient
with oculomotor paresis (Abel et al., 1978; but see Optican and Robinson,
1980), spinocerebeller patients (Zee et al., 1976) as well as with patients
under the effects of alcohol and drugs (Carpenter, 1977). Although these
effects have usually been observed with large saccade amplitudes. Abel et
al. (1978) conclude from their observations that the central balance
mechanism responsible for saccades can 'selectively alter the operation of
different aspects of the saccadic system'. These results provide further
support for this view by demonstrating that the duration of saccades 5 degs.
and smaller can be lengthened, particularly with extended feedback practice.

Nevertheless, the results also make the point that trial by trial feedback
is notsufficient to disturb the normal levels of saccade peak velocities.
Therefore, in Experiment 2, continuous on-going feedback was used in an
attempt to modify saccadic peak velocities.

EXPERIMENT 2

Extensive research in the field of psychophysiology has shown that direct
feedback is particularly effective in the modification of many biological
systems. The method of continous auditory feedback indicated a possible
useful method for an attempt to induce adjustments in the peak velocity
of a saccade and might therefore provide further insight into the possibilities of plasticity in the saccadic system.

Eight subjects were tested for their ability to increase or decrease the
speed of their saccadic eye movements. Subjects were divided equally into
the BF-fast and BF-slow tasks. Subjects were asked to saccade between two
point targets spaced at either 5 or 20 degrees of visual angle. In each
condition 2 subjects were tested at each target eccentricity. Saccadic
velocities were measured immediately before, during and after auditory
feedback information trials. This feedback procedure used a variable
analogue tone whose frequency was modulated with the position of the eye.
The range of tone variation depended to some extent on the gain setting of
the eye movement recording system which could differ slightly for each
subject. A 20 degree saccade corresponded approximately to a change in
tone frequency from 1 KHz to 3.5 KHz. Velocity information was given by the
contingent relationship of the eye velocity and the gradient of the tone
frequency modulation. Each subject was also tested at different eccentrici-

ties in an equivalent control condition, without feedback, when the instructions were to saccade normally. Horizontal eye movements were recorded
with a photoelectric device, (based on the system described by Findlay,
1974). Eye movements were recorded in the dark to minimize the effect of
any distracting visual stimuli.

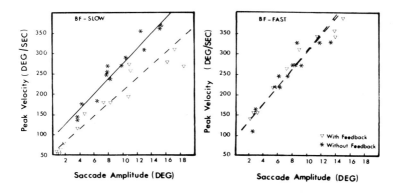

Figure 2   Typical Results from the BF-slow and BF-fast tasks
                with continuous feedback.

RESULTS

No consistent effects were observed under the instruction to speed up the
saccade, only 1 of 4 subjects produced saccades which showed any enhancement in the peak velocities. However, 3 out of 4 subjects were able to
slow saccades by the end of the BF-slow condition. Two of these subjects
were in the 20 degree target amplitude condition. These modified saccades
had peak velocities which fell below the normal saccade amplitude/velocity
relationship (Fig. 2). Saccade durations were also lengthened in these
BF-slow trials, suggesting that the saccades may have been controlled in a
feedback loop in accordance with the model proposed by Jurgens et al. (1981)
Inspection of the velocity profiles of slowed saccades showed that the usual
single peak was occasionally replaced by a double peak trajectory. Subsequent tests immediately after feedback training showed that normal velocity
saccades were evident within 10-15 responses.

SUMMARY AND CONCLUSIONS

These results demonstrated that continuous auditory feedback facilitates
the slowing of saccade peak velocities whereas discrete trial by trial
feedback does not. The data also indicated that saccades in the 20 degree
target condition were more consistently slowed than saccades in the 5
degree condition. Further unpublished experiments on saccadic trajectories
have shown that the trajectories of 20 degree saccades are in fact more open
to modification than 5 degree saccadic eye movements. This is consistent
with Posner (1978) and others who have suggested that the processing of
small saccades may be characterised by processing limitations that do not
apply to larger saccades.

Experiment 1 showed that the saccadic system is able to modulate the

duration of the signals generating a saccade depending on the demands
imposed by a saccadic movement task. This mechanism apparently has the
capacity to modify a saccade duration, while permitting the saccade amplitude to be less restricted by the saccade duration when it does so.

The finding that it is possible to lengthen the duration of a saccade,
while inducing no general variations in the peak velocities, implies that
these saccade parameters may be determined by different components of the
neurophysiological input. This is consistent with neurophysiological
research on this question. Saccade peak velocities can apparently be
modified with the aid of feedback methods, although the possibility of
adjustments in arousal states etc. cannot be ruled out. Further experiments will be necessary to examine the precise nature of strategies used
in the control of saccadic characteristics.

THEORY

Robinson (1975) proposed that the neural information used to move the eye
in a saccadic movement is composed of step-pulse components, in systems
analysis terms. The height of the pulse determines the impulse for
acceleration of the eye, while the duration of the pulse determines the
duration of the response. Jurgens et al. (1981) argue that the trajectory
of the saccade is controlled in a continuous feedback loop in which both
the pulse intensity and pulse duration may vary indpendently of the amplitude of the saccade. According to this formulation, during a saccade the
eye is continuously driven by a pulse generator until it arrives at the
target location. If saccades are controlled in this on-line mode, it is
reasonable to suppose that modifications of the pulse input could produce
almost immediate effects on a saccadic movement. This model would make
the possibility of adjustments in the flight of a saccade particularly
plausible (Zee et al., 1976, Jurgens et al, 1981).

It remains to be determined why saccades appear to be more easily slowed
than enhanced in normal subjects. (A similar phenomenon has been found in
the control of human heart rate). One possibility is that saccades may
be already produced at speeds close to the limits of the system. Faster
saccades would leave less time for the operation of the feedback loop
responsible for the control of a saccade trajectory.

ACKNOWLEDGMENTS

I would like to thank Dr. J. Findlay and Dr. L. Harris for their valuable
comments on earlier drafts of this paper. The work was supported by
Science and Engineering Research Council, Grant no. 80309579.

REFERENCES

Abel, L.A., Schmidt, D. and Dell'Osso, L. Saccadic System Plasticity in
Humans. Ann. Neurol. Vol. 4 (1978) 313-318.

Bahill, A.T., Adler, D. and Stark, L. Most naturally occurring saccades
have magnitudes of 15 degrees or less. Invest. Ophthal. 14 (1975)

Carpenter, R.H.S. Movement of the Eyes. (Pion, London, 1977).

Findlay, J.M.   A simple apparatus for recording microsaccades during
fixation. Quart. J. Exp. Psychol. 26 (1974) 167-170.

Frost, D. and Poppel, E.   Different Programming Models of Human Saccadic
Eye Movements: Indications of a functional subdivision of the visual
field. Biol. Cybernetics. 23. (1976) 39-48.

Fuchs, A.E.:   The Saccadic System, in: Bach-y-Rita, P. and Collins, C.
(eds.).   Eye Movements and Psychological Processes (Lawrence Erlbaum
Associates, Hillsdale, New Jersey, 1976).

Jurgens, R., Decker, W. and Kornhuber, H.   Natural and drug induced
variations of velocity and durations of saccadic eye movements:
evidence for a control of the neural pulse generator by local feedback.
Biol. Cyber. 39 (1981) 81-96.

Optican, L.M. and Robinson, D.A.   Cerebellar dependent adaptive control
of primate saccadic system. J. Neurophysiol. 44 (1980). 1058-1076.

Posner, M.I.   Chronometric Explorations of Mind (Lawrence Erlbaum Associates
p. 192-197. 1978).

Robinson, D.A.   Oculometer control signals in: Basic Mechanisms of Ocular
Motility and Their Clinical Implications (Pergamon Press, Oxford,
1975)

Shakhnovich, A.R.   The Brain and the Regulation of Eye Movements (Plenum
Publications, 1977).

Van Gisbergen, J.A.M., Ottes, F.P. and Eggermont, J.J.   Responses of the
saccadic system to sudden changes in target direction in: Roucoux, A
and Crommelinck, m. (eds.) Physiological and Pathological Aspects of
Eye Movements (W. Junk Publishers, 1982).

Viviani, P and Berthoz, A.   Voluntary deceleration and perceptual activity
during oblique saccades, in: Baker, R. and Berthoz, A. (eds.)   The
Control of Gaze by Brainstem Neurons (Elsevier/North-Holland Biomedical
Press, New York.

Westheimer, G.   Eye Movement responses to a horizontally moving visual
stimulus. Arch. Ophthalmol. 52 (1954) 932-94=.

Yarbus, A.L.   Eye Movements and Vision (Plenum/New York, 1967).

Zee, D.Ss, Optican, L.M., Cook, J.D., Robinson, D.A. and Engel, W.K.
Slow saccades in spinocerebellar degeneration.   Arch. Neurol. Vol.33
(1976) 243-25=.

Theoretical and Applied Aspects of Eye Movement Research
A.G. Gale and F. Johnson (Editors)
© Elsevier Science Publishers B.V. (North-Holland), 1984

THE EFFECTS OF TASK VARIABLES AND PROLONGED PERFORMANCE ON SACCADIC EYE MOVEMENT PARAMETERS

Tayyar Sen and Ted Megaw

Department of Engineering Production
University of Birmingham
Birmingham
ENGLAND

The extent to which prolonged work at a VDU influences a variety of saccadic eye movement parameters was investigated in two experiments. In the first, trials on a step-jump and number comparison task were presented to subjects over a period of one hour. No consistent effects were observed of the type of task or their duration on saccadic peak velocities or durations. In the second experiment, subjects performed a Neisser type search task for one hour in between completing trials on the number comparison task. Peak velocities and duration of saccades were unchanged as a result of performing the search task. Some changes in the variability of peak velocities and in the frequencies of dynamic overshoot and glissades were observed although their characteristics were specific to individual subjects.

INTRODUCTION

With the rapid introduction of VDUs into several working environments there has been a noticeable revival of interest in research into the definition and measurement of visual fatigue. Whether or not this concern is justified is debatable. However, it cannot be denied that many studies, reviewed by Dainoff (1982), have indicated that operators report a significant increase in complaints of fatigue following prolonged VDU work. A problem in assessing visual fatigue is that many of the measures that have been used to quantify it are themselves influenced by factors other than purely visual ones. These include the extent of pacing, the amount of repetitiveness in the task and the working memory load. Measures which are less likely to be affected by these confounding factors are those concerned with ocular motor control. Of these, the most important relate to the control of the extra-ocular muscles over saccades and vergence movements and to the control of the ciliary muscles over accommodation. Although changes in the accommodative state during prolonged VDU work have been reported (Ostberg, 1980), there is the possibility that these are due to changes in the pupil size and, therefore, to changes in depth of field. While vergence movements are one of the easiest ocular-motor systems to fatigue (Luckiesh and Moss, 1935), it is difficult to assess to what extent such movements are involved during normal VDU work.

Because of a large reserve force, the execution of saccades involves a relatively small proportion of the extra-ocular muscle fibres (Young, Zuber and Stark, 1966) and it is, therefore, unlikely that any observed changes in saccadic parameters are due to muscular fatigue. One of the earliest studies to report a slowing of saccades from fatigue was made by Dodge (1917). More recently, and with much improved recording and analysis methods, Bahill and Stark (1975) have shown that operators display an increasing tendency to execute overlapping saccades and saccades with reduced peak velocities when performing a continuous series of large amplitude step-jumps. These so-called non-main

sequence saccades could be identified by their position in relation to
unfatigued saccades on plots of peak velocity or duration against amplitude.
Bahill and Stark also reported a similar increase in the frequency of closely
spaced saccades. On the other hand, Schmidt et al (1979) who employed a similar
task to induce fatigue were unable to demonstrate reduced peak velocities as a
result of prolonged performance. On the contrary, some subjects showed an
increase. In the same study, Schmidt et al obtained data from one subject who
was tested under two different states of mental fatigue. It was clear that peak
velocities were reduced for all amplitudes of movement when the subject was in
the more fatigued state.

In addition to a reduction in peak velocities, Bahill and Stark (1975) have
reported an increase in the frequency of occurrence of glissades with fatigue.
Glissades, like non-main sequence saccades, occur monocularly. According to
Bahill and Stark (1979) they are caused by a mismatch between the pulse and step
components of the saccadic control signals. If the step component itself was
inappropriate, a binocular corrective saccade would follow (Weber and Daroff,
1972). A further parameter that might be influenced by fatigue is dynamic
overshoot. Bahill, Clark and Stark (1975) have found that 70% of main saccades
are accompanied by dynamic overshoot caused by small pulses of antagonist muscle
activity towards the end of the saccade resulting from neurological control
signal reversals.

Some criticisms can be made of the studies on fatigue to the saccadic system.
All of them have required subjects to track a simple step function. Such a task
is unnatural in that it does not include a cognitive element but places a large
emphasis on accuraccy of movement. The studies have tended to involve a high
proportion of large amplitude saccades. This conflicts with the usual finding
that saccades greater than 15 deg are rarely executed (Bahill, Adler and Stark,
1975). None of the studies provided any statistical evidence to corroborate the
conclusions reached. To overcome these criticisms, the experiments described
in this paper use a variety of tasks to induce fatigue without resorting to
require subjects to execute large amplitude saccades at high frequencies.

EXPERIMENT 1

Method
Subjects were presented with alternating blocks of trials on a step-jump and
number comparison task. Both tasks were presented on a VDU with a P31 phosphor
and involved the execution of horizontal saccades never greater than 18 deg
within + or - 9 deg either side of the mid position. Screen luminance was
approximately 0.25 cd m^{-2}. A headrest and chin support were arranged so that
the viewing distance was approximately 30 cm. The method of recording eye-
movements was similar to that described by Bahill, Brockenborough and Troost
(1981).

Step-jump task (SJ). Each block of trials included 10 discrete horizontal step-
jumps, the amplitude of which varied randomly from between 2.7 deg and 17.4 deg.
A trial began with the presentation of an X on the screen. When this changed to
a horizontal dash subtending 0.27 deg the subject knew the dash would then move
500 ms later to a new horizontal position. The dash remained displayed for 1100
ms before changing to an X. The next trial was initiated after the eye movement
recording was checked for artefacts. Only the first trial was guaranteed to
start from the mid-position. Typically, one main saccade was required to
complete each trial, although this was frequently followed by a corrective
saccade.

Number comparison task (NC). Each block of trials included 10 number
comparisons. The subjects initiated each trial by pressing a key. This caused
a vertical dash to appear on the screen and one s later the dash was replaced by

two 3-digit number strings, one either side of the fixation point. The separation between the middle digits of the two number strings varied randomly between 9 and 16 deg. On half the trials the digit strings differed by one digit and subjects were required to press the appropriate key as quickly as possible depending on whether or not the two digit strings were the same or different. The digits were displayed for 1280 ms. Typically, a minimum of two main saccades was exececuted during each trial, one fixation on each of the number strings, although sometimes a third main saccade was made.

Experimental design. Three subjects with normal vision were tested on 12 blocks of trials. Each block was preceded by a calibration trial. Blocks of SJ and NC trials were alternated. The experiment took a minimum of one hour to complete.

Eye movement recording. The horizontal movements were recorded independently using standard photoelectric techniques. Pairs of infrared emitters (TIL24) and receivers (LS618) were mounted on a pair of trial frames. Following amplification, the signals were passed through a 300 Hz low-pass analog filter to avoid the Nyquist frequency resulting from the sampling rate of 1000 Hz. used by the 12-bit A-D converter. The position data were then stored on floppy discs with an Altos 8000-2 microcomputer.

Recording program. This controlled the presentation of the stimulus material, the recording and storage of the position data and the presentation and storage of calibration trials. The program enabled the experiment to be run interactively. This ensured that trials were presented only if the preceding calibration was acceptable and that individual trials were stored only if they were free from artifacts including blinks. To overcome storage limitations, data were recorded at certain periods. These were 640 ms from when the dash moved to a new position for the SJ trials and 1280 ms during which the number strings were presented for the NC trials.

Calibration procedure. To calibrate the position data, subjects performed four sweeps of horizontal pursuit tracking. The target was a horizontal dash whose position was generated by the output of a joystick moved by the subjects at their preferred rate. Each sweep was 25 deg, 12.5 deg either side of the mid-position, and provided eye position data for 34 discrete character positions on the screen. For each eye, a linear regression was performed on the relationship between the output from the sensors averaged over the four sweeps and the angular separation of the 34 character positions. In calculating the angular separation, the distance between the two eyes was taken into account. An algorithm removed the effects of blinks. If the resulting F-ratios were highly significant ($p < 0.0001$), the calibration values in units deg^{-1} were stored and the appropriate block of trials was initiated.

Analysis program. Before applying routines to identify saccades and glissades, velocity and acceleration data were calculated. The velocity was obtained by applying a two point central difference algorithm to the position data and then passing the data through a zero-phase 21-point digital filter with a 80 Hz cut-off. To obtain acceleration data, a two-point central difference algorithm was applied to the velocity data before passing it through a 60 Hz cut-off zero-phase digital filter.

To register a saccade, the peak velocity must have exceeded 20 deg s^{-1} for a minimum of 10 ms. Using the calibration data, the amplitude and duration of the saccades were calculated from the position data. The same routine was used to identify dynamic overshoot. Originally, it had been hoped to identify overlappping saccades by the presence of more than one peak in either the acceleration or deceleration profiles of the saccades. However, the noise levels were too high to permit this method of identification. Following the detection of a saccade, a glissade detection routine was applied. A glissade

T. Sen & T. Megaw

was registered provided the velocity remained above 2 deg s^{-1} for longer than 30 ms and started within 25 ms from the end of the preceding saccade. Saccadic latencies to the presentation of the step-jumps and the number strings were calculated.

The second phase of the analysis program performed two sets of linear regressions on the data from individual or combined blocks of trials. The first of these was between log PV (where PV = peak velocity) and log A (where A = amplitude) provided A exceeded one deg. Dynamic overshoot was excluded from the regression. Thus the data were fitted by a power function of the form PV = k.A^a The second linear regression was between Duration and A. To ensure a good fit, saccades less than 5 deg were excluded. From these two sets of regressions it was possible to obtain estimates of peak velocities and durations for saccades of 5, 10 and 15 deg with 95% confidence limits.

Results

A summary of the results on peak velocity and duration from the right eye are given in Table 1. No obvious changes can be observed as a function of task type or block number. Had there been an increase in the frequency of either overlapping or low velocity saccades as performance on either task continued one would have expected to find a decrease in peak velocity and an increase in duration. A close look at individual saccades failed to reveal a single example of overlapping saccades. The same results were obtained from the left eye. Generally these observations were confirmed by the results from statistical analyses. Although the application of Bartlett's test for homogeneity of variance often indicated significant differences between the regression lines obtained from the 6 blocks of trials for a particular subject and task type, these differences did not reflect a consistent trend. This conclusion also applied to other comparisons made between pairs of regression lines obtained from equivalent blocks on the two different tasks.

		Block number			
		1+2	3+4	5+6	Mean
Peak velocity, deg s^{-1}					
5 deg	SJ	215	209	213	212
	NC	210	205	208	208
10 deg	SJ	342	335	352	343
	NC	350	349	348	353
15 deg	SJ	450	442	475	456
	NJ	471	466	461	466
Duration, ms					
5 deg	SJ	38.7	37.3	39.0	38.3
	NC	40.3	40.7	41.0	40.7
10 deg	SJ	48.0	49.0	48.3	48.4
	NC	47.7	48.0	47.0	47.8
15 deg	SJ	57.7	60.7	57.3	58.6
	NC	54.7	55.3	55.0	55.0
Latency, ms					
	SJ	255	264	262	263
	NC	211	196	215	207

TABLE 1. Estimates of saccadic peak velocity, duration and latency for the right eye from the first experiment, results averaged over the three subjects.

No obvious trends were apparent when the frequencies of dynamic overshoot and glissades were examined. However, the system noise level was much higher than anticipated so that detection sensitivity was not very high. The mean noise level in the velocity channel was + or - 7 deg s^{-1} at 50 Hz. This meant that a glissade had to maintain a velocity above this level for at least 30 ms for it to be identified. Despite the claim by Weber and Daroff (1972) that the velocity of glissades is in the region of 20 deg s^{-1}, the evidence from this and other experiments suggests that this value is overestimated. Although the low frequency of dynamic overshoot may also have been a result of the noise level, it should be noted that their frequency was highest in the noisiest channel.

EXPERIMENT 2

Rather than attempting to induce fatigue by having subjects perform the same task which was used to measure the fatigue, in this experiment fatigue was induced by having them perform a Neisser type search task (Neisser, 1963) for a period of one hour. Any changes in saccadic parameters resulting from this could be assessed by comparing the saccades obtained from number comparison trials given before and after the search task.

Method
Number comparison task (NC). This was the same as before except that a delay of 100 ms was introduced between the presentation of the two number strings. This was to discourage subjects from adopting a strategy of always first looking at the left or right string.

Neisser search task. For each trial the screen was filled with 10 rows of four 6-letter strings. Each string was in upper case and was accompanied by a reference number displayed in relatively low contrast. When the screen was filled, one of four target letters appeared at the top and subjects could then begin to search the array for the target using a horizontal scanning strategy. Only 50% of the trials included a target and there was never more than one target on each page. Two keys were provided for subjects to indicate whether or not they found the target. If they did find one they were required to give the reference number of the relevant letter string. If they missed the target, the relevant letter string flashed on and off for 5 s. Subjects were free to move their head during the search task.

Experimental design. Five subjects with normal vision were given two blocks of NC trials followed by one hour on the search task. This was immediately followed by a further two blocks of NC trials. Before and after the experiment, subjects were given a 16-point self-assessment questionnaire which required them to rate symptoms of visual and general fatigue.

Recording and analysis of eye movements. The sensors were changed with the result that the mean noise level in the velocity channel was reduced to + or - 3 deg s^{-1}. Some new analyses routines were developed in order to assess the correspondence between the characteristics of pairs of saccades from the left and right eyes.

Results.
The data summarised in Table 2 show no obvious effects of the intervening search task on either saccadic velocity or duration. These results were confirmed by comparing the regression lines from NC trials before and after the search task for individual subjects. Only in the case of one subject (GB) was there a significant difference (p < 0.01) reflecting an increase in the slope of the relationship between duration and amplitude from 1.66 ms to 2.11 ms deg^{-1}. On the other hand, three subjects (EL, LC, AA) showed a significant increase in the variability of peak velocity (p < 0.01) following the serach task reflected by an increase of approximately 50% in the 95% confidence limits for the

estimated values for 15 deg saccades. In agreement with the results of Robinson (1964), two subjects (AA and EL) were found to have temporal saccades with higher peak velocities and shorter durations than nasal saccades.

		Peak velocity, deg s^{-1}		Duration	
		Before	After	Before	After
5 deg	LE	214	210	38.0	38.6
	RE	219	217	37.8	38.6
10 deg	LE	354	343	47.2	48.0
	RE	356	351	46.8	47.2
15 deg	LE	475	455	56.0	57.6
	RE	472	467	55.8	55.8

TABLE 2. Estimates of peak velocity and duration from the second experiment before and after performing the search task, results averaged over 5 subjects.

The frequencies of dynamic overshoot, glissades and static corrections are given in Table 3. Except in the case of static corrections, the frequencies are shown separately for each eye and direction of movement (nasal or temporal). Clearly the pattern of results is very idiosyncratic. For example one subject (TM) showed no dynamic overshoot while in the case of subject EL, all dynamic overshoot was restricted to temporal saccades of the left eye. On the other hand, for subject GB the dynamic overshoot was restricted to nasal saccades. There are some indications of changes following the search task, but again there are no general trends. Subject AA showed a marked increase in glissadic overshoot for temporal saccades of the left eye, while subject LC showed a general increase in dynamic overshoot, glissadic overshoot and static corrective undershoot. Subject GB, on the other hand, showed reductions in dynamic overshoot, glissadic overshoot and glissadic undershoot. There were no changes for the other two subjects.

		Number of saccades		dynamic overshoot		glissadic overshoot		glissadic undershoot		static correction	
		N	T	N	T	N	T	N	T	Under	Over
Subject		LE,RE	LE,RE	LE,RE	LE,RE	LE,RE	LE,RE	LE,RE	LE,RE	shoot	shoot
GB	Before	21,21	21,21	4,15	0,0	3,21	0,0	0,0	0,12	11	6
	After	21,20	20,21	4,0	0,0	1, 9	1,3	0,0	0,0	9	0
AA	Before	21,22	22,21	0,0	0,0	1,4	0,0	3,7	1,0	5	3
	After	23,23	23,23	0,0	0,2	1,0	15,0	3,10	1,0	10	0
LC	Before	23,21	21,23	1,1	3,8	0,1	6,9	0,0	0,0	9	1
	After	22,20	20,22	3,7	16,14	9,8	17,14	0,0	0,0	29	0
TM	Before	21,26	26,21	0,0	0,0	0,0	0,0	5,17	12,3	5	1
	After	21,30	30,21	0,0	0,0	0,0	0,0	3,11	13,3	5	1
EL	Before	26,28	28,26	0,0	23,0	0,15	21,0	19,0	0,3	9	2
	After	28,35	35,28	0,0	27,0	0,9	27,2	23,0	0,0	8	2

TABLE 3. Frequencies of dynamic overshoot, glissadic overshoot, glissadic undershoot and static corrections (binocular) for the five subjects from the second experiment.

The results shown in Table 4 concern difference between some parameters of the main saccades of the left and right eyes. Essentially, they show no reduction in the correspondence between the responses of the two eyes following performance on the search task. Only in the case of subject LC was there an increase in the discrepancy between the amplitudes of pairs of saccades together with an increase in the tendency for left eye saccades to have larger amplitudes than the right eye.

All subjects reported an increase in symptoms of fatigue following the search task. Most of these symptoms related to general fatigue such as over-tiredness, boredom, suffering from headaches and difficulty in maintaining attention. However, four subjects reported aching behind the eyes and the onset of double vision. The number of pages of letter strings searched during the course of the experiment varied between 84 and 130.

		First main saccade (LE-RE)			All saccades > 1 deg	
Subject		Latency,ms	Amplitude,deg	PV, deg s^{-1}	Absolute	Relative (LE-RE)
GB	Before	-0.3	0.00	-8.0	8.8	-1.0
	After	-0.7	0.36	15.9	5.0	3.5
AA	Before	-0.1	-0.48	-3.4	10.0	-7.7
	After	-0.9	0.04	10.4	13.9	0.9
LC	Before	0.2	0.22	14.8	7.3	1.4
	After	0.3	0.53	25.4	13.0	8.9
TM	Before	-0.5	0.22	12.7	3.5	3.0
	After	-0.5	-0.05	0.2	2.8	-0.1
EL	Before	0.3	0.26	10.2	10.3	7.0
	After	-0.8	0.44	20.2	10.5	9.1

TABLE 4. Mean differences between pairs of left and right eye saccades. The last two columns refer to percentage differences in amplitude.

DISCUSSION

The results from both experiments generally support the conclusion that prolonged viewing at a VDU does not dramatically affect saccadic parameters irrespective of whether or not the attempt to induce fatigue is made by the same task used to measure the fatigue or by an additional search task, although results from the second experiment do suggest that the variability of peak velocity is influenced by fatigue. It could be argued that the task demands including task duration were insufficient to produce fatigue. Haider, Kundi and Weibenbock (1980) have indicated that temporary myopization only occurs to any significant extent after 3 hours of continuous VDU work. On the other hand, subjects from both experiments reported symptoms of fatigue on completing the experiments, although it is open to question whether such fatigue is visual in origin. Of the various saccadic parameters investigated in this study, there is no doubt that estimates of peak velocity and duration are the most reliable because they are relatively unaffected by the noise level in the recording system. From neither experiment could any examples of monocular overlapping saccades be found, although binocular closely spaced saccades were observed. However, these tended to occur when subjects were anticipating the stimulus onset. Such staircase-like responses have been reported previously by Megaw and Armstrong (1973).

In contrast to these negative results, other studies have demonstrated that certain factors can exert a considerable influence over the dynamic characteristics of main saccades. Both Wilkinson, Kime and Purnell (1974) and Guedry et al (1975) have reported a reduction in velocities ranging from 13 to 17% following the intake of alcohol. Unfortunately, neither study stated whether or not account was taken of any changes in saccadic amplitude. On the other hand, Gentles and Thomas (1971) have indicated that the 11% reduction in peak velocity following the intake of diazepam could be partially if not completely explained by reduced amplitudes of movement. The effect of age on saccadic velocities is inconclusive. Spooner, Sakala and Baloh (1980) report a 12% reduction in asymptotic values of peak velocity for older subjects. In contrast, Henricksson et al (1980) and Abel, Troost and Dell'Osso (1983) report the absence of any effects of age.

If the effect of fatigue was a reduction in the amplitude of main saccades, one might expect an increase in the frequency of corrective saccades to compensate for undershoot. Two of the subjects (AA, LC) from the second experiment did show such an increase. However, it should be pointed out that what are classified as corrective saccades in Table 4 includes examples where the inter-saccadic interval was much longer than if they were small saccades compensating for undershoot. They could reflect re-fixations on a different digit within the 3-digit string.

Results on dynamic overshoot and glissades do suggest some influence of the intervening search task. The differences observed following the search task were always much greater than the differences between the two pairs of blocks of trials given before or after the search task. Two subjects showed an increase in glissadic frequency following the search task, and one subject a reduction. Only one subject showed an increase in the frequency of dynamic overshoot. There is no doubt that the results on glissades and dynamic overshoot should be treated cautiously. They are influenced both by the design of the detection routines and by the system noise levels. In this study, the application of the main saccadic detection routine to the detection of dynamic overshoot meant that the minimum size of glissade that could be identified was 0.25 deg. However, visual inspection of the records showed that dynamic overshoot as small as 0.1 deg could easily be identified. Unfortunately, if the minimum velocity in the saccadic detection routine is reduced much below 20 deg s^{-1}, glissades are identified as saccades. At the same time, to prevent glissades from being masked by 30-60 Hz noise, it would be necessary to introduce a further low-pass filter to the velocity data.

REFERENCES

Abel, L.A., Troost, B.T. and Dell'Osso, L.F., The effects of age on normal saccadic characteristics and their variability. Vis. Res. 23 (1983) 33-37.

Bahill, A.T., Adler, D. and Stark, L., Most naturally occurring human saccades have magnitudes of 15 degrees or less, Invest. Ophthalmol. 14 (1975) 468-469.

Bahill, A.T., Brockenborough, A. and Troost, B.T., Variability and development of a normative data base for saccadic eye movements, Invest. Ophthalmol. Vis. Sci. 21 (1981) 116-125.

Bahill, A.T., Clark, M.R. and Stark, L., Dynamic overshoot in saccadic eye movements is caused by neurological control signal reversals, Exp. Neurol. 48 (1975) 107-122.

Bahill, A.T. and Stark, L., Overlapping saccades are produced by fatigue in the saccadic eye-movement system. Exp. Neurol. 48 (1975) 95-106.

Bahill, A.T. and Stark, L., The trajectories of saccadic eye movements, Sci. Am. 240 (1979) 85-93.

Dainoff, M., Occupational stress factors in visual display terminal (VDT) operation, Behav. & Inf. Tech. 1 (1982) 141-176.

Dodge, R., The laws of relative fatigue, Psych. Rev. 24 (1917) 89-113.

Gentles, W. and Llewellyn Thomas, E., Effect of benzodiazepines upon saccadic eye movements in man, Clin. Pharm. Therap. 12 (1971) 563-574.

Guedry, F.E., Gilson, R.D., Schroeder, D.J. and Collins, W.E., Some effects of alcohol on various aspects of oculomotor control, Aviat. Space & Env. Med. 46 (1975) 1008-1013.

Haider, M., Kundi, M. and Webenbock, M., Worker strain related to VDUs with differently coloured characters, in: Grandjean, E. and Vigliani, E. (Eds), Ergonomic Aspects of Visual Display Terminals (Taylor & Francis, London, 1980).

Henricksson, N.G., Pyykko, I., Schalen, L. and Wennmo, C., Velocity patterns of rapid eye movements, Acta Otol. 89 (1980) 540-512.

Luckiesh, M. and Moss, F.K., Fatigue of the extrinsic ocular muscles while reading under sodium and tungsten light, J. of Exp. Psychol. 25 (1935) 216-217.

Megaw, E.D. and Armstrong, W., Individual and simultaneous tracking of a step input by the horizontal saccadic eye movement and manual control systems, J. of Exp. Psychol. 100 (1973) 18-28.

Neisser, U., Decision time without reaction time: experiments in visual scanning. Am. J. of Psychol. 76 (1963) 376-385.

Ostberg, O., Accommodation and visual fatigue in display work, in Grandjean, E. and Vigliani, E. (Eds.), Ergonomic Aspects of Visual Display Terminals (Taylor and Francis, London, 1980)

Robinson, D.A., The mechanics of human saccadic eye movement, J. Physiol. 174 (1964) 245-264.

Schmidt, D., Abel, L.A., Dell'Osso, L.F. and Daroff, R.B., Saccadic velocity characteristics: intrinsic variability and fatigue, Aviat., Space & Env. Med. 50 (1979) 393-395.

Spooner, J.W., Sakala, S.M. and Baloh, R.W., Effect of ageing on eye tracking, Arch. Neurol. 37 (1980) 575-576.

Weber, R.B. and Daroff, R.B., Corrective movements following re-fixation saccades: type and control system analysis, Vis. Res. 12 (1972) 467-475.

Wilkinson, I.M.S., Kime, R. and Purnell, M., Alcohol and human eye movement, Brain 97 (1974) 785-792.

Young, L.R., Zuber, B.L. and Stark, L., Visual and control aspects of saccadic eye movements, (NASA CR-564, Washington, 1966)

Theoretical and Applied Aspects of Eye Movement Research
A.G. Gale and F. Johnson (Editors)
© Elsevier Science Pulishers B.V. (North-Holland), 1984

SACCADES DURING ACTIVE HEAD MOVEMENTS:

INTERACTIVE GAZE TYPES

W. H. Zangemeister

G. Huefner

Neurological University Clinic Hamburg
D 2000 Hamburg 20
Martinistr. 52
F. R. G.

Eye saccades that occured with special gaze
shift types where eye and head movements are
executed at the same time,so that saccade and
vestibulo-ocular reflex (VOR) act in opposite
directions,were analyzed.The resulting saccade
VOR interaction was often nonlinear,and saccadic fine structure was more strongly influenced than expected from actual head movement
dynamics.Since our subjects performed in a
highly predictable situation,we suppose that
under this condition interaction took place
on a higher level of control,so that the dynamics of saccades would be influenced beyond
the limits of the VOR.

INTRODUCTION

In recent years,several studies in humans and monkeys have been
reported concerning the additivity of two different types of
eye movements,like a saccade and a VOR-generated eye movement.
From earlier studies we know (6) that saccades and nystagmus
quick phases might be generated in the same neural circuitry.
Chun and Robinson (4) postulated an off-switching of the slow
command during the execution of VOR quick phases in the monkey.
However,interaction of saccades as compared to quick phases
with the VOR -an interaction that can occur in both directions
for saccades- appears to differ from each other (9).This is not
very surprising,since in everyday's life saccades serve quite
different tasks than quick phases: precise foveation as compared to fast repositioning of an involuntary slow eye movement.
Consequently,Juergens and Becker(9) found goal directed saccades,
with precision being the optimized criterion,evidently differ
in their dynamics from normal ones in that they follow a linear
summation with the ongoing VOR only up to 70% of that sum.
Morasso et al.(13) demonstrated the adjustment of saccadic dynamical characteristic,when goal directed eye-head movements to

predictable targets were unexpectedly braked in monkeys.They
explained this phenomenon by a local feedback mechanism of the
VOR in addition to preprogramming of the ongoing saccade.Fur-
ther evidence for this hypothesis came from their studies in
labyrinthectomized monkeys (Dichgans et al.(8)),where the dynamic characteristics of saccades did not adapt to these rapid
unexpected changes before a long time of additional training.
Nam et al.(14),on the other hand,reported that a linear summation can be found using a different experimental paradigm with
relatively slow selfpaced sinusoidal head movements.Kenyon et
al.(10,11) could show that on a more peripheral level of control nonlinear interaction of state variables in the neuromuscular
plant could account for deviations of Hering's law of equal in-
nervation.Their explanation,based on modeling evidence in addition to experimental evidence,did hold for nonlinear summation
of amplitudes of saccades and vergence movements -saccadic in-
equalities being usually greater at the start of the vergence--
as well as pursuit movements.Also,saccades reported by (13) with
free head movements could be simulated by their model.

From these studies it follows that the specific interaction that
might take place between a fast and a slow eye movement clearly
depended upon the experimental conditions that were used.Conse-
quently, four explanations of effects of interaction on different levels of control have been put forward: (a)peripheral linear summation (14),(b)peripheral nonlinear interaction (10,11,
15),(c) nonlinear interaction at the oculomotor neurons' i.e.
brainstem level, (d) higher level preprogrammed changes in saccadic and compensatory eye movement (CEM) amplitude and/or velocity in addition to (c) (8,13).
In this study,we examined saccades that occured simultaneously
with an early head movement,analyzed and described as gaze type
III in our previous work (18,19,20).Only in highly predictive
situations using larger amplitudes than 15°--since most naturally occuring saccades lay below this upper limit (1)-- and fast
active head movements this gaze type will occur most often --
43% on the average (19).The gaze type III saccades being executed almost simultaneously with the head movement occur together
with a relatively fast rising head velocity.Since the target in
this situation is highly predictable in time and in amplitude,
the saccade as well as the head movement generated VOR with its
compensatory eye movement could be preprogrammed.Our questions
then were: Do saccades and VOR interact linearly in this condition? If not,is there another 'rule' for attenuation of saccade
dynamics by the VOR --in addition to peripheral interactions?
Do these changes depend upon --supposedly higher level-- latency differences in generating fast eye and head movements?
We did not yet try to analyze in how far eye position   change
in space ('gaze') might be the dominant variable that is influenced and possibly taken into account.

METHODS

In five healthy volunteers with normal visual acuity we studied
interactive gaze types.The subject sat in a dark room with one
eye closed to avoid interference arising from vergence movements

looking at a semicircular screen.They were asked to perform fast active head movements towards small green target lights,that
were lit alternatively with a frequency of 1Hz and 0.5Hz,subten-
ding an angular amplitude of 10,20 and 30° around the center.
Eye and head movements were calibrated within a range of +-5°
and 30° ,for eye saccades with the head tightly fixed. Each
subject had to execute 100 head movements (50 to each side) for
each amplitude.Since we were looking for the saccadic fine structure that occured with the interactive gaze type, we selected
typical examples of saccadic time functions (Figs.2,3),and we
also averaged the data for each particular amplitude that occured in each subject (n=10).As our subjects showed slightly lower velocity and acceleration values,and longer duration of head fixed saccades,than those reported from (14,21,22) using the
same recording method,the latter values were plotted additionally on the main sequence graphs.
Eye movements were recorded by the well known IR-technique (16)
with the photocells firmly linked to the head.Head movements
were recorded by use of a high precision ceramic potentiometer
linked through a set of universal joints to a frame that was
tightly strapped onto the subject's head,so that no slip would
occur.Data were recorded on a high precision chart recorder (0-
3Khz),and could be analyzed offline from a FM-recorder (0-4Khz)
on a microcomputer using especially developed software (OFFLINE
9000) based on differentiating algorithms suggested by Bahill
et al.(2) for obtaining peak positive and negative acceleration
(A^+,A^-),peak velocity (V^+),the times of these extrema as well
as amplitude and duration of eye and head movements.
To help the reader understand the specific gaze type that has
been experimentally produced through favouring certain experimental conditions (like e.g.increased amplitude,intent and predictability,see Tab.1), figure 1 (from (19)) shows the time
functions of eye and head movements (right column),and of target
and gaze (eye in space) movements including gaze error (left).
Gaze type III is favoured by the above noted conditions,having
an overall frequency of occurence of 43%.The arrows in Tab.1 relate to the numbers that are ratios between probabilities of
gaze type III for two variants of the same conditional variable
(like e.g. intent), shown in the lefthand column.

GAZE TYPE CON-DITION	I SYNCHRONOUS EYE NECK EMG (34%)	II LATE HEAD MOVEMENT (4%)	III EARLY HEAD MOVEMENT (43%)	IV LATE EYE SACCADE AND EARLY HEAD. MOVEMENT (19%)
AMPLITUDE: 60° / 15°	← 0,90	↓ 0,09	↑ 3,05	↑ 2,10
INTENT: forced / natural	← 0,73	↑ 3,80	↑ 2,20	↑ 3,50
PREDICTABILITY: high / low	← 0,71	← 0,85	↑ 1,60	↓ 2,90
ACCELERATION TYPE				

Tab. 1

RESULTS

Normal saccade dynamics with head fixed (Fig.2a;3a(from Cook (7)) are compared with eye and head dynamics for gaze type III in the original recording (Fig.2b) and a digitally processed one (Fig.3b).The basic differences between both conditions (head fixed vs. simultaneous movements of eye and head) can be abstracted from these figures:
Markedly lowered peak velocity and positive acceleration,and a distinctly increased peak negative acceleration with a too far compensating CEM that does not bring the fovea precisely back to target (about 3° off target in Fig.2b) as compared to head fixed saccadic velocity,and the ratio between peak positive and negative acceleration.
In addition,saccadic positive acceleration and rising velocity traces clearly show an infraction that happens to occur quite often together with gaze type III --with corresponding changes of velocity and acceleration. Perhaps,this is an indication of a kind of'switch' during this mode of movement.

EYE AND HEAD COORDINATED GAZE TYPES

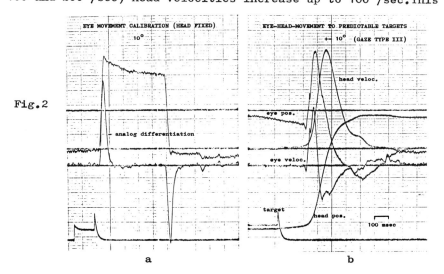

Fig.1

Finally,and most importantly,the synchronous contribution of head velocity through the VOR is evidently very low,as head velocity values are around 40 to 50°/sec (Fig.3b) during the very much slowed eye saccade --and only later together with the CEM (CEM-velocities ranging between 100 and 200°/sec) head velocities increase up to 100°/sec.This

Fig.2

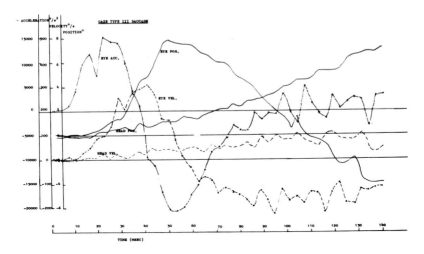

Fig.3b

points to an obvious discrepancy between the possibly very limited influence of the head movement initiated VOR onto the
saccade,and the marked changes that show up in the saccadic fine structure.

The double logarithmic main sequence plots (Fig.4a,b) demonstrate the deviations of the saccades of one subject (continous
line) as well as from other main sequence data (14,21,22)(dashed line).Slow velocities that differ even for very similar amplitudes and prolonged durations are evident (Fig.4a).The respective decrease of peak positive acceleration and increase
of peak negative acceleration shows up in Fig.4b even more str-
ikingly than in the time functions.

The phase plane plots (Fig.5a,b)
illustrate the deviations for
another subject from a different point of view.Fig.5a,up-
per and lower part,also shows
the noted acceleration deviati
ons when plotted as function
of peak velocity.The differences between gaze type III saccades (dots;continous,fitted
line in Fig.5) and the subject's main sequence is significant (p < 0.01).

The same holds for the ratio of
peak positive and negative acce
leration,that is normally fitted

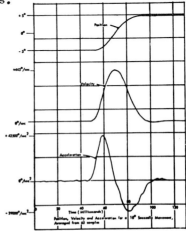

Fig.3a

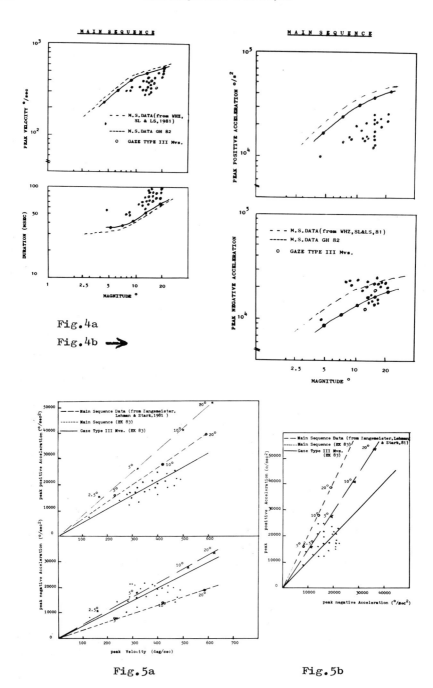

Fig.4a

Fig.4b ➡

Fig.5a   Fig.5b

by the dashed line $(21,22)$ with a slope of about 60°,meaning that A^{+} is about one third higher than A^{-} .In the gaze type III saccades this ratio is significantly $(p < 0.01)$ reversed.

When we look for the relationship between timeof the head movement begin with respect to eye saccade in our subjects (Fig.6), the early head movement's influence on peak velocity (decrease) and peak negative acceleration (increase) is obvious.In case the head began to move before about half of the saccade was finished ,this becomes more and more evident.(Fig.6,abscissa:ratio between time of eye saccade after and before head movement begin;90% means a head movement during 90% of the saccade,0% means no head movement during the saccade.Ordinate: ratios of $V^{+}$and A^{-} respectively,with head moving and with head fixed).The earlier the head movement occurs the more are V^{+} and A^{-} respectively de- and increased.

Comparision of the linear summation hypothesis (Fig. 7,continous lines for 3 am plitudes) with the data of our subjects clearly shows the marked difference that occured using our experime ntal paradigm: (a)all sac cades lie off the linear summation line,(note:each amplitude is the actually occuring saccadic,i.e.eye in orbit,amplitude),(b)for smaller saccades this dif ference appears to be grea ter than for larger ones, (c)for the same amplitude (e.g.16°)and the same head velocity peak eye saccade velocities can differ mar kedly.Note the different scale calibrations in Fig. 7 with abscissa scale being spread by a factor of 5,the

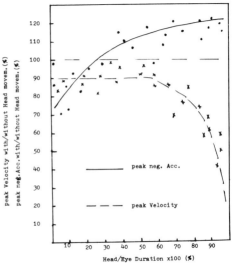

Fig.6

negative foresign meaning that the head movement initiated eye movement is always directed oppositely to the saccade.

Possible types of --neurologically-- ballistic movements are il lustrated in figure 8 (from Stark(17)).Since in our experimental paradigm highly predictable targets were used,to elicit fast ac tive head movements in conjunction with eye saccades,only the re sponse on the lefthand side appears to be used by the subject.Re flex influences maybe play a minor if not negligible role --for muscle stretch as well as the VOR-- as indicated in figure 8 by a voluntary,optimal,force output,that was preprogrammed by the subject.

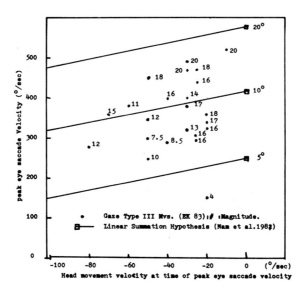

Fig. 7

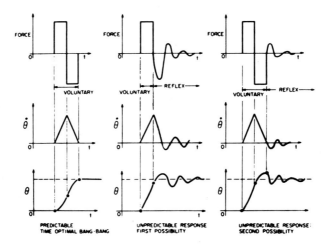

Possible types of ballistic movements.

Fig. 8

CONCLUSIONS

Because of load differences head movements are much slower than eye movements (21,22;20),so that head position change occurs only when the fast eye saccade is about finished --given simultaneously initiated control signals as seen in the begin of agonistic EMG.With high prediction of time and a large amplitude the latency for head movement control signals is reduced significantly more than for eye movements (18).Particularly,in this condition gaze type III occurs, leading to saccade/CEM interaction. How does this interaction take place? As noted above,there are basically four possibilities that reflect the different levels of control: peripheral at the plant level, the brainstem level including local feedback, preprogramming on a higher level (i. e. cortical,frontal or parietal, or subcortical,e.g.colliculus superior level) or combinations of these.When considering our here reported results in conjunction with evidence from other studies we feel that the kind of interaction that takes place strongly depends upon the experimental conditions.

Our experiments show that the dynamical fine structure of saccades that occured with the interactive gaze type III varied more than expected,often nonlinearly.Detailed analysis of initial and concurrent conditions of these saccades with respect to the ongoing head movement (and its velocity and acceleration) failed to show a linear summation of saccade velocity and head velocity.Rather,even with similar amplitudes this 'summation' resulted in slowing of peak eye saccade velocity and positive acceleration as well as an increase of peak negative acceleration. Typically,the duration of these saccades was increased - and not slightly decreased as it happens physiologically with saccades that show dynamic overshoot and that also have increased peak negative accelerations (5,12).Considering the uneven ratio between a relatively low head velocity and acceleration at the time of saccadic peak positive acceleration and velocity it appears reasonable to assume that an important part of eye head interaction has been executed in advance as an account for the found data.In addition,often occuring changes during the rise of initial eye acceleration could point to some 'switch' that is used under this condition.

As the preprogramming of the fast head movement is more efficient than of the eye movement (18;note again Fig.8 for possible types of ballistic movements), eye saccades could be influenced on different levels of control: peripherally,where the summation of saccade and VOR velocity could be both nonlinear or linear, and by preprogramming of slow saccades of unusually long duration,generally with too high negative accelerations;the height of negative acceleration depends maybe only in part onto the VOR in this case,that might have a rather low ,suppressed gain because of the imagined i.e. anticipated fixation situation that is intended by the subject (Barr et al.(3)) - and it might differ depending upon the subject's intent (see Tab.1).Therefore ,when we find nonlinearly and sometimes seemingly irregularly interacting saccades in gaze type III,the asymmetry in preprogramming of eye and head movements,with the higher efficiency

of the head movement,with time as the criterion to be optimized,
could well explain our results.In this way the subject's intent
becomes the main influencing variable.
Simulations are under way that might explore and help explain
the variety of interactions relating different kinds of neural
control signals with the latency shifts that occur with gaze
type III.

REFERENCES

1. Bahill,T.,Adler,D.,Stark,L. Invest.Ophthalm.14,468(1975)

2. Bahill,T.,McDonald,J. IEEE Trans Biomed.Engin.30-3,191(1983)

3. Barr,C.,Schultheis,L.,Robinson,D.A. Acta Otolaryngol.81,
    365(1976)

4. Chun,K.,Robinson,D.A. Biol.Cybern.28,209(1978)

5. Clark,M.,Stark,L. Math.Biosci.20,239(1974)

6. Cohen,B.,Henn,V. Bibl.Ophthalmol.82,36(1972)

7. Cook,G. Control Systems Study of the Saccadic Eye Movement
    Mechanism.Sc.D.Diss.Dept.Electr.Engin.MIT (1965)

8. Dichgans,J.,Bizzi,E.,Morasso,P.,Tagliasco,V. Exp.Brain Res.
    18,548(1973)

9. Juergens,R.,Becker,W.,Rieger,P. Progr.Oculomot.Res.,ed.by
    A.Fuchs and W.Becker,Elsevier North Holland,33 (1981)

10. Kenyon,R.,Ciuffreda,K.,Stark,L. Am.J.Optom.&Physiol.Optics
    57-9,586(1980)

11. Kenyon,R.,Stark ,L. Math.Biosci. 63,187(1983)

12. Lehman,S.,Stark,L. J.Cybern.Inf.Sci. 4 ,21(1979)

13. Morasso,P.,Bizzi,E.,Dichgans,J. Exp.Brain Res.16,492(1973)

14. Nam,M.,Winters,J.,Stark,L. JPL Publication 81-95, 537(1981)

15. Ono,H.,Nakamizo,S.,Steinbach,M. Vision Res.18,735(1978)

16. Stark,L.,Vossius,G.,Young,L. Human Fac.in Electron.HFE-3,
    52(1962)

17. Stark,L. Neurological Control Systems.Plenum Press N.Y.1968

18. Zangemeister,W.H.,Stark,L. Exp.Neurol.75,389(1982)

19. Zangemeister,W.H.,Stark,L. Exp.Neurol. 77,563(1982)

20. Zangemeister,W.H.,Stark,L.,Meienberg,O.,Waite,T. J.Neurol.
    Sci. 55,1(1982)

21. Zangemeister,W.H.,Lehman,S.,Stark,L. Biol.Cybern.41,19(1981)

22. Zangemeister,W.H.,Lehman,S.,Stark,L. Biol.Cybern.41,33(1981)

Theoretical and Applied Aspects of Eye Movement Research
A.G. Gale and F. Johnson (Editors)
© Elsevier Science Publishers B.V. (North-Holland), 1984

AIMING PRECISION AND CHARACTERISTICS OF SACCADES

Zoi A.A. KAPOULA

Laboratoire de Psychologie Expérimentale,   Paris V
Visitor at the Department of Psychology, Durham University, England

This experiment was designed to test the existence of a
speed-accuracy trade-off for saccades of 2-10 degrees.
A simple two choice discrimination task was used to elicit
saccades towards a single target which appeared in peripheral
vision. Task requirements and precision were manipulated
a) by using two different sets of targets; b) by verbal
instructions emphasizing either accuracy or speed; c) by
arranging a saccade contingent display in which the target
was only visible for 150 msec after saccade triggering.
The results show that when high accuracy is demanded
variability of saccade amplitude decreases with a concomitant
increase in saccade latency.

INTRODUCTION

Ideas on the saccadic system have been considerably revised in recent years
(see Bach-y-Rita, Collins, Hyde (1971); Lennerstrand, Bach-y-Rita (1975);
Fuchs, Becker (1982)).

One rather neglected aspect however is the extent to which voluntary
attentional and cognitive factors may affect saccadic characteristics. This
form the purpose of the present study.

The method used is to look for the existence of a speed-accuracy trade-off
for the saccadic system. This trade-off is a fundamental characteristic of
motor skills. Its investigation could provide a more satisfactory understanding of the mental and physiological processes underlying motor control.
It has been intensively investigated for other types of movement such as
manual reaction time (Pachella, Pew (1968); Pew (1969); Schmidt (1969);
Keele (1968); Klapp (1975), etc.).

For hand movements the speed-accuracy trade-off is now known to apply not
only to the final correction phase using visual feedback, but even in the
initial impulse phase, that is the ballistic part of the movement (Meyer,
Smith (1982)). Thus although saccadic eye movements can not be controlled
by a continuous visual feedback (Vossius (1960); Young (1962)) the speed-
accuracy trade-off could nevertheless exist for the oculomotor system.

Several authors had invoked the hypothesis of a trade-off for the saccadic
system (Festinger (1971); Bouma (1978); Russo (1978)), but there is no clear
experimental demonstration. The few earlier studies which deal with it
(Leushina (1965); Cohen, Ross (1978); Viviani, Swensson (1982) either don't

manipulate explicitely the accuracy requirements or they don't use
appropriate estimations of the accuracy. However, in my previous work
(Kapoula (1982)) I showed that a form of speed-accuracy trade-off occurs
in scanning situations resembling reading: fixation accuracy was found to
increase as fixation duration increases. Following up this work the present
investigation looks for the speed-accuracy trade-off in situations in which
only one saccade at a time is produced, and accuracy and task requirements
are explicitely manipulated.

METHOD

Apparatus

The experiment as well as data processing was carried out by a CAI alpha
ISI 2/20 microcomputer. The visual display consisted of a Hewlett Packard
1321 A display oscilloscope (P31 phosphor).

Subjects were seated at eye level 57 cm from the display scope and head
position was held constant by a head and bite board system.

Horizontal eye movements were recorded by a photoelectric limbus tracker
(modified by using two probes, Findlay (1974)). The system was linear to
within a few percent over central region of 10 degrees.

Calibration data were obtained for three screen positions, the central
position and 4.4 degrees of visual angle to either side. The resolution of
the system was better than 10 min. arc.

Procedure

Six subjects,members or students in the department of Psychology at Durham
University, took part in the experiment. The experiment was conducted in
two sessions on different days.

The experimental paradigm I used was a simple discrimination task derived
from that used by Findlay (1982) [1] . On every trial a small square (0.5
degree of angular dimension) was presented, inside which was a small number
of additional spots. The subject's task was to report the number of spots.
In the beginning of each session the subjects received some training trials
in which the target appeared at the central position. In the main experiment
the square appeared at various eccentricities (2.7°-9.5° left or right), to
which the subject had to make a saccade and then make a simple two-choice
discrimination.

Two conditions were used. In the first, subjects were told to attempt to
move their eyes very precisely to the target. In the second they were
instructed to saccade normally by fixating the target as soon as possible
after it appeared. Two different sets of targets were used which had similar
physical characteristics (density, luminosity and identical external contour).
But in the first condition discrimination was between four and five spots
(see fig.1). Preliminary experiments showed that this was only possible if
the subject fixates the target precisely. In the second condition the
discrimination was between two and five spots; this can be made in near
periphery and so didn't require a precise adjustment.

In addition a saccade contingent display was programmed so that the target
was only visible for 150 msec after the saccade triggering (see fig.1). Thus
I rendered corrective saccades ineffective since target was visible only

110-120 msec after saccade execution (30-40 msec). The subject was informed of this method and thus appreciated the need for high accuracy in the first condition.

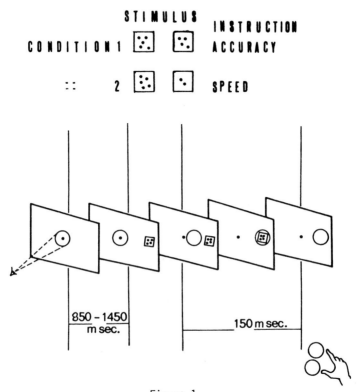

Figure 1
Subject fixated a central point (eye's position is indicated by circles). After a variable interval 850, 1050, 1250, 1450 msec the target appeared in the peripheral visual field. The program sampled eye's position every 7 msec. Once the saccade triggered and detected by the computer the target was only visible for 150 msec.

Subjects were run in blocks of 80 trials in each condition lasting 20 minutes. Order of conditions was counterbalanced. Calibration recordings were taken every 10 trials.

Subsequent off-line analysis programs were used to estimate saccade amplitude, latency, duration and peak velocity.

RESULTS

Two different measures were used to estimate saccade accuracy: the mean saccade amplitude of all the responses to a given target eccentricity and standard deviation.

Figure 2 illustrates the results pooled together for 6 subjects. Full symbols show the data from the high accuracy condition and empty symbols the data obtained in the speed condition. Vertical bars indicate the mean individual dispersions. The dashed line illustrates the performance expected if saccade amplitude fits perfectly with target eccentricity. As can be seen

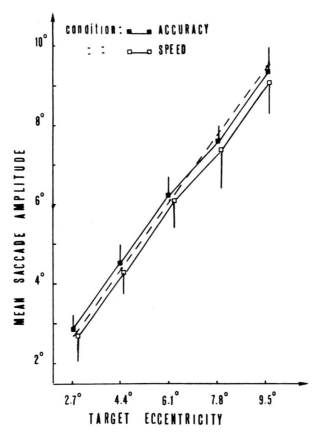

Figure 2
Mean saccade amplitude for every target
eccentricity and each condition

in the accuracy condition the saccade amplitudes are very close to the ideal fit line, while in the speed condition they systematically undershoot target eccentricity especially beyond 6°. However this difference is not significant. Analysis of variance with condition and eccentricity as factors gives: condition effect : $F(1,5)=2.26$, $p=.19$; the interaction between eccentricity and condition is not significant either: $F(4,20)=0.17$, $p=.94$.

The results on the variability of saccade amplitudes are more interesting. As can be seen in figure 2 the random error, indicated by the length of the bars, is smaller in the accuracy condition. Further, while in the speed condition this error increases with eccentricity, in the accuracy condition if remains almost constant for every eccentricity. These effects are highly significant: (condition effect: $F(1,5)=8.24$, $p=.03$; interaction between condition and eccentricity: $F(4,20)=4.14$, $p=.01$).

The fact that accuracy requirements act mainly upon saccade amplitude variability and not so much on mean saccade amplitude has been already observed by Leushina (1965), although the task used was not a task of oculomotor precision.

The reduction of the random adjustment error is accompanied by an increase of saccade latency. Figure 3 illustrates the mean saccade latency across subjects. As can be seen, saccade latencies are longer in the accuracy condition. This difference is statistically significant ($F(1,5)=15.46$, $p=.01$). Latencies variability is also significantly higher in the accuracy

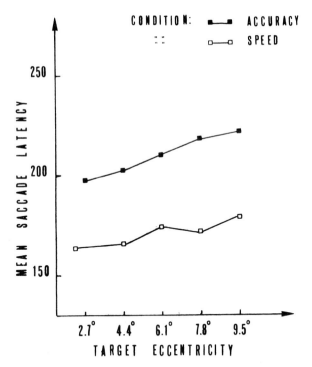

Figure 3
Mean saccade latency for every target
eccentricity and each condition

condition ($F(1,5)=16.79$, $p=.01$). These results show that indeed a trade-off between saccade accuracy and latency can occur.

It is interesting to follow up this data by analysing the dynamic character-istics of saccades.

It is generally agreed that saccade duration and peak velocity increases approximately linearly with amplitude. Effects on these characteristics can be studied by comparisons across conditions of the values of correlations between amplitude-duration or amplitude-peak velocity. Table 1 contains these correlations for every subject and every condition, as well as means and standard deviations. As it was expected high level correlations were obtained for both, duration and peak velocity with corresponding amplitude of the saccade.

Table 1
Values of correlations between : saccade amplitude
and duration; saccade amplitude and peak velocity

Condition :	ACCURACY				SPEED			
				Amplitude-duration				
	r	$\overline{m}$	σ	N	r	$\overline{m}$	σ	N
M.H.	.90	30	10	46	.87	37	9	66
Z.K.	.86	40	8	77	.84	39	8	73
C.M.	.90	41	10	73	.92	48	11	53
A.D.	.79	40	11	64	.84	43	10	61
J.F.	.92	33	10	61	.92	38	9	71
J.K.	.84	27	9	56	.92	25	8	40
				Amplitude-peak velocity				
M.H.	.90	199	55	46	.96	200	57	66
Z.K.	.96	217	64	77	.93	209	60	73
C.M.	.93	186	57	73	.95	145	55	53
A.D.	.90	200	49	64	.92	201	48	61
J.F.	.80	217	58	61	.97	210	62	71
J.K.	.87	246	65	56	.93	257	84	40

The amplitude duration correlations are basically similar in both conditions. The amplitude peak velocity correlations however tend to be lower in the accuracy condition. This difference attains a significant level for two subjects (subject M.H.: X^2=8.88, p<.001, subject J.F.: X^2=30.81, p<.01). Figure 4 shows the peak velocity saccade amplitude scatters from subject J.F. It is clear that more variability on peak velocities occurs in the accuracy condition than in the speed condition. This could be seen as reflecting an effort to improve accuracy during saccade execution. Although these results are weak they seem interesting and deserve further research.

CONDITION

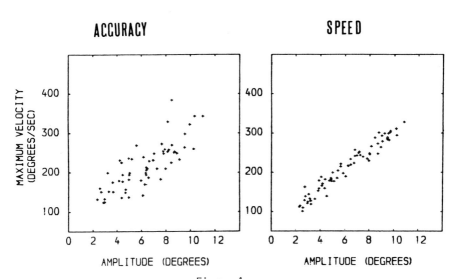

Figure 4
Relation between saccade amplitude
and its maximum velocity for each condition

The main finding of this experiment however is the reduction of the random adjustment error with a concomitant increase on saccade latency, which demonstrates that the speed-accuracy trade-off exists for the saccadic system.

DISCUSSION

What can this trade-off teach us about oculomotor functioning ?

First of all, the reduction of the random adjustment error in the accuracy condition can be seen as reflecting an effort to achieve high aiming consistency over successive trials. Since the physical stimulus characteristics are similar in both conditions the improvement of accuracy when verbal instructions and task requirements emphasize it must be the outcome of a specific generating mechanism. It seems that in the accuracy condition saccades are produced under a voluntarily controlled mode of functioning of the oculomotor system. Some aspects of the results seem to favour this hypothesis.

It is well known that in general saccade accuracy decreases with target eccentricity. Now we have seen that in the accuracy condition the random adjustment error remains low and constant for all eccentricities. This can be seen as a voluntary effort to keep the error small throughout the experiment. This mechanism doesn't operate in the speed condition leading this way to the more normal and generally observed increase of the error with eccentricity.

The increase of latencies and their variability in the accuracy condition
is another result favouring the distinction between natural and voluntarily
controlled mode of functioning. Although it is difficult to know exactly
the specific mechanism involved on saccade programming when subject aims
for high accuracy, I will present some hypotheses: the changes in accuracy
condition could arise from the fact that the command to trigger saccade
programming comes from some central pathways increased elaboration and not
only from the retinal signal error. But it is equally possible that central
factors act during the elaboration of the efferent motor command. In the
former case the increase of latencies can be produced by a more precise
analysis of the peripheral visual information. Recent data reported by
Becker and Jürgens (1979) support the idea of the existence of a temporal
window, according to which saccade calibration is based on the mean of
several samplings of visual information taken over a period of about 60 msec.
Thus it is possible that the accuracy improvement is made by delaying the
moment the temporal window occurs. But it is also possible that the increase
of latencies arises from attentional influences in a hypothetical spatial
map; some kind of internal verification would take longer than an approximate
adjustment. Such a map is becoming an important feature of much work in
oculomotor control (Zee, Optican, Cook, Robinson, King Engel (1976); Mays
and Sparks (1981)). These hypotheses incite several potentially productive
directions for further research.

FOOTNOTES

1 This work was done with the collaboration of J. Findlay in his laboratory
   at Durham University, while I held a fellowship from the European Science
   Foundation.

REFERENCES

(1) Bach-y-Rita, P.P., Collins, C.C., Hyde, J.E. (eds.), The control of Eye
    Movements (New York, London, 1971).

(2) Becker, W., Jürgens, R., An analysis of the saccadic system by means
    of double-step stimuli, Vision Research 19 (1979) 967-983.

(3) Bouma, H., Visual search and reading: eye movements and functional visual
    field, in: Requin, J. (ed.), Attention and Performance (Hillsdale,
    New Jersey, Erlbaum, 1978).

(4) Cohen, M.E., Ross, L.E., Latency accuracy characteristics of saccades
    and corrective saccades in children and adults, J. Exp. Child Psychol.
    26 (1978) 517-527.

(5) Festinger, L., Eye movements and perception, in: Bach-y-Rita, P.P.,
    Collins, C.C., Hyde, J.E. (eds.), The control of eye movements
    (New York, London, 1971).

(6) Findlay, J.M., A single apparatus for recording microsaccades during
    visual fixation, Q. J. exp. Psychol. 26 (1974) 167-170.

(7) Findlay, J.M., Global visual processing for saccadic eye movements,
    Vision Res. 22 (1982) 1033-1045.

(8) Fuchs, A.F., Becker, W. (eds.), Progress in Oculomotor Research
    (Elsevier/North Holland, 1982).

(9) Kapoula, Z., Le contrôle des mouvements des yeux : information périphérique et modes d'exploration d'alignements graphiques, Thèse de 3ème cycle, Univ. Paris V (July 1982).

(10) Keele, S.W., Movement control in skilled motor performance, Psychol. Bull. 6 (1968) 387-403.

(11) Klapp, S.T., Feedback versus motor programming in the control of aimed movements, J. of Exp. Psychol. 104 (1975) 147-153.

(12) Lennerstrand, G., Bach-y-Rita, P., Basic Mechanisms of Ocular Motility (Oxford, 1975).

(13) Leushina, L.I., On the estimation of position of photostimulus and eye movements, Biofisika 10 (1975) 130-136.

(14) Mays, L.E., Sparks, D.L., Saccades are spatially, not retinotopically, coded, Science 208 (1980) 1163-1165.

(15) Meyer, D.E., Smith, J.E.K., Models for the Speed and Accuracy of Aimed Movements, Psychological Review 89 (1982) 449-482.

(16) Pachella, R.G., Pew, R.W., Speed-accuracy trade-off in reaction time: effect of discrete criterion times, J.of Exper.Psychol. 76 (1968) 19-24.

(17) Pew, R.W., The speed-accuracy operating characteristics, Acta Psychologica 30 (1969) 16-26.

(18) Schmidt, R.A., Movement time as a determiner of timing accuracy, J. of Exper. Psychol. 79 (1969) 43-47.

(19) Vossius, G., Das system der Augenbewegung, Z. Biol. 112 (1960) 27-57.

(20) Young, L.R., A sampled data model for eye tracting movements, Ph.D. Thesis, Dept. of Aeronautics and Astronautics, Cambridge Univ. (1962).

(21) Zee, D.S., Optican, L.M. Cook, J.D., Robinson, D.A., King Engel, W., Slow saccades in spinocerebellar degeneration, Arch. Neurol. 33 (1976) 243-251.

Theoretical and Applied Aspects of Eye Movement Research
A.G. Gale and F. Johnson (Editors)
© Elsevier Science Pulishers B.V. (North-Holland), 1984

THE RELATION BETWEEN VISIBILITY SPAN AND EYE MOVEMENTS IN VARIOUS SCANNING TASKS

Ariane Lévy-Schoen, J. Kevin O'Regan,
Arthur M. Jacobs & Christian Coëffé

Groupe REGARD, Laboratoire de Psychologie Expérimentale,
Université Paris V, C.N.R.S., E.P.H.E.
28, rue Serpente, 75006 PARIS, France

We define static visibility span in a character
string as the number of letters which can be identified
on each side of the fixation point with better than a
certain percent correct.

Our purpose here is to understand how this static
visibility span determines dynamic oculomotor behaviour. By manipulating viewing distance and lateral
masking, we induce variations in static visibility span
which are measured for a given group of Subjects. We
then study the effect of these same manipulations on
the same Subjects' eye movements in several different
scanning tasks, using the same typographic material.

In reading, we find that visibility span is not the
main determiner of eye movements. It appears that
other factors such as word length and linguistic context play a stronger role. However in a search task
where these factors are not active, visibility span is
also not the main determiner of eye movements. Again
it appears that cognitive factors are more important
than visibility factors in determining eye movements.

1 INTRODUCTION

While reading a text or scanning the visual field in search of an object,
the saccades and fixations of our eyes are governed by efficient rules which
neither theoretical nor experimental research have yet clearly elucidated.
Among the factors which play a role in determining the exploration strategy
of the eyes, the ones which will concern us here are those which act at a
very elementary sensory level. The question we shall ask is: How do the
sensory limits of peripheral vision, acting within each fixation, determine
the way eye movements are controlled? The experiments that we shall report
are part of a research program devoted to studying the static and dynamic
functional organization of human vision during everyday behaviour.

1.1 EXTENSION OF STATIC VISUAL CAPACITIES

Since Javal (1905) and Tinker (1958), the visibility of print has interested
psychologists studying reading. However, up until now, no satisfactory
theory has been available to predict the behaviour of readers as a function
of print characteristics. In an attempt to make such a theory, O'Regan
(1983) has developed a VISIBILITY MODEL, based on the notion of VISIBILITY
SPAN.

Static VISIBILITY SPAN is defined as the amount of visual material which can
be easily processed while the eyes are fixating some point of the field. It
depends on the peripheral acuity curve of the subject, on the size of the
critical features which must be resolved in order to recognize the elements
in the field, and on the size of their projection upon the retina (i.e. on
the viewing distance). Calculations predict how visibility span for letter
strings should vary as a function of changes in viewing distance and character spacing. We have tested these predictions in several experiments.

1.2 VISIBILITY EXPERIMENT

In an experiment performed by Coëffé (1983), Subjects had to identify two
letters which simultaneously appeared on a display screen to the right and
left of a central fixation point, and which disappeared before the eyes
could move (i.e. after 150 ms). The two letters had the same eccentricity,
but this eccentricity was varied within a block of trials in successive
increasing or decreasing series. The target letters were taken from the
set of ten letters <A,P,T,Z,X,C,L,E,K,M>. Performance curves were plotted
and adjusted using a Probit fit. The maximum eccentricity within which
characters could be seen with 50% or better probability was deduced from the
curves. We will refer to this as the "unilateral visibility span". Because
it is more relevant for reading, we will measure the span in number of
letters visible, not in degrees.

The effects of two factors were studied: 1) viewing distance was varied from
30 cm to 120 cm; 2) lateral interference due to neighbouring characters was
varied by introducing two digits flanking each letter on both sides, at
spacings of zero (no more space than in normal printing of the letters in a
word) to 2 (two blank character-spaces between letter and digit). The visibility model predicts a slight but regular decrease of the size of the
visibility span (measured in number of letters) when viewing distance increases, meaning that about two more letters on each side of the fixation
point should be recognizable in a string when it is viewed from 30 cm than
when it is viewed from 120 cm.

This is what was found in the experiments, whose results are plotted in Fig
1. It is seen that as distance increases, span diminishes. The additional
drop in span at 30 cm can be attributed to the fact that at small distances,
the eccentric letters on the flat or slightly convex screen appear compres-
sed because they are viewed more from the side. The effect of increasing
spacing from 0 to 2 character spaces is to diminish by one or two the number
of characters recognized.

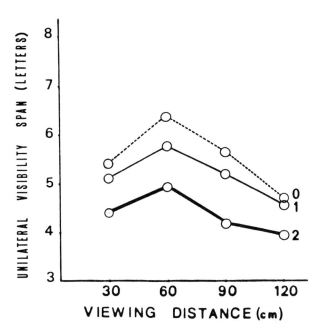

Figure 1

VSIBILITY limits for
the three spacing
conditions (0,1,2) as
a function of viewing
distance.

2 PREDICTIONS FOR DYNAMIC VISUAL CONTROL

Starting from this theoretical knowledge and these experimental data, we
attempted to understand how changes in visibility span affect oculomotor
behaviour when text is scanned and processed.

Intuitively, we expect that in reading or scanning, each saccade should
bring the eyes to a point where further information is to be found. If this
is true, saccade amplitude variations should parallel visibility variations.
The number of letters spanned by saccades should decrease when viewing
distance decreases and also when inter-letter spacing increases. A further
prediction concerns fixation durations. On the one hand these are expected
to depend on saccade sizes, since the distance spanned by saccades presumably determines the amount of material processed at each fixation. On the
other hand fixation durations should also increase as viewing distance
increases, since characters become harder to distinguish.

2.1 READING EXPERIMENT

To test these predictions we performed a READING EXPERIMENT (Jacobs, 1981;
O'Regan, Lévy-Schoen & Jacobs, in press) in which we measured saccade amplitudes and fixation durations for four Subjects reading simple texts displayed using the same typography as in the visibility experiment, and in
which the same distance and spacing manipulations were performed in order to
induce changes in visibility. A peculiar pattern of results was found in
which saccade amplitudes did not parallel the variations of visibility span
when viewing distance is changed, but did on the other hand follow character

spacing changes in the expected way. Fixation durations did follow the
expected pattern.

We interpreted the results as showing that under normal conditions of visibility, visual span is not what mainly limits eye movements. Rather, eye
behavior appears to be determined by factors operating within the visibility
span, which are related to the cognitive processing involved in reading,
possibly including factors such as word boundary isolation, and linguistic
processing.

3 LETTER SEARCH EXPERIMENT

The absence of an effect of changing visibility span in the reading experiment prompted us to find a task in which visibility is the main determiner
of saccade sizes. We therefore used a situation where no linguistic processing or word boundary effects could have been active, but which nevertheless
involved reading-type eye movements. The task we chose required the Subject
to search for a target letter within lines of random letters. To ensure
that the VISIBILITY experiment data would apply to the results, we used the
same typography and display conditions as there.

3.1 EXPERIMENTAL PROCEDURE

The letter "B" was arbitrarily chosen as the target letter, against a set of
nine non-targets which were the next consonants of the alphabet. Twelve sets
of 14 lines of 40 characters each were generated so as to respect the
following conditions: The first and last line of each set always contained
one or two targets, as well as two other lines selected at random among the
12 remaining ones. The position of the targets within these "target-lines"
was randomized, as well as the order of the background (non-target) letters.
In order to avoid clusters, repetition of adjacent letters was forbidden.
Two additional sets of 14 lines served as training material at the beginning
of the experimental session for each subject. In order to vary the visibility conditions in the same way as for our VISIBILITY study, 4 viewing distances crossed with 3 character spacings determined twelve presentation conditions. For technical reasons, we could not use spacing 2, so we replaced it
with a spacing of 1/2. Figure 2 gives examples of target lines.

Spacing 0:
             CJDKFMLBCDHJKJFHJHDGMKDJLJCDLDCMFHGFDGMC

Spacing 1/2:
             C J D K F M L B C D H J K J F H J H D G M K D J L J C D L . . .

Spacing 1:
             C J D K F M L B C D H J K J F H J H D G M K D . . .

Figure 2: Examples of target lines in
the three spacing conditions.

The order of presentation of the four viewing distances followed a latin
square design for the four subjects. The three spacing levels were presented

consecutively at each of the viewing distances, in an increasing order for two of the subjects and decreasing for the two others. Each Subject saw all 12 sets of 14 lines. Taking the four Subjects together, the distribution of the sets was counter-balanced across viewing distance and spacing factors.

The Subject was instructed to scan the lines of characters in order to count the number of target letter "B"'s that occurred. He sat in a dental armchair with his head resting on lateral supports. His eye movements were recorded using a photo-electric spectacle-mounted device measuring scleral reflection.[1] Automatic detection of saccades was provided by the computer sampling every 10 ms. A saccade was considered to have occurred when the computer detected an eye-position change of more than one character space, occuring in less than 50 msecs, and followed by a fixation of more than 50 msecs.

The visual material was displayed on a television monitor one line at a time. This allowed better calibration to be maintained across the whole series of lines scanned. The sampling program detected when the S's eyes made a large return movement from the right to the left of the line, and this triggered the change of the line into the next one. Since this change occurred during the Subject's ocular return sweep, it remained invisible to him. The line change was therefore accompanied by an audible beep emitted by the computer.

Before the presentation of each set of lines, calibration of the eye movement recording device was done using a feedback method. Accuracy during a session was guaranteed by the computer continually checking the first and last fixations made in each line. Under good conditions, saccade sizes were accurate to about one character space.

The four Subjects were the same as those who had previously performed the VISIBILITY experiment.

3.2 RESULTS

Since we are interested in the ocular behaviour while scanning the lines in search for an expected target, and not in the adjustment of the eyes when a target falls within the peripheral field of view, only the lines containing no target were selected for analysis.

Figure 3 shows the mean amplitude of progression saccades, averaged over the four Subjects. The amplitude distributions of the individual Subjects are well behaved and allow us to consider these means as representative of the oculomotor behaviour.

Two main points clearly appear:
- saccade amplitudes do not decrease when viewing distance is increased, even though the visibility span, as measured under the same distance and spacing conditions in the VISIBILITY experiment, did decrease.
- saccade amplitudes in all conditions are always much smaller than the unilateral visibility span: no more than about four letters are skipped in the mean.

------

[1]The equipment was developed in our laboratory by Y. Barbin.

A. Lévy-Schoen et al.

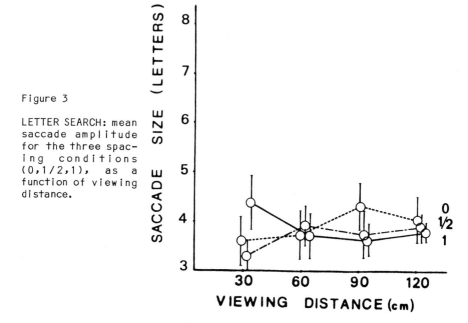

Figure 3

LETTER SEARCH: mean saccade amplitude for the three spacing conditions (0,1/2,1), as a function of viewing distance.

An analysis of variance confirms that neither of the two factors (distance and spacing) has any significant effect on mean saccade amplitude.

The data for fixation durations are given in Figure 4. When viewing distance increases, there is always a systematic increase in fixation durations. Also we observe that the smallest spacing gives rise to the longest fixation durations. However, for triple spacing, fixations are not shorter than for double spacing. Both factors are significant at the .01 level, and their interaction is not significant.

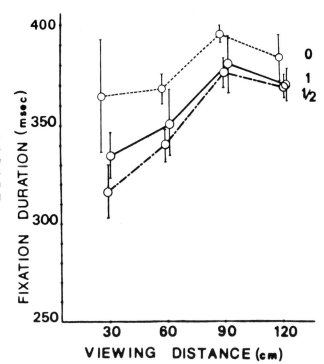

Figure 4.

LETTER SEARCH: mean fixation durations for the three spacing conditions (0,1/2,1), as a function of viewing distance.

3.3 DISCUSSION

First it clearly appears that something other than the limits of visibility governs the eyes in this scanning task, restricting saccades to cover short-er distances within the zone of visibility: In a given fixation, not all the characters visible from the fixation point are processed.

An alternative possibility is that instead of being guided on the basis of the spatial limits of sensory capacities, saccades adapt to processing constraints, even in such an apparently simple task as searching for a target in a nonsignificant background. Two hypotheses come to mind. Saccades can be determined on the basis of TIME LIMITS: decide to move when a stand-ard fixation time is over (for instance around 300 ms) and jump to the next group of letters not yet processed during that time. But if this is cor-rect, it implies that fixation durations should remain constant across conditions of variable visibility. Since we observe that fixation durations show systematic variations as a function of viewing distance, this hypothe-sis seems not to hold. The second hypothesis is that saccades are deter-mined on the basis of a LIMITED COGNITIVE EFFORT to expend for each fixation (for instance: process no more than about four different letters in a single chunk and jump to next four letters).

This 'piecemeal' strategy may also have another advantage. In the present task of visual search along a repetitive spatial structure, it is difficult to know where the gaze is situated within the line. It may be necessary to proceed by small regular steps so that at each step the eye is anchored relative to the preceding one. The situation is similar to using one's finger to help count closely spaced identical dots.

If this limited cognitive effort hypothesis is correct, it explains why no variation in saccade amplitude is observed in any of the conditions in our experiment, in which the same kind of letter processing has always to be done. If this is true, we would expect that systematic changes in oculomotor regulation should be induced by varying the processing load. Another letter search experiment presently in progress investigates the effects of reducing processing load by using fewer nontarget characters. We also are studying the effect of varying the similarity of target and nontargets. It appears that saccade amplitudes do indeed vary in a systematic and stable fashion as a function of such factors. When searching for a target B embedded in lines consisting of two alternating letters (QRQRQR), mean saccade amplitude increases to about five characters. When searching for the same B target among a similar line whose letters (CDCDCD) are more easily distinguished from the target, mean saccades further increase to around seven letters. However still no effects of viewing distance appear, probably because the visibility span of the B within these backgrounds has also increased to more than seven letters. In order to show up the expected change of saccade size with visibility changes, it is necessary to use background stimuli which entail only very weak cognitive load, and targets which are not easily visible in peripheral vision. This will be attempted in further experiments.

4   CONCLUSION

The size of the zone around each fixation point in which visual processing can be done plays less of a role than first imagined in determining the eye's scanning behavior. In reading and even in simpler search tasks, it appears that the progression of the gaze along the line is more strictly constrained by processing load limitations than by sensory limitations related to the sharp drop of acuity in periphery.

For a given set of critical features to analyse, if sensory discrimination of these critical features is easy (moderate viewing distance i.e. large apparent size of the letters), no more letters are processed at each fixation but they are processed faster (shorter fixation durations than for larger viewing distances). It is only when cognitive load diminishes because fewer features are to be analysed, that processing span increases and saccade amplitudes become larger.

It is important to note the   basic   role played in the present studies by the VISIBILITY model and experiment. Because of the precise predictions made by the VISIBILITY model about the sensory effects of changing viewing distance and character spacing, we were able to partial out the sensory from the cognitive constraints imposed on eye movements in reading and scanning.

REFERENCES

COEFFE, C. L'effet de la distance d'observation sur l'empan de la zone de visibilité. Le "grain" dépend-il de l'espace inter-lettres? Mémoire de Maîtrise. Psychologie Expérimentale, Paris V, 1983.

JACOBS, A. M. L'effet de la distance de lecture et de l'espacement entre les caractères sur le comportement oculomoteur. Mémoire de Maîtrise. Psychologie Expérimentale, Paris V, 1981.

JAVAL, E. Physiologie de la lecture et de l'écriture. Paris, Alcan, 1905.

McCONKIE, G. W. Eye movements and perception during reading, in RAYNER, K. (Ed.), Eye movements in reading: Perceptual and Language processes, N.Y., Academic Press, 1983, pp. 65-96.

O'REGAN, J. K. Elementary perceptual and eye movement control processes in reading, in RAYNER, K. (Ed.), Eye movements in reading: Perceptual and language processes, N.Y., Academic Press, 1983, pp. 121-139.

O'REGAN, J. K., LEVY-SCHOEN, A., JACOBS, A. M. The effect of visibility on eye movement parameters in reading, Perception & Psychophysics, in press.

TINKER, M. A. Recent studies of eye movements in reading. Psychological Bulletin, 1958, 55, 215-231.

Theoretical and Applied Aspects of Eye Movement Research
A.G. Gale and F. Johnson (Editors)
© Elsevier Science Pulishers B.V. (North-Holland), 1984

SACCADES DIRECTED AT PREVIOUSLY ATTENDED LOCATIONS IN SPACE

Jonathan Vaughan

Department of Psychology
Hamilton College
Clinton, New York 13323
U.S.A.

Having previously fixated a location increases the latency
of subsequent saccadic eye movements towards that location.
In a saccadic pursuit task, the last saccade of each trial
was directed directly at, or up to three deg away from the
location of a previous fixation on that trial. Saccades to
a previously fixated location were longer by 8 to 15 msec
than those directed 3 deg away from any previously fixated
location, even 1700 msec after the prior fixation. The
elevation in saccade latency is consistent with an
attentional bias in favor of fresh sources of information.

INTRODUCTION

Attending to a location in space has two effects in sequence. First,
responding is faster if attention is directed to a parafoveal location where
targets may appear [7]. Second, having temporarily directed attention to a
location can interfere with later processing of signals originating from
that location or nearby [5, 8, 9]: Posner and Cohen [8] had subjects
maintain central fixation, then cued one side or the other with an
attention-eliciting cue to which the subject was not to overtly respond. If
attention was caused to be withdrawn from the cued location (either
directly, by the presentation of a cue elsewhere in the visual field, or
indirectly, by making targets highly probable elsewhere) then manual
reaction time to later targets in the first cued location was longer,
beginning a few hundred msec after cue presentation and persisting for up to
two sec [5, 8, 9]. This time course would seem to rule out purely sensory
explanations for the effect, such as forward masking.

These experiments address how the latency of eye movements is affected by
previously having attended to and fixated a location in space. Like
reaction time, saccade latency is also elevated by prior cuing at a location
in space [5]. The inhibition is mapped in spatial coordinates (rather than
being mapped retinotopically) and it can bias the direction of voluntary
saccades [10]. However, much is yet to be learned about the spatial and
temporal distribution of inhibition, as it affects saccadic eye movements,
and the conditions under which inhibition may be observed.

EXPERIMENT I: SPATIAL AND TEMPORAL DISTRIBUTION OF INHIBITION

A fixation point was moved about a display so that its last location (the
target of the last saccade) coincided with or was a few degrees above or
below a previously fixated location, that is, a presumably inhibited
location in space.

144   *J. Vaughan*

Method

<u>Subjects, Apparatus, and Procedure</u>. Eight subjects were run for one
practice and one data session each. Each session consisted of about 500
trials run in blocks of 30 each with a short break between blocks. A PDP-9
computer drove an oscilloscope display system, refreshed at 60 Hz, with a
P11 phosphor. Screen background luminance was 0.05 cd/m^2. Eye orientation
was recorded using the electrooculogram, and saccade onset was measured to
millisecond accuracy. In each trial the subject was instructed to foveate a
fixation point (a plus sign subtending about 12 min of arc) which was
instantaneously displaced from one location to another on the oscilloscope
screen, and to manually report the onset of an occasional target stimulus
(clearly discriminable from the fixation point), which could replace it at
any time during the course of the trial. To ensure that each fixation point
was attended, the manual response target appeared on some trials during the
first, second or third fixation. The potential locations of the fixation
point are shown in Figure 1. Only one fixation point was illuminated at any
time.

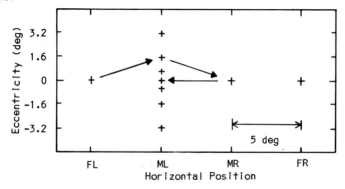

Figure 1. Potential locations of fixation points during a trial.
(Arrows indicate sequence of saccades made on a 1.6 deg
eccentricity experimental trial. Two points omitted for clarity.)

There were two kinds of trials, experimental and control trials (in addition
to those which involved detection of the manual response target and catch
trials, which will not be considered further). In each trial four fixations
occurred. The first fixation was on the far left (FL) location (Figure 1).
After fixation had been maintained for 700, 1200, or 1700 msec, the point
was moved to one of 9 locations, coincident with or up to 3.2 deg above or
below the middle left (ML) location. The second fixation was maintained at
this location for 1000 msec, measured from the beginning of the saccade
which brought the gaze to it. Then the point was extinguished and the point
at the middle right (MR) location was illuminated; the third fixation was
maintained for 700, 1200, or 1700 msec, measured from the beginning of the
saccade to that location.

There were two possible locations for the fourth fixation of each trial. On
experimental trials, it was to ML (on the horizontal). This point was
nearer or farther from a previously fixated location, depending on the
vertical distance of the second fixation location from the horizontal. On
control trials, the location of the fourth fixation was the far right (FR)
point. Latency of this third ("last") saccade, from MR to ML or FR, was

the major dependent measure for the experiment. For convenience, I will refer to the distribution of inhibition (if any) around the second fixation location as its inhibition field. The strength and extent of the inhibition field was assessed by observing the latency of the third saccade on experimental trials (always from MR to ML on the horizontal) as a function of the distance between the locations of the second and fourth fixations. (Note that the last saccade on experimental trials is directed towards a cue whose retinotopic location places it in the left visual field; all other saccade targets fall in the right visual field).

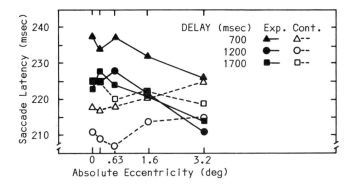

Figure 2. Latency of the last saccade as a function of the vertical distance between its destination and the location of the second fixation, Experiment I.

Results

The latency of the last saccade can be used to measure the inhibition field around the presumably inhibited location. Figure 2 indicates this latency, as a function of the distance between the location of the fourth fixation from the second fixation, for each of the delays after the end of the fixation on the inhibited location. Saccade latency is longer on zero-eccentricity trials, when the saccade is directed to the middle left point after a prior fixation on it, compared with the latency of control saccades directed at the far right point. Analysis of variance showed a significant effect of delay (a non-specific warning effect), $F_{(2,14)} = 5.43$, $p < .02$, and a delay X type interaction, $F_{(2,14)} = 20.03$, $p < .0002$. There was also a significant type X eccentricity interaction, $F_{(4,28)} = 6.09$, $p < .002$: the latency of only experimental saccades varied with eccentricity (Figure 2). The inhibition field appears to extend about 1.5 deg, in that there is no difference between experimental and control saccade latencies when the distance of the second fixation location from the destination of the last saccade is 3.2 deg.

Discussion

The elevation of saccade latency persists up to 1700 msec after the end of the fixation on the inhibited location. This cannot be due to retinal aftereffects of prior stimulation, since the cue which elicited the last saccade fell in the opposite visual field from the prior cues in that trial. Rather, it maps spatiotopically.

Elevation of saccade latency is qualitatively similar to the effect of prior
peripheral stimulation on manual reaction time to peripheral targets [8],
though it is smaller in absolute terms (10-15 msec, rather than 40).

EXPERIMENT II: FACTORS AFFECTING MEASURED INHIBITION

The second experiment assessed the robustness of the effect observed in
Experiment I, by replicating it with a number of procedural variations, such
as ambient illumination level, the pattern of locations in space fixated,
and the size of the saccades made. Though it differed in only seemingly
minor respects, it had quite different results.

Method

Subjects, Apparatus, and Procedure. Seven subjects served for four sessions
(after one practice session), each comprising 224 trials. The experiment
was controlled by a PDP-8 computer that refreshed an oscilloscope display
(P31 phosphor with a green filter added to minimize persistence after
nominal stimulus offset) at 100 Hz. Saccade latency timing was synchronized
with scope refresh. Background luminance was 1.0 cd/m^2. Each trial began
with a fixation at a central location. One second after this first fixation
began, the fixation point moved left or right to a lateral point (3.6 deg
horizontally from the center) that was the same height as the first point or
up to 3.6 deg above or below the horizontal. One second after the beginning
of the second fixation, the point was moved back to the center by the
simultaneous extinction of the lateral point and reillumination of the
center point. When the third fixation had lasted for one second, the point
moved left or right of fixation 3.6 deg on the horizontal, either on the
same side (experimental trials) or on the opposite side (control trials) as
the prior excursion. As in Experiment I, latency of the last saccade was
recorded as a function of the vertical distance of the last saccade from the
location of the second fixation.

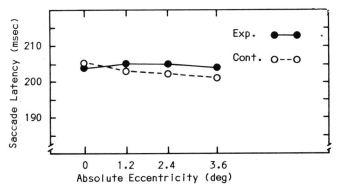

Figure 3. Latency of the last saccade as a function of the
vertical distance between its destination and the location of the
second fixation, Experiment II.

Results

There was no increase in saccade latency when saccades were to be directed
near a location previously fixated either in the overall data (Figure 3) or
in the data of individual sessions.

Discussion

There were several seemingly minor differences between Experiment I and
Experiment II, including saccade size (5 deg in Experiment I, 3.6 deg in
Experiment II); visual field stimulated (always right vs. alternately right
and left prior to the last saccade, respectively); number of sessions;
variability of the time before saccades; and level of ambient illumination
(in the scotopic range, vs. in the lower photopic range). One or more of
these factors must have contributed to the difference in the results. One
that comes readily to mind is a possible difference in the persistence after
offset of the image of the fixation points due to a difference in ambient
illumination or phosphor used. Even though phosphor decay is initially
rapid (typically 20-30 msec to .1% of full brightness) a small but
persistent trace may remain for some hundreds of msec, clearly visible to
dark-adapted subjects, which might have had more effect at the low luminance
levels. To test this a brief control experiment (the "lights out"
experiment) repeated Experiment II with the room illumination off, for five
subjects, one session each. The subjects spontaneously reported strong and
annoying persistence under these conditions, but there was no evidence of an
inhibitory field around the location of the second fixation.

The next experiment considered another difference between the first two
experiments, the ease with which saccades could be initiated. In Experiment
I, for the last saccades, the delay before the signal to move the eye was
variable. In Experiment II, by contrast, this delay was constant for all
saccades, 1000 msec after the beginning of the fixation. Because the time
of movement was always predictable, and because the stimulus for movement
was nearly optimal (simultaneous offset of fixation and onset of the
peripheral cue) the conditions of Experiment II may have made saccade
latency insensitive as an indicator of spatially localized inhibition. By
hypothesis, the existence of an inhibition field around a previously fixated
location might be due to the aftereffects of releasing fixation of that
point, and correlated with the difficulty of such release.

EXPERIMENT III: EFFORTFUL VS. AUTOMATIC SACCADES.

Saslow [12] has shown that the latency of a saccadic eye movement to the
onset of a peripheral cue is longer if the fixation point remains on, rather
than being extinguished simultaneously with the onset of the peripheral
cue. This manipulation was used in Experiment III to impede the rapidity of
movement away from the inhibited location.

Method

Subjects and Procedure. Ten subjects were run under conditions similar to
Experiment II (except as noted below) for four sessions of 224 trials each.
Half of the trials in each session (non-persistence trials) were identical
to those of Experiment II. The remaining trials (persistence trials)
differed only with respect to the second fixation point. On a random half
of the trials, this point was not extinguished when the cue came on for a
return saccade to the center; rather, the point remained on until the
saccade to the center began. Thus, the second and third fixation points
were simultaneously illuminated during the latency of the second saccade.
The simultaneous offset of the fixation point when a new cue to move was
illuminated was maintained for all other saccades in the trial.

Results

So that the number of trials analyzed would correspond roughly with those in
Experiment I, the data from the first two and from the last two sessions
were analyzed separately. During the first session, the latency of the
saccade away from the second fixation location was elevated from 189 msec
(on non-persistence trials) to 247 msec (on persistence trials) by the
continued illumination of the fixation point (as Saslow [12] had observed).
During the second session these saccade latencies were 186 and 229 msec,
respectively. The latency of the last saccade was, by contrast, not
different between nonpersistence trials and persistence trials. However,
there was in the first two sessions a difference in the latency of last
saccades as a function of eccentricity, F (3,27) = 3.15, p < .05, though not
in the second two, F (3,27) = .93, p > .4 (Figure 4).

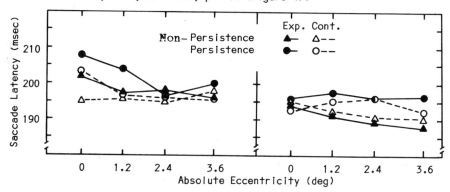

Figure 4. Last saccade latency as a function of the distance of
its target from the location of the second fixation, Experiment
III.

Discussion

Persistence of the fixation point increased the "effortfulness" of making a
saccade away from a location (as indicated by the increase in the latency of
the saccade after onset of the peripheral cue). This manipulation also
transiently increased the latency of saccades towards a previously fixated
location. Thus, merely fixating to detect a target is not a sufficient
condition to maintain inhibition in highly practiced subjects.

GENERAL DISCUSSION

The failure to observe inhibition in Experiment II and after the first two
sessions of Experiment III helps to rule out purely sensory (retinal,
masking, or oscilloscope persistence) or mechanical (direction reversal)
hypotheses for the inhibition observed in Experiment I. Similarly, these
results show that having to make saccades in a direction opposite to the
immediately prior one is not, per se, sufficient to lengthen saccade
latency. It is possible that the factors which made Experiment I show an
inhibition field around a previously fixated location differ from those
responsible for Experiment III. Nevertheless, both these experiments have
some condition (unpredictability of cue movement time, or stimulus

persistence) for at least some of the fixations that might be expected to increase the degree to which attention or effort is required in initiating the saccade. Jonides [3] has argued that there is an inherent difference in the attentional effects of different types of cues. Peripheral cues with sudden onset automatically elicit shifts of attention in space. Central cues (such as a right- or left-point arrow at fixation) require more time and more "attentional" resources to produce equivalent facilitation in responding to peripheral targets.

This kind of automatic/attention-demanding distinction might serve to distinguish conditions (Experiment II) that do not demonstrate aftereffects of having fixated a location from those conditions (Experiments I and III) that do. If this were consistently the case in further experimentation, it would support a general argument that inhibition around previously fixated locations is an attentional, rather than a purely sensory (e.g. retinal), phenomenon. Inhibition might be attenuated, then, as practice made performance more automatic. Such a hypothesis would be consistent with the transitory effect observed in Experiment III (see, for example, [13]).

There is another way in which the conditions of Experiment II and Experiments I and III can be distinguished. Because of the low level of ambient illumination in Experiment I, and by design in Experiment III, there was a persisting stimulus at the second fixation point of some trials that showed inhibition. (The lights off experiment showed that persistence per se is not sufficient to produce evidence of inhibition). The subjective perceptual experience when there is a persisting stimulus is that of moving the eyes from one object with continuous existence to another, separate object. In Experiment II, on the other hand, where there was no persistence, one subjective description of the task might be that it is to follow a single, individual object from one location to another. These two situations might invoke different mechanisms in attention, appropriate for marking the location of a persisting object in space in the first instance and following a moving object in the second, as a higher-level representation of the stable external world is synthesized and modified in perception [2]. Ecologically, it might be more useful to remember the location of once-attended objects than it is to recall the various locations traversed by a single moving object. Most prior studies on spatially localized inhibition [e.g. 5, 8] employed continuing peripheral stimuli. It remains to be seen if this distinction will prove to be a useful one.

An explanatory application of the present results concerns the asymmetry of the perceptual span in reading. Rayner, Well, and Pollatsek [11] have shown that, when eye movements are made across text, this span is larger to the right of fixation, and a similar study by Pollatsek, Bolozky, Well and Rayner [6], and one by Chang [1] using reaction-time methods, have shown that the span is larger to the left when reading Hebrew or upside-down English, when the reader scans right to left. Posner and Cohen [8] have interpreted these results as indicating an analog internal scanning process that goes from left to right for normally oriented English and right to left for Hebrew or upside-down English. We may certainly argue a priori that reading requires active attention to the locations fixated. An inhibition field localized around each previously attended or fixated part of the text might provide, in normal reading, the mechanism producing this asymmetry in perceptual span.

Spatially localized inhibition around previously fixated locations does,

150   *J. Vaughan*

then, affect the latency of saccades. It does so in a manner similar to the
effect on reaction time described by Posner and Cohen [8]. In both
contexts, the effect is consistent with a bias against sampling visual
information from previously attended locations, in favor of getting visual
information from new areas, those not recently fixated or attended.

REFERENCES

[1] Chang, F. Distribution of attention within a single fixation in
     reading: Studies of the perceptual span. Unpublished doctoral
     dissertation, University of Oregon, 1981.
[2] Feldman, J. A. Four frames suffice: a provisionary model of vision
     and space. Technical Report No. 99, Computer Science Department,
     University of Rochester, 1982.
[3] Jonides, J. Voluntary versus automatic control of the mind's eye's
     movement. In J. Long and A. Baddeley (Eds.) Attention and Performance
     IX. Hillsdale, N.J.: Erlbaum, 1981.
[4] Maylor, E. Unpublished observations, University of Durham.
[5] Maylor, E. Effects of prior activation on the processing of signals.
     Paper delivered to the Experimental Psychology Society, Cambridge,
     England, March, 1982.
[6] Pollatsek, A., Bolozky, S., Well, A. D., and Rayner, K. Asymmetries in
     the perceptual span for Israeli readers. Brain and Language, 1981,
     14, 174-180.
[7] Posner, M. I. Orienting of Attention. Quarterly Journal of
     Experimental Psychology, 1980, 32, 3-25.
[8] Posner, M. I., and Cohen, Y. Components of visual orienting.
     To appear in H. Bouma and D. Bouwhis (Eds.), Attention and Performance
     X. Hillsdale, N.J.: Erlbaum, in press.
[9] Posner, M. I., Cohen, Y., Choate, L., Hockey, R., and Maylor, E.
     Sustained concentration: passive filtering or active orienting? In
     Kornblum, S., and Requin, J. (Eds.) Preparatory Processes (in press).
[10] Posner, M. I., Choate, L., and Vaughan, J. Unpublished obervations,
     University of Oregon.
[11] Rayner, K., Well, A. D., and Pollatsek, A. Asymmetry of the effective
     visual field in reading. Perception & Psychophysics, 1980, 27,
     537-544.
[12] Saslow, M. G. Effect of components of displacement-step stimuli upon
     latency for saccadic eye movement. Journal of the Optical Society of
     America, 1967, 57, 1024-1029.
[13] Schneider, W. and Shiffrin, R. M. Controlled and automatic human
     information processing: I. Detection, search and attention.
     Psychological Review, 1977, 84, 1-66.

Acknowledgements: The work reported here was supported by National Institute
of Mental Health National Research Service Award 1 F32 MH 08716 and by a
faculty research grant from Hamilton College. I appreciate the assistance of
Debra Leder, Julie Ross, and Elizabeth Walton in conducting the experiments,
and Lisa Choate, Yoav Cohen, Wendy Kellogg, Elizabeth Maylor, Michael
Posner, and Douglas Weldon for valuable discussion and comments. I
especially thank the University of Oregon Psychology Department for its
hospitality during a sabbatical year and subsequent summer.

Section 3

READING AND PERIPHERAL VISION

Theoretical and Applied Aspects of Eye Movement Research
A.G. Gale and F. Johnson (Editors)
© Elsevier Science Pulishers B.V. (North-Holland), 1984

READING AND PERIPHERAL VISION

INTRODUCTION

Geoffrey Underwood and Elizabeth Maylor
Department of Psychology, University of Nottingham
Nottingham, England.

In recent years there has been much interest in the relationship between
skilled reading and the concurrent patterns of eye movements. Kennedy
(1978) argued that such investigation is important "since any analysis of
reading demands a clear description of the manner in which information is
gathered" (p. 218). During both oral and silent reading, characteristic
patterns of eye movements are observed. Approximately one tenth of the
total reading time is taken up by predominantly horizontal saccades of
up to 15 letter spaces with fixations of about 250 ms. duration occupying
the remaining time. A number of models of eye movement control have been
proposed ranging from random to strong control (see Kennedy, 1978, for a
review). From the work of Rayner and his colleagues which demonstrated
that fixation duration is influenced by the type of information being
processed (for example, Rayner, 1977; Rayner and McConkie, 1976), it is
clear that eye movements are to some extent under the control of cognitive
information.

A provocative study by Willows (1974) reported that young good readers were
more impeded than young poor readers by words printed between the lines of
text which was to be read and understood. This suggests that one aspect
of skilled reading is the ability to make use of information which is not
fixated. This question has been addressed in recent studies of word and
sentence processing concerned with the specific issue of whether or not useful information is extracted from parafoveal vision in order to guide subsequent eye movements. The question has not received an unequivocal reply,
with, in general, single fixation studies suggesting the plausibility of the
notion of peripheral control, and continuous eye-monitoring studies failing
to support the notion. Tachistoscopic single-fixation studies have demonstrated that semantic information is available from parafoveal vision (for
example, Bradshaw, 1974; Underwood, 1976; Underwood and Thwaites, 1982) and
that unattended words in the parafovea influence the processing of attended
foveal words not only when they appear in insolation but also when embedded
amongst other material (Underwood, 1981). Support for the notion of the
use of peripheral vision in the cognitive control of eye movements has also
come from eye-monitoring experiments. A study by Kennedy (1978) using a
free-reading task provided evidence that eye movements can be influenced by
the meanings of words ahead of fixation, and Balota and Rayner (1983) have
also reported an influence of the meanings of peripherally presented words
upon reading behaviour. In order to demonstrate effects of peripherally
presented words, we have so far found it necessary to use experimental tasks
which investigate components of reading, and whereas these tasks establish
the plausibility of the use of peripheral information, our conclusions will
remain tentative until the influences can be demonstrated in free-reading
tasks.

The following chapters illustrate current attempts to investigate further
the relationship between reading, peripheral vision and eye movements. The
experiments reported by O'Regan and by McConkie, Zola and Blanchard employ-

ing highly sophisticated recording techniques, are concerned with the moment
-to-moment variability in eye movement behaviour. In particular, O'Regan
confines his chapter to the scanning behaviour of single words and addresses
the problems of both identifying the sources of information used to control
eye movements and the time at which such sources become effective. Although
his previous work demonstrated that there is an optimum place to fixate
within a word, this could be attributed not to lexical information but merely to knowledge about the position of the eyes within a word. From his
present experiments, however, it is possible to conclude that lexical processing does influence the behaviour of the eyes during word recognition.
In addition, the effect is rapid as it influences characteristics of both
the current and subsequent fixation. An important result from this experiment is that, at least for the identification of isolated words, there is
no evidence that parafoveal information potentially available on the previous fixation is of use on the current fixation (see, for example, the short
145 ms fixation at the 'wrong' end of 'protagoniste' followed by a fixation
of 356 ms at the 'right' end compared to a single fixation of 352 ms at the
'right' end).

McConkie et al. similarly begin by noting the well-established finding that
there is considerable variability in fixation duration during reading. They
then provide an excellent summary of the many experiments conducted in their
laboratory designed to investigate possible eye movement control processes
that produce this variability . They are specifically concerned with identifying the 'triggering event' for saccades during reading and first consider two extreme positions, namely (1) that saccades are initiated solely on
the basis of information obtained prior to the current fixation and (ii)
that saccades are initiated on the completion of the processing of information from the current fixation. Both positions are rejected in favour of
the view that the decision to move the eyes is influenced by information
from the current fixation but that the processing of the information is not
always completed before the decision is made. In addition they propose a
'variable utilisation time hypothesis' which attributes some of the saccade
variance to differences from fixation to fixation in the amount of processing carried out before the information from the current fixation is utilised
- a highly plausible hypothesis, but one which may prove difficult to test
without the eye movement recording techniques available to McConkie and his
colleagues.

The chapter by Jennings and Underwood provides an example of the way in
which tachistoscopic single-fixation tasks can be used in order to investigate eye movements and reading. In this case the question is that of the
possible integration of parafoveal information across fixations. The study
was able to separate effects (on naming latency) due to prior context and
those due to information from parafoveal vision. They demonstrated that
both contextual information derived from previously processed words and information extracted from the parafovea facilitate identification of single
words. In addition, prior contextual information increases the pertinence
of related words presented subsequently to the parafovea. From these results Jennings and Underwood claim that contextual constraint may have two
independent effects during normal reading. First, it facilitates the recognition of and thus fixation duration on predictable words (which is also
true of information extracted from the parafovea). Secondly, by increasing
the pertinence of predictable words in the parafovea such items may then
have an increased effect upon the characteristics of the following saccade.

The influence of cognitive factors on eye movements during reading is demonstrated more directly in the three chapters by Kennedy and Murray, Vonk and Kerr and Underwood. For example, in the latter study subjects read sets of three contextually-related sentences in which the third sentence contained a pronoun referring incongruously ('she' to 'engineer'), congruously ('he' to 'surgeon') or ambiguously ('he' to 'student') to the subject of the first sentence. Fixation duration and inspection duration (that is, including regressive fixations) of the pronoun and total reading time of the third sentence were all longer for the incongruous condition compared to the congruous and ambiguous conditions. Thus pragmatic information is an important factor influencing the control of eye movements during reading.

The study reported by Vonk, like that of Kerr and Underwood, concerns the comprehension difficulties imposed by the introduction of a pronoun into a sentence. At least when inspecting sentences with a fixed syntactic structure, readers are able to select rapidly the parts of the sentence which convey critical information. From this Vonk argues that her data provide evidence for an immediacy notion of the processing of text in which at least the lexical properties of words are identified during the first fixation. The sensitivity of inspection behaviour to ambiguous text was also investigated in the Kennedy and Murray experiment, and they emphasise the importance for comprehension of the availability of prior text. Additionally, the spatial positions of words within sentences appear under some circumstances to be of use when deriving the meaning of a sentence.

Further insight into the relationship between eye movements and reading can be gained from the study of changes in behaviour as a result of practice, as in the experiment reported by Menz and Groner, of increased task difficulty, as in the experiment reported by Netchine, Greenbaum and Guihou, or of the addition of a secondary task, as described by Schroiff. First, Menz and Groner are concerned with changes in two aspects of ocular behaviour during the learning of a novel typography over the course of twenty one-hour sessions. A distinction is made between the width of processing (defined as saccade length) and depth of processing (defined mainly in terms of fixation duration). Although the width of processing remained stable throughout training (at approximately 1.5 degrees visual angle corresponding to a single character), changes were observed in the depth of processing, that is, a decrease in the mean fixation duration over time. It would be very surprising if the width of processing did not increase with further extensive training at the task, raising the interesting possibility that an increase in the width of processing may occur when the depth of processing reaches an upper limit. It might be interesting to compare these results with those obtained from studies in which adults are monitored while learning novel but real orthographies (e.g. arabic or an oriental writing system, rather than using non-sense symbols).

Second language learning was of more specific interest to the study reported by Netchine et al., and in addition to monitoring childrens' and adults' eye movements during reading they consider the role of head movements. Young children, particularly, prefer to alter their direction of gaze by moving the head. Such movements are relatively slow compared to saccades and are initiated before and finish after the saccade. Although the exact function of head movements is unclear, it is argued that they reflect the difficulty of the text and indeed evidence is presented of increased head movements in adult second language compared to first language reading. The extent to which eye movements can be trained is raised by Pugh's discussion, which presents an overview of some preliminary studies of the use of eye movements in the investigation of comprehension.

The chapter by Schroiff is also concerned with changes in overt behaviour as a function of task difficulty, in particular with the effects of a secondary task on eye movements during both oral and silent reading. Total reading time, forward and backward fixations and reading errors all increased with the addition of the secondary task indicating both a general decrease in processing rate and a breaking-up of the text into smaller processing units.

Thus by manipulating the difficulty of the reading task, changes in overt behaviour can be observed in terms of the pattern of head and eye movements. The final chapter, by Stein and Fowler demonstrates the reverse, in that by controlling one aspect of ocular behaviour, reading ability can be affected. They identify a significant problem associated with the very early stages of learning to read, that is, the problem of knowing precisely where the eyes are pointing. Accurate visual localisation is facilitated by the development of a fixed leading eye (the reference eye chosen for calibrating the visual direction of objects). The results indicated that this is associated with reading ability so that the development of a fixed leading eye may help to resolve the confusion of ocular motor and retinal signals produced during reading. Moreover, important evidence is presented to suggest that if dyslexics (who are more likley to have unfixed leading eyes than normal readers matched for age and performance IQ) are forced to develop a fixed leading eye by means of monocular occlusion, their reading ability improves.

We now have rich descriptions of the patterns of eye movements made during the reading of sentences, and these descriptions have provided valuable information about the comprehension difficulties faced by readers of different abilities. In turn, the descriptions are being used in the development of cognitive theories of the processing of text. Eye movements do not reveal cognitive processes, of course, but they are sensitive to processing, and can be used as an indication of the focus of attention during reading. The current availability of sophisticated eye movement recording equipment ensures that these valuable measures will continue to aid the development of cognitive theory.

Acknowledgement

Preparation of this report was supported by the Science and Engineering Research Council (grant GRC/02259).

References

Balota, D.A. and Rayner, K. Parafoveal visual information and semantic contextual constraints. Journal of Experimental Psychology: Human Perception and Performance, 9. 1983, 726-738.

Bradshaw, J.L. Peripherally presented and unreported words may bias the perceived meaning of a centrally fixated homograph. Journal of Experimental Psychology, 103, 1974, 1200-1202.

Kennedy, A. Reading sentences: some observations on the control of eye movements. In Strategies of Information Processing (G. Underwood, ed.). Academic Press: London, 1978.

Rayner, K. Visual attention in reading: Eye movements reflect cognitive processing. Memory and Cognition, 4. 1977, 433-448.

Rayner, K. and McConkie, G.W. What guides a reader's eye movements? Vision Research, 16. 1976, 829-837.

Underwood, G. Semantic interference from unattended printed words. British Journal of Psychology, 67, 1976, 327-338.

Underwood, G. Lexical recognition of embedded unattended words: Some implications for reading processes. Acta Psychologica, 47, 1981, 267-283.

Underwood, G. and Thwaites, S. Automatic phonological coding of unattended printed words. Memory and cognition, 10, 1982, 434-442.

Willows, D.M. Reading between the lines: Selective attention in good and poor readers. Child Development, 45, 1974, 408-415.

Theoretical and Applied Aspects of Eye Movement Research
A.G. Gale and F. Johnson (Editors)
©Elsevier Science Pulishers B.V. (North-Holland), 1984

HOW THE EYE SCANS ISOLATED WORDS

J.K. O'Regan
Groupe Regard, Laboratoire de Psychologie Expérimentale,
Université Paris V, C.N.R.S., E.P.H.E.,
28 rue Serpente, 75006 Paris

There is increasing evidence in the literature that
eye movement patterns reflect ongoing processing in
reading. However linguistic processing is insufficiently understood for precise predictions to be made
about how the eye should behave. The present paper
looks at the simpler activity of recognizing words
and letter strings in an attempt to understand the
eye's scanning behavior there. Experiments are described in which the position of information within a
string or a word is either mainly at the beginning or
not. A satisfactory account of the data can be put
forward in which it is assumed that the eye's behavior is entirely determined by the amount of processing being done in the vicinity of the location
fixated.

INTRODUCTION

Much of the work that has been done over the past decade on eye movements in
reading is centered on the question of what guides the eye (c.f. the excellent article by McConkie, 1983). It seems to me that the most useful way of
posing the problem is to distinguish two fundamental questions.

The first question is: What sources of information are being used to determine the eye's behavior? The sources of information can be classified
according to the amount of processing that is required to obtain them. For
example, the physical layout of the text, the lengths of the words, the
presence of capital letters and punctuation marks, are features available in
the visual field which can be made use of by the eye guidance system without
very much processing having to be done. More processing is required however
to determine the identity of individual words, and still more is required to
fit them together into a syntactic or semantically coherent whole.

The second fundamental question concerning eye guidance is: How quickly do
these sources of information have an effect on eye guidance? There are two
constraints which determine the answer to this question, one imposed by the
sources of information, and one by the characteristics of the oculomotor
control system. A source of information cannot influence eye guidance
before it becomes available: the more processing that has to be done to
obtain information from the visual field, the longer it will take following

pickup of the information before it can affect eye movements. On the other
hand, the oculomotor system cannot react arbitrarily fast, and even if
information is available on a given fixation, it is not clear whether this
can be used to determine the immediately following eye behavior.

Several studies have shown that physical aspects of the text such as spacing
(c.f. Abrams & Zuber, 1972) and word length (O'Regan, 1979; Kliegl, Olson, &
Davidson, 1983) affect the eye very rapidly, that is, on the fixation where
the information is gathered and on the immediately following saccade. On
the other hand, studies investigating the effect of linguistic variables
have only shown up slower forms of control (O'Regan, 1980; Holmes & O'Regan,
1981; Carpenter & Just, 1983). An explanation for this slowness could be
related to the complexity and distributed nature of linguistic processing.

There is however one aspect of lingustic processing which occurs on a local
scale, and probably proceeds comparatively fast: word recognition. Little
attention has been paid in the literature to the way the eye scans words,
though it has been observed that the eye tends to prefer to fixate certain
positions within words (Dodge, 1907; Dunn-Rankin, 1978; Rayner, 1979; O'Regan 1981). Recently in the Groupe Regard in Paris we have shown that there
is an optimum place to fixate within words, where processing time is shortest (O'Regan, Pynte, Levy-Schoen & Brugaillere, submitted for publication).
These facts suggest that ongoing lexical processing may influence the eye's
behavior in scanning words. In the present Chapter we look more closely at
the evidence for this, and we attempt to determine whether the information
being used to control the eyes is really lexical information and not merely
information about the position of the eye within the word, which is easier
to extract. We also attempt to determine how quickly the information is
being put to use.

Before investigating the question of the influence of ongoing word processing on eye movements, it is useful to consider first a task involving not
word recognition but digit recognition, where it is clearer what kind of
processing must be occurring.

DIGIT EXPERIMENT

I constructed stimuli consisting of 12 o's with a random digit embedded
either at the second or second from last position: e.g. o3ooooooooooo or
ooooooooooo4o. In each trial, such a stimulus appeared at the center of the
screen, and another one appeared 10 degrees in peripheral vision to the
right. The subject had to look at the first, move his eyes to the second,
and decide whether there was the same digit in both. Now the first stimulus
appeared in one of two positions with respect to the Subject's initial eye
position. Either it appeared so that the digit was precisely where the
initial fixation point was, or it appeared shifted horizontally so that the
eye was fixating at the "wrong" end of the string. I recorded 5 Subjects'
eye movements in this task.

Under the hypothesis that ongoing processing immediately determines the
eye's behavior, we expect the following. If the eye is exactly fixating the
digit in the string, it should stay the time necessary to process the digit,
and then move onwards to the comparison string. If the eye starts fixating
at the "wrong" end of the string, then (unless the digit can be seen clearly

in peripheral vision, which was not the case) it should stay the time
necessary to realize that it is in the wrong place, move to the digit, and
then stay same time it would have stayed if it had arrived there in the
first place. This is exactly what happens:

EYE FIXATES THE "RIGHT" PLACE

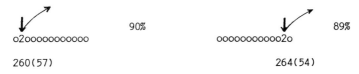

260(57)   264(54)

EYE FIXATES THE "WRONG" PLACE:

265(50)   174(21)   170(23)   299(50)

The arrows in the diagram indicate the position of the eye, and the numbers
indicate the mean duration of the fixations (with 95% unilateral confidence
intervals calculated between Subjects) and the mean percentage of cases
where the associated behavior was observed. When the eye is fixating in the
"right" place, it makes a 260 msec fixation and moves to the comparison
string with a probability of about 90% (in 10% of the cases a second fixation is made before leaving the test word -- this will not be discussed
here). But when it is fixating in the "wrong" place, the fixation there is
very short, around 170 msec, and the eye quickly moves on (92% of the time)
to the "right" place, where it makes a fixation of about the same length as
if it had landed there in the first place. The "ongoing processing" hypothesis therefore perfectly explains the data.

What sources of information are involved in generating this behavior? Since
trials were randomly mixed, Subjects had no way of knowing before each trial
whether their eyes were going to be at the digit position or not. The
different eye movement patterns found could only have been based on information gathered at the fixation point at each individual trial, that is by
recognizing the presence or absence of the digit at the fixation point.
This recognition probably could be done without great difficulty because
target digits were slightly higher than the lowercase o's, thereby providing
a simple visual clue.

How quickly did the information act? Since fixation durations differ significantly according as the fixation was or was not on the digit, we conclude
that the information acted on the very fixation at which it was gathered.
It also immediately influenced the probability of making a second saccade in
the string.

I will note here an aspect of the data which will be useful to later discussion: the 170 msec spent in looking at the "wrong" end of the string
appears to be entirely lost, not contributing at all to shorten the time

subsequently to be spent on the "right" end of the string. This is presumably because seen from the "wrong" end of the string, information about the target digit was too poor to be of any use.

The present data show that in scanning the simple strings used here, the eye can be controlled in a moment to moment way. The source of information being used must at least involve the extraction of a simple visual clue from the location fixated, although it need not involve actual recognition of the digit. However we have done a further experiment where the task can only be accomplished if processing goes far enough for the target digit to be recognized, and similar results are found. This suggests that there may be time for even higher forms of processing to affect eye movements in scanning strings. The following experiment is an analogy of the digit experiments extended to a situation where the information to be extracted is lexical.

WORD EXPERIMENT

In analogy with the digit experiment, we attempted to construct two word classes having identical physical characteristics, such that words in one class had their "information" at the beginning, and words in the other class had their "information" at the end. Not knowing what "information" is used in word recognition, we did the following. Each of the 20 words in the "information at beginning" class was such that it was the only word in the dictionary of about its length (10 to 13 letters) having the same first six letters. Tests showed that 20 Subjects given these six letters were always able to correctly guess the remaining 4-7 letters. Examples of such words were: "protagoniste", "sarcastique", "coccinelle". The "information at end" class consisted of 20 words whose last six letters gave more information than their beginnings. Subjects given these last 6 letters could correctly complete the word 89.5% of the time. Given the first six letters of such words Subjects were only able to correctly guess the word 18.5% of the time. Examples of such words were: "supercherie", "interpreter", "extravagance". Note that the information-bearing part of these words were not as effective in enabling the Subject to guess the word as was the case for the other word class. It seems that in lexical processing less use can be made of word endings, even if these endings are perfectly constraining when measured by dictionary counts. What is clear in any case is that given the beginnings of these words, Subjects were unable to determine their endings: it would be most accurate to have called this class of words the "information not at beginning" rather than the "information at end" words.

The experiment (c.f. O'Regan, Lévy-Schoen, Pynte, & Brugaillère, submitted for publication) was exactly analogous to the experiment with digits. The words were presented in a randomly mixed list, shifted horizontally so that on appearance of the word the Subject was either fixating the beginning (third letter) or the end (third letter from end) of the word. For each position, half the time the "information" was at the eye position, and half the time it was not. The Subject had to compare the test word to a comparison word on the right, and we recorded his eye movements on the test word. The 14 Subjects did the experiment twice. In this way we could examine their eye movements on the second reading, when they were more familiar with the words used, and presumably lexical access was facilitated owing to a more restricted effective vocabulary.

The table below gives the mean durations and 95% unilateral between Subject confidence intervals for 14 Subjects for the first and second readings of the words (upper and lower numbers, respectively). The arrows show the successive eye positions, and saccade sizes (in letter-spaces) are indicated with their associated confidence intervals. Only the eye movement strategies involving one or two fixations on the word will be discussed.

EYE FIXATES THE "RIGHT" PLACE

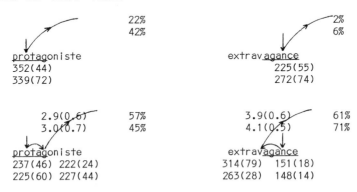

EYE FIXATES THE "WRONG" PLACE:

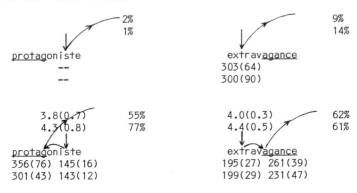

The pattern found is different from that of the digit experiment. First, there is not the same perfect symmetry between the information at beginning and information at end cases. Second, there is more variety in the eye behavior. However the observed fixation durations and positions have small confidence intervals (for durations in some cases these are about 15 msec; eye position estimates have confidence intervals of between 0.3 and 0.7 letters). Also there is striking stability across the two readings of the words. I think the data should be taken seriously.

Consider first the case where the information is at the beginning (e.g. "protagoniste"). In the case where the eye fixates the word at the end,

i.e. in the "wrong" place, it shifts over to the "right" place after a very
brief 143-145 msec fixation, and then makes a long fixation of 301-356 msec.
This latter duration is comparable with the 339-352 msec fixation required
if the eye is in the "right" place to start with. This pattern of a short
fixation followed by a long one is identical to that found in the digit
experiment. It is reasonable to assume, as we did there, that the short
143-145 msec fixations, being in the "wrong" place, could not contribute to
processing. This is also supported by the small inter-subject variability
observed (confidence intervals of about 15 msec), to be expected if the
decision to move is taken at a very low level of processing. It is also
supported by the fact that unlike most of the longer fixations, the 140 msec
fixations do not shorten on second reading when presumably processing is
facilitated.

The case where "protagoniste" is fixated twice in the "right" place, is a
pattern not found in the digit experiment, but which we can reasonably
suppose corresponds to distributed processing. Instead of a single 339-352
msec fixation, processing is distributed over two shorter (225-237 msec)
fixations separated by a comparatively small (2.9 letter) saccade.

Now consider the case where information is at the end of the word (e.g.
"extravagance"). If we retain the idea that a fixation around 140 msec with
small inter-subject variability means no processing is done, then we must
conclude that no processing is done at the first fixation on "extravagance"
when this falls at the end, precisely where supposedly the information is.
In fact the pattern of fixations is very similar to "protagoniste" when the
eye falls at the end, i.e. where there is no information. This suggests
that the word count method of determining information content does not
correspond to the way humans process words. Note however that when the eye
falls at the beginning of a word like "extravagance", fixation duration is
195-199 msec. With a confidence interval of 27-29, this is significantly
different both from the "no information" 143-145 msec fixations (confidence
interval 12-16), and from the "much information" 339-352 msec fixations
(confidence interval 44-72) found for a single fixation on the words with
information at the beginning. This suggests that though there is not as
much information to be got from the first part of the "extravagance" type
words as from normal "information at beginning" words, there is something to
be got. This small amount of processing might also explain why the second
fixation in this case is only 231-261 msec, comparatively shorter than the
301-356 msec found for the "information at beginning" words.

The above account of the pattern of results appears satisfactory. But what
firm proof have we that the behavior is truly being generated by use of
lexical processing? Could it not be explained by an eye movement strategy
which makes use only of information concerning the physical aspect of the
words, without necessitating any lexical processing? There is one aspect of
the data which goes against such an explanation: when the eye starts by
fixating at the beginning of a word, the pattern of fixations depends on the
word's lexical structure: the probability of making only one fixation is 22-
42% when information is at the beginning, and only 9-14% when it is not;
when two fixations are made, their durations are different and the saccade
leading from one to the other is different across the two lexical categories. These differences, though not large, are very stable across the two
repetitions of the experiment. They must have been caused by the effect of
lexical processing on the eye movement strategy, and cannot have been generated by information about position or length. In addition the effect is
rapid, because it affects the duration of the fixation where the information

is picked up, and the position the following saccade goes to.

The situation is different when the eye fixates at the end of a word. The eye behaviors observed are not significantly different between the two lexical categories. They could be explained by the action of a simple eye movement strategy which depends only on information about the position of the eye in the word. In particular the very short 140 msec fixations, instead of being caused by the fact that only very little lexical processing is being done, could have been caused by a strategy such as: "always shift rapidly leftward if you first land on the right". We cannot distinguish between these two alternatives unless a class of words is found for which we can be sure that processing occurs via the end; and unfortunately such words may not exist at all, since it may be that lexical processing always goes from left to right!

ANOTHER WORD EXPERIMENT

Partial resolution of these problems comes from detailed analysis of an experiment reported in O'Regan, Lévy-Schoen, Pynte & Brugaillère (submitted for publication), similar to the above word experiment, but in which only normal "information at beginning" words were used, and where there were a greater variety of initial fixation points.

We found that the behavior of the eye in recognizing the words depended strongly on the place where the eye was fixating when the words appeared. In some cases the eye made only a single fixation on the word before leaving it, and in some cases, it made two or more fixations. The probability of making a single fixation is given in the Figure below.

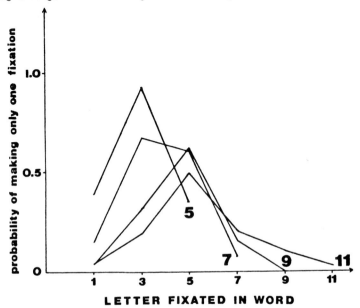

Each curve corresponds to the data for words of a given length, ranging from
five to eleven letters. When the eye is initially fixating the third letter
(for 5 and 7 letter words) or the fifth letter (for 9 and 11 letter words)
in the word, the probability is high that only one fixation will be made.
This is precisely what would be predicted from the idea that processing will
be most efficient if the eye is fixating where the "information" in the word
is. As we move away from this "optimum" position, the probability of making
a single fixation falls off quickly.

The next Figure concerns the fixation durations found in the experiment.
The lozenges at the top give the durations of the fixation made when only
one is made in the word. Data is only reliable in the cases near the
optimum position. The durations are all around 400 msec.

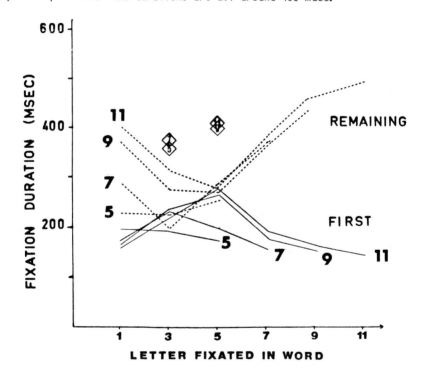

The remaining data in the Figure correspond to the cases when two or more
fixations are made in the word. The solid lines give the durations of the
first fixation as a function of where it was. There is a separate curve for
words of length 5, 7, 9, and 11 letters. When the eye starts by fixating at
the end of a long word, that is in the "wrong" place, fixation durations are
about 140 msec, just as in the experiment reported earlier. Now, as the
first fixation occurs closer to the "optimum" position, its mean duration
increases (just as we found 190 msec fixations for the "extravagance" type
words), and reaches a maximum at precisely the optimum point, where the most
processing is presumably done.

While this pattern is exactly what is expected from ongoing lexical processing, it could also be explained by a scanning strategy which says: "rapidly make a leftward saccade if you are near the end of a word, but slow down the closer you are to center."

Let us look at the further fixations. The dashed lines in the Figure give the total remaining time taken by the eye on the second and further fixations, as a function of where the first fixation was. When the first fixation is in the "right" place, remaining time taken is short. When it was in the "wrong" place, time is long. What this means is that when a word is analyzed by several fixations, this can happen in a variety of ways: either there is a first very short fixation in the "wrong" place, followed by a long one in the "right" place, or there are two shorter fixations near the "right" place. As can be seen by taking the sum of the two graphs, the total time taken is however always greater than the approximately 400 msec required if a single fixation is made, and is also longest when processing is unequally distributed. It is better to make two medium length fixations near the "right" place, than a short one in the "wrong" place followed by a long one in the "right" place. It is best of all to make only a single fixation in the "right" place.

Again the pattern of the eye behavior is exactly what would be expected from ongoing lexical processing, but again, a strategy depending only on position within the word cannot be excluded. However the strategy we require to completely explain the fixation pattern would have to be unrealistically complicated: "1. If you land at the third to fifth letter of a word, make a single long fixation with probability X, else two medium length fixations. 2. If you land near the end, make a short fixation, followed by a long one. 3. If you land further inwards, lengthen the first and shorten the second...". I think this is asking too much of a preprogrammed scanning strategy! The complexity of the program approximates what would be required to do the lexical processing that it is trying to avoid!

CONCLUSION

The evidence from the above discussion is in favor of the idea that lexical processing can influence the moment-to-moment behavior of the eyes in scanning words, and that this influence is very rapid, affecting the fixation at which information is picked up and the characteristics of the following saccade. We cannot however exclude the possibility that some of the time, in particular when the eye lands at the end of a word, its behavior is determined by a simple strategy dependent not on lexical processing, but on information about the eye's position in the word. However several arguments go against this idea. One is that since lexical processing is determining the behavior of the eye when it lands on the beginning of the word, why should it not also control the eye's behavior when it lands on the end of the word? The second argument is that in the digit experiment, where physical layout and task were very similar to those of the word experiment, no such a strategy was apparent. Finally, the exact strategy that would have to be postulated in order to account for the data of the last experiment is quite complicated. It appears more parsimonious to assume that lexical processing is what is determining the eye's behavior at every moment.

The implications of these conclusions are interesting. It has always been the psycholinguist's dream that eye movements might reflect moment to moment processing. It may be that the reason why up to now evidence for this has

been hard to come by is that the eye in fact so reliably reflects processing, both at the sensory, lexical, and linguistic level, that the resulting
complexities have resisted understanding!

REFERENCES

Abrams, S.G., & Zuber, B.L. (1972). Some temporal characteristics of information processing during reading. Reading Research Quarterly, 8, 42-51.

Carpenter, P.A. & Just, M.A. (1983). What your eyes do while your mind is
reading. In: K. Rayner (Ed.). Eye Movements in Reading. Perceptual and
Language Processes. New York, Academic Press, 275-307.

Dodge, R. (1907). An experimental study of visual fixation. Psychological
Review Monograph Supplement, 35 (8).

Dunn-Rankin, P. (1978). The visual characteristics of words. Scientific
American, 238, 122-130.

Holmes, V., & O'Regan, J.K. (1981). Effects of syntactic structure on eye
fixations during reading. Journal of Verbal Learning and Verbal Behavior,
20, 417-430.

Kliegl, R., Olson, R.K., & Davidson, B.J. (1983). On problems of uncon-
found-ing perceptual and language processes. In: K. Rayner (Ed.). Eye Movements in Reading. Perceptual and Language Processes. New York, Academic
Press, 333-343.

McConkie, G.W. (1983). Eye movements and perception during reading. In: K.
Rayner (Ed.). Eye Movements in Reading. Perceptual and Language Processes.
New York, Academic Press, 65-96.

O'Regan, J.K. (1979). Eye guidance in reading: Evidence for the linguistic
control hypothesis. Perception & Pychophysics, 25, 501-509.

O'Regan, J.K. (1980). The control of saccade size and fixation duration in
teading: The limits of linguistic control. Perception & Psychophysics, 28,
112-117.

O'Regan, J.K. (1981). The convenient viewing position hypothesis. In D.F.
Fisher, R.A. Monty, & J.W. Senders (Eds.), Eye movements: Cognition and
visual perception. Hillsdale, New Jersey: Erlbaum.

O'Regan, J.K., Lévy-Schoen, A., Pynte, J., & Brugaillère, B. Naming latency
and fixation duration in word recognition depend strongly on the position
the eye fixates in the word and on its lexical structure. Submitted for
publication in Journal of Experimental Psychology, Human Perception &
Perfor-mance.

Rayner, K. (1979). Eye guidance in reading: Fixation locations within
words. Perception, 8, 21-30.

Theoretical and Applied Aspects of Eye Movement Research
A.G. Gale and F. Johnson (Editors)
© Elsevier Science Pulishers B.V. (North-Holland), 1984

WHAT IS THE BASIS FOR MAKING AN EYE MOVEMENT
DURING READING?

George W. Mcconkie, David Zola, & Harry E. Blanchard

Center for the Study of Reading
University of Illinois at Urbana-Champaign
Champaign, Illinois
U. S. A.

At some time during the period of a fixation a decision
is made to move the eyes. This paper considers possible
bases for this saccade initiation decision during reading. Two extreme theoretical positions seem unlikely:
that the decision is made without consideration of
information from the current fixation or that the eyes
are moved when processing of information from that fixation is finished. Alternative explanations which suggest that the eyes are moved after some but not all processing is complete are considered and tested against
recent data.

INTRODUCTION

At some time during every fixation a decision is made to move the eyes,
directing them to a new location in the stimulus array. From fixation to
fixation, there is a great deal of variability in the time at which a saccade begins. In one set of eye movement records made as people were reading, 92% of the time the eyes were moved between 100 and 500 msec after the
beginning of the fixation. The median fixation duration was 230 msec.
Variability in fixation duration has been an aspect of behavior that has
intrigued psychologists, since it seems likely to reflect differences in
processing time required to deal with the information acquired on different
fixations. A number of studies have demonstrated relationships between
characteristics of the stimuli being fixated and the times, or durations, of
these fixations. For reviews of this literature see McConkie (1983), Levy-
Schoen and O'Regan (1979), and Rayner (1978).

The purpose of the present paper is to raise an issue which underlies these
investigations: namely, on what basis does the mind decide that it is time
to move the eyes? Thus, our purpose here is not to account for the variability in fixation times. Instead, the purpose is to understand the eye
movement control processes that produce this variability. To do this
requires that we investigate the relation between the cognitive processes
taking place and the decision to move the eyes. This issue will be discussed within the context of the eye movements of skilled adult readers.

We will consider three general hypotheses concerning the cognitive basis for
deciding to move the eyes: (a) the saccade initiation time is determined on
the basis only of information obtained prior to the current fixation; (b)
the saccade initiation occurs only after all processing permitted by the
presently viewed information has been completed; and (c) the saccade initia-

tion occurs when some processing event takes place prior to completing the
full processing of the presently viewed information. The first two state
positions at opposite extremes and the third takes an intermediate position.

SACCADE INITIATION BASED ONLY ON PRIOR INFORMATION

It has been suggested that during reading there is a considerable lag
between the time a part of the text is fixated and the time the information
is processed (Shebilske, 1975). Results from studies of the eye-voice span
have supported this view, since as one word is being vocalized during oral
reading the eyes are usually fixating words further along in the text
(Levin, 1979). However, recent eye movement studies have demonstrated that
characteristics of a word may influence the length of time it is fixated,
thus arguing against a processing lag (Just & Carpenter, 1980). Fixation
durations are correlated with several aspects of the words being fixated,
including frequency of the word in the language (Kliegl, Olson, & Davidson,
1983), part of speech of the word (Rayner, 1977; Wanat, 1971), predictability of the word in its context (Rayner, McConkie, & Ehrlich, 1978; Zola,
1981), whether or not the word is a reasonable continuation of the sentence
as it is being understood (Frazier & Rayner, 1982), and whether or not the
word is spelled correctly (Zola, 1981).

The fact that characteristics of words influence the amount of time they are
fixated argues against the prior information hypothesis. However, this fact
can be explained in either of two ways. First, prior context provides sufficient information about the characteristics of the next word to be fixated
that an appropriate fixation time could be allocated on that basis alone.
Second, on the prior fixation a peripheral preview of the word to be fixated
next could provide sufficient information to determine how long the eyes
should remain on that location for the next fixation. Both of these alternatives are ruled out by studies in which a word is changed during the saccade in which the eyes are moving to it. In this case, characteristics
unique to the word present on a fixation have been found to influence the
duration of that fixation (Rayner, 1975; and two recent unpublished studies
from our own laboratory). Characteristics of these words could not have
been picked up on previous fixations. Thus, while the time of saccade initiation may be influenced by information from earlier fixations, control is
not based exclusively on that information. Currently perceived words have
an effect.

SACCADE INITIATION OCCURS WHEN ALL PROCESSING IS FINISHED

A common assumption about the relationship of cognitive processes and eye
movement control suggested in recent literature is that the period during
which the eyes are centered on some stimulus unit, such as a word or an
object, is the period of time during which it is processed (Just & Carpenter, 1980). Processing is assumed to start at the beginning of this
period and to terminate when the eyes are moved. Thus, the triggering event
for an eye movement is assumed to be the completion of processing of the
presently viewed information. However, this assumption has implications
which do not seem reasonable. These are raised by considering the response
time of the eyes.

In several studies we have masked the text or made other display changes at
a certain time after the onset of selected fixations. When we compare the
frequency distributions of the time durations of these fixations with simi-

lar frequency distributions for fixations in which no mask occurred, the
distributions are identical up to a point 80 to 100 msec following the onset
of the mask. At that point, however, there is a sharp dip in the distribution for the mask condition, followed by a large hump in the distribution at
a later time. This indicates that for most fixations that would normally
have terminated with a saccade more than 80 to 100 msec after the occurrence
of the mask, the time of initiation of that saccade was postponed by the
experimental manipulation. Fixations that would normally terminate prior to
that time were unaffected by the mask. Thus, it appears that the minimum
time it takes the eyes to respond to retinal stimulation is about 80 to 100
msec. This is in agreement with results from studies of the saccadic system
in which the target to which the eyes are to be moved is shifted prior to
the initiation of that movement (Becker & Jurgens, 1979).

The response time of the eyes can be divided into three periods: stimulus
transmission time, the time required for transmission of neural activity
from the retina to the visual cortex; response transmission time, the time
required for transmission of neural activity from the motor cortex to the
muscles of the eyes and their latency in beginning to respond; and the processing period intervening between these two. Russo (1978), after reviewing
the relevant neurological literature, estimated the stimulus transmission
time to be about 60 msec and the response transmission time to be about 30
msec, which is in reasonable agreement with our own estimate of the total
response time. Hence, it is assumed that the earliest point at which a
stable neural pattern representing the stimulus array is established in the
brain is approximately 60 msec following the beginning of each fixation.
Prior to that time, the brain is stimulated by the rapidly changing pattern
on the retina during the saccade. Likewise, there is a point 30 msec prior
to the initiation of a saccade at which the motor signal for that saccade
was sent. Brain activity following that point is assumed to occur too late
to have any affect on the time of initiation of that saccade.

An implication of this chronology is that if a command is sent to move the
eyes only after the processing of the presently available information is
complete, then the actual processing time would not be the period of a fixation, but rather would be about 80 to 100 msec less than this (e.g., about
140 msec for a 230 msec fixation.)Furthermore, following the point at which
a signal is sent to move the eyes, an average period of 125 msec would pass
prior to the stimulation of the brain by input from the next fixation,
assuming an average saccade duration of 35 msec. An information processing
system which is idle, waiting for new input, almost half the time, is certainly inefficient. It is unlikely that this is an appropriate description
of brain activity during reading or other visual tasks where the necessary
information is continuously present. Instead, the brain probably initiates
a saccade prior to the completion of processing associated with a particular
stimulus unit. Evidence in support of this conjecture will be presented
shortly.

SACCADE INITIATION BASED ON SOME INTERMEDIATE EVENT

We have argued that neither of the two extreme hypotheses accurately
describe the nature of the decision to move the eyes during reading; that
is, this decision is not based entirely on information obtained prior to the
present fixation, nor does it occur only following completion of all the
processing permitted by the presently available visual information. We
believe that the remaining alternative is closer to the correct position:

the decision to move the eyes is influenced by visual information available during the current fixation, but is made prior to the completion of the processing of that information. The challenge is to identify the basis on which this decision is made. What perceptual or cognitive event indicates that it is time to initiate a saccade? For purposes of present discussion, we will refer to this as the triggering event during the processing sequence (i.e., the processing event which triggers the onset of the next eye movement). We will discuss two examples of possible triggering events: failing to obtain needed visual information and completing word identification.

A left-to-right processing sequence during fixations. One proposal for the nature of the triggering event is based on the assumption that there is a left-to-right processing sequence during a fixation (McConkie, 1979). It was assumed that during the period of a fixation there is actually a left-to-right progression in the region of text being attended, dealing with successive units of some sort (e.g., letters, orthographic units, syllables or words). This attending of the text is in response to the needs of the language processing activities, providing visual information as required to support those activities. A saccade is then initiated when visual information is needed from a retinal region where acuity is too poor to provide that information. The triggering event in this case would be the failure of the visual system to provide data with sufficient clarity to support the choices required in lower level language decisions, although higher levels of processing may continue beyond that time.

This proposal has the implication that visual information is used from the regions to the left and central parts of the visual field prior to those from the right. Three studies have now sought evidence for this hypothesis, and all have failed (Blanchard, McConkie, Zola, & Wolverton, 1983; McConkie, Underwood, Zola, & Wolverton, 1983; and Slowiaczek & Rayner, cited in Rayner, 1983). Thus, at present there is no evidence for such a progression of attention during fixations in reading.

Saccade initiation at the completion of word identification. Another reasonable hypothesis is that there is some point during a fixation at which the processing of visually provided information is completed. At that point, the visual system is ready to deal with new input and the eyes are advanced, positioning them to deal with the next visual region. In the meantime, higher processing continues using the visually provided information to advance the understanding of the text. If this were the case, a reasonable candidate for the triggering event would be the completion of word identification. Evidence for this possibility comes from studies, mentioned earlier, which indicate that characteristics of words which are known to influence their ease of identification, such as their cultural frequency, also affect the time they are fixated during reading.

While this is an appealing possibility, recent results from work in our laboratory are not in harmony with it. In one study, as subjects were reading, there were certain fixations on which all letters to the left of the directly fixated letter, or all letters more than three to the right of it, had been replaced by other letters. Normal text returned during the following saccade. Thus, the study investigated the effect of having erroneous letters present at specific retinal regions on individual fixations during reading. We assumed that the presence of erroneous letters would disrupt the word identification process and that the difficulties produced would increase processing time. This, in turn, should delay the saccade onset if it is normally dependent on successful word identification, resulting in longer fixations when the errors are present (McConkie, Underwood, Wolverton, & Zola, 1983).

When the erroneous letters lay to the left of the fixated letter, the duration of that fixation was increased, as expected. However, further examination indicated that the increased fixation time occurred only in the 21% of the cases in which subjects made a regression on the following saccade. In the cases where the following saccade was a forward movement, there was no change in the fixation duration, as compared to fixations on which there were no errors. The following fixation was longer than normal, and the frequency of regressions on the following saccade was greatly increased, indicating that the erroneous letters had been perceived.

Even more striking were the results when the erroneous letters were in the right part of the visual field. Here we found no effect on the duration of the fixation on which the errors were present, although the following saccade was shortened. Again, effects on later fixations and saccades indicated that the subjects had perceived the errors.

In this study subjects encountered errors on certain fixations in the words they were reading, and these errors clearly disrupted their processing. Yet in most cases the duration of the fixation on which the errors occurred was not affected. These results do not support the hypothesis that the eyes are moved only after word identification occurs. If this were the case, we would expect to see a change in the frequency distribution of the durations of the fixations on which errors were present. The errors should create processing difficulties which would delay the following saccade. Such effects were found only under limited circumstances. In the great majority of cases only later fixations were affected.

What we seem to be dealing with, then, is a system in which the characteristics of the words being identified influence, at least sometimes, the time of initiating the following saccade when processing proceeds smoothly. However, when processing is disrupted by errors lying three or more letter positions to the right of the fixated letter, or in many cases by errors to the left of the fixated letter, then the decision to move the eyes is not postponed in order to provide additional time dealing with the problems arising. Rather, the eyes are advanced at the time they would have been without the errors. Then additional processing time is gained either by lengthening the following fixation or by later returning the eyes to the region where the problems had occurred.

There is another finding which relates to the hypothesis being considered here. Not only do the characteristics of the fixated word influence the duration of the fixation on it, as described earlier, but if the following saccade sends the eyes beyond the next word, then that next word also affects the fixation (Hogaboam, 1983; Kliegl, Olson, & Davidson, 1983).Thus, there is evidence that in cases where a saccade skips over a word, that word was identified during the prior fixation. This being the case, we might expect the system to require more processing time in the case where the following word was skipped than in the case where the eyes were sent to it, since an additional word was apparently read in the former case. This suggests that there should be a correlation between the duration of a fixation and the length of the following saccade. Neither of these predictions is correct, however. Skipping a word does not increase the prior fixation

duration (Hogaboam, 1983) and the correlations between fixation duration and
the length of the following saccade is near zero (Andriessen & deVoogd,
1973; Rayner & McConkie, 1976). Thus, another reasonable prediction from
the word identification hypothesis is not correct.

As yet, we have no direct support for the hypothesis that the completion of
word identification is the triggering event for a saccade. It will be
necessary to conduct studies designed more specifically to test this
hypothesis, but at the present time it does not look promising.

Further considerations. The data from another recent study were analyzed to
determine how long a fixation had to be in order to be influenced by erroneous letters lying to the left of the fixated letter (McConkie, Underwood,
Zola, & Wolverton, 1983). As in the study described earlier, on certain
fixations all letters to the left of the directly fixated letter were
replaced by erroneous letters. The frequency distribution of durations of
these fixations was compared to that of fixations on which there were no
errors. It was found that the distributions were identical in the lower
fixation duration range but separated in the 140-160 msec range. Thus, it
appears that a fixation must be at least 140 to 160 msec long in order to be
long enough to be influenced by the orthographic characteristics of the
text.

One implication of this result is that fixations which are shorter than 140
msec are too short for even the orthographic aspects of the text to have an
influence on when the saccade occurs. In one set of data of college students reading a 417 word passage describing the early history of Alaska, 10%
of the fixations were less than 140 msec and some were even shorter than the
response time of the eyes. In these cases the saccade initiation times must
be determined without input from the present fixation. Thus, for some
fixations, the triggering event has nothing to do with the processing of
information obtained on that fixation. However, this occurs on relatively
few instances.

In another study, as subjects read the text was masked with a line of X's
either 50, 80 or 120 msec after the onset of each fixation (Blanchard,
McConkie, Zola, & Wolverton, 1983). The mask lasted for 30 msec, and then
the text returned. There were certain word locations in the text that could
be occupied by either of two words which differed by only a single letter.
On fixations in the region of these words, the distinguishing letter was
changed following the mask, and then changed back during the following saccade. Thus, during the first part of each fixation one word was in that
word location, and during the latter part of the fixation the other word was
present. We wanted to find out whether subjects would perceive this change,
and, if they did not, whether they would report having seen the word present
at the beginning or end of the fixation.

The results indicated that subjects reported having seen only one of the two
words 65% of the time, and that in these cases, the fixation durations were
the same as those from control instances in which the text was masked but
the word was not changed, thus giving no evidence that the word change was
disruptive. Furthermore, in these cases subjects sometimes reported having
seen the word present at the beginning of the fixation, and sometimes the
word present at the end. One interpretation of the results is that the time
during the fixation at which the visual information from a word is utilized
in the reading process varies from fixation to fixation. This is referred

to as the <u>Variable Utilization Time Hypothesis</u>. It suggests that in reading there is a specific time during the fixation at which the information provided by a word is brought into play in the language processing, and that this time can occur at any time during the fixation when the information is needed. If this hypothesis is accurate, then much or even most of the variability in fixation times is variability in the amount of time that elapses prior to the utilization of the fixated word, rather than variability in the processing time of the word itself. This contrasts with the notion that processing during reading proceeds in stages time-locked to the beginning of each successive fixation, as with Just and Carpenter's (1980) initial stage, entitled Get Next Input (McConkie, 1983). While this does not clarify exactly what serves as the triggering event for a saccade, it may contribute to an understanding of why the time of that event is so variable.

CONCLUSIONS

At some point during a fixation a message is sent to the ocular muscles to initiate an eye movement which will cause the eyes to move to some new location in the display. The purpose of the present paper is to consider the nature of the processing event which triggers this message. Present research suggests the following statements.

1. The time of initiating a saccade can be influenced by visual information acquired on the fixation which terminates in that saccade. Cultural frequency and orthographic characteristics are two aspects of the text stimulus which can have this influence.

2. However, processing of the information acquired during a fixation is not necessarily completed by the time of the decision to move the eyes. If it were, the system would spend much of its time in waiting for the next visual input.

3. At least some characteristics of the stimulus can only influence the saccade initiation time if a saccade has not occurred by a certain time. For instance, the presence of orthographically irregular letter strings on a fixation only affect the durations of fixations lasting at least 140 msec; shorter fixations show no effect.

4. As a corollary to the point just made, some fixations appear to be too short to be affected by any visual information acquired during those fixations. In these instances, then, the initiation of the next saccade must be based entirely on information available prior to the fixation.

5. With the exception of the shortest fixations, the saccade initiation time is probably determined to some degree by information acquired on that fixation, but not by the completion of the processing of that information. What, then, is the intermediate processing event which is linked to moving the eyes? Present data do not support either of the two possibilities considered here: namely, that a left-to-right consideration of the text is completed, or that word identification is completed.

6. Part of the variability in the saccade onset time may be associated with the amount of processing which takes place before the information available from the present fixation is utilized.

These suggestions fall far short of specifying the cognitive events accompanying the initiation of a saccade. However, they do question several reasonable alternatives and place constraints on future theorizing.

Finally, it may seem reasonable to postulate simply that the eyes are moved when there is a shift of attention to a new region of text. While this may well be true, it still leaves open the original question, which would now be stated as: What is the cognitive event which indicates that it is time to shift attention to a new region, and hence initiate an eye movement?

It is our hope that the above discussion will stimulate research and theorizing on the nature of the triggering event for saccades in on-going tasks like reading. Further knowledge on this issue will increase our understanding of what fixation time data can tell us about processing time in these tasks.

REFERENCES

Andriessen, J. J., & deVoogd, A. H. Analysis of eye movement patterns in silent reading. *IPO Annual Progress Report*, 1973, *8*, 30-35.

Becker, W., & Jurgens, R. An analysis of the saccadic system by means of double step stimuli. *Vision Research*, 1979, *19*, 967-983.

Blanchard, H. E., McConkie, G. W., Zola, D., & Wolverton, G. S. *The time course of visual information utilization during fixations in reading*. Unpublished paper, 1983.

Frazier, L., & Rayner, K. Making and correcting errors during sentence comprehension: Eye movements in the analysis of structurally ambiguous sentences. *Cognitive Psychology*, 1982, *14*, 178-210.

Hogaboam, T. W. Reading patterns in eye movement data. In K. Rayner (Ed.), *Eye movements in reading: Perceptual and language processes*. New York: Academic Press, 1983.

Just, M. A., & Carpenter, P. A. A theory of reading: From eye fixations to comprehension. *Psychological Review*, 1980, *87*, 329-354.

Kliegl, R., Olson, R. K., & Davidson, B. J. On problems of unconfounding perceptual and language processes. In K. Rayner (Ed.), *Eye movements in reading: Perceptual and language processes*. New York: Academic Press, 1983.

Levin, H. *The eye-voice span*. Cambridge, MA: The MIT Press, 1979.

Levy-Schoen, A., & O'Regan, K. The control of eye movements in reading. In P. A. Kolers, M. E. Wrolstad, & H. Bouma (Eds.), *Processing of visible language*. New York: Plenum Press, 1979.

McConkie, G. W. On the role and control of eye movements in reading. In P. A. Kolers, M. E. Wrolstad, & H. Bouma (Eds.), *Processing of visible language*. New York: Plenum Press, 1979.

McConkie, G. W. Eye movements and perception during reading. In K. Rayner (Ed.), Eye movements in reading: Perceptual and language processes. New York: Academic Press, 1983.

McConkie, G. W., Underwood, N. R., Wolverton, G. S., & Zola, D. Encountering erroneous letters at different retinal locations during reading. Unpublished manuscript, University of Illinois, 1983.

McConkie, G. W., Underwood, N. R., Zola, D., & Wolverton, G. S. Some temporal characteristics of processing during in reading. Unpublished manuscript, University of Illinois, 1983.

Rayner, K. The perceptual span and peripheral cues in reading. Cognitive Psychology, 1975, 7, 65-81.

Rayner, K. Visual attention in reading: Eye movements reflect cognitive processing. Memory & Cognition, 1977, 4, 443-448.

Rayner, K. Eye movements in reading and information processing. Psychological Bulletin, 1978, 85, 616-660.

Rayner, K. The perceptual span and eye movement control during reading. In K. Rayner (Ed.), Eye movements in reading: Perceptual and language processes. New York: Academic Press, 1983.

Rayner, K., & McConkie, G. W. What guides a reader's eye movements? Vision Research, 1976, 16, 829-837.

Rayner, K., McConkie, G. W., & Ehrlich, S. Eye movements and integrating information across fixations. Journal of Experimental Psychology: Human Perception and Performance, 1978, 4, 529-544.

Russo, J. E. Adaptation of cognitive processes to the eye movement system. In J. W. Senders, D. F. Fisher, & R. A. Monty (Eds.), Eye movements and the higher psychological functions. Hillsdale, NJ: Erlbaum, 1978.

Shebilske, W. Reading eye movements from an information-processing point of view. In D. Massaro (Ed.), Understanding language. New York: Academic Press, 1975.

Wanat, S. Linguistic structure and visual attention in reading. Newark, DE: International Reading Association, 1971.

Zola, D. The effect of redundancy on the perception of words in reading (Tech. Rep. No. 216). Urbana: University of Illinois, Center for the Study of Reading, 1981.

Theoretical and Applied Aspects of Eye Movement Research
A.G. Gale and F. Johnson (Editors)
© Elsevier Science Pulishers B.V. (North-Holland), 1984

THE INFLUENCE OF PARAFOVEAL INFORMATION
IN A SIMPLE READING TASK

Graham D.J. Jennings and Geoffrey Underwood

Department of Psychology, University of Nottingham
Nottingham, England

This paper briefly reviews previous experimental
results bearing upon the issue of the integration
of parafoveal information across fixations in
reading and attention is drawn to the problem of
separating effects due to contextual constraint from
those due to information extracted from parafoveal
vision. An experiment is described which employs a
single fixation task to examine the influence of
these factors upon word identification. Prior
parafoveal exposure was found to facilitate
subsequent identification of fixated words
independently of a facilitative effect of contextual
constraint. Semantic context however did increase
the pertinence of related words in parafoveal vision
and it is argued that context has separate effects
upon the processing of words at fixation and those
in parafoveal vision.

INTRODUCTION

How is parafoveal vision ahead of fixation used to aid the extraction of
information during reading? A number of recent studies have employed
sophisticated eye movement techniques to investigate this issue. Ehrlich
and Rayner (1981) observed eye movements as subjects read passages
differing in the degree to which a particular word was predicted by the
preceding context. When the critical word was highly constrained then the
number of fixations on the critical word was reduced. This effect is
similar to the THE skipping reported by O'Regan (1979). If an item is
sufficiently constrained then it seems that enough information may be
available from parafoveal vision for the word to achieve semantic analysis
and thus be influential in altering the pattern of eye movements. However,
when the eyes did alight upon the critical word in this condition, fixation
durations were significantly shorter than those in the low constraint
passage. There is a degree of ambiguity about the interpretation of these
results. It is not clear whether the effects upon fixation duration are
due to the influence of contextual constraints upon the critical word when
it is being fixated, or to the availability of information from parafoveal
vision affected by the preceding context. Only the latter possibility
would require integration of information across fixations to influence
fixation durations on the critical word.

A similar difficulty is encountered in the interpretation of Kennedy's
(1978) results. Subjects in this experiment were instructed to read a

short series of passages as naturally as possible in order to understand
them. Each paragraph consisted of three sentences which were presented one
at a time on a CRT screen. In one condition a word in the first sentence
was a high associate of one of the words embedded in the final sentence,
this "priming" stimulus being replaced by an alternative word in the
control condition. Subjects' eye movements were monitored to determine
whether the critical word in the third sentence would be fixated more
quickly when the item had been primed. The eye movement data showed that
mean latency to the first fixation on the critical word after presentation
of the final sentence was 372 ms when the word had been primed and 482 ms
when there was no prime. This is an important result because in order to
fixate an item more quickly than would be expected, the meaning of this
item must have been appreciated prior to that fixation. That is, the word
must have been semantically processed whilst it was still in parafoveal
vision. Fixation duration on the target word, however, was longer when the
prime had been present in the first sentence. This is perhaps a surprising
result, but Kennedy accounts for the finding by arguing that extra
processing is necessary under these circumstances to accomplish richer
cross-referencing.

Rayner, McConkie and Ehrlich (1978) employed a simplified reading task to
demonstrate that some information from parafoveal vision can be integrated
across fixations. A letter string was displayed in parafoveal vision and
while subjects moved their eyes to fixate it, the stimulus was changed into
a word which was to be named. Naming latencies to this word were
facilitated if it was visually similar to the initially presented item
suggesting that visual information extracted while the stimulus was in the
parafovea had aided its identification. There was also some evidence to
suggest that if the initial item was displayed only 1 deg. visual angle
beyond fixation then semantic information was available to influence
subsequent word identification. The latter result however was not
replicated in a following study (Rayner, McConkie and Zola, 1980) which
suggested that the critical aspect of a word in parafoveal vision is the
initial few letters. There are some aspects of these studies, however,
which urge caution in the interpretation of their results. Paap and
Newsome (1981) and McClelland and O'Regan (1981) both note that in all of
these experiments the same words were repeatedly presented to each subject.
Furthermore, in some experiments, the first letter of the initially
presented string was the same as that of the named word on a large
proportion of these trials (up to 80% in Rayner et al., 1980, Experiment
6). This is an important bias in a study which claims to demonstrate
readers' reliance upon the initial letters of parafoveal words during
normal reading.

Paap and Newsome (1981) used a simulated eye movement procedure similar to
that with which Rayner et al. (1978) successfully replicated their own
results. They found no evidence that subjects were able to derive any
benefit from parafoveal presentations except when the subjects were highly
familiar with the stimulus set. McClelland and O'Regan (1981) using a
similar procedure also found an advantage for parafoveal cues which were
constrained in some way. This effect was observed both when the parafoveal
string was selected from a small population of alternatives and when it was
constrained by preceding sentence context. Both the Paap and Newsome (1981)
and the McClelland and O'Regan (1981) studies suggest that the
predictability of parafoveal items is influential in determining the extent
to which it may aid identification of the word on a subsequent fixation.

One inconsistency in the results of these investigations, however, is that
while McClelland and O'Regan always found an advantage when the initially
presented item was the same as the target word, even in the absence of any
constraining information, this was not the case in Paap and Newsome's study.
It is interesting therefore to note that in the latter case subjects were
processing another item while the parafoveal stimulus was being presented.
This is a departure from the designs of both McClelland and O'Regan and the
Rayner et al. studies, and is potentially an important one. It is very
possible that the processing of fixated items may influence the processing
of material in parafoveal vision. A further simplification of these
designs involves the presentation of the parafoveal string in isolation.
It could be that this results in processing of the parafoveal item that
does not occur in normal reading due to the increased complexity of the
visual array. An experiment by Underwood (1981), however, suggests that
this is not the case. A number of studies have shown that a single
parafoveal word can influence the processing of a semantically related item
presented to fixation at the same time (e.g. Bradshaw, 1974; Craddock,
1983; Underwood, 1976). Underwood (1981) extended this result to show an
effect of a related word embedded in a list of stimuli displayed to the
parafovea of vision. At presentation durations of both 50 and 200 ms a
single related word, whether embedded in a list of other words or a list of
non-words, produced a significant effect upon response latencies to the
fixated word. Whereas these results alone do not demonstrate that
parafoveal words influence reading behaviour, they do establish the
plausibility of the notion. The processing of fixated stimuli can be
influenced by the semantic qualities of non-fixated stimuli.

It seems then that information is available from parafoveal vision, and
that it may aid identification of a word on a subsequent fixation. A
single fixation task was employed to examine this possibility further with
respect to the integration of information derived from the parafovea and
that already extracted from earlier in the passage. The design was
modelled upon that of Kennedy (1978). Three critical moments can be
identified in the reading of Kennedy's sentences. First there is the
fixation in the first sentence upon the possible associate of the target
word. Secondly there is a fixation prior to that on the target in which
the critical word is in parafoveal vision, and finally there is the
fixation upon the target word itself. Each trial in the present
experiment consisted of three presentations corresponding to these three
fixations. Thus the presence of "contextual" information and parafoveal
information could be manipulated independently to investigate their
individual and joint contributions to the identification of a target word.

METHOD

Subjects

Twelve members of the Department of Psychology (7 males) acted as subjects.
All were right handed and had normal or corrected to normal vision.

Materials and Design

A trial in the experiment consisted of the presentation of three slides.
The final slide on each trial bore a single word (the target word) placed
to appear at fixation. The main feature of the design was the manipulation
of the relationship between the target word and the words appearing on the

two slides displayed immediately prior to each target. Two factors were
varied. The word on the first slide was either a high associate of the
target or a filler item matched with the associate for frequency (Kucera
and Francis, 1967), word length and number of syllables. This corresponds
to the presence or absence of a target associate in the first sentence of
Kennedy's paragraphs. A filler word was displayed to fixation on the
second slide of each trial. This slide also bore a letter string placed to
the right of the first such that its initial letter appeared 3.75 deg.
visual angle beyond fixation. The item projecting to parafoveal vision
could either be the target word or a pronounceable non-word. These factors
combine to give the four priming conditions: a related prime to fixation;
an identity prime to parafoveal vision; both the related prime and the
parafoveal prime; and a no-prime control.

The critical stimuli were 32 word pairs selected from the Postman and
Keppel (1970) word association norms. These were used together with 64
filler words in the construction of 32 sets of slides. These slide sets
were divided into 4 groups of 8 sets, the groups being matched as closely
as possible for associative frequency between the related pair (Postman and
Keppel, 1970), frequency (Kucera and Francis, 1967), word length and number
of syllables. For every subject each of these groups was assigned to one
of the four priming conditions, the groups being rotated through the design
across subjects. The slides were back projected on to a white screen using
a Kodak S-RA 2000 projector modified by Forth Instruments Ltd. for
tachistoscopic display. The fixation field, consisting of a high spatial
frequency random dot mask and a fixation cross to coincide with the centre
of the left-hand word, was displayed using a Kodak S-AV 2000 projector
mounted directly below the test field projector. Exposure of the fixation
field was terminated by operation of a Forth Instruments Ltd. pulse-
generator (model FI 272) which simultaneously exposed the test field and
started a Camden Instruments Ltd. timer (model 565). The fixation field
mask was re-presented upon offset of the test field. A microphone placed
in front of the subject was connected to an Electronics Developments Ltd.
voice key, operation of which stopped the timer. An R.D.P. Electronics
Multi-X was used to produce a brief tone as a warning signal immediately
prior to each slide presentation.

Procedure.

Subjects were instructed to fixate directly above the cross and were told
that words would be briefly projected on to the screen in that position
immediately after they heard a warning tone. The words were displayed for
75 ms and the presentation rate was one slide every 5 seconds. Having
been told to read aloud the words appearing above the cross as quickly as
possible, the subjects were given 5 practice presentations followed
immediately by the rest of the series.

RESULTS AND DISCUSSION

Errors were recorded on less than 1% of occasions and were removed from the
means and analysis. Mean naming latencies to the target words in each
condition are presented in table 1.

Table 1. Mean naming latencies (in ms) to the target in
          the four priming conditions.

| | | unattended identity prime | |
		present	absent
semantically related prime	present	505	515
	absent	519	543

A mixed model analysis of variance was run to examine the effects of
priming upon target naming latency. The factors were (i) subjects, (ii)
presence or absence of the associative prime and (iii) presence or absence
of the parafoveal identity prime. The analysis showed reliable main
effects of both the related prime, $F(1,11)=6.41$, $p < 0.05$, and the
parafoveal identity prime, $F(1,11)=6.04$, $p < 0.05$, but the interaction
between these two factors did not approach significance ($F < 1$). This
indicates that these sources of facilitation are independent, and more
specifically that contextual constraint does not increase the benefit
derived from parafoveal vision.

This is not to say that the presence of the associate did not increase the
pertinence of the parafoveal word. A second analysis examined naming
latencies to the word presented to fixation on the middle slide of each
trial. These words were unrelated to any other words in the experiment,
but on half of the trials the target word was presented parafoveally on
this slide. On half of these occasions a high associate of this word had
been presented to fixation on the preceding slide. The mean naming
latencies for each condition are presented in table 2.

Table 2. Mean naming latencies (in ms) to the fixated word
          on slide 2 in the four priming conditions.

| | | unattended identity prime | |
		present	absent
semantically related prime	present	531	525
	absent	508	524

An analysis of variance was run to examine naming latencies to these words
as a function of the type of letter string in the periphery and the
presence of the related word on the preceding slide. The analysis showed
no main effect of the type of parafoveal stimulus ($F < 1$), but there was a
significant influence of the presence of the associate, $F(1,11)=6.26$,
$p < 0.05$. This effect, however, is complicated by a significant
interaction between these two factors, $F(1,11)=5.81$, $p < 0.05$. Planned
comparison t-tests showed, as expected, that when a non-word was in the
periphery the presence of an associate of the target word had no effect
upon response latency ($t < 1$). When the target was present in parafoveal
vision however response times were slower if the related word had been
presented ($t=3.56$, $df=11$, $p < 0.01$).

Taken together then these results suggest that information extracted from parafoveal vision can aid identification of the word when it subsequently appears to fixation. Contextual information derived from previously processed words also facilitates naming, but the increased pertinence of a parafoveal word when preceded by a related word was not transferred across the simulated fixations. Although the present results alone do not establish that these mechanisms operate during normal reading the findings are consistent with those from studies employing free reading tasks. The findings also offer an interpretation of the influence of context upon eye movements. Contextual constraint, it is suggested, has two independent effects which influence different aspects of eye movement behaviour. First, it facilitates recognition of, and thus fixation duration upon, predictable words as does information available from parafoveal vision on prior fixations. Secondly, it increases the pertinence of predictable words in parafoveal vision thus affecting the influence these items have upon the onset (Kennedy, 1978) or length (O'Regan, 1979; Ehrlich and Rayner, 1981) of the following saccade.

This research was supported by a studentship to the first author from the Social Science Research Council and a Project Grant No. GRC/02259 to the second author from the Science and Engineering Research Council.

REFERENCES

1. Bradshaw, J.L., "Peripherally presented and unreported words may bias the perceived meaning of a centrally fixated homograph", Journal of Experimental Psychology 103, pp.1200-1202 (1974).

2. Craddock, K., Visual Attention, Unpublished PhD Thesis, Nottingham University (1983).

3. Ehrlich, S.F. and Rayner, K., "Contextual effects on word perception and eye movements during reading", Journal of Verbal Learning and Verbal Behaviour 20, pp.641-655 (1981).

4. Kennedy, A., "Reading sentences: some observations on the control of eye movements", in Strategies of information processing, ed. Underwood, G., Academic Press, London (1978).

5. Postman, L. and Keppel, G., Norms of word association, Academic Press, New York (1970).

6. Kucera, H. and Francis, W.N., Computational analysis of present-day American English, Brown University Press, Providence, R.I. (1967).

7. McClelland, J.L. and O'Regan, J.K., "Expectations increase the benefit derived from parafoveal visual information in reading words aloud", Journal of Experimental Psychology: Human Perception and Performance 7, pp.634-644 (1981).

8. O'Regan, K., "Saccade size control in reading: Evidence for the linguistic control hypothesis", Perception and Psychophysics 25, pp.501-509 (1979).

9. Paap, K.R. and Newsome, S.L., "Parafoveal information is not sufficient to produce semantic or visual priming", Perception and Psychophysics 39, pp.457-466 (1981)

10. Rayner, K., McConkie, G.W. and Ehrlich, S., "Eye movements and integrating information across fixations", Journal of Experimental Psychology: Human Perception and Performance 4, pp.529-544 (1978).

11. Rayner, K., McConkie, G.W. and Zola, D., "Integrating information across eye movements", Cognitive Psychology 12, pp.206-226 (1980).

12. Underwood, G., "Semantic interference from unattended printed words", British Journal of Psychology, 67, pp.327-338 (1976).

13. Underwood, G., "Lexical recognition of embedded unattended words: Some implications for reading processes", Acta Psychologica 6, pp.267-283 (1981).

Theoretical and Applied Aspects of Eye Movement Research
A.G. Gale and F. Johnson (Editors)
© Elsevier Science Pulishers B.V. (North-Holland), 1984

READING WITHOUT EYE MOVEMENTS

Alan Kennedy and Wayne S. Murray

Psychology Department
University of Dundee
Dundee, Scotland

Subjects were asked to read sentences containing local
structural ambiguity. The material was presented word-
by-word, under subject control, in three different display
modes.   In the Cumulative condition words previously read
remained visible.   In the Sequential condition only one
word was visible at a time but was presented in its appropriate spatial location.   In the Central condition all words
appeared in the same physical location.   The results suggest that subjects may be less sensitive to ambiguity if
prior text is not available for inspection, and that removing information on differential spatial location may influence comprehension.

It is well established that the time spent inspecting words varies for normal
readers as a function of several text properties.   It is much less clearly
established that such variation in inspection time is *necessary*.   Similarly,
in normal reading, saccade extent varies systematically as a function of
properties of the word about to be fixated.   It is also possible to question
the degree to which this variation is necessary for adequate comprehension.
In fact, the possibility that readers can fully comprehend text without
making eye movements at all has been the subject of recent debate (Juola,
Ward & McNamara, 1982; Just, Carpenter & Wooley, 1982).   The issue turns
on the fact that it is technically possible to present text to the reader,
word-by-word, in such a way that each word occupies the same physical
location.   Reading in this case may take place without the execution of eye
movements.   This paper addresses the question of whether in such a situation
readers are sensitive to the structure (and hence the meaning) of what they
read.   The question is far from being settled, despite a considerable amount
of experimentation, largely because different studies have used different
methods of assessing comprehension and have manipulated different variables.

It is possible to identify two major variables which may influence performance when text is presented word-by-word.   The first relevant factor is
whether the display time for individual words is under subject- or
experimenter-control.   Obviously, if presentation is under the control of
the subject, words will be displayed for variable amounts of time.   Such
variation may be random, or may relate in an interesting fashion to the
processing demands which are placed on the reader by the text.   Under
experimenter-control, words are generally, but not necessarily, presented for
fixed periods of time.   These may vary from the extremely brief intervals
used in the RSVP procedure to times equivalent to the average fixation durations found in normal reading.   The second factor relates to the reading

material itself.   Clearly, different texts impose different demands on the
reader.   The fact that some may be processed without eye movements does not
necessarily imply that this will be true for all types of reading material.

It is important to note that single-word presentation in a fixed location
differs from normal reading in two major respects.   First, words in normal
text can be mapped on to distinct spatial coordinates by the reader.   In
contrast, the single-word presentation mode mimics the presentation of words
in speech, where only temporal coding is possible.   When reading normal
text, words can be scanned in different temporal sequences, yet perceived
appropriately as a consequence of their spatial organisation.   When faced
with a single-position display the possibility of such spatial organisation
is removed.   The consequences of this for comprehension are uncertain.
Secondly, the reader of normal text is free to engage in reinspection.
What has been read before is available to be read again.   This possibility
is denied the reader presented with a single-word display, and again, the
consequences of this for adequate text-processing have yet to be fully
ascertained.   We have sought to assess the influence of these two factors
by using different types of single-word display to present text to the
reader.   We employed three presentation modes.   (1) CENTRAL - each word
of the text appears singly in the same physical location on a computer-
controlled display.   (2) SEQUENTIAL - words again appear singly, but
under this condition they are located in appropriate positions on the
display.   The effect, from the reader's point of view, is of a window, one
word wide, moving from left to right across the sentence.   (3) CUMULATIVE -
this condition is similar to sequential, in that words appear in their
appropriate spatial locations, but in this case previously presented words
remain on view.   A number of measures have been used to assess the effects
of different presentation procedures.   However, if text is presented under
subject-control, viewing-time itself should provide a satisfactory measure
of performance.   The degree to which the reader varies presentation-rate
can be used to assess sensitivity to properties of the material being read.

The question of the necessity of eye movements in reading is of considerable
theoretical consequence.   Two issues arise.   The first relates to the fate
of previously-inspected text.   Several studies (Carpenter & Just, 1977;
Ehrlich, 1983; Ehrlich & Rayner, 1983; Frazier & Rayner, 1982; Kennedy,
1978) suggest that certain types of reading material may induce the reader
to make reinspections.   Such regressive eye movements appear to serve a
function in resolving ambiguity and in determining the antecedents of
anaphor.   Obviously, for reinspection to take place, the text must be displayed in a spatially extended format and must be available to the reader.
Both Sequential and Cumulative display modes provide this necessary spatial
extension, but only under the Cumulative display is prior text actually
available for inspection.   To the extent that reinspection is *necessary* for
adequate comprehension, we would expect an advantage to be shown for the
Cumulative method of display.   The contrast between performance under these
two presentation modes, therefore, allows an assessment of the effects of
the availability of prior text on reading.

The second theoretical issue relates to the question of spatial extension
itself, quite apart from the visibility of prior text.   The fact that words
appear in distinct spatial locations in normal reading may be of functional
significance.   There is evidence, for example, that readers preserve information as to the location of words for some period of time (Christie & Just,
1976; Kennedy, 1983).   This locative code serves to allow the reader to

direct reinspections to appropriate informative regions of text.   Thus, if
differential spatial location cannot be encoded as text is processed, the
effect may be deleterious, whether or not reinspections are in fact made.
While in both Central and Sequential display modes only one word of the
sentence is visible at any one time, it is only in the Sequential condition
that differential spatial information is available as each word is encoded.
Thus, a contrast between these two conditions provides an appropriate means
for determining the effects of the availability of differential spatial
coding *per se* on reading performance.

In summary, we have carried out an experiment which has assessed the influence of prior text (by contrasting Sequential and Cumulative modes) and the
influence of spatial code (by contrasting Central and Sequential modes) on
reading, using a measure of on-line reading-time to assess performance.   We
examined these two theoretical questions by making use of a type of sentence
which was already known to generate patterns of reinspection under normal
reading conditions.   The materials were based on those used by Frazier and
Rayner (1982).   These were sentences containing local structural ambiguity
such as 1(a) and 1(b).

1(a) *While the teacher was reading the book it fell off the table.*

1(b) *While the teacher was reading the book fell off the table.*

If sentences like these are presented one word at a time they are temporarily
ambiguous.   Only when the reader encounters the word *it* or *fell* is it
possible to assign an unambiguous structural description to the sentence.
A common intuition is that sentences like 1(a) are more easily understood
than 1(b).   Frazier and Fodor (1978) propose a theory which accounts for
this intuition.   It is assumed that readers adopt a *late closure* strategy -
attaching incoming words, wherever possible, to the clause or phrase
currently being analysed.   This strategy leads to the reader assigning the
wrong structure temporarily in the case of early-closure sentences such as
1(b).   (It will be noted that in normal usage such ambiguity is often,
though not always, resolved by the use of punctuation.   We attempted to
avoid sentences where such punctuation appears to be *demanded.*)   The theory
also predicts that the longer the ambiguous phrase, the more difficulty the
reader will have in dealing with the eventual violation of the assigned
structure.   Such a prediction arises whether or not it is assumed that
readers normally compute alternative structures simultaneously for such
sentences or adopt a single preferred analysis which must later be revised.
In either case a longer ambiguous phrase will impose a greater load on memory.
Thus it is predicted that sentences such as 1(d) should cause even more
problems than 1(b) when compared to an appropriate control such as 1(c).

1(c) *While the teacher was reading the huge old book it fell off the table.*

1(d) *While the teacher was reading the huge old book fell off the table.*

In presenting sentences of this kind, together with appropriate filler material, we were concerned with the degree to which the reader's inspection-times
would reflect their awareness of structural ambiguity.   Our assumption here
is that this provides a sensitive index of comprehension, since appropriate
syntactic analysis must be completed before the sentences can be properly
understood.   Different groups of subjects were asked to read the materials

under one of the three presentation conditions.   Subjects were told to
pace through the words of each sentence as quickly as possible consistent
with understanding it.   The sentence was then followed by a question,
presented as a full line of text on the display.   Subjects were asked to
respond yes or no to these questions as quickly and accurately as possible.
The questions served as an additional incentive for subjects to read the
stimulus materials carefully.

Two predictions arise from the study.   First, if prior text must be available for readers to carry out reanalysis, a greater sensitivity to closure

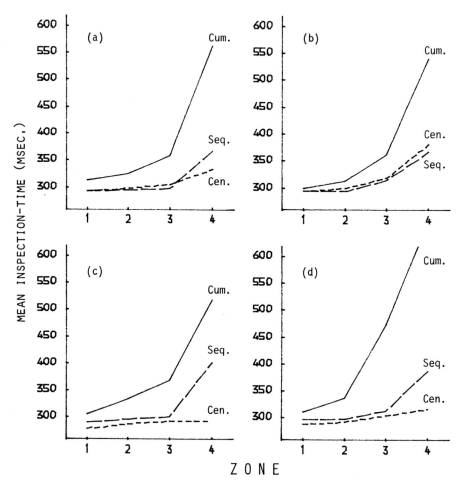

FIGURE 1:   Mean inspection-time for words in four zones of (a) short
late-closure; (b) short early-closure; (c) long late-closure;   and
(d) long early-closure sentences, under the Cumulative, Sequential and
Central presentation conditions.

should be evident under the Cumulative than under the Sequential mode of display. Further, this presentation-mode effect should be most apparent in sentences with long ambiguous regions. The second prediction relates to the availability of differential spatial information. If the reader suffers a disadvantage as a consequence of an inability to code locative information, this should be apparent in poorer sensitivity to structural ambiguity under Central presentation than under Sequential.

A summary of the results of this study is shown in Figure 1. The four zones of the sentence illustrated comprised: (1) the initial, unambiguous, region (e.g. *While the teacher was reading*); (2) the temporarily ambiguous noun phrase (*the huge old book*); (3) the disambiguating region, excluding the final word (*it fell off the*); and (4) the final word (*table*). These results are discussed in greater detail elsewhere (Kennedy & Murray, 1983), but may be summarised quite simply. Comparing the Sequential and Cumulative presentation modes, highly significant differences in inspection-time are found in different regions of the sentences under both conditions. However, sensitivity to the effects of early closure (which is indexed by an increase in inspection-time in the third zone) was found *only* in the Cumulative condition, and even there, only for sentences with long ambiguous regions. These differences between Sequential and Cumulative modes clearly suggest that the presence of prior text has a significant effect on the reader's mode of inspection. It would appear that reinspections serve a function: readers may be unable to engage in successful reanalysis if they cannot be made. Further, it appears that such reanalysis takes place at the point where the ambiguity is resolved, rather than being deferred until reading of the sentence has been completed. In fact, the data suggest that under the Cumulative condition readers may actually be aware of having *entered* a temporarily ambiguous zone: the divergence of Cumulative and Sequential data at zone 2, apparent in Figure 1, is statistically reliable.

It is interesting, however, that the data show little apparent sensitivity to closure for short sentences. We conclude from this either that subjects may be able to sustain two simultaneous parsings for short periods, or that there is little cost associated with revising the structural interpretation of relatively short phrases.

Turning to the question of the availability of differential spatial information, the outcome of the study is more equivocal. There are small, but highly significant, differences in inspection-time between Sequential and Central presentation modes, but they are confined to data derived from the final word. Here we find a significant divergence between the two presentation modes, but again only in the case of sentences with long ambiguous regions. The most that can be said is that if these long final inspection-times index sentence wrap-up processes or reanalysis, then such strategies differ following Sequential and Central modes of presentation. Direct records of eye movements will be needed to ascertain if subjects actually engage in 'reinspections' at this point.

While this study did not provide particularly clear evidence in support of the second hypothesis, other results from our laboratory do suggest more clear-cut differences in inspection-time under Central and Sequential presentation conditions. If subjects are presented with material containing more potent structural ambiguity, such as 'garden-path' sentences of the form, *the students heard the brilliant lecture was cancelled*, clear differences become apparent between Sequential and Central conditions.

Inspection-times under the Sequential condition are strongly influenced by
sentence-structure, whereas no effects of structure are apparent under
Central presentation.

This study also provided one further finding of relevance to the present
discussion.   The results showed instances, for all three display modes,
where inspection-time did not increase as the subject paced through the
sentence.   In fact, times actually decreased in some zones for certain
display modes.   Thus, it appears that the overall increase in inspection-
time through the sentence, apparent in Figure 1, is not simply an artifact
of the procedures employed, but reflects real changes in processing load.

In summary, our data suggest that reading is possible without eye movements,
but that the reader may be unable to carry out a complete analysis of
certain types of sentence under conditions where previously-read text is
not visible or where differential spatial information cannot be encoded.

REFERENCES

Carpenter, P.A. and Just, M.A., Reading comprehension as the eyes see it.
in M.A. Just and P.A. Carpenter (Eds.), Cognitive processes in
comprehension (Hillsdale, New Jersey: Erlbaum, 1977).

Christie, J. and Just, M.A., Remembering the location and content of sentences in a prose passage, Journal of Educational Psychology 68 (1976) 702-
710.

Ehrlich, K., Eye movements in pronoun assignment: a study of sentence
integration, in K. Rayner (Ed.), Eye movements in reading: perceptual and
language processes (New York: Academic Press, 1983).

Ehrlich, K. and Rayner, K., Pronoun assignment and semantic integration
during reading: eye movements and immediacy of processing, Journal of
Verbal Learning and Verbal Behavior 22 (1983) 75-87.

Frazier, L. and Fodor, J.D., The sausage machine: a new two-stage model
of the parser, Cognition 6 (1978) 291-325.

Frazier, L. and Rayner, K., Making and correcting errors during sentence
comprehension: eye movements in the analysis of structurally ambiguous
sentences, Cognitive Psychology 14 (1982) 178-210.

Juola, J.F., Ward, N.J. and McNamara, T., Visual search and reading of
rapid serial presentations of letter strings, words, and text, Journal of
Experimental Psychology: General 111 (1982) 208-227.

Just, M.A, Carpenter, P.A. and Wooley, J.D., Paradigms and processes in
reading comprehension, Journal of Experimental Psychology: General 111
(1982) 228-238.

Kennedy, A., Eye movements and the integration of semantic information
during reading, in M.M. Gruneberg, R.N. Sykes and P.E. Morris (Eds.),
Practical aspects of memory (London: Academic Press, 1978).

Kennedy, A., On looking into space, in K. Rayner (Ed.), Eye movements in reading: perceptual and language processes. (New York: Academic Press, 1983).

Kennedy, A. and Murray, W.S., Inspection times for words in syntactically ambiguous sentences under three presentation conditions. MS submitted for publication 1983.

Theoretical and Applied Aspects of Eye Movement Research
A.G. Gale and F. Johnson (Editors)
© Elsevier Science Pulishers B.V. (North-Holland), 1984

Fixation Time on Anaphoric Pronouns Decreases
with Congruity of Reference

John S. Kerr and Geoffrey Underwood

Department of Psychology,
Nottingham University,
Nottingham,
England.

Subjects' eye movements were recorded while they read
sets of 3 contextually related sentences from a slide
projector. The subject of the first sentence was a
ROLE (eg Engineer), this was followed by a filler
sentence, and the third sentence contained an anaphoric
PRONOUN (He/She). Each set either matched the sex
stereotype ("He" referring to "engineer"), did not
match ("She" referring to "surgeon") or was neutral
(reference to a "gender-unconstrained" role like "patient" or "student"). Measures were taken of fixation
duration and inspection duration upon the pronoun in
the third sentence of each set of sentences, together
with total reading time. All three measures were longer
for sentences containing incongruous (non-matching)
pronouns. These results suggest that pragmatic information is used as part of the process of comprehension.

INTRODUCTION

The semantic/pragmatic division is usually expressed, when considering
comprehension, as a two-process system: the first a "process aimed at
extracting the literal meaning of a sentence via a sequential parsing
system, while the second process determines the significance of the
analysed meaning in terms of the context in which the sentence was
encountered."(Johnson-Laird 1977). An example given by Miller (1977)
illustrates this:

"The Smiths saw the Rocky Mountains while they were flying to California."

This sentence would be ambiguous without the knowledge that mountains do
not fly; this knowledge is practical as opposed to lexical (semantic).

Some of those investigating comprehension processes consider this distinction artificial: Johnson-Laird (1977), Sanford and Garrod (1981) and others
have pointed out that comprehension should be considered as a unified
process which uses frame-like structures (or scenarios) as a model of the
discourse. In the strongest form of this view reading proceeds by mapping
the text on to the mental model. In this way pragmatic knowledge becomes an
integral part of understanding, occurring as part of the reading process

and not as some "final stage".

The present experiment examines the extent to which practical knowledge is used in the resolution of pronominal reference; specifically reference involving gender, ie using he/she (the referent) to refer to a previously mentioned entity (the antecedent), in this case a role such as "nurse", "student" or "engineer".

This experiment is partially based on work by Simmons (1981): using sentence reading time as an indication of processing time, she showed that it took significantly longer to resolve a reference by a pronoun whose gender was "atypical" of its antecedent, eg "The surgeon.....She.....". Simmons explained this effect in terms of a slower "referential mapping" by the parser ie the semantic feature "she" is somehow less available to the reader than "he" is, in the case of surgeon.

According to the Semantic Marker model (e.g. Clark 1975), reference resolution proceeds by, on reading an antecedent, reducing it to semantic components eg:

> Man(x)   male(x)
>   adult(x)
>   singular(x).

When a possible referent is read this is decomposed in the same way:

> He(x)   male(x)
>   singular(x).

Resolution is made in this case by matching the features of He with those of Man.

To account for any extra processing time observed in incongruous reference this model would have to postulate eg a male feature for nurse which was less accessible than the female feature:

> Nurse(x)   female(x)
>   singular(x)
>   adult(x)
>   ....
>   ....
>   male(x).

In this case the theory would have imported "real-world" information to an extent which would make it indistinguishable, in terms of prediction, from a frame theory.

Sanford and Garrod (1981) involve pragmatic knowledge in a unified comprehension process which utilises what they call a scenario. When a noun phrase such as "The nurse" is encountered, not only do semantic features become available but also a suitable scenario is accessed. A scenario is a frame-like structure which holds information about what to expect in a given situation. This includes information about characters, props, relationships etc., in a similar way to Schank's scripts (Schank and Abelson 1977). Part of the scenario activated by "nurse" will hold the information "female", since most nurses are female. Therefore when the reader encounters the pronoun "she" there is little processing load since the information already exists as a default value in the scenario: on reading "he" this is "new" (as she is "given") and so takes longer to process, resulting in a longer reading time. In Minsky's frame theory it is the difference between walking into your bedroom and seeing a bed or a plough.

The present experiment was designed to distinguish between the two models described above, to replicate Simmons' finding of longer processing times for "incongruous" reference resolution, and to investigate the processing of reference to a "gender unconstrained" role such as pupil or patient.

The scenario model predicts that processing time for a "gender-neutral" reference (Student.....S/He) will be longer than "instantiation" time (Nurse.....She), since the information is neither "new" nor as greatly predicted by the scenario, but shorter than that for the incongruous case where the information is opposite to that expected, and must be actively integrated with the current scenario (Nurse.....He).

It was decided that eye-movement recording would be a more sensitive technique than the sentence reading time used hitherto, since the time spent processing the pronoun only could be ascertained from the fixation time (Just and Carpenter 1980).

METHOD

Design and Materials

Materials were taken from a previous experiment conducted in 1981, and were similar to those used by Simmons. They consisted of 17 sets of 3 sentences of the form (1) The ROLE...verb phrase. (2) Filler sentence related to sentence (1). And (3) He/She...verb phrase. For example:

The surgeon examined the X-rays.
It was a bad case.
He would have to operate immediately.

Of the 17 sets, 6 contained gender ambiguous roles, ie roles which were thought not to be strongly predictive of gender (eg "patient", "pupil"). Half of the pronouns in these sets were male, and half were female. The remaining sets contained roles which do have a strongly associated gender ("engineer", "nurse"). These were divided into two groups: congruous sets, in which the pronoun agreed with the antecedent's expected gender; and incongruous sets in which it did not (nurse....she; nurse....he).These were reversed for half the subjects. Thus all subjects did all conditions, and materials were rotated for half the subjects.

The dependent variables were:
1 The length of the first fixation on the pronoun.

2 The "inspection" time (the length of the first fixation on the pronoun plus any regressive fixations)

3 The reading time for the sentence containing the pronoun.

4 The number of regressive saccades to the pronoun.

Apparatus

Horizontal movements of the left eye were recorded using a device developed
in the Psychology Department at Nottingham University, which monitors the
position of the iris-sclera border by means of reflected infra-red (Wilkinson 1976). The device consists of emitters, a simple optical arrangement
for focus, and detectors all mounted on an adjustable band placed on the
subjects head. A chin rest was used to reduce head movement. Output from
the eye movement device operated a Watanabe WTR 211 pen recorder running at
10mm/sec. Overall accuracy is approximately +/-20' of arc in a range of 60
degrees.

Stimuli were mounted on slides and presented on a Bell and Howell Auto-
Focus back projector modified so that when subjects pressed a key this
advanced the projector and sent a pulse to the pen recorder enabling easy
analysis of sentence reading time.

All the apparatus was situated in a sound-proof cubicle.

Procedure

Subjects were seated 40 cm in front of the projector with the chinrest
adjusted to a height such that the subject had to look slightly upwards to
see the stimuli. After calibration subjects were instructed that they were
to read silently, as they would normally, sets of three related sentences.
They were told to go at their own pace, pressing the button for the next
slide when they were ready. The end of the experiment was signalled by a
marker slide. Ss were then debriefed. The average time to run a subject,
including setting up, was about eight minutes.

Subjects

Sixteen subjects with adequate uncorrected vision were randomly assigned to
one of the two parts of the experiment.

RESULTS

	Relationship of Pronoun to Referent		
	Ambiguous	Incongruous	Congruous
Duration of First Fixation (ms)	232	261	231
Inspection Duration (ms)	255	319	250
Sentence Reading Time (ms)	2261	2406	2239
No. of Regressives	8	16	7

Table 1. Mean values for each dependent variable in each condition.

The mean results for each of the dependent variables in each of the
conditions are shown in Table 1. The means for each subject were calculated
and an analysis of variance was carried out on these data with subjects
treated as a random factor. In the incongruous condition, subjects spent
significantly more time on the first fixation of the pronoun, $F(2,26)=3.49$,
$p<0.05$, also that, including regressive fixations (inspection), they spent
a longer total time fixating the pronoun, $F(2,26)=8.31$, $p<0.002$, and a
longer time reading the whole sentence, $F(2,26)=3.8$, $p<0.04$; all compared
to the other two groups. Comparing the ambiguous and congruous groups,
there was a slight, non-significant effect of congruity in the expected
direction. As can be seen from Table 1, the number of regressive fixations
was small throughout but double for the incongruous condition ($X = NS$).
These results indicate that it takes significantly longer to process an
"unexpected" pronoun.

DISCUSSION

The present results support the hypothesis that pragmatic information is
used in the resolution of pronominal reference, and therefore in reading.
The results are consistent with Sanford and Garrod's (1981) scenario model
of comprehension, and with the results of the experiment by Simmons (1981)
mentioned above.

Gender is an important factor when making a pronoun assignment: Caramazza
et al.(1977), investigating implicit causality, a property of verbs, found
that, in sentence pairs such as
a) Ronald scolded Joe because he was annoying.

b) Barry scolded Pete because he was annoyed.

subjects prefer to assign "he" in a) to Joe. In b) assignment is in the
opposite (contrary to implicit causality) direction, and thus a longer
reading time is found, and more errors. The finding relevant to this
experiment is that if the gender of one of the characters is changed, eg
"Susan scolded Mark", reading time is significantly reduced, along with
errors.

Ehrlich (1979) found that the implicit causality effect (a syntactic
property) was less important than pragmatic information, which itself
became unnecessary if gender was unambiguous ie other pragmatic information
is used only if gender does not yield a unique referent.

It is therefore likely that gender information is readily available to the
reader (foregrounded, see Chafe, 1976), and if this is the case then a
frame type processor would be more effective since in the usual case it
needs only instantiation, with recourse to bridging only when an unexpected
entity is encountered. It would not be economical for a reader to wait
until a pronoun is encountered, and then use this "new" information to make
an assignment.

A model which predicts expectancy in reading must give an account of how this may reduce reading time. The best theoretical account of a possible mechanism is that of Morton (1970) who would say that an expected word has a lower threshold for recognition. This also accounts for simpler lexical priming effects (Stanovich and West 1983) which have been regularly demonstrated in lexical decision, naming, phoneme-monitoring and other tasks (eg Meyer and Schvaneveldt 1971) but are only now being established in tasks closer to normal reading (eg Reder 1983; Ehrlich and Rayner 1981), usually by use of eye movement recording techniques.

It is unclear to what extent lexical priming is used in everyday reading: Kennedy (1978) failed to find a priming effect using a reading task and eye movement recording. However, recently, we have demonstrated a priming effect using materials similar to Kennedy, and which also showed priming in naming and lexical decision tasks. For example:

One of the guests spilt drink on the carpet.
It made a bad stain.
A rug was bought to cover it.

In this set of sentences, a shorter fixation time on the target "rug" was recorded when the high associate synonym "carpet" was included in the first sentence, compared to the control condition where "carpet" was replaced by "sofa". Similarly, in naming and lexical decision tasks, reaction time to words like "rug" was significantly reduced when preceded by a prime (carpet), as compared to a control (sofa). This demonstrates that materials used in single word priming tasks do show lexical priming effects in sentence processing, and may serve both to increase reading speed and to aid in the integration of word meanings during comprehension.

The results of the present experiment show no significant difference in processing time for the ambiguous and congruous conditions, although the means are in the predicted direction (ambiguous > congruous). However it may be that the roles used in this condition are in fact predictive of one sex only and since equal numbers of each gender pronoun were used this resulted in a mixture of congruous and incongruous reference causing the appearance of an intermediate processing time when the means were calculated.

In conclusion this experiment demonstrates that pragmatic information is available to, and used by, the reader as part of the process of comprehension, and not merely as a final stage. This supports the findings of Simmons (1981), and provides evidence in support of the scenario account of text comprehension.

REFERENCES

Caramazza, A., Grober, E.H., Garvey, C., and Yates, J.B., Comprehension of anaphoric pronouns, Journal of Verbal Learning and Verbal Behaviour, 16 (1977) 601-9.

Chafe, W., Givenness, contrastiveness, definiteness, subjects, topics and point of view, in: Li, C. N. (ed.), Subject and Topic (Academic Press, New York, 1976).

Clark. H. H., Bridging, in: Schank R., and Nash-Webber, B., (Eds.),

Theoretical Issues in Natural Language Processing, Proceedings of a conference at the Massachusetts Institute of Technology, (June 1975).

Ehrlich, K., The comprehension of pronouns, Quarterly Journal of Experimental Psychology, 32 (1980) 247-255.

Ehrlich, S. F., and Rayner, K., Contextual effects on word perception and eye movements during reading, Journal of Verbal Learning and Verbal Behaviour, 20 (1981) 641-655.

Johnson-Laird, P. N., Psycholinguistics without linguistics, in: Sutherland, N. S. (ed.), Tutorial Essays in Psychology Vol. 1 (Erlbaum, New Jersey, 1977).

Just, M. A., and Carpenter, P. A., A theory of reading: from eye fixations to comprehension, Psychological Review, 87 (1980) 329-354.

Kennedy, A., Reading sentences: Some observations on the control of eye movements, in: Underwood, G. (ed), Strategies of Information Processing, (Academic Press, London, 1978).

Meyer, D. E., and Schvaneveldt, R. W., Facilitation in recognising pairs of words: Evidence of a dependence between retrieval operations, Journal of Experimental Psychology, 90 (1971) 227-234.

Miller, G. A., Practical and lexical knowledge, in: Johnson-Laird, P. N. and Wason, P. C. (eds.), Thinking: Readings in Cognitive Science, (C.U.P., Cambridge, 1977).

Morton, J., A functional model for memory, in: Norman, D. A. (ed.), Models of Human Memory, (Academic Press, New York, 1970).

Reder, L. M., What kind of pitcher can a catcher fill? Effects of priming in sentence comprehension, Journal of Verbal Learning and Verbal Behaviour, 22 (1983) 189-202.

Sanford, A. J., and Garrod, S. C., Understanding Written Language, (Wiley & Sons, New York, 1981).

Schank, R., and Abelson, R., Scripts, plans, goals and understanding: An enquiry into human knowledge structures, (Erlbaum, New Jersey, 1977).

Simmons, J., The effect of gender on reference resolution, Undergraduate thesis, Dept. Psych., Glasgow Univ. (1981).

Stanovich, K. E., and West, R. F., On priming by a sentence context, Journal of Experimental Psychology: General, 112 (1983) 1-36.

Wilkinson, H. P., Wide range eye and head movement monitor, Quarterly Journal of Experimental Psychology, 28 (1976) 123-124.

Acknowledgements

This work was supported by a research studentship to the first author from the Social Science Research Council, and Project Grant No.GRC/02259 to the second author from the Science and Engineering Research Council. We are grateful to C. I. Howarth, A. Kennedy and E. A. Maylor for their comments on this work, and to H. P. Wilkinson for his technical expertise.

Theoretical and Applied Aspects of Eye Movement Research
A.G. Gale and F. Johnson (Editors)
© Elsevier Science Pulishers B.V. (North-Holland), 1984

EYE MOVEMENTS DURING COMPREHENSION OF PRONOUNS

W. Vonk

Max-Planck-Institut für Psycholinguistik
and
University of Nijmegen
Nijmegen, The Netherlands

This study deals with inference processes in anaphora
resolution. Subjects had to read sentences such as
Alex lied to Andy because he smelled trouble and to
indicate the referent of the pronoun. The gender of
the pronoun and of the referents was varied so as to
create conditions with and without a gender cue.
Additionally, the conjunction was varied. The main
data are the eye fixation durations on the conjunction,
on the pronoun and on the verbphrase of the second
clause in the first pass through the sentences as well
as the number of fixations in later passes through the
sentences. The results give evidence for the immediacy
of processing as far as the lexical properties of the
pronoun is concerned and can be interpreted in terms
of a rational selection of information in reading sentences.

In order to understand a text, a reader has to construct a coherent representation: he/she has to integrate the incoming information with the current
mental representation of the text that is read so far. There are several
means in the language that signal the necessity to make connections. Anaphora is one of them. In encountering a pronoun, for instance, the reader has
to look for the appropriate antecedent.

Several factors are reported in the literature that may influence pronoun
assignment. One such factor is the position of the pronoun in the surface
structure of the sentence. For instance, the pronoun in the second clause
of sentence (1)
    (1) John went to Pete and he asked him ....
is preferably identified with the antecedent that has the same grammatical
function in the first clause. *He* in sentence (1) refers to John, *him* to
Pete. This strategy is called the parallel function strategy (Sheldon, 1974;
Grober, Beardsley, & Caramazza, 1978). The use of contrastive stress can
change the assignments.

A second factor is related to the verb. Caramazza c.s. introduced the notion
of "implicit causality" to refer to a property of the verb that determines
the selection of the referent (Garvey & Caramazza, 1974; Garvey, Caramazza,
& Yates, 1976; Caramazza, Grober, Garvey, & Yates, 1977). In sentence (2)
    (2) Mary won the money from Helen, because she ....
she refers to Mary and not to Helen. When subjects have to complete sentence

(2) they continue by describing an action of Mary. The reverse is true for a sentence like (3)
    (3) Mary punished Helen, because she ....
In this case, *she* clearly refers to Helen. If the main clause contains the verb *like*, *confess*, *win*, *sell*, *lie* .... the pronoun in the subordinate clause will be assigned to the subject of the main clause. These verbs are called NP1 verbs. On the other hand, with verbs like *punish*, *congratulate*, *envy*, *fear* .... the object of the main clause will be perceived as the antecedent. These verbs are called NP2 verbs. It is not quite clear from the literature what underlying linguistic or other features differentiate between these verbs (cf. Cowan, 1980). But at least it is possible to categorize these verbs as NP1 or NP2 verbs on empirical grounds.

Closely related to the so called "implicit causality" is the role of the conjunction. Replacing *because* by *but* can change the pronoun assignment (cf. Ehrlich, 1980). In sentence (4)
    (4) John lied to Pete because he smelled trouble
he refers to John, but in sentence (5)
    (5) John lied to Pete but he smelled trouble
he has to be assigned to Pete.

Another factor is the relation of the pronoun to the larger configuration of the discourse. The pronoun assignment is influenced by what the topic of the discourse is and by what is foregrounded (Frederiksen, 1981). Furthermore, there are other kinds of inferences, pragmatic or otherwise, that affect the pronoun assignment. These inferences may determine whether the characteristics predicated of the pronoun are consistent with the characteristics of the available antecedents in the previous discourse. What is called the *experiencer constraint* is one example. In sentence (6)
    (6) Bill told Harry that John bored him
him refers to Bill, because Bill will not inform Harry about Harry's own feelings. On the other hand, in sentence (7)
    (7) Bill told Harry that he bored John
he is unconstrained; it can refer to Bill as well as to Harry.

Finally, an obviously important factor in the pronoun-referent mapping are the lexical properties of the pronoun, e.g., gender, number and case.

It is an empirical question what the relative importance of the different factors are in the mapping process and how the different factors operate in real time. Does the reader simply use the information on gender for determining the coreference whenever the assignment unambiguously can be made on the basis of the gender of pronoun and available antecedents? Or is it the case that even in this situation other factors play a role as well. This would not be the most parsimonious procedure, of course. Inferences are supposed to be time and effort consuming processes. But Springston (1976) showed that the experiencer constraint still had an effect when the pronoun can be assigned on the basis of the gender alone.

So, one may assume that several factors influence the process of pronoun assignment. The question now is: what is the on-line process of the assignment of the pronoun. Does the reader process the sentence in a selective way, paying more attention to words if they contain information bearing on the pronoun assignment? How quickly can the reader decide what the relevant information is? To study this, two kinds of sentences were constructed in which pronoun assignment can be made on different pieces of information. Examples are the sentences (8) to (11):

(8) Alex lied to Andy because he smelled trouble
(9) Alex lied to Anna because he smelled trouble
(10) Alex lied to Andy but he smelled trouble
(11) Alex lied to Anna but she smelled trouble
Sentences (9) and (11) contain a gender cue. This cue is sufficient for the pronoun assignment. In all sentences the pronoun can be mapped onto an antecedent on the basis of inferences with respect to the meaning of the verbs and of the conjunction and in sentences (8) and (10) it has to be done this way.

Ehrlich (1980) used this kind of sentences with the conjunctions *because*, *but*, and *and* in an experiment, in which the task of the subjects was to indicate who *he* or *she* is. The sentences in which the gender of the pronoun unambiguously determined the antecedent were processed more quickly than the other sentences. Moreover, in the sentences without a gender cue there was a difference in the speed of processing between the *because*, *but*, and *and* sentences. This effect disappeared in the sentences with a gender cue. A likely interpretation is that the reader uses the gender of the pronoun in case the first clause contains two names of different gender. If both names are of the same gender, the sentence has to be processed more deeply, allowing for differences in processing time between the *because*, *but*, and *and* sentences.

If it is indeed the case that more sources of information are used, then the question is what kind of information is used in a particular situation. In answering this question, eye movement registration was used, because this method is more sensitive to on-line processes than, for instance, registration of reading times. With respect to the question of *when* the information from a region of a particular fixation is being processed, and *what kind* of processing (perceptual, lexical access, syntactic parsing, semantic integration) is taking place, one can find several claims, depending on the different theories about eye fixations: claims from cognitive lag theory (Bouma & De Voogd, 1974; Kolers, 1976), from the process monitoring hypothesis (Rayner, 1977; Ehrlich & Rayner, 1981), and from the strong version of the eye-mind assumption (Just & Carpenter, 1980). A claim that seems safe enough - and one that is sufficient - is: If an effect of a particular variable can be measured at a particular moment, that variable has had influence at least at that moment.

The task in the experiment was similar to the task in Ehrlichs (1980) experiment. Subjects had to vocalize as quickly as they could the name of the antecedent of the pronoun. Vocalisation latency was measured from the onset of the sentence on the screen to the onset of the answer of the subject. The eye movement data were obtained with a corneal-reflectance eye tracking system (Gulf & Western) that samples every 20 msec. Subjects read with two eyes, but eye movement data were recorded from one eye, the left one, only. The eye movements were measured while subjects read the sentences on a display until the answer was given. Before a sentence was displayed, a fixation point was presented at the point where the first word of the sentence would appear. The sentence appeared when the subject pushed a button. After the subject had named the referent, he/she had to push the button again, and the fixation point came on the screen again. The sentences were written in lower and upper case. Three characters equalled about 1.5 degree of visual angle. Fixations were determined according to an algorithm that takes into account the distance between succeeding sample points, the distance between the current sample point and the running mean, a look ahead on the next sample

point, and finally a minimum fixation duration of 60 msec. The outcome of
this algorithm allows for two succeeding fixations to be as near as one
character. A fixation on a word was defined to be a fixation on the characters of the word or a fixation on the two blank spaces before the word. If
on a word there were more than one fixation, the fixation duration of that
word is the sum of the durations of the fixations on that word *in the first
pass* of that word.

Twelve sentences were constructed. In order not to confound the assignment
of the pronoun with the position of the referent in the first clause, half
of the verbs of the main clause were NP1 verbs, half NP2 verbs. NP1 verbs
and NP2 verbs were selected according to the results of a sentence completion test using the following sentence frames:
        John *verbphrased* Ann, because ....
        John *verbphrased* Ann, but ....
The verbphrase of the second clause was chosen in congruence with the NP1/
NP2 bias of the main verb. The appropriatedness of these selections was
tested in a separate experiment. It was checked carefully that the bias to
NP1 in the *because* version of the sentence was accompanied with the bias
to NP2 in the *but* version of the sentence. Three conjunctions were used:
because, *but*, and *and*. There were four gender configurations: both names
were male, both names were female, the first name male and the second one
female and the reverse. These configurations form two conditions, the
condition with gender cue (names in main clause are of different gender) and
the condition without gender cue (names are of the same gender). There were
10 subjects, all university students. To each subject was presented the
complete set of 144 sentences preceeded by 18 practice sentences of the
same format, but with different content words.

The responses of the subjects indicated that the material was indeed constructed in an unambiguous way. (table 1).

Table 1
Percentage of correct answers as a function of verbtype, gender
cue and conjunction condition.

Conjunction	NP1 verb gender cue		NP2 verb gender cue	
	+	−	+	−
Because	98	96	98	93
But	100	93	99	97
And	100	*	100	*

In the *and* sentences the implicit causality of the verb does, presumably,
not determine what the correct antecedent is; the partitioning of the
answers between NP1 and NP2 in the condition without gender cue is given in
table 2. There was no complete bias towards either NP1 or NP2. Data of the
and sentences will not be reported in this paper.

Table 2
Percentage of answers on nounphrase 1 (NP1) and nounphrase 2
(NP2) as a function of verbtype for sentences without gender
cue with the conjunction *and*.

Conjunction	NP1 verb		NP2 verb	
	NP1	NP2	NP1	NP2
And	48	52	88	12

The vocalisation latencies (see table 3) for the sentences with a gender cue
are smaller than for the sentences without a gender cue (p < .001 for both
F_1 and F_2). This is in agreement with the results reported by Ehrlich (1980).
But there was no effect of the conjunctions ($F_1 < 1$; $F_2 < 1$), nor was the
interaction between gender cue and conjunction conditions significant
(p = .13 for F_1; $F_2 < 1$). This is in conflict with the results of Ehrlich.

Table 3
Mean vocalisation latencies (msec) as a function of gender cue
and conjunction condition.

Conjunction	gender cue	
	+	−
Because	3763	4109
But	3658	4289
mean	3711	4199

Ehrlich found a difference between the *because*, *but*, and *and* sentences only
if there was no gender cue. She attributes this difference to a more carefully processing of this kind of sentences. It has to be noticed, however,
that the percentage correct answers for sentences without a gender cue in
the experiment of Ehrlich is not very high, and, moreover, is lower for the
but sentences than for the *because* sentences (63% vs. 78%). This difference
may be reflected in the longer latencies for the *but* sentences in that experiment. This interpretation is in agreement with the latency data of the
and sentences in the present experiment: In the condition without gender cue
the mean vocalisation latency was a very long 4698 msec.

The mean fixation duration in the first pass of the sentence on the conjunction (*omdat (because)* or *maar (but)*) and on the pronoun (*hij (he)* or *zij
(she)*), and the mean of the sum of the fixation durations in the first pass
on the part of the sentence after the pronoun (second part of second clause)
are presented in table 4. In this table are also included the mean duration
of the first fixation after the pronoun and of the last fixation of the
sentence, both in the first pass. The 0 msec observations that appear if a
reader does not fixate words are not included in the mean fixation durations.
Henceforth, these 0 msec observations will be called nonfixations.

Table 4
Mean fixation duration on conjunction and pronoun, mean total
fixation duration on second part of second clause, and mean
duration of first fixation after pronoun and last fixation of
sentence (msec) in first pass of sentence as a function of
gender cue and conjunction condition.

	Conjunction			
	because		but	
	gender cue		gender cue	
	+	-	+	-
Conjunction (*omdat, maar*)	227	235	205	217
Pronoun (*hij, zij*)	255	198	255	219
2nd part 2nd clause	448	498	436	500
1st fix. after pronoun	245	233	226	219
Last fix. of sentence	224	243	230	240

The fixations on the pronoun will be discussed first. In the gender cue
sentences the pronoun assignment can be made on the basis of the gender of
the pronoun. So, an efficient procedure would be to utilize the information
in the pronoun as soon as the reader fixates it. Therefore, in contrast
with the fact that the vocalisation latencies for the sentences with a gender cue is *smaller* than for sentences without a gender cue, it can be expected that the fixation duration for the pronoun in the sentences with a gender cue is *larger* than in the sentences without a gender cue. This was found
indeed (p < .01 for F_1 and p < .001 for F_2). As in the vocalisation latency
data, there was no main effect of conjunction (p > .20 for both F_1 and F_2),
nor an interaction effect (p = .08 for F_1, p > .20 for F_2). There was no
effect, whatever, of the conjunction condition on any of the dependent measures. The effect of the gender cue suggests that the reader is selective in
analysing the information. It is when the pronoun is informative for the
assignment that the reader more carefully processes the pronoun. This suggests that the reader is selective in a very rational way. The reader behaves
according to what will be called the "principle of rational selection of
information".

The data give some indication about the time course in which this process
takes place. First, it is only when the reader has processed the second name
and has discovered that the two names differ in gender, that he/she can
decide that the crucial information for the assignment can be derived from
the pronoun. How quickly does the reader make this decision? The maximum
time he has available is the time from some moment after the onset of the
fixation of the second name to somewhere during the fixation of the pronoun.
A rough estimate on that basis yields a maximum of 500 msec. That this decision is taken is evident also from the nonfixations on the pronoun. The
number of nonfixations should be greater in the condition without gender cue
than in the gender cue condition. This was indeed the case (table 5).

Table 5
Percentage nonfixations on conjunction and on pronoun in
first pass of sentence as a function of gender cue condition.

	gender cue +	gender cue -
Conjunction (*omdat, maar*)	37	18
Pronoun (*hij, zij*)	17	40

This confirms the idea that after having encountered the second name, the
reader very quickly knows where he has to get the information for the pronoun assignment. As soon as possible, the subject decides *where* he has to
acquire the information and then he processes that information immediately.
Another piece of evidence confirming a form of the immediacy hypothesis
comes from the fixation after the pronoun (table 4). When there is a gender
cue, the fixation of the pronoun is longer than when there is no gender cue,
but for the first fixation after the pronoun there is no difference, whether
there is a gender cue or not (p = .17 for F_1, F_2 < 1).

There is, at least, one objection one could raise against this interpretation, namely, that nothing is processed at all except the gender of the
names in the first clause and the gender of the pronoun. This objection
comes down to the following strategy: The reader bets that the sentence has
a gender cue which will enable him to perform the task. Only if it turns out
that there is no gender cue, then he will regress to the first clause -
after having read the second clause - in order to process the sentence more
fully. The data, however, reject this objection, because there appears to be
an effect of the verb in the first clause on the fixation duration of the
NP2. The fixation on the NP2 is longer after an NP2 verb than after an NP1
verb (324 msec vs. 293 msec). So, subjects in fact do process the first
clause and not only the gender of the names.

What can be concluded from the fixations on other parts of the second clause?
In the condition without gender cue the pronoun can only be assigned by
using the information from the conjunction and from the verbphrase of the
subordinate clause. So it is expected that the fixation duration on the conjunction is longer in the non gender cue sentences than in the gender cue
sentences. Similarly, it is expected that the fixation durations on the
second part of the second clause are longer in the condition without gender
cue than in the gender cue condition. The results for the conjunction (table
4) are in the predicted direction, but the difference is not significant
(p <.10 for F_1, p =.17 for F_2). The total fixation durations on the verbphrase, on the other hand, give a clear result. When the information for the
pronoun assignment has to be deduced from the verbphrase, the verbphrase
requires indeed more processing time (p < .01 for both F_1 and F_2). This
result is confirmed by the number of nonfixations for the conjunctions
(table 5). In the gender cue condition, the conjunction is not the appropriate place for extracting the information concerning the anaphor resolution.
In the condition without gender cue, however, the conjunction contains useful information. The number of nonfixations in this case is indeed smaller
than in the gender cue condition. This, in fact, exactly mirrors the nonfix-
ation on the pronoun.

To summarize the differences between gender cue condition and condition without gender cue: In the gender cue condition the reader fixates the pronoun 47msec longer and the verbphrase of the second clause 57 msec shorter than in the condition without gender cue. So, the total durations are about the same. It should be noted, that these data refer only to the fixations in the first pass through the sentence.

After the first pass, the reader makes quite a lot of regressions and new fixations (table 6). These are more frequent and take more time in the sentences without gender cue than in the gender cue sentences. It is the difference in processing time after the first pass that makes up for the difference in the vocalisation latencies between the two conditions.

Table 6
Mean number of fixations on conjunction and pronoun, and mean number of NP1 - NP2 switches after first pass of sentence as a function of gender cue condition.

	gender cue		
	+	-	Δ
Conjunction (*omdat*, *maar*)	174	297	123
Pronoun (*hij*, *zij*)	207	249	42
NP1 - NP2 switch	218	292	84

It appears that the regressions and fixations after the first pass are not arbitrary. The number of regressions and fixations in the condition without gender cue is larger than in the gender cue condition. This is true in particular for the number of regressions and fixations on the conjunction and for the number of switches between the first and second nounphrase. In comparison with these differences, the difference with respect to the pronoun is small: The number of regressions to the pronoun in the condition without gender cue is relatively low. This suggests that, also in the additional pass(es), those words are more fully processed that carry the information for the pronoun assignment. The additional pass is a real processing routine and not only the check of the outcome by fixating only the answer.

A remark should be made with respect to the task. Eye fixations in reading tasks normally exhibit a long fixation at the end of the sentence. Just and Carpenter (1980) introduced the term sentence wrap-up: At the end of a sentence some extra time is spent to integrate the information. The data on the last fixation of the sentence show two things (table 4). There is a tendency for longer fixations in the condition without gender cue. This is in agreement with the data on the verbphrase. The point, however, is that the last fixation is not longer than other fixations. So, there is no evidence for the sentence wrap-up that normally takes place in reading sentences.This, probably, is the case because the task is not only a reading task: Much of the processing is done after the first pass of the sentence. (This, of course, is true for all other studies that used this task, e.g. Caramazza c.s., 1977; Springston, 1976; and Ehrlich, 1980.) It is an open question whether the results of the first pass will be replicated when the reader has to process a sentence in a normal context and without a vocalisation task. A small pilot study on *because* sentences in such a reading situation suggests that the fixations on the pronoun are longer in the gender cue condition.

In summary, the data lead to the following conclusions. First, the reader is able to decide very fast which words or parts of sentences are crucial for the adequate understanding. The reader is able to selectively attend to words to a certain degree depending on where he can extract the information he needs. This is the principle of rational selection of information. Secondly, the data support a qualified version of the immediacy claim (cf. Ehrlich & Rayner, 1983). As soon as a word is fixated, it is immediately processed at least as far as the lexical properties are concerned. Although the processing is immediate, it is not completed during the first fixation.

Acknowledgements.
P. Wittenburg and P. Rahner developed the programs for analyzing the data, R. de Wijk conducted the experiments, and L.G.M. Noordman gave many comments and suggestions on an earlier version of this paper. Their contributions are gratefully acknowledged.

REFERENCES

Bouma, H. and De Voogd, A.H., On the control of eye saccades in reading, Vision Research 14 (1974) 273-284.
Caramazza, A., Grober, E., Garvey, C. and Yates, J., Comprehension of anaphoric pronouns, Journal of Verbal Learning and Verbal Behavior 16 (1977) 601-609.
Cowan, J.R., The significance of parallel function in the assignment of intrasentential anaphora, in: Kreiman, K.J. and Oteda, A.E. (eds.), Papers from the Parasession on Pronouns and Anaphora (Chicago Linguistic Society, Chicago, 1980).
Ehrlich, K., Comprehension of pronouns, Quarterly Journal of Experimental Psychology 32 (1980) 247-255.
Ehrlich, K. and Rayner, K., Pronoun assignment and semantic integration during reading: Eye movements and immediacy of processing, Journal of Verbal Learning and Verbal Behavior 22 (1983) 75-87.
Ehrlich, S.F. and Rayner, K., Contextual effects on word perception and eye movements during reading, Journal of Verbal Learning and Verbal Behavior 20 (1981) 641-655.
Frederiksen, J.R., Understanding anaphora: Rules used by readers in assigning pronominal referents, Discourse Processes 4 (1981) 323-347.
Garvey, C. and Caramazza, A., Implicit causality in verbs, Linguistic Inquiry 5 (1974) 459-464.
Garvey, C., Caramazza, A. and Yates, J., Factors influencing assignment of pronoun antecedents, Cognition 3 (1976) 227-243.
Grober, E.H., Beardsley, W. and Caramazza, A., Parallel function strategy in pronoun assignment, Cognition 6 (1978) 117-133.
Just, M.A. and Carpenter, P.A., A theory of reading: From eye fixations to comprehension, Psychological Review 87 (1980) 329-354.
Kolers, P.A., Buswell's discoveries, in: Monty, R.A. and Senders, J.W. (eds.), Eye Movements and Psychological Processes (Erlbaum, Hillsdale, N.J., 1976).
Rayner, K., Visual attention in reading: Eye movements reflect cognitive processes, Memory and Cognition 4 (1977) 443-448.

Sheldon, A., The role of parallel function in the acquisition of relative
   clauses in English, Journal of Verbal Learning and Verbal Behavior
   13 (1974) 272-281.
Springston, F.J., Verb-derived contraints in the comprehension of
   anaphoric pronouns, Paper presented at the Eastern Psychological
   Association (1976).

Theoretical and Applied Aspects of Eye Movement Research
A.G. Gale and F. Johnson (Editors)
© Elsevier Science Pulishers B.V. (North-Holland), 1984

THE ACQUISITION OF A NEW LETTER SYSTEM:
EFFECTS OF WORD LENGTH AND REDUNDANCY

Christine Menz and Rudolf Groner

Department of Psychology
University of Berne
Berne
Switzerland

Adult Subjects (=Ss) had to learn a new letter system by reading aloud prose texts, isolated words and pronounceable nonwords. During the whole learning process, speech and eye movements were recorded. In the data analysis a distinction was made
between 'width of processing', defined by the saccade size and
'depth of processing', defined by the temporal factors fixation
duration and pause duration. During the entire training, saccade
length did not exceed the size of one single letter for words and
nonwords. With respect to the depth variables we found strong individual differences indicating different information processing
strategies of the Ss.

INTRODUCTION

Reading is a highly complex skill which involves a broad range of perceptual
and cognitive activities. From methodological reasons reading is difficult to
investigate in its full complexity. For this reason reading research is concentrated either on the study of isolated subskills of the reading process
(e.g. word recognition) or on the experimental manipulation of aspects affecting reading performance. Our study can be assigned to the latter approach. We
were interested in the acquisition of a complex skill, involving the encoding
of several short stories and some specially constructed material printed in
a new typography. During the entire learning process, eye movements and speech
signals were recorded. Of course we do not claim this task to be a simulation
of children's reading acquisition. However, one cannot deny that both situations involve the acquisition of a few subskills of the reading process
(for a detailed comparison of the two situations see Menz and Groner (1981)
p.202-206).

The idea this experimental procedure is based on is the hypothesis that a
hierarchy of processing units is built up during the learning process. The
existence of such a hierarchy would imply an information organization on
different levels (Rozin and Gleitman (1977)). That means, the organization
of information processing becomes more efficient during the reading training.
This should become manifest in a difference of size of the units available
during reading. The units can be single letters, syllables or words. Two
aspects of information processing activities are directly observable and can
be related to processing units: The perceptual input operationalized by means
of eye movements and the resulting output registered by voice utterances.

With the experimental task of learning to read a new letter system we tried
to tackle two questions:

1) How does the efficiency of information processing activities depend on the
   Ss' training with the new letter code ?
2) How does the semantic and syntactic structure of the text influence the
   availability of different processing units ?

In order to shed some light on these questions we made a distinction between
width of processing and depth of processing. Width of processing is related
to the concept of perceptual span (McConkie and Rayner (1975), Rayner (1975),
O'Regan (1980), Rayner, Inhoff et al. (1981)), and it is operationally defined
by saccade length. Depth of processing is defined by the temporal parameters
fixation duration and pause duration. Depth of processing refers to the amount
of central processing and bears some relation to Craik and Lockhart's (1972)
levels of processing. Other aspects of eye movement and speech recordings
will not be considered here.

EXPERIMENTAL PROCEDURE

We ː̃d six volunteer Ss but since always pairs of them were trained in the same
expeⁱ . ̃ntal set-up, the following discussion will be concerned with just two
of them. Every S took part in at least 20 experimental sessions, each lasting
for about one hour. In this time the Ss had to read text that was presented
in a standardized form. Figure 1 shows an example of a display unit containing a code list and four lines of text with a maximal line length of 15
symbols. The displays were generated on a graphic system and displayed on a

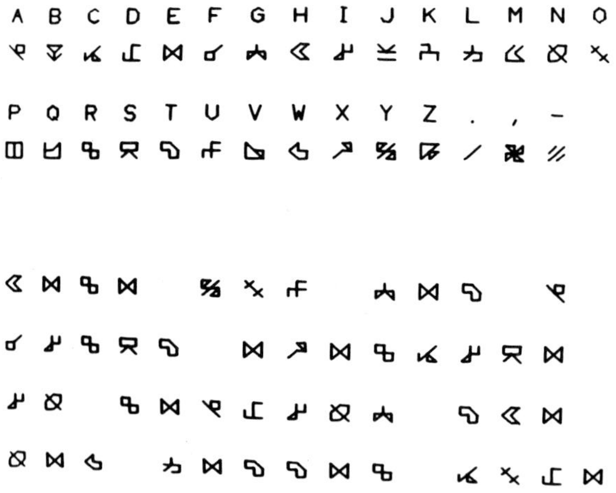

Figure 1
Example of a display unit with a code list (upper part) and
four lines of text (lower part).

cathoderay tube. The Ss had a hand button for controlling the display changes corresponding to turning the pages in a book. The stimulus material was displayed in a field of 40x30 cm corresponding to a visual angle of 23 deg in the horizontal and 17 deg in the vertical axis. Single symbols were separated by approximately one degree of visual angle. The strict hyphenation rules of German orthography were not obeyed in order to obtain a relatively constant number of symbols per display unit (50 ± 2).

Eye movements were recorded with a TV-based infrared system at a frequency of 50 Hz. The voice signal was digitized into utterances and pauses at the same sampling rate.

The texts consisted of novels, presented in order to have a natural reading situation involving careful reading of continuous text for the purpose of comprehension (McConkie (1983)). This material was presented in a total of 850 display units. In addition to the novels, some specially constructed reading material was presented in various learning phases. The experiments discussed next concern this special material.

EXPERIMENT 1
Here single words were shown, matching in statistical frequency (equal frequencies of 100-200 per 50'000 words (Meier (1967))) but differing in length. The words were either four (=4LW), seven (=7LW) or fifteen (=15LW) letters long. Figure 2a shows an example of each condition. Four different sets of

(a)	(b)
FLAU HAIN KOHL	PEVEKEMUNES EH-
SAME FLUG KITT	AHE TAREN ETOL-
ZINS RIND EFEU 4 LETTER WORDS	OGUTATU SANERUS 1. ORDER APPROXIMATION
BUDE GIER HEMD ---------------	AMERELOEHUNIK ----------------------
ANDRANG MONARCH	ZUMELARIS PUND-
STOLLEN BAUSTIL	AU GENELL ERATE
ADMIRAL ZAPPELN 7 LETTER WORDS	UNDURDE WAN 'JS- 3. ORDER APPROXIMATION
HABHAFT DENKART --------- ------	EDESO GA R D- ----------------------
UNBEWEGLICHKEIT	LITER ABGERING-
FRONTAUSDEHNUNG	ETRIESTAL BESE-
HAUPTGEGENSTAND 15 LETTER WORDS	SORTENE LEHRE 8. ORDER APPROXIMATION
REICHHALTIGKEIT ---------------	UNDE KEINE MIL- -----------------------

Figure 2
Transcripted examples of four-, seven- and fifteen-letter words used in Experiment 1 (a) respectively examples of first, third and eighth order word approximations used in Experiment 2 (b).

EXPERIMENTAL SET-UP

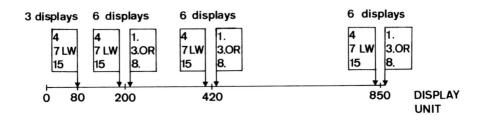

Figure 3
Schema of the experimental set-up of the entire study

items were presented after reading 80, 200, 420 and 850 display units. The set that was presented after 80 displays consisted of three displays for each wordlength. Later we presented six displays of each condition (Fig. 3).

EXPERIMENT 2
For a second experiment, pronounceable nonwords were constructed by scrambling letter sequences of different length. Figure 2b shows an example of each of the three approximation levels used. First order approximations (=1.OR) are nonwords composed of consonant-vocal sequences with their actual probabilities in written language. Third order approximations (=3.OR) are nonwords composed of letter triplets. Eighth order approximations (=8.OR) are nonwords composed of randomly selected letter octets. Three sets of items were presented after 200, 420 and 850 displays coinciding with the presentation times of the single words. Every time the Ss had to read six displays per approximation level (Fig. 3).

The following discussion will be concentrated on the analysis of these two experiments. They are part of a more extensive study discussed elsewhere in more detail (see Menz and Groner (1982)). With respect to the two experiments our initial questions must be reformulated:

1) Do ocular and voice related parameters change during the learning process ? Under what experimental conditions and at what training stage does this change take place ?
2) How does the experimental manipulation of the reading material affect the availability of different processing units ?

RESULTS

Separate analyses of variance were carried out for both experiments. A mixed model factorial design was submitted for the dependent variables saccade length, fixation duration and pause duration. In contrast to a simple ANOVA which treats the Ss as within-cell replications we used a randomized block design with the Ss as block factor. For the wordlength experiment the factors were Wordlength, Subject, Learning Phase and Replication. The factors of the nonword experiment were Approximation Level, Subject, Learning Phase and Replications.

Everyone would expect the rather trivial finding that with increasing reading practice the saccade length would certainly increase. Surprisingly this result did not come out. The saccade length remained stable over all learning phases in both experiments with an average length of approximately 1.5 deg of visual angle. That means perceptual processing units were single letters for all conditions in both experiments. Apparently width of processing stays unaffected by our learning conditions.

The evaluation of the depth variable showed a steady decrease of the mean fixation duration with increasing reading practice in the nonword experiment[1] (Fig. 4). The wordlength experiment showed a similar tendency but was not significant.

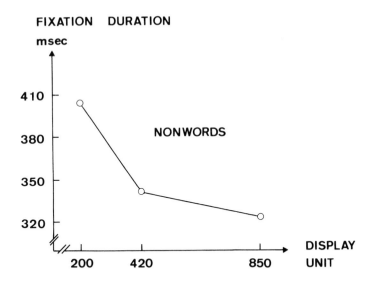

Figure 4
Fixation durations as a function of reading experience with pronounceable nonwords

[1]However, this result cannot be substantiated by inferential statistics $(F(2,2)=9.78, p=.09)$ because the error term of the appropriate F-ratio (i.e. the interaction Subject x Learning Phase) is larger and also close to significance $(F(2,90)=2.62, p=.08)$.

For the voice related depth variable the picture becomes more complex. The
results of both experiments are shown in Figure 5. In the nonword condition
the two Ss differ significantly from each other (main effect Subject F(1,90)=
118.4, p ◀ .0001). The pause duration for S2 is almost twice that of S1 independent of the approximation level. With meaningful words, however, there is a
strong interaction between Subject and Wordlength (F(2,48)=18.64, p ◀ .0001).
While pause durations are identical for both Ss in the short word condition,
they double for S2 and cut in half for S1 in the long word condition. This
result can be explained by a difference in the information processing strategy: S2 processes a complete word before he utters it, while S1 segments longer
words during reading and utters word parts.

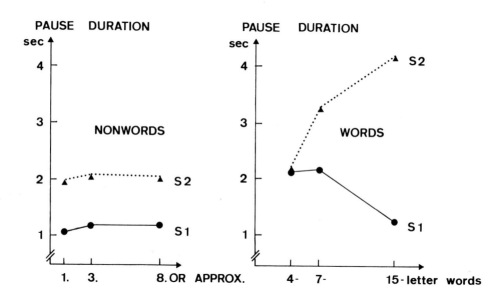

Figure 5
Pause durations as a function of approximation levels of nonwords
respectively as a function of wordlength (Experiment 1)

Besides the individual differences the performance did not differ whether
the reading material consisted of meaningful words or pronounceable nonwords.
This finding is in accordance with results Baron and Thurston (1973) found in
their experiments on the word superiority effect.

The next analysis concentrates on the question of how the experimental manipulation of the reading material affects the organization of information processing. As dependent variable we used fixation duration since it refers to
the temporal decisions made in the reading process (McConkie (1983)). Four
gaze areas were defined by dividing the display in a way that each line is

subdivided in four equal parts. The time spent in each of these four gaze areas was computed per display and defined as gaze duration. This was done for all three conditions in both experiments. The data were submitted to a mixed model factorial design. The factors were Subject, Wordlength respectively Approximation Level, Learning Phase, Replication and Gaze Position.

For the wordlength experiment a significant second-oder interaction was found between Wordlength, Learning Phase and Gaze Position ($F(18,18)=2.37$, $p=.037$) and is shown in Figure 6. Each graph representing one of the wordlength conditions contains separate curves for the four different learning phases (T1-T4) The four data points in each curve are the means of the gaze durations spent within the four different gaze positions.

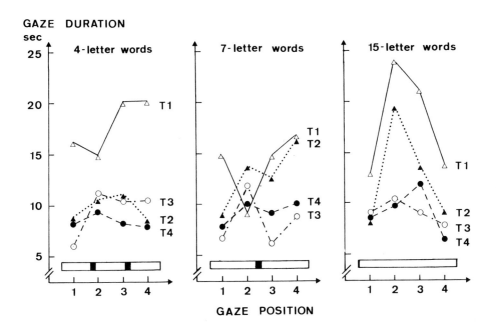

Figure 6
Second-order interaction between Wordlength, Learning Phase and Gaze Position. Gaze durations in the four learning phases T1-T4 are plotted as a function of gaze positions separately for each word length. Bars above the abscissas represent the four gaze areas. Black squares in the bars show the word boundaries in the three wordlength conditions

Comparing the curve shapes over the wordlength conditions one finds remarkable differences. Except for T1 gaze durations for four-letter words are about the same for all gaze positions. This finding might be explained by the fact that

in the short word condition three words were presented per line which were in
no way related to their neighbours. So the reader was unable to guess about
the meanings of neighbour words. With seven-letter words more fixation time
was spent on the second half of the words independent of their position in a
line. This means that word ends become more important for the encoding process
with increasing training. With long words all four curves have an inverted
u-shaped form. If one has to explain these data one has to mention how long
words are built in German. Long words are made out of several shorter words
concatenated together (i.e. Pferdebahnwagen, Affenbrotbaum). Thus the Ss are
able to make quite good guesses about the rest of a word once they have read
part of it.

The ANOVA for the wordlength experiment showed the interaction between Subject,
Wordlength and Gaze Position to be significant ($F(6,192)=2.1$, p=.05). Figure 7
shows the mean gaze durations for the wordlength conditions separate for each
S. The two Ss differ most in the middle gaze position (2,3) of the long word
condition. Here S2 spent much more time than S1. This finding can be brought

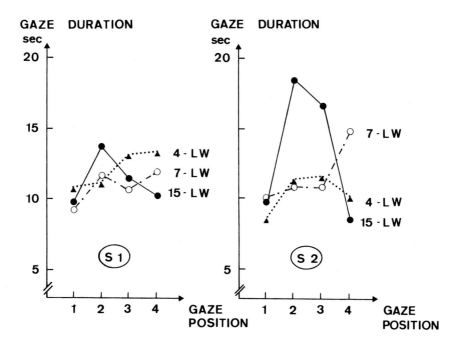

Figure 7
Second-order interaction between Subject, Wordlength and Gaze Position.
Separate graphs for the two Ss with gaze durations for all three word-
length conditions as a function of gaze position.

into connection with the observation mentioned earlier how Ss verbalized
during reading. S2 operates on the word as a whole before he utters it and

while encoding he does not much care about information in the first and fourth gaze position. This means that hypothesis testing seems to be an important constituent of his information processing strategy. Apparently guessing and hypothesis testing is more important for S2 than for S1.

While word length did influence the learning process from the beginning, observing not fixation duration but gaze duration, significant effects of the different approximation levels could not be found besides individual differences. Several interactions involving Subject as factor were significant (Subject x Approximation Level ($F(2,360)=46.07$, p$<$.0001); Subject x Learning Phase ($F(2,360)=15.9$, p$<$.0001) and their higher order interactions). All these results show that S2 is able to improve reading performance by making use of the redundancy of orthographic spelling patterns, while S1 does not. This again confirms the differing organization of cognitive units of the Ss during reading.

CONCLUSIONS

The results of the two experiments can be summarized by noting that saccades did not traverse the span of one single letter for all conditions on the average. With respect to the depth variables a general although not significant decrease of fixation duration and pause duration was stated with more training. The evaluation of fixation durations in gaze areas showed that wordlength has an effect on the reading performance in all learning phases. Nonwords composed of regular spelling patterns differing in length did not affect the learning process in any phase. There were some systematic differences between our Ss concerning the way they encoded words and nonwords. In view of these differences it seems questionable to pool the results of different Ss who might have used different information processing strategies. Furthermore, the research method of this study is probably not an optimal one, mainly for two reasons. First one might get into problems with the interpretation of results because each S is considered as a complete replication of the experiment and second it may be difficult to disentangle the many variables that are confounded within a single S. But on the other hand the tradition of classical experimental psychology, putting together all Ss in the data analysis, certainly does not solve the problem either.

REFERENCES

(1) Baron, J. and Thurston, I., An analysis of the word-superiority effect, Cognitive Psychology 4 (1973) 207-228.

(2) Craik, F.I.M. and Lockhart, R.S., Levels of processing: A framework for memory research, Journal of Verbal Learning and Verbal Behavior 11 (1972) 671-684.

(3) McConkie, G.W., Eye movements and perception during reading, in: Rayner,K. (ed.), Eye Movements in Reading (Academic Press, New York, 1983).

(4) McConkie, G.W. and Rayner, K., The span of the effective stimulus during a fixation in reading, Perception and Psychophysics 17 (1975) 578-586.

(5) Meier, H., Deutsche Sprachstatistik (Olms, Hildesheim, 1967).

(6) Menz, Ch. and Groner, R., "Zweitlesenlernen" - die experimentelle Analyse der okulomotorischen und artikulatorischen Koordination bei einer komplexen Dekodierleistung, in: Foppa, K. and Groner, R. (eds.), Kognitive Strukturen und ihre Entwicklung (Hans Huber, Bern, 1981).

(7) Menz, Ch. and Groner, R., The analysis of some componential skills of reading acquisition, in: Groner, R. an Fraisse, P. (eds.), Cognition and Eye Movements (Verlag Deutscher Wissenschaften, Berlin, 1982).

(8) O'Regan, K., The control of saccade size and fixation duration in reading: The limits of linguistic control, Perception and Psychophysics 28 (1980) 112-117.

(9) Rayner, K., The perceptual span and peripheral cues in reading, Cognitive Psychology 7 (1975) 65-81.

(10) Rayner, K., Inhoff, A.W., Morrison, R.E., Slowaczek, M.L. and Bertera, J.H., Masking of foveal and parafoveal vision during eye fixations in reading, Journal of Experimental Psychology: Human Perception and Performance 7 (1981) 167-179.

(11) Rozin, P. and Gleitman, L.R., The structure and acquisition of reading II: The reading process and the acquisition of the alphabetic principle, in: Reber, A.S. and Scarborough, D.L. (eds.), Toward a Psychology of Reading (Erlbaum, Hillsdale, New Jersey, 1977).

Theoretical and Applied Aspects of Eye Movement Research
A.G. Gale and F. Johnson (Editors)
© Elsevier Science Pulishers B.V. (North-Holland), 1984   223

CEPHALIC AND OCULAR COMPONENTS OF GAZE DISPLACEMENT DURING ORAL AND
SILENT READING IN CHILDREN AND ADULTS

Serge Netchine, Connie Greembaum and Marie-Claude Gihou
Laboratoire de Psycho-Biologie de l'Enfant, Ecole Practique
des Hautes Etudes, (Laboratoire associé au C.N.R.S.), Paris
France.

Cephalic displacement during reading evidences a specific
mode of appropriation of two dimensional space. The
quantitative importance of this mode of gaze displacement
constitutes an indicator of difficulty of access to text
meaning. Results of an experiment on young French readers
shows a significant increase of cephalic participation
in gaze displacement from silent to oral reading. A
similar experiment on adult second language reading shows
head movement present in silent reading and a significant
jump in quantity of cephalic displacement in oral reading.

In most studies of gaze displacement during reading, the spatial and
temporal characteristics of information processing are entirely derived
from recordings of ocular saccades and fixations. One of the main reasons
for this is linked to the mechanical requirements of monitoring instruments
which necessitate head immobilization in order to obtain precise measures
of the eye as regards a specific portion of text. However, observation of
subjects reading in their habitual manner - without being constrained by
laboratory equipment - reveals a variety of means of obtaining gaze shifts
(Salel and Gabersek, 1975). Although the great proportion of gaze shift
mechanisms remains entirely ocular in experienced adult readers reading
easily accessible material, this is not the case for other categories of
readers: in particular, children obtain gaze displacement for the most
part through horizontal head movements (Netchine, 1981, Netchine et al.
1981; 1983). Furthermore, the quantitative importance and the incoercive-
ness of cephalic movements in young children has been reported as an experimental liability, and not in terms of its role as an inherent characteristic of children's visual exploration of the world.

We would like to suggest that it is indeed possible to record conjugated
head and eye movements during reading, and that there are interesting
reasons for doing so,

A quantitative estimate of the relative proportion of oculo-motor and ceph-
alo-motor components of gaze displacement can be obtained through the
combined use of an EOG recording device and a potentiometer strapped to the
vertex for horizontal cephalic movements. It is to be noted, however, that
this recording does not allow for a high degree of precision in the analysis
of local and/or elementary reading activities (see Figure 3).

The characteristics of gaze displacement during reading observed under these
conditions differs from those observed in the case of rapid head movements

(Zangemeister and Hufner, this volume). From the beginning to the end of
a line, cephalic movements are relatively slow and only rarely require compensatory eye movement. In the majority of cases, the eyes and the head
move in the same direction. Given this form of relationship, it is possible
to express head movement and positioning in terms of saccades and fixations
which are normally used to describe ocular movements. However, cephalic
"saccades" differ from ocular saccades by their lower velocity. During
reading, when they appear conjointly with ocular saccades, the are initiated before the ocular saccade and finish after it. Furthermore, the curve
of cephalic movements shows either a smooth slope, or a series of steps
(Netchine, 1981; Netchine et al. 1981; 1983). When head movements govern
a sizeable and irreducible part of gaze displacement, it seems reasonable
to ask whether other complementary or substitute mechanisms are at stake,
rather than those habitually referred to in the literature. These mechanisms are associated with developmental aspects of spatial orientation, and
the relationships between mode or gaze displacement and text difficulty.

Ocular and cephalic displacements differ as to the source of information
governing their control and regularity. The child uses head movements to
calibrate reading space by reference to his/her own body. The shifts from
the beginning to the end of a line and the return sweep are thus guided by
partially proprioceptive sources of information. The reduced ocular saccade size in young readers, coupled with their limited useful field of view
during fixations are compensated for by a certain proportion of cephalic
participation in gaze displacement. This change in modality has direct
repercussions on the process of information itself. Furthermore, the small
ocular saccade size observed when gaze displacement is accompanied by head
movements testifies to the head's role in foveal recentering (Howard, 1982).

When gaze displacement is at least partially ensured by head movements, the
role of peripheral vision is necessarily reduced. The mode of gaze displacement thus influences cognitive processing and vice-versa, in that obsta-
cles met during reading influence the mode of gaze shift used. The choice
of a particular mode of gaze displacement may in itself be a means of
increasing possible information extraction. Head movement in adults may
serve as a possible trade-off procedure for spatial processing ensured by
the eyes whenever the ocular modality must deal with text difficulty.

Modifications in gaze displacement can be elicited in adults by introducing
specific variants into the reading task, such as non-standard spellings.
In this case, adult performance takes on characteristics found in young
beginning readers. Increasing difficulty of access to meaning by heighten-
ing syntactic, semantic or pragmatic features, notably in the case of
foreign language reading, are all accompanied by a corresponding increase
in the amount of gaze shifts showing cephalic participation, and simultaneously by changes in ocular displacements including the type of return
sweep - this last theoretically being considered as purely spatial in nature
(G. Englander, 1981; C. Greenbaum, 1983). However, it should be stated that
the cephalic component in gaze displacement in adults is small as compared
to children. Nevertheless, measurement of the amount of cephalic participation in gaze shifts seems to be an observable method of evaluating text
difficulty and is complementary to the observation of ocular gaze displacement both in children and in adults.

To test this hypothesis, we have conducted a series of experiments on children and adults. Two types of difficulty were measured: oral vs. silent

reading in children and adults, and foreign language reading in adults. We selected oral reading as a form of difficulty since we believe that regardless of the amount of automaticity involved in reading (Laberge and Samuels, 1974), oralization requires a necessary passage via phonological coding and thus represents a supplementary task (Foucambert, 1976; Neville and Pugh, 1982). If indeed this is the case, we should observe a greater proportion of cephalic participation in gaze displacement in oral reading as compared to silent reading, in texts of equal difficulty. Foreign language reading was introduced in adults because it allows specific manipulation of elements of the foreign language for a particular population of readers. In our case, we used twenty middle-proficiency French adults reading two types of text controlled to be "easy" and "difficult" in both French and English. The adults also read successive paragraphs of the text silently and orally. The population of children was composed of 19 native French seven-to-eight year olds, and 21 nine-to-ten year olds. Head movements were recorded by the technique described above. Eye movements were recorded in the form of an EOG. Texts, controlled for their homogeneity, (Greenbaum, 1981) were 10-13 lines long, each line measuring 27 cm.

RESULTS

At all ages, mean amplitude of cephalic participation in gze displacement per line is significantly greater in oral reading than in silent reading. In both groups of children, cephalic components of gaze shifts are the general rule in both modes of reading. The amplitude of cephalic participation in gaze shifts is lower in the older groups, suggesting that ocular control of gaze displacement is lined to progressive acquisition of reading skill. In both age groups, however, the switch from silent to oral reading constitutes a change in gaze shift modality of equal magnitude. In adults, there are practically no occurrences of cephalic participation in gaze shifts in silent reading of the native language. In English, however, the cephalic component is present for silent reading and evidences a steep rise in oral reading, as can be seen in the chart below.

Population	N	Order of reading mode	Silent Reading X̄	σ	Oral Reading X̄	σ	F	P
7-8 year olds	9	Silent-Oral	4°8	2°8	9°4	3°7	$F_{(1,8)}=$ 22.34	.01
"	10	Oral-Silent	3°7	2°4	6°4	2°5	$F_{(1,9)}=$ 7.98	.025
9-10 year olds	11	Silent-Oral	2°5	2°0	6°3	2°3	$F_{(1,10)}=$ 73.65	.001
"	10	Oral-Silent	4°3	3°7	6°9	3°4		
French adults	20	Sil-Oral Easy English	0°3	0°6	1°7	1°6	$F_{(1,19)}=$ 25.8	.001
French adults	20	Sil-Oral Difficult English	0°3°	0°7	2°0	1°4	$F_{(1,19)}=$ 37.46	.001

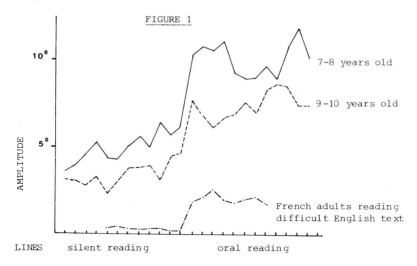

AMPLITUDE OF HEAD GOVERNED GAZE SHIFTS PER LINE
IN SILENT AND ORAL READING IN THREE AGE GROUPS

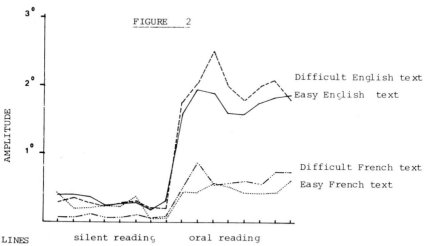

AMPLITUDE OF HEAD GOVERNED GAZE SHIFTS PER LINE
IN ADULTS' SILENT AND ORAL READING OF FRENCH AND
ENGLISH TEXTS

We have stressed results concerning changes in gaze displacement modality. An analysis of the number of regressions shows a greater number in oral reading, although evaluation and interpretation of regressions observed in conditions with cephalic participation are subject to caution: they may be related to an eye-voice span effect, or to text verification, as in classic studies, or be caused by compensatory eye movements following cephalic overshoots. Secondly, an examination of mean reading time per line conforms to standard observations and shows slower reading times for oral reading. However, these results should also be treated with caution since trends may be associated with ongoing task processes in the youngest children. It seems essential to take temporal factors into account for all the variables used, in order to confirm global results obtained in each reading mode.

As concerns amplitude of cephalic participation in gaze displacement, line per line, results show that the change in reading mode is accompanied at all ages by notable differences, which cannot be attributed to a trend effect (figures 1 and 2), as is confirmed by the application of an algorithm designating the most pertinent point of dichotomization of the values into a time series having two periods (Netchine, 1981).

$$p = 1/ \frac{\sum (_i - \bar{x}_i)^2 + (x_j - \bar{x}_j)^2}{\sum (x - \bar{x})^2}$$

DISCUSSION

Reading has been taken here to involve all the possible means of exploration used by the subject to gain access to text. The reader has at his/her disposal varied combinations of cephalic or ocular components of gaze displacement. Extreme cases involve predominant use of ocular gaze displacement or predominant use of cephalic displacement, in which case the ocular components function in relation to their location within the head, and are only involved in limited exploration.

Gaze displacement with a predominant ocular component has the advantage of allowing programming of saccades and fixations via anticipatory linguistic control. Cephalic gaze displacement, on the other hand, is dependent upon physical parameters such as the inertia of the head, and a guidance system based on reafferences intervening concurrently with visual exteroceptive information.

Thus the greater the participation of cephalic movement in gaze displacement, the slower and less flexible these shifts are, and the more dependent on point to point recalibration. When text processing requires a mode of gaze displacement limited to small portions of text, this mode of gaze displacement is suitable, as has been seen in the case of adults reading in a foreign language and oral reading in general.

More broadly speaking, a relationship can be drawn from mode of ocular gaze displacement and level of acquisition, and can be expressed in terms of the perceptuo-motor scheme solicited. The predominantly ocular-based system of gaze displacement found in adults thus should not be used as the unique reference for the study of reading. Rather it seems more appropriate to study the different situations that produce different combinations of ocular and cephalic modes of gaze displacement in terms of the needs

S. Netchine et al.

FIGURE   3

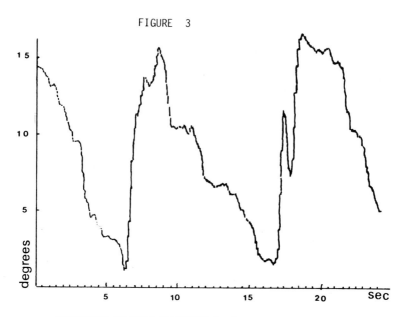

RECORDING OF HEAD MOVEMENT DURING READING OF 2.5 LINES
BY A CHILD 7YRS 6 MOS.

basic to reading strategies according to age and proficiency.

REFERENCES

Englander, G. L'organisation des mouvements oculaires et céphaliques au
    cours de la lecture d'un texte français et d'un texte hébreu chez des
    enfants de 11 à 12 ans. Diplôme d'Études Approfondies, Dept. of Psych.
    Université de Paris V, 1981.

Foucambert, J. La manière d'être lecteur, O.C.D.L. Sermap, Paris, 1976.

Greenbaum, C. Construction et analyse de textes de lecture utilisés dans
    une recherche sur les modalitiés proprioceptives et oculomotrices chez
    l'enfant d'âge scolaire, Master's Thesis, Dept. of Linguistics,
    Université de Paris X-Nanterre, 1981.

Greenbaum, C. Stratégies d'organisation des déplacements du regard témoig-
    nant des micro-genèses chez l'adulte au cours de la lecture dans une
    langue étrangère, Diplôme d'Etudes Approfondies, Université de Paris V,
    1983.

Howard, I.P., Human Visual Orientation, Wiley, Chichester, 1982.

Laberge, D. and Samuels, S.J., Toward a Theory of Automatic Information
    Processing in Reading, Cognitive Psychology, 6, 1974, 293-323.

Netchine, S., Evolution au cours de l'enfance des relations entre
    proprioception et oculomotricité dans des activitiés perceptives
    appliquées à un plan bi-dimensionnel: lecture, copie de textes,
    et formes géometriques. Rapport d'A.T.P. du CNRS, 1981.

Netchine, S., Guihou, M.C. Rôle des mouvements céphalogyres dans l'organis-
    ation des déplacements due regard chez les jeunes lectures: paper
    delivered at the Forum Espace III - Position et Mouvement, Marseille,
    1981.

Netchine, S., Solomon, M., Guihou, M.C. Composants oculaires et céphaliques
    de l'organisation des déplacements du regard chez les jeunes lecteurs.
    Psychologie Françzise, 26, 1981, 110-124.

Neville, M.H. and Pugh, A.K. Towards Independent Reading, London, Heine-
    mann, 1982.

Salel, D. and Gabersek, S.J. Cinq aspects due comportement de l'enfant
    en cours d'apprentissage de la lecture, Revue d'EEG et de Neurolo-
    physiologie clinique, 5, 1975, 345-350.

Theoretical and Applied Aspects of Eye Movement Research
A.G. Gale and F. Johnson (Editors)
©Elsevier Science Pulishers B.V. (North-Holland), 1984   231

READING UNDER NORMAL CONDITIONS:
RECENT STUDIES USING EYE MOVEMENTS AND OTHER OBSERVATIONAL METHODS

A. K. Pugh

School of Education
The Open University
Milton Keynes
U.K.

Information about reading under normal conditions is of
relevance to the development of reading abilities and to the
design and structuring of written information. For the study
of reading under normal conditions a direct observational
method was devised and more recently an eye movement recording
system has been used in conjunction with it. This paper
describes the principles of the present system and presents
some preliminary findings. These are from studies of reading
subtitles and other text on television as well as of printed
text.

INTRODUCTION

The influence on theories and practices in the teaching of reading from
the early literature on eye movements was very considerable if not always
well acknowledged or fully appreciated. The background is discussed in
e.g. Kolers (1976) and Pugh (1978, 1983a), where reasons are also
considered for the lesser impact of more recent eye movement studies. In
part at least this is due to a realisation that a good many of the studies
made earlier in this century were so constraining of the reader, often
physically as well as in terms of what he could read and in what order,
that generalisation from the findings to more normal reading tasks was
not so obviously warranted as once it seemed. Also, the training of the
eyes, which seemed to be supported by eye movement studies, deteriorated
to arid drills which were as obviously quackish as their justification in
the reading courses (usually for adults and older schoolchildren) was
pseudoscientific. Eye training represented a confusion common in
education whereby a useful research procedure becomes adopted as a teaching
technique, neatly short circuiting the problems involved in discerning
and applying the implications of the findings of research rather than its
surface methodology. Another example in this field is cloze procedure, a
valuable research tool with questionable applications in teaching and
testing, where it has nevertheless been widely adopted (Neville and Pugh,
1982).

Theoretical issues, such as the basic one of the extent to which eye
movements are voluntary, are important for the design of courses and
materials. Indeed, this particular issue, itself part of the more basic
question of what guides search for meaning in reading, is very important
for assessment of the importance of eye movement training just mentioned.
Recent views, summarised e.g. by Clifton (1983) seem to accord with ours
(Pugh 1978) that the reader can do very little directly to change his own
eye movements while reading, except perhaps for reducing major regressive

sweeps, simply because he has no way of knowing what the movements are.
Whether it can prove worthwhile to provide feedback using the more
sophisticated eye movement apparatus now available is another question,
which seems worth exploring for some special applications at least. It
has seemed, however, that larger strategies might be more important for
the reader to know about.

PREVIOUS OBSERVATIONAL STUDIES

Given the limitations of the equipment then available and a concern with
obtaining information about students' use of long texts, it was decided
some years ago to concentrate on observational studies of reading
behaviour, backed by careful specification of tasks and assessment of their
achievement. The method used has been described elsewhere (e.g. Pugh,
1978, 1979) where results are also given. Basically the approach involves
obtaining a direct image of the text from above, using a video camera
pointed downwards, while a reflected image of the subject's head is
obtained in the same 'frame' since he reads at a reading stand consisting
of wood and, in the upper part, semi-silvered glass. This glass is
preferred to a mirror,because with suitable lighting, it may appear
transparent to the student while, from another angle, the camera obtains
a good image without the subject being distracted by either the camera or
his own movements. Depending on how the camera is focussed, it is possible
to make judgements about the reader's general direction of gaze, although
this is an apparatus for general surveillance of reading and related
activities and not for the measurment or recording of detailed movements
of the eyes.

Analysis of data from the video recordings requires judgements to be made
and is not capable of automisation. The analysis can be informed by the
subject giving a commentary on the recording, both to clarify actions and
to attempt to discern intentions. Indeed, it is to this kind of case
study work that the apparatus is particularly suited although samples of
over 20 subjects have been studied in several investigations of the use
of books to obtain information. The studies have been exploratory of
reading styles, in order to contribute to our knowledge of how readers
at different levels approach certain reading tasks in real books, but they
have also involved an element of assessment. The criteria are difficult
to establish for good performance, but with university students and sixth
formers (Pugh, 1978, 1979) and with middle school children (Neville and
Pugh, 1982) it has been found that the demands and expectations of teachers
with regard to reading are disconsonant overall with the observed level of
skill of the readers studied. That university reading lists are often
unrealistic is, of course, a common student complaint. What is less in
accord with the commonplace view is that, within admittedly homogeneous
samples, the correlation between reading ability as measured by a
standardised test and ability to use books (in our tasks) was almost always
very low.

CURRENT EYE MOVEMENT SYSTEM

The studies of book use pointed to a need for better descriptions of
readers' strategies and text structures, preferably obtained by obtaining
data about the interaction of the two. Recently, numerous proposals for
analysing text structure have become available, although few of those in
e.g. Spiro, Bruce and Brewer (1980) have been tested against actual reading

behaviour, as opposed to what can be inferred about that behaviour from the reader's recall. There have been studies of eye movements in the reading of longer texts (see e.g. Shebilske and Fisher, 1983 for some recent work) but in general rather short texts under carefully controlled conditions are still usual in eye movement research reading judging from the papers in Rayner (1983)and those given at this conference.

To obtain better information on readers' strategies, in relation to the texts they read, eye movement apparatus was seen to be necessary despite the fact that the emphasis on realistic texts and conditions was a little out of accord with the mainstream of eye movement research. In part it was acknowledged that in any study a trade off must be involved, since even with recent advances in the design of equipment and with the benefits of microcomputing development, there remain nevertheless some formidable technical problems in the way of obtaining accurate results from equipment built at low cost, with portability for use in the 'field' (schools, etc.) an essential attribute and with insistence on the need to be able to record movements of both eyes. The importance of measuring binocular vision has, perhaps for technical reasons, been insufficiently acknowledged. Evidence from Bedwell, Grant and McKeown (1980) and in Clark and Crane (1978) suggests that it is not safe to assume a constant dominance during reading.

The equipment which will be described has been used in studies briefly reported in the latter part of this paper, but is also still under development as a major component of a comprehensive system for monitoring reading behaviour. That system would include possibilities for measuring head movement, rather than constraining it. Netchine et al. (1983 and this conference) have evidence to show that movement of the head appears to increase with difficulty of text, foreignness of the language or inexperience on the part of the reader; accuracy of return sweep is similarly related to these factors. As discussed more fully elsewhere (Pugh, 1983b) there are interesting parallels here with what we know about the incidence of inner speech in reading, and a means of monitoring subvocal activity would be a useful part of the comprehensive system whose development is currently being planned in collaboration with Dr. Netchine's team.

The equipment so far uses scleral reflection following principles presented in detail in Haines (1980) who was responsible for building the analog part of the apparatus. Spectacles on which are mounted light emitting diodes and photosensors are connected to a box where the signals can be balanced as well as being amplified for output to a chart recorder. In addition to this output, use of a dedicated Acorn microprocessor permits digitalisation of the signals from the spectacles so that data can be stored for analysis using a main frame computer (on-line or subsequently). A graphics output from the microprocessor to a videorecorder shows point of gaze for either or both eyes on a television monitor, while input from a camera may be mixed with these graphics if required. A facility is available for adjusting what distance is to count as a movement within the 256 x 256 digitalised matrix; between one and ten squares may be selected.

Sampling is at 20 millisecond intervals. Calibration follows principles similar to those given by McConkie (1980), whose plea for fuller details of equipment and procedure has been heeded here, at least in proportion to the space available. It is indeed difficult to assess findings in some

reports about eye movements in reading, as so little information is given
on how the eye movements were recorded. Nevertheless, as this is not a
technical paper, but one concerned with applications of eye movement
study to normal or realistic reading tasks, it is to three examples of our
preliminary studies that we now turn.

STUDIES OF READING IN TELEVISION PROGRAMMES

It is not intended to be typical that two of the studies referred to here
are of reading written information which forms part of television
programmes, rather than reading printed text. This is accidental in the
sense that these areas, in which the equipment is being tried out, as
it is in other work, have been the first to yield results. They do not
suggest that the main thrust is to be diverted away from the reader/
text interaction in books which has been mentioned earlier, though there
are nevertheless interesting and important applied areas opened up by the
study of the reading of subtitles on foreign language television as in
the study of programmes intended to help young children to read. .However,
despite the pervasiveness of television, it remains a medium which has
been submitted to relatively little hard scrutiny; even its overtly
educational output tends to be evaluated by questionnaires sent to
teachers, to the extent that it is ever   examined at all.

Exceptions to this have not generally included the subtitling versus
dubbing controversy which affects many people outside the English speaking
countries in the world. Considerations of national language policy,
cultural and political influence, verisimilitude of the programme (as
opposed to the 'spaghetti western' look), time available for translation
before broadcasting, and cost all enter into the debate. Once a country
does opt for extensive use of subtitling, rather than dubbing, a good
many other questions arise about the graphic design of the titles, the
language used in them, the duration for which they remain on the screen,
and so on.

Some of these questions lend themselves to investigation using eye
movement equipment, since it is possible to judge the attention value
of the subtitles in relation to what else is on the screen. From a
point of view of using the present apparatus, there are also advantages
in watching a vertically mounted display at a fixed distance, as a
television screen is, since head movements in adults seem to be much less
in these conditions than in reading.

As a new technical system for actually mounting subtitles on the screen
was at that time being implemented, Belgian Radio Television supported a
small-scale study in 1982 of the design and utlilisation of subtitles.
A specially prepared version of Dallas, chosen because it afforded a
preview of the next episode which some viewers eagerly await, was given
some normal subtitles and others, in seven groups, which were in some way
different. For example some were shown for a shorter duration than the
usual (and standard) six seconds and others for longer; some contained
words judged to be hard, some were on two lines with an unusual break
point, and some on two lines where one would suffice. Subjects were 16
adults chosen from the panel of people who help Belgian Radio Television
in research (usually of the audience survey type). They were selected
to represent differing age groups and educational levels.

For various reasons only ten records could be analysed, using pen recorder

output since temporary difficulties were experienced at this stage with
the microprocessor part of the equipment. Records were manually
examined, with attention to the horizontal and vertical, and a judgement
was made about whether the subject was looking in the direction of the
subtitles at any given time. A comparison with a record of when the
subtitles were actually shown produced a figure, to the nearest ¼ second,
for the time spent reading a subtitle. Although the procedure was
onerous, slow and did not yield fine measures, nevertheless some
information was obtained on questions which have not been examined in eye
movement research, except for some little work on the reading by the deaf
of specially prepared subtitles (e.g. O'Bryan, 1976; Baker, 1981).

From our study,fuller details of which are given in Muylaert, Nootens,
Poesmans and Pugh (1982a and b), the most clear finding is that almost
all the subjects looked towards a subtitle as soon as it appeared and
spent most of the time for which it was displayed in looking at it,
probably reading it. Indeed from our records of 141 different subtitles
available to the ten different subjects, only ten cases occurred where a
subtitle was not looked at. These were all in the record of one person
whose overall reading was much the fastest and whose knowledge of English
appeared to be excellent. Other findings were that the different types
of subtitle were read similarly from person to person, in that inter-
correlations between time spent in looking towards the subtitles of
different types were high in most cases, although there were some low
relationships which would reward further examination with a larger sample.
Within the normal subtitles, between which the others were interspersed
in groups, correlations were high suggesting both satisfactory testing
and analysis and a consistency in approach by the viewers to their use of
subtitles. Comparisons of means were also made, the clearest finding
being that a significantly greater proportion of available time was spent
on the subtitles of the shortest duration. This is not surprising in
itself, though it might be expected from subjective evidence collected
by audience research that some of the other kinds of subtitle would also
cause difficulties which would be shown in these comparisons of differences.

The main theoretical issue of psychological interest arising from this
study has to do with the complex monitoring task carried out by the
viewer who has not only to attend to image changes and sound signals,
usually designed together by the producer, but also subtitles which are
rarely well synchronised with either the original changes of camera shot
or with the sound. Some similar questions are raised by programmes
designed to teach children to read by watching television, except here, of
course, deliberate attempts are made to capture the attention, at certain
points, by the text presented on the screen.

For our study of learning to read with the help of television we took one
of the BBC Look and Read series, entitled The Boy from Space. A part of
programme 5, The Hold-up, was selected as containing examples of several
different kinds of presentation of written language. The extract, of
about 8 minutes duration, was broadcast in February, 1982. It is,
therefore, current even though the series (and this programme) seem to
originate in 1971. The series is accompanied by booklets for teachers
and for pupils. They were not used for this study, in which records for
analysis were obtained from 12 children of mean age 7.4 years and mean
reading age 8.6. In fact a stratified sample by sex and reading age
was taken and the high mean is a reflection of the very high ability of
children in this particular school.

Procedure was to show children a part of the film in order to allow them
to settle down and relax before any attempts were made to calibrate.
Although only three records had to be discarded, and only one of these
due to problems in calibration, nevertheless these children had the
greatest of difficulty in keeping still for much longer than the
calibration period of, normally, less than a minute. Accordingly a
simple chin rest was devised as an aid for them, although the problem
was not entirely overcome, In fact, it was only when their attention
appeared to be strongly held by the screen that they kept their heads and
bodies quite still.

The method of recording was by mixing the digitalised signals from each
eye, shown as a dot, with the image from a camera focussed on the screen.
This is not good practice for various reasons, but a suitable mixer was
not available. Analysis was by reviewing these videotapes which were
thus made of the programme with the eye movements superimposed. Note
was taken of the periods spent in looking at the various titles, some of
which were subtitles, some superimposed titles, some on placards held by
actors in the story and others in the form of animated sequences.

Results of the analysis show that these children also spent a good deal
of the time when text was on the screen in looking at it. However,
perhaps because they were so much less accustomed to reading titles than
the Belgian adults, they tended to take their time before turning to the
title whereas the appearance of a subtitle had almost always virtually
immediately caught the attention of the adults. It was also noted at
certain points that children seemed not to be looking at any particular
part of the screen but that the dot would be stuck at one of the boundaries.
At first this was thought to be due to a fault in calibration or then to
movement of the head, as indeed it may well have partly been since
the registration usually returned when there was animation or text on
the screen. Recent papers by Rice (1983) and by Pezdek and Hartman (1983)
both refer to the sophisticated viewing behaviour of children and the
latter in particular suggest, from their own findings, that children
can apportion attention very selectively between watching and listening.
However, it was our impression that the poorer readers, albeit in a group
of able children, were those whose attention wandered most. This warrants
further investigation with different materials to ascertain whether
literacy at a certain level is associated with certain types of viewing
activity, or, if our impression is confirmed, whether what is at issue is
attention span which is in turn related to reading ability.

On the more immediate question of what type of text gained attention most
readily and for most of the time, it was found that animated sequences
were given attention by all viewers, as were the placards held by an
actor. However, these were looked at for different proportions of time
available which is not as expected. Also the correlation between time
spent on the animated sequences and reading ability is very high
(rho = .80, p < .01) whereas it is much less for the placards
(rho = .53, p < .05). It is, of course, rather depressing that it is
the better readers who spend more time on the animated sequences which
are intended, at least in this case, to give phonic help. This finding
may well be related to our observation above about the different approaches
to viewing which seemed to be associated with different reading abilities.
The other types of title appeared to receive less consistent attention
from the group, whether they were normal subtitles, freestanding or

superimposed a quarter or so of the way up the picture. These latter
seemed to be least successful, very possibly simply because they were
so feint.

These are suggestive findings from preliminary work which seems well
worth pursuing both because of practical implications and because
of the light findings could cast on television viewing related to the
development of literacy, and vice versa, for example. A fuller report
of this study, made with E. Fisher, is in preparation and plans have
been made for further data collection in this area.

COMPARISON OF CLOZE AND NORMAL READING

Cloze texts are used for teaching, testing and in research. They are
prepared by deleting words which the reader has to supply. In some
of our earlier work (see, for a review, Neville and Pugh, 1982) we
examined errors made by children in several types of cloze task in
order to make inferences about the strategy they used in reading, as
opposed to listening for example. This work is continuing with adults
and a system for computer analysis has been developed (Pugh and Blenkhorn,
forthcoming). However, there is contention about how close to normal
reading the reading of cloze texts is, and despite the very considerable
literature on cloze procedure there seems to be little evidence
(introspection and analysis of responses apart) about how people read
cloze texts.

A further preliminary study was, therefore, made of how middle school
children read cloze texts compared with their reading of undeleted text.
Sixteen pen recorder records were collected from a sample selected by
reading ability (high, R.Q. circa 108 and low, R.Q. circa 90) and sex.
Mean age was 10 years, 3 months at the time of testing. The two forms
of the GAP test (McLeod and Unwin, 1970) were used, one in the original
and one in the undeleted version. In order to minimise the effects of
differences in difficulty, and number of no responses especially, only
the first two items of the tests were given. Order of administration
by form and version were systematically varied.

The spectacles proved suitable for this purpose and the binocular
records are capable of interpretation to test issues raised by, for
example, Bedwell, Grant and McKeown (1980) about problems of convergence
and divergence related to reading difficulty. Because of the writing
involved, it is not reasonably possible for subjects to keep their head
still so calibrated digital data could not be collected. Inspection of
the pen recorder records shows the cloze reading to be much slower, as
expected, and to be interrupted on many (though not all) occasions
when writing occurs. More detailed analysis should show whether divergence
was associated with points of difficulty and whether the different types
of reading observed are in any way related to level of reading ability.
However, careful preliminary inspection without quantification suggests
that the reading of the poorer readers is more akin to the cloze reading
of themselves and of the better readers than to the better readers'
ordinary reading. If borne out by the detailed analyses and by further
study this would confirm some of our earlier findings and suggest that
cloze procedure should be used only with the greatest caution as a
teaching device.

238   *A.K. Pugh*

CONCLUSION

The system so far developed appears to be capable of collecting usable
data without unduly restricting the reader. The findings given here are
tentative, and from preliminary studies which were exploratory of the
applications of the apparatus as well as the questions examined.
Nevertheless, there seem to be valuable leads for further studies of
reading under normal conditions.

REFERENCES

1.   Baker, R.G. Guidelines for the Subtitling of Television Programmes
     for the Deaf and Hard of Hearing. London: IBA, 1981.
2.   Bedwell, C.H., Grant, R. and McKeown, J.R. Visual and ocular control
     anomalies in relation to reading difficulty. British Journal of
     Educational Psychology, 50, 1978, 61 - 70.
3.   Clark, M.R. and Crane, H.D. Dynamic interactions in binocular vision.
     In J.W. Senders, D.F. Fisher and R.A. Monty (eds.) Eye Movements and
     the Higher Psychological Functions. Hillsdale, N.J.:Erlbaum, 1978.
4.   Clifton, C. Psycholingustic factors reflected in the eye. In K. Rayner
     (ed.), 1983.
5.   Haines, J. Eye movement recording using optoelectronic devices.
     In I. Martin and P.H. Venables (eds.) Techniques in Psychophysiology.
     Chichester: Wiley, 1980.
6.   Kolers, P.A. Buswell's discoveries. In R.A. Monty and J.W. Senders
     (eds.) Eye Movements and Psychological Processes. Hillsdale, N.J.:
     Erlbaum, 1976.
7.   McKonkie, G.W. Evaluating and Reporting Data Quality in Eye Movement
     Research. Urbana, Ill.: Center for the Study of Reading, 1980.
8.   McLeod, J. and Unwin, D. GAP Reading Comprehension Test. London:
     Heinemann, 1970.
9.   Muylaert, W., Nootens, J., Poesmans, D. and Pugh A.K. The Perception
     of Subtitles on Foreign Language Television Programmes. (In Dutch).
     Brussels: Belgian Radio Television Research Department, 1982a.
10. Muylaert, W., Nootens, J., Poesmans, D. and Pugh A.K. Design and
     utilisation of subtitles on foreign language television programmes.
     In P.H. Nelde (ed.) Theorie, Methoden und Modelle der
     Kontaktlinguistik. Bonn: Dümmler, 1982b.
11. Netchine, S., Guihou, M-C., Greenbaum, C. and Englander, G. Retour à
     la ligne, age des lecteurs et accessibilité au texte. Le Travail
     humain, 46, 1983, 139 - 153.
12. Neville, M.H. and Pugh, A.K. Towards Independent Reading. London:
     Heinemann, 1982.
13. O'Bryan, K. Captioning Television Programs for the Deaf. Boston,
     Mass.: WGBH Caption Center, 1976.
14. Pezdek, K. and Hartman, E.F. Children's television viewing: attention
     and comprehension of auditory versus visual information. Child
     Development, 54, 1983, 1015 - 1023.
15. Pugh, A.K. Silent Reading: an introduction to its study and teaching.
     London: Heinemann, 1978.
16. Pugh, A.K. Styles and strategies in silent reading. In P.A. Kolers,
     M.E. Wrolstad and H.Bouma (eds.) Processing of Visible Language 1.
     New York: Plenum, 1979.
17. Pugh, A.K. Eye movement studies of readers and text. In A.K. Pugh and
     J.M. Ulijn (eds.) Reading for Professional Purposes: studies in
     native and foreign languages. London: Heinemann, 1983a.

18. Pugh, A.K. Development of fluency and strategy in native and foreign language reading: some comparisons and contrasts. Paper for 3rd European Conference on Reading, Vienna, July 1983, for publication, 1983b.
19. Pugh, A.K. and Blenkhorn, P. Computer analysis of cloze responses: a comprehensive system and some findings from its application,forthcoming.
20. Rayner, K. (ed.) Eye Movements in Reading: perceptual and language processes. New York: Academic Press, 1983.
21. Rice, M. The role of television in language acquisition. Developmental Review, 3, 1983, 211 - 224.
22. Shebilske, W.L. and Fisher, D.F. Eye movement and context effects during reading of extended discourse. In K. Rayner (ed.), 1983.
23. Spiro, R.J. Bruce, B.C. and Brewer, W.F. (eds.) Theoretical Issues in Reading Comprehension. Hillsdale, N.J.: Erlbaum, 1980.

Theoretical and Applied Aspects of Eye Movement Research
A.G. Gale and F. Johnson (Editors)
© Elsevier Science Pulishers B.V. (North-Holland), 1984

SECONDARY TASK EFFECTS ON OCULOMOTOR BEHAVIOUR
IN READING

Hans-Willi Schroiff
Department of Psychology
Rheinisch-Westfälische Technische Hochschule, Aachen
West-Germany

The effects of simultaneous performance of a secondary task on silent and oral reading for comprehension
were investigated in two experiments - the first study
also included text difficulty as an independent variable. The results indicate a pronounced effect of the
secondary task. To a lesser extent text difficulty
also influenced total reading time and the number of
forward and backward fixations. In silent reading
densities of progressions and regressions were increased compared to reading aloud. The results are
interpreted within a reader's strategy to overcome
capacity limitations imposed by text difficulty and
simultaneous performance of a secondary task.

INTRODUCTION

Viewing man as an information processing system led to numerous efforts
which tried to determine what is called his "processing capacity". The term
usually stands for the amount of resources or effort by means of which
cognitive events are initiated or maintained (NORMAN & BOBROW, 1975;
NAVON & GOPHER, 1979; EYSENCK, 1982). If a task is performed poorly according to NORMAN & BOBROW (1975) two reasons are possible: either the process
is data limited (poor quality of the data) or resource limited. Exceeding
the upper limit of the available resources becomes more probable when two
tasks (a primary and a secondary task) have to be performed simultaneously.
The experimental paradigm based on this assumption (the "secondary task"
paradigm) offers a way to describe a defined primary task in terms of its
demands on processing resources: when control performance on the secondary
task alone is better than secondary task performance in combination with a
primary task the latter is said to consume processing capacity (provided
that the resource demands of the primary task are fulfilled first).
The basic assumption of this approach is the so-called "principle of complementarity" according to which several tasks compete for resources from a
general pool which can be allocated in a flexible way. In this context,
several complicating issues deserve to be mentioned:

(1) Resources may not be constant under different instructional and situational conditions (KAHNEMANN, 1973). Therefore processing capacity allocated is not only a function of task demands, but also of the individual's
allocation strategy which may be influenced by external and internal
factors.

(2) The process of allocating resources between tasks itself requires
capacity. This leaves the possibility that the sum of demands when

performing two tasks simultaneously is larger than the amount when the same tasks are performed separately. In a similar vein qualitative changes may occur when additional mental operations are required which are not needed in single task performance.

(3) EYSENCK (1982) has pointed out that the cognitive system is not perfectly free to distribute its resources as is made believe by the "principle of complementarity".Behaviour in a STROOP-test or the orienting reaction are examples of mandatory processes not under complete control of an allocation strategy.

(4) Subjects may differ in the relative importance they place on one of the tasks though receiving the general instruction to perform the secondary task without letting it interfere with the primary task.

We have tried to incorporate these comments in a simple non-steady state model of secondary task performance which also provides the theoretical basis for our present experiment. Total workload L_{12} is regarded as the ratio of the joint demands of task 1 and task 2 plus the demands placed by their interference to the capacity available. Each task is thought to be composed of a number of elementary information processes. The total workload of the task can be seen as the sum of workloads placed by its elementary information processes. The amount of workload placed by the interference between the tasks is represented by their product. Maximum capacity C_{max} is weighed by an index S_{12} which symbolizes situational and motivational factors involved in performing tasks 1 and 2. For a detailed description of the model we refer to SCHROIFF & KOCH (in preparation). Equation 1 summarizes the model:

$$L_{12} = \frac{\sum_{n=1}^{N} D_{1n} + \sum_{m=1}^{M} D_{2m} + \sum_{n,m}^{N,M} W_{nm} D_{1n} D_{2m}}{S_{12} C_{max}} \qquad (1)$$

If $L_{12} > 1$, allocation strategies are needed which place a demand D_s on total capacity. There are several allocation strategies possible. Assuming that highest priority is given to task 1 an allocation strategy can be described as follows: given the amount of resources $S_{12} C_{max} - D_s$, allocate the partial amount $\sum_{n=1}^{N} D_{1n} + \sum_{m=1}^{M} D_{2m}$ to task 1 and the remaining partial amount to task 2.

We have employed the secondary task paradigm using reading as a controllable primary task in order to get a better understanding of the reading process in general and of the use of processing capacity in that domain. A number of studies in this field have been done by BRITTON and his associates (BRITTON & PRICE, 1981; BRITTON & TESSER, 1982). In these studies a secondary task technique was used to assess the cognitive demands of reading various text passages. The main results - as they are important here - can be summarized as follows:

(1) Cognitive capacity was used up more completely in reading passages with more discourse level meaning.

(2) Difficult passages take longer to read than easy passages.

(3) Prior knowledge used in a cognitive task also uses up resources of the same limited capacity system that is used to perform the reading task by
- retrieving knowledge from inactive memory
- change of state from an inactive to an active status
- maintenance of the active state
- requiring additional cognitive operations

(4) More complex surface forms used more cognitive capacity to process. This is probably due to additional mental activities required.

It might now be interesting to take a closer look at the allocation strategies adopted by the subjects in this situation. In our study we tried to investigate the effects of additional cognitive load on the allocation of processing resources as indicated by eye fixations. As a second independent variable we varied text difficulty mainly in order to assess possible in-interaction effects.

Before discussing some expected effects of our independent variables we explain the secondary task employed. The selection of the secondary task was partially determined by the experimental setting which ruled out visual stimuli and verbal responses. BROWN (1978) has advised to use discrete stimuli of constant load and a forced-paced performing schedule. The locus of interference should occur within the processing mechanisms rather than at the sensory input or motor output. A motor reaction to randomly presented acoustical signals seemed to meet these criteria when reading is used as a primary task. NAVON & GOPHER (1979, 1980), however, have argued in favor of a "multiple resources" approach stating that "... each channel (auditory or visual) may have its own capacity" (1979, p. 233). An experiment reported by BRITTON & PRICE (1981) was decisive in choosing the auditory task. The authors varied the allocation of processing resources for the primary task (reading) and the secondary task (manual reaction to occasional random clicks presented during reading) by appropriate instructions. A resulting performance operating characteristic confirmed the hypothesis that for this task combination the principle of complementarity obviously holds.

Effects of text difficulty on fixation behavior have been extensively studied. Though a precise definition of text difficulty is still lacking (HELLER, 1982) one may generally conclude that a lower fixation density is associated with "easy" text paragraphs. Explanations are partly based on linguistic considerations: JUST & CARPENTER (1980) attribute differences in text difficulty to factors like word encoding, lexical access, assigning semantic roles, and relating information in a given sentence to the content of previous sentences and previous knowledge. It has been argued that increased processing load may result in a shrinkage of the "useful field of view" (MACKWORTH, 1965) which in turn implies a higher fixation density per line of text. VANECEK (1972) varied text difficulty by using different orders of approximation to normal text. His basic assumption was that hypotheses about text elements are based on information acquired via peripheral vision. The higher the semantic and syntactic redundancy the higher is the probability that hypotheses are confirmed and the less capacity is needed for a correct reading of the paragraph. Low fixation density found with easy texts was explained by a broader useful field of view (see also IKEDA & SAIDA, 1975).

Concerning the effects of text difficulty we expect to confirm the known findings that more difficult texts are associated with an increased number of fixations per line indicating a higher processing load placed by the text (see HELLER, 1982).

Which effects are to be expected when a secondary task is introduced ? According to our model we should expect an increase in the manual reaction times to an acoustical signal during reading compared to the response latency when the secondary task is performed alone ("free" reaction). Furthermore we hypothesize that response latencies for a more difficult text are increased compared to an easier text. If the subject follows the allocation strategy specified above no deterioration in performance on the reading task should occur (longer reading times, reading errors etc.). In a number of test trials, however, we observed that subjects obviously tried to use a "trade-off" strategy which distributed resources more equally on the tasks - contrary to instructions given. The limitations of processing capacity assigned for the primary task probably result in a further shrinkage of the functional visual field which in turn should imply an increased number of forward and backward fixations. We hold that interference occurs despite instructions to give priority to the reading task which is not predicted by the allocation strategy mentioned above.

In this study our analysis of fixation behaviour will be restricted to the number of progressions and regressions which could be reliably scored by a frame-by-frame video tape analysis. Further results concerning fixation times, saccade durations, and saccade length will be reported elsewhere (SCHROIFF & KOCH, in preparation).

METHOD

Experiment I

Subjects
28 students of the RWTH Aachen participated in the experiment. Applicants wearing glasses or contact lenses could not be considered due to restrictions imposed by the registration system. Subjects participated for a total of about 15 min.

Apparatus
Registration of eye-movements was obtained via the DEBIC-80 system which we have described elsewhere (SCHROIFF, 1983; SCHROIFF & SOMMER, in preparation). The texts were photographed on slides and projected via a random access projector (ZAK-EPta 5a) on a display about 190 cm away from the subject. The distance between the first and the last letter of a line of text was 20°. The acoustical signal (2500 Hz tone) was generated by a Massey-Dickinson wave generator, reaction times were measured in msec by a self-built device.

Materials
The texts used in this study were selected from a pool of alternatives on the basis of their "cloze"-values (DICKES & STEIWER, 1977) as well as independent ratings (SCHROIFF & KOCH, in preparation). The text refered to in the following as "easy" is part of a German reading aptitude test by ANGER, BARTMANN & VOIGT (1971), the text refered to as "difficult" is the slightly modified first paragraph of MUSIL's novel "Der Mann ohne Eigenschaften" (1965). "Cloze"-values for the easy text were C= 27.07 and C= -19.09 for the difficult text, respectively. The values indicate a facilitated readability of the easy text which was confirmed by the results of the individual ratings. The texts were identical with respect to word number, but differed in text length (161.7 cm for the easy text, 182.9 cm

for the difficult text). To "equalize" texts on this level dependent
variables for the easy text were corrected when performing statistical
analyses between texts.

Design
Our experimental design is a 2 x 2 factorial design with repeated measurements on the text factor (two texts differing with respect to their readability). Combined with a task factor (secondary task: yes/no) the four
experimental conditions are labeled E (reading easy text), D (reading
difficult text), ES (reading easy text plus reaction to tone), and DS
(reading difficult text plus reaction to tone).
According to their sequence of registration the subjects were assigned to
the experimental group with its conditions ES/DS and to the control group
with conditions E/D. Since 28 subjects were tested there were 7 cases per
cell in the experiment. Dependent variables were
(a) "free" reaction time (performing the secondary task alone)
(b) total reading time
(c) number of forward fixations
(d) number of backward fixations
(e) number of reading errors
(f) response latencies for the secondary task during reading
(g) number of items correct on a multiple choice test on the contentsof the
passage read
(h) total time spent on the multiple choice test

Procedure
During a dark adaption period of about 5 min the subjects were instructed
to read the texts aloud in their normal reading speed and that several
questions on the contents of the text had to be answered after reading.
The experimental group received instructions to react to the acoustical
signal by a keypress without letting it interfere with the simultaneous
reading task. The control group was instructed to ignore the signal.
After the calibration routine had been completed the texts were presented
successively. Presentation order was completely balanced in all experimental conditions. The acoustical signal was triggered when the subject
reached the position of the words No. 15, 45, and 107 in both texts. It
could be stopped within a period of 2000 msec by a keypress of the subject
only.
50% of the subjects performed the "secondary task alone" before reading the
texts, the other 50% after having read the texts. "Free" reaction times
were scored in 10 trials and then averaged.
Finally the subjects were timed when answering the multiple choice questions on the contents of the texts.

Results
Before performing statistical analyses on fixation behaviour and other
dependent variables we had to confirm our assumption that the secondary
task indeed did compete for processing resources. Mean reaction times
under secondary task conditions and "free" reaction times are plotted in
figure 1a. RT-differences between the "free" reaction and the text reading
conditions are highly significant (χ^2 = 48.73, df=3, p < .001).
RT-difference between the two texts did not reach statistical significance
though mean reaction times lie in the direction expected. There were no
substantial correlations between "free" reaction time and other dependent
variables, systematic differences between the experimental and control
group were not observed.

Mean reading times for the experimental conditions are shown in figure 1b:

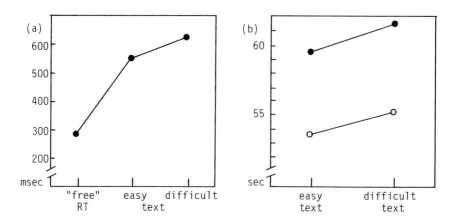

<u>FIG. 1</u>: (a) Mean reaction times for the secondary task under various experi-
mental conditions (b) Mean reading times for the control group
(O——O) and the experimental group (●——●).

An analysis of variance showed a significant effect of the secondary task
on total reading time $[F(1,28) = 11.23, p(F) < .01]$. The secondary task also
produced more reading errors $[F(1,28) = 9.09, p(F) < .01]$, but also text
difficulty explained part of the variance $[F(1,28) = 5.25, p(F) < .05]$.

Fixation behavior was analyzed separately for progressions and regressions.
The results are plotted in figure 2:

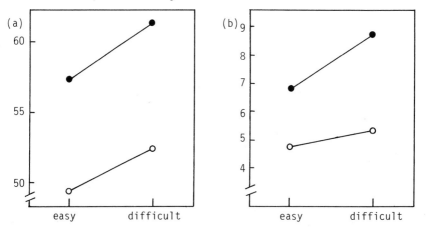

<u>FIG. 2</u>: (a) Mean number of total progressions (b) Mean number of total
regressions

The simultaneous performance of a secondary task increases average fixation density for forward fixations $[F(1,28) = 8.85, p(F) < .01]$. Text difficulty and interaction effects are not significant. The same result is obtained for the mean number of regressions: the secondary task is responsible for an increase in regressions $[F(1,28) = 6.53, p(F) < .01]$.
Finally, the analysis of the time to work on the multiple choice test showed an effect of text difficulty $[F(1,28) = 15.09, p(F) < .01]$.

Experiment II

On the basis of these results a follow-up study was carried out where we tried to replicate our findings and introduced some changes in the experimental conditions. These changes consisted in (a) silent reading of a text and (b) analysis of the number of text propositions recalled as a measure of text comprehension.

When reading aloud it may be possible that a reader allocates a large amount of his processing capacity to reading without an audible mistake (correctness of pronounciation, no stuttering etc.). If this assumption holds, text comprehension is more likely to suffer under "reading aloud" conditions. Both the relatively low numbers of correct answers to our multiple choice questions and the forced-paced character of oral reading seem to indicate that in this reading mode integration of previously read material may be impaired.
Furthermore we observed that multiple choice questions partly cued correct answers to them. So in experiment II we switched to "free recall" by assessing the number of text propositions recalled as a dependent variable (KINTSCH, 1974; TURNER & GREENE, 1979).
All other experimental conditions and the experimental setting remained the same as in the first experiment. A total of 14 subjects was investigated in only two of the experimental conditions (D and DS).

Results

Compared to oral reading mean reading times in the silent reading condition were generally increased (tD = 55.07 sec; tDS = 68.72 sec). Figure 3 shows mean number of progressions and regressions compared to the corresponding values for oral reading from experiment I.

Comparison between silent and oral reading of the difficult text showed that for both conditions differences in reading time are attributable to the simultaneous performance of task 2 $[F(1,14) = 5.08, p(F) < .05]$, while number of forward fixations are influenced by the reading mode $[F(1,14) = 26.92, p(F) < .01]$ and the secondary task $[F(1,14) = 10.09, p(F) < .01]$. The same holds for regressions: silent reading $[F(1,14) = 9.74, p(F) < .01]$ as well as the secondary task $[F(1,14) = 11.29, p(F) < .01]$ increase the number of backward fixations. No significant interaction effect was found in these analyses.
Mean latencies for the keypress showed no significant differences between silent and oral reading though in the silent reading condition reaction time was slightly increased. Analyses of propositions were abandoned due to the minor degree of coherence in the difficult text: only very few propositions were recalled by the subjects indicating a very strong recency effect.

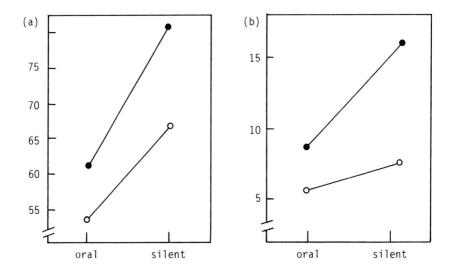

FIG. 3: (a) Mean number of total progressions for oral and silent reading
        (b) Mean number of total regressions for oral and silent reading
        (●——●) secondary task/ (○——○) no secondary task.

DISCUSSION

We have tried to demonstrate in these two studies the effects of a secondary
task on the reading of paragraphs which varied in text difficulty. One may
generally conclude that the two selected tasks share a common pool of
resources: response latencies for the keypress were significantly increased
in all experimental conditions compared to a "free" reaction. The more
difficult text entailed the longer latency expected. On the other hand
interference effects were also found in the reading task despite the instructions to give highest priority to reading the text. Subjects obviously
used a "trade-off" allocation strategy which provided less than $D_1 + D_{12}$
for the primary task: compared to a control group the secondary task increased reading time, number of forward and backward fixations, and the
number of reading errors. These effects can be interpreted as a general
slow-down of the processing rate in this self-paced task combined with a
breaking up of the line of text into smaller processing units. A possible
explanation may be based on a further shrinkage of the useful field of view
due to additional cognitive load (WILLIAMS, 1982). This hypothesis will be
tested explicitly by artificially narrowing down the text area around the
fixation point. This reduction of the normal field of view should then have
no influence on reading time and fixation behaviour in secondary task conditions. Whether a slow-down of the processing rate is a crucial point
could be tested by using a strictly forced-paced reading task.
These effects are considered to be parts of a reader's strategy to get
along with the additional constraints on processing capacity imposed by the
secondary task.In spite of this strategy, however, errors in reading are
more probable as indicated by the number of reading errors.

When reading silently we found a further increase in reading time under
secondary task conditions. The same holds for the number of progressions
and regressions. We interprete these results as the effect of a totally
self-paced reading mode which may be confirmed by the observation that in
silent reading readers frequently jumped back for one or more lines in
order to search for referents. These jumps rarely occured when reading
aloud. Our doubts that subjects in the reading aloud condition were not
reading for comprehension were corroborated by the post-experimental in-
quiries: Subjects in the silent reading condition more often reported that
they had tried to comprehend the contents by going back in text and read
a sentence again. This also explained the higher densities in progressions
and regressions found in the silent reading condition. Subjects reading
aloud claimed that being tied to consecutive reading prohibited to jump
back and integrate previous parts of the text.

Apart from the necessity to replicate our findings we clearly feel the need
for several modifications:
(1) Text parameters (word number, word length, number of text lines etc.)
should be kept constant when varying text difficulty. A promising approach
by BRANSFORD & JOHNSON (1972) varied comprehension difficulty by providing
a number of seemingly unrelated sentences with an integrating headline.
(2) Propositional analysis of free recall protocols seems to make more
sense when using texts with a certain degree of coherence. The analysis
of free recall protocols had to be abandoned due to the difficulty of the
MUSIL paragraph. So our hypotheses that reading for comprehension is more
effective under silent reading conditions and consequently placing more
demands on general capacity still await verification.

REFERENCES

(1) Anger,H., Bartmann,R. & Voigt,M.   Verständiges Lesen VL 5-6, Schul-
    und Begabungstest für 5. und 6.Klassen (bearbeitet von Raatz, U.,1971).
    Weinheim: Beltz, 1971.
(2) Bransford, J.D. & Johnson,M.K.   Contextual prerequisites for under-
    standing: some investigations on comprehension and recall. Journal of
    Verbal Learning and Verbal Behavior, 1972,11, 717-726.
(3) Britton,B.K. & Price,K.   Use of cognitive capacity in reading:
    a performance operating characteristic. Perceptual and Motor Skills,
    1981,52, 291-298.
(4) Britton,B.K. & Tesser,A.   Effects of prior knowledge on the use of
    cognitive capacity in three complex cognitive tasks. Journal of Verbal
    Learning and verbal Behavior, 1982,21, 421-436.
(5) Brown,I.D.   Dual task methods of assessing work-load. Ergonomics,
    1978,21, 221-224.
(6) Dickes,P. & Steiwer,L.   Ausarbeitung von Lesbarkeitsformeln für die
    deutsche Sprache. Zeitschrift für Entwicklungspsychologie und Pädago-
    gische Psychologie, 1977,9, 20-28.
(7) Eysenck,M.   Attention and arousal. Springer: Heidelberg, 1982.
(8) Heller,D.   Eye movements in reading. In. R.Groner & P.Fraisse (eds.)
    Cognition and eye movements. Amsterdam: North-Holland Publishing
    Company, 1982.
(9) Ikeda,M. & Saida,S.   Span of recognition in reading. Vision Research,
    1978,18, 83-88.
(10) Just,M.A. & Carpenter,P.A.   A theory of reading: from eye fixation to
    comprehension. Psychological Review, 1980,87, 329-354.

(11) Kahnemann,D.   Attention and effort. Englewood Cliffs, N.J.: Prentice-
     Hall, 1973.
(12) Kintsch,W.   The representation of meaning in memory. Hillsdale, N.J.:
     Erlbaum, 1974.
(13) Mackworth,N.H.   Visual noise causes tunnel vision. Psychonomic
     Science, 1965,3, 67-68.
(14) Musil,R.   Der Mann ohne Eigenschaften. Hamburg: Rowohlt, 1965.
(15) Navon,D. & Gopher,D.   On the economy of the human processing system.
     Psychological Review, 1979,86, 214-255.
(16) Navon;D. & Gopher,D.   Task difficulty, resources, and dual-task per-
     formance. In: R.S.Nickerson (ed.) Attention and performance VIII.
     Hillsdale, N.J.: Erlbaum, 1980.
(17) Norman,D.A. & Bobrow,D.G.   On data-limited and resource-limited pro-
     cesses. Cognitive Psychology, 1975,7, 44-64.
(18) Schroiff,H.W.   Experimentelle Untersuchungen zur Reliabilität und Va-
     lidität von Blickbewegungsdaten. Unpublished doctoral dissertation,
     RWTH Aachen, 1983.
(19) Schroiff,H.W. & Sommer,D.   The DEBIC-80 eye movement registration
     system: a user's report. (manuscript in preparation).
(20) Schroiff,H.W. & Koch,M.   Doppeltätigkeiten und Fixationsparameter
     beim Lesen. (manuscript in preparation).
(21) Turner,A. & Greene,E.   The construction and use of a propositional
     text base. JSAS Catalog of Selected Documents in Psychology, 1979.
(22) Vanecek,E.   Fixationsdauer und Fixationsfrequenz beim stillen Lesen
     von Sprachapproximationen. Zeitschrift für experimentelle und ange-
     wandte Psychologie, 1972,19, 671-689.
(23) Williams,L.J.   Cognitive load and the functional field of view.
     Human factors, 1982,24, 683-692.

ACKNOWLEDGEMENTS

The author wishes to thank Dr.D.Heller for helpful comments on an early
version of the paper and Miss M.Koch for her assistance in collecting the
data. Thanks are also due to Norbert A. Streitz for his discussions on the
workload model.

Theoretical and Applied Aspects of Eye Movement Research
A.G. Gale and F. Johnson (Editors)
©Elsevier Science Pulishers B.V. (North-Holland), 1984

OCULAR   MOTOR   PROBLEMS   OF   LEARNING   TO   READ

J. F. Stein and S. Fowler

University Laboratory of Physiology,
Parks Road,
OXFORD OX1 3PT
ENGLAND.

Learning to read makes higher demands on a child's ocular motor
control system than he has ever faced before. We wish to examine here
how children solve a classical perceptual problem for this purpose, the
problem of how to maintain a stable view of the world even though the
eyes are moving all the time. How do we use information about the
direction in which the eyes are pointing to fix the location of observed
objects even after the eyes have moved on?

In order to identify the position of an object in the outside world
it is necessary to make rapid and reliable associations between retinal
signals about its nature, and extraretinal signals indicating the
direction in which the eyes are pointing at the time it is inspected.
Determination of visual direction can not be achieved by retinal signals
alone. Information about the direction of gaze with respect to the
centre of a subject's visual world, his 'ego-centre', is essential as
well. ·

Extraretinal cues indicating the direction of gaze are of two kinds.
Proprioceptive signals are fed back from the ocular muscles and orbital
tissues. These report the position of the eyes within the orbits at any
moment. The second source of extraretinal information is derived from
the ocular motor control system itself. In addition to despatching
commands to the ocular muscle motoneurones the control centres direct the
same signals, known as efferenzkopie or corollary discharge, to visual
processing areas. This information is essential for the visual system
to be able to ignore retinal image movements which are purely a
consequence of the eyes moving, and do not indicate that anything is
really moving in the outside world.

Under most circumstances both proprioceptive feedback and corollary
discharge tell the same story about the position of the eyes. Corollary
discharge informs the visual system where the subject intended to
position his eyes, and proprioceptive feedback confirms that he
succeeded. However sometimes confusion can arise. High level ocular
motor control centres issue separate commands for lateral deviation of
the eyes, and for vergence movements to deal with the third dimension
(1), distance. For these movements both the eyes are yoked together in
different ways. But proprioceptive signals come from each eye muscle
individually. Hence feedforward and feedback signals can occasionally
disagree with each other, especially during inaccurate vergence; this
may cause great confusion.

Association of extraretinal and retinal signals for the purpose of determining the location of objects probably takes place in the posterior part of the parietal lobe and angular gyrus. Here neurones respond not only to visual stimuli but are modulated by the direction in which the eyes are pointing at the time (2). Lesions of the posterior part of the parietal lobe disturb accurate localisation of objects (3). In humans it is probable that it is the right parietal lobe which is particularly important for such visuospatial localisation (4).

When a child begins to learn to read accurate visual localisation becomes more important than ever before. Small letters viewed at the normal reading distance subtend only $1/4^{\circ}$ of visual angle. Hence a child must learn to control and register the position of his eyes with this degree of precision if he is to learn to read properly. If, when inspecting the 'd' in the word 'dog', a child does not know whether his eyes were to the left or right of the 'o', he can easily misread the word as 'god'. This type of missequencing error is characteristic of beginning readers and children suffering from dyslexia (5).

$1/4^{\circ}$ ocular motor precision is particularly difficult to achieve when starting to read for a number of reasons. The first is that in order to read small print, the eyes must be converged. Under these circumstances corollary discharge and proprioceptive feedback signals may disagree. In the corollary discharge information about lateral deviation of the eyes is combined with a single vergence signal, but proprioceptive feedback indicates that the two eyeballs are orientated differently in the orbits. Moreover neither eye's orbital angle corresponds to the direction of a letter with respect to the egocentre, since this lies between the two eyes. Hence to recover the precise direction of gaze from this mixture of signals must be very difficult.

Furthermore, particularly in young children, convergence is not always absolutely exact (6). Not only does this confuse still further the ocular motor cues as to the location of what is being inspected , but also the two foveae may often end up inspecting different letters. Dissimilar letters are usually fusable; so the child rarely experiences frank diplopia. But of course the results of this fusion are uninterpretable, unless the child develops a means for temporarily disregarding the foveal signals of one eye.

A third reason why learning to read presents especially onerous problems for the visual system arises from the fact that both sides of the foveae are represented in both hemispheres. Some axons derived from ganglion cells situated in the temporal part of the retina project not to the ipsilateral lateral geniculate nucleus (LGN) as usual, but take an aberrant course across the optic chiasm and pass to the contralateral LGN (7). Furthermore neurones at the border between primary (V1) and secondary visual cortex (V2) project via the splenium of the corpus callosum to the homologous cortical area in the opposite occipital lobe (8). Thus points representing up to 5° out in the left and right hemifields are probably connected together homotopically. Cells at the V1/V2 border may therefore end up receiving retinal information from mirror image points to the left and right of the centre of the fovea. This convergence from both halves of the fovea on to a single region of cortex clearly offers great potential for confusion; it may well help to explain why beginning readers and dyslexics tend to make so many mirror image (b for d type) confusions.

Nevertheless, since most children do finally learn to read it seems that they must find some way of overcoming this plethora of problems. We suggest that the solution which most children adopt is to develop a reliable 'leading' or 'reference' eye (9). We have preferred not to use the term 'dominant' in this context in order to avoid any implication that the eye which is selected has anything particular to do with the dominant left hemisphere. Indeed, as implied earlier, it seems quite likely that the right rather than the left hemisphere may control this process.

By 'leading' or 'reference' eye we mean the eye whose ocular motor and foveal signals are selected to be treated as references, and used for calibrating the visual direction of objects. Developing a reference eye simplifies the task of unravelling the confusion of ocular motor and retinal signals described earlier, in several important ways:

In the first place labelling the ocular motor signals provided by one of the eyes helps to solve the problem of corollary discharge and proprioceptive signals which disagree with each other during lateral deviation and convergence. The subject now only 'attends' to the proprioceptive signals of one eye. Hence the direction of the object fixated by the reference eye can be specified simply and uniquely. Similarly the foveal signals supplied by the reference eye now become the only ones which receive attention. Therefore when convergence is not precise the conflicting retinal information supplied by the other eye is simply ignored.

Developing a reference eye probably enhances the contribution to visual processing of ocular motor and retinal signals relating to that eye alone. This enhancement may take place in either hemisphere since each eye supplies retinal and ocular motor signals to both hemispheres. Once it is established however, confusion of the mixture of signals arriving from both the left and right sides of the fovea need no longer occur. The signals provided by the reference eye can be interpreted unambiguously, and conflicting mirror image signals supplied by the other eye can safely be discounted, when it is known that the reference eye did not supply them.

Thus there are good reasons for supposing that the development of a stable leading eye is important for helping to resolve the potential confusions engendered by binocular convergence, bilateral representation of the foveae and homotopic interhemispheric transfer of retinal and ocular motor signals. However although the idea that one eye is dominant has been around for a very long time (10), there has never been very good empirical evidence in its favour. Tests dependant upon retinal signals alone, such as those based on retinal rivalry, disregard the important contribution of ocular motor cues. Similarly most of those that take into account ocular motor information (e.g. the sighting eye test) are monocular; and therefore unreliable and inconclusive.

Demonstration that a subject has attained a fixed reference eye requires evidence that the ocular motor and retinal signals relating to foveal vision provided by one eye are habitually associated with each other, whilst those coming from the other eye are not. Hence a satisfactory test for stable reference eye needs to employ foveal vision, since it is the foveae which are bilaterally represented; and it should involve binocular vergence movements, so that conflicting eye position signals are sent to the two hemispheres. Such conditions are found in a visual illusion of movement, which is employed in an orthoptic test introduced by Dunlop (11).

The subject is asked to view through a synoptophore two slides of a macular sized house with central front door. In the slide viewed by the right eye a large post is placed on the left of the front door whilst that viewed by the left eye has a small post on the right of the door. This arrangement is designed to project retinal signals about the posts from each eye to the primary visual cortex on the same side. Most children are able to fuse the two images of the house and they see a large post to the left of the door and a small one to its right. The synoptophore tubes are then diverged. When this is done, for most subjects either the large or small post appears to move towards the front door. Ocular motor signals relating to each eye are projected first to the contralateral hemisphere (12). The apparent movement of one post implies that one eye's ocular motor signals are successfully associated, by transfer across the corpus callosum, with the retinal signals projected to the ipsilateral hemisphere that denote the post which that eye sees. For each child the test is repeated ten times. Normal children always see the same post move towards the door. Their 'leading' or 'reference' eye is said to be 'stable' or 'fixed'.

Whether the left or right eye is selected as the reference is not strongly correlated with the hand they use for writing. Thus 'crossed dominance', reference eye opposite to the preferred hand, does not seem to be particularly significant (9). What does seem to be important is that whichever eye is chosen as reference, that selection should be stable. We have studied several hundred normal and dyslexic children using our version of the Dunlop test. At all age levels from 5 - 11 if children have a stable reference eye on the Dunlop test their reading is significantly better than those who have not yet achieved a reliable leading eye (13). Also 2/3rds of children with specific developmental dyslexia have unfixed reference eyes compared with less than 1% of age/sex and I.Q. matched normal readers (14). Furthermore a simple procedure designed to encourage the development of a stable leading eye – occlusion of the other for reading and writing – helps dyslexic children to learn to read if it gives them a fixed reference (9).

We performed the Dunlop test on over 650 primary school children aged between 5 and 11 in order to see how stable ocular motor/retinal associations developed with age. We found that about 50% of children have achieved a stable leading eye by the age of 6 and about 7% more children achieve stability in each successive year, so that among 10 year olds over 80% have achieved a fixed leading eye. We were of course also interested to find out whether those children who had achieved a fixed leading eye were better readers than those who had not. Fig. 1 shows that this is the case; it supports our hypothesis that development of a fixed leading eye helps to resolve the confusion of ocular motor and retinal signals encountered during reading.

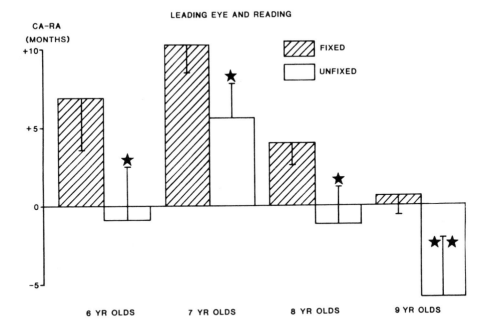

Fig. 1.   Reading age of children with fixed leading eye is significantly
higher than those without.   * = P<0.05;   ** = P<0.01.

We went on to study children with specific reading retardation. We
performed the Dunlop test in 80 dyslexics and 80 normal readers matched
for age and I.Q. The orthoptist performing the test did not know which
children were normal and which dyslexic. Nearly 2/3rds of the dyslexics
were found to have unstable leading eyes on the test whereas only 1
normal child was unfixed (Fig. 2). We found that the dyslexic children
with unfixed leading eyes made typically 'visual' errors when attempting
to read. These included mis-sequencing, substitution and reversal of
letters when writing.   To some extent, judgement of which errors are
specifically visual is bound to be arbitrary however.

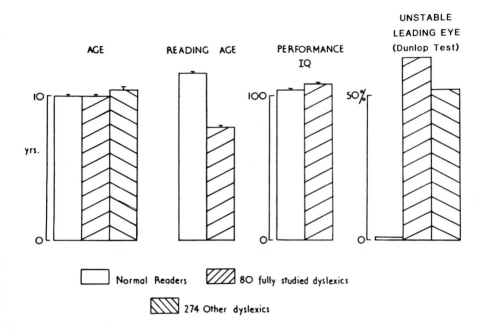

Fig. 2. Over 50% of dyslexics, matched with controls for age and
performance I.Q., have unstable leading eye.

        Algernon Charles Swinburne was probably dyslexic. When reading
poetry he would stand in front of his audience with a hand clamped over
one eye to help his reading.   We found many dyslexics with unstable
leading eyes had learnt similar tricks. Some would hold their hands over
one eye; others got so close to the page that only one eye could see the
print. Some developed a manifest squint when they first started to read,
with visual suppression of the squinting eye. We have even been referred
a number of memorable cases with the alarming symptom of sudden loss of
vision in one eye presenting when they started to learn to read (15).
All these manoeuvres have the common effect of blanking out the retinal
signals of one eye; this simplifies the retinal/ocular motor problems
faced by the child by removing confusing retinal signals coming from the
occluded eye and thereby encouraging the development of the other as the
leading eye. We therefore decided to try similar treatment on our
dyslexics with unstable leading eyes, by occluding their left eyes when
reading, in order to encourage the seeing eye to form stable associations
between its retinal and ocular motor signals, and thus to become the
fixed leading eye.

The results of our initial study were most encouraging (9). A group of 15 dyslexics with unfixed leading eyes were given plain glasses with the left lens occluded with opaque material, to wear when reading. 15 other dyslexics, matched for age and I.Q., were not so treated. The treated dyslexics achieved a fixed leading eye which persisted after the glasses were removed. At the same time their average reading age improved by over 12 months in the 6 months of observation. Those without treatment remained unfixed, and their average reading age barely improved by 6 months.

We have now almost completed a further double blind controlled study involving nearly 200 dyslexic children. All have been fully assessed by a reading specialist using the British Ability Scale tests together with writing and phonetic exercises of our own devising. They have all been examined also by an ophthalmologist and orthoptist who were not given the reading specialist's reports. A fourth person with no knowledge of the results of any of the tests then gave alternate children either plain spectacles or spectacles with the left lens occluded. We have thus obtained four randomly selected groups - children with unstable leading eyes who were given occluded spectacles; those who were not; children with fixed leading eyes who were given occluded spectacles and those who were not. The children have been followed up by the orthoptist at 2 monthly intervals. At 6 months reading, writing, phonetic and the Dunlop tests were repeated. Those who had a stable leading eye then had their glasses removed. Those who were still unstable were given occluding spectacles. We will therefore finally have a group of over 50 children with initially unfixed leading eyes who started off with no treatment for a 6 month control period and then were switched over to occlusion. The reading, writing and Dunlop tests were repeated at the end of one year and in some children they were repeated again 18 months after the children were first seen.

The results of this double blind study fully support those of our preliminary trial of occlusion in dyslexics with unstable leading eyes. Nearly 50% of children who were initially unfixed on the Dunlop test became fixed after 6 months occlusion compared with approximately 20% who learnt the trick spontaneously wearing plain spectacles. The percentage who achieved a fixed leading eye without occlusion is similar to the proportion of the 650 normal children still unfixed at these ages whom we found to become fixed spontaneously. The average reading age of 30 dyslexic children who started with unfixed leading eyes but achieved stability after 6 months monocular occlusion increased by 13.1 months whereas that of 38 children not given occluding spectacles who remained unfixed increased by only 6.4 months. The difference between these two groups is significant at better than the 1% level. When allowance is made for the I.Q. of each child by calculating the regression of their reading retardation (the difference between their chronological and reading ages) against non-verbal I.Q., for successfully treated and untreated children, the difference between the groups still remained highly significant (Fig. 3).

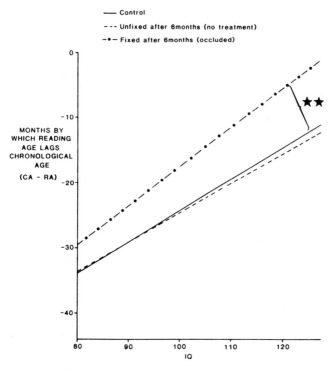

Fig. 3.   Regression of reading retardation (Reading age - Chronological
            age) performance I.Q. (B.A.S.) for all children before treatment
            (----------) 38 children who received no treatment (——————)
            and 30 children who became fixed after 6 months occlusion.
            ** = P<0.01.

    Thus monocular occlusion helps a substantial proportion of dyslexics
with unfixed leading eyes to achieve stable ocular motor/retinal
associations and thereby helps them to learn to read. This provides
further evidence in favour of our basic hypothesis, namely that many
children are dyslexic because they fail to achieve reliable associations
of ocular motor with retinal signals. We conclude therefore that many
dyslexics do indeed fail to solve the classical problem of perception
with which we started. They do not know precisely where their eyes are
pointing when they are trying to read. Normal children probably simplify
the ocular motor problems of learning to read by developing a reliable
reference eye. They learn to make stable ocular motor/retinal
associations with one eye in order to remove the confusion of conflicting
signals when both eyes compete with each other.

REFERENCES

1. Kenyon, R.V., Cinfreda, K. & Stark, L. (1978) Binocular eye movements during vergence, Vis. Res., 18, 545.

2. Mountcastle, V.B. & Motter, B.C. (1981) Light sensitive neurones in posterior parietal cortex. J. Neurosci., 1, 3-26.

3. Stein, J.F. (1978) "Effects of posterior parietal lobe cooling on reaching in the conscious monkey". In: "Active Touch" ed. G.Gordon, Pergamon Press.

4. De Renzi, E. (1983) Gaze paresis and posterior parietal lobe. Arch. Neurol., 39, 482.

5. Naidoo, S. (1969) Specific dyslexia, Pitman Medical.
6. Ashton, H. & Haines, J. (1982) Binocular coordination in severely disabled readers. Symposium on Vis. Dyslexia, Worthing, U.K.

7. Bunt, A.H., Minckler, D.S. & Johanson, G.W. (1978) Bilateral projections of the central retina of the monkey. J. Comp. Neurol., 171, 619.

8. Choudhury, P.B., Whitteridge, D. & Wilson, M.E. (1965) The function of the callosal connections of the visual cortex. Q.J. Exp. Physiol., 50, 214.

9. Stein, J.F. & Fowler, S. (1981) Visual dyslexia. Trends in Neuroscience, 4, 77-80.

10. Porta, L.B. (1593) De refractione. J. Carlinus, Naples.

11. Dunlop, P. (1972) Dyslexia. The orthoptic approach. Australian Orthoptic Journal, 12, 16-20.

12. Westheimer, S. & Mitchell, D.E. (1969) The sensory stimulus for disjunctive eye movements. Vis. Res., 9, 749-755.

13. Stein, J.F. & Fowler, S. (1982) Ocular motor dyslexia. Dyslexia Review, 5, 25-28.

14. Stein, J.F. & Fowler, M.S. (1982) Diagnosis of dyslexia by means of a new indicator of eye dominance. Br. J. Ophthalmol., 66, 332-336.

15. Fowler, M.S. & Stein, J.F. (1983) Ocular motor dominance as a factor in some orthoptic problems. Brit. Orthoptic J., 40, 433-45.

Section 4

EXAMINING DISPLAYS

Theoretical and Applied Aspects of Eye Movement Research
A.G. Gale and F. Johnson (Editors)
© Elsevier Science Publishers B.V. (North-Holland), 1984

EXAMINING DISPLAYS

INTRODUCTION

Dr. E. Megaw,
Department of Engineering Production,
University of Birmingham, England.

While it is true to say that all three of the contributions to this session
involved studies where subjects looked at displays, there the similarity
ends. The contributions of both Goillau and Hainline are concerned with
aspects of the saccadic control system while Mehta et al, investigated vergence movements. Goillau recorded the search strategies of experienced
adult observers as they examined either artificial or real scenes while
Hainline was interested in the main-sequence characteristics of saccades
recorded from young infants as they examined displays of simple geometric
forms of Gibsonian texture gradients. In the study by Mehta et al.,
practiced subjects observed random dot stereograms and their fixation disparity was monitored as they tracked a target which appeared to move in
depth either towards or away from them. The authors stress the advantages
of their method of assessing disparity over the more traditional paradigm
of nonius alignment. By introducing +0.5D lenses, the authors were able
to confirm predictions from their initial results concerning the accommoda-
tion/disparity conflict. Altering the spatial frequency of the random dot
targets did not affect the fixation disparities and the authors mention the
implications of this result for the Marr/Poggio model of stereopsis. The
eye movement data confirmed that subjects could tolerate a large retinal
image disparity while maintaining binocular fusion.

Hainline was surprised to find that, at least for some of the time, the
characteristics of infants' saccades closely resembled those of adults.
This suggests that the saccadic system is one of the first motor systems to
become co-ordinated. An interesting result for the infants is that higher
properties of slow saccades and so-called oscillations occurred when they
viewed the geometric forms as opposed to the texture gradients. The author
suggests this may have been due to the higher level of reticular arousal
induced by the gradients. It is worth noting the obvious problems experienced in obtaining clean recordings of eye movements from the young infants.

When observing real scenes, the results from Goillau's experiments indicate
that the probability of target detection is higher and search times shorter
when the display subtense is equal or greater than 45 degrees. Naturally,
it is not possible to conclude whether this is due to the size of the
display per se or to the magnification factor. In further experiments, the
effects of a grid superimposed over the display were found only to have a
small beneficial effect. It appears that there are several task factors
which determine whether or not a grid is likely to improve search performance.

Theoretical and Applied Aspects of Eye Movement Research
A.G. Gale and F. Johnson (Editors)
© Elsevier Science Publishers B.V. (North-Holland), 1984

THE EFFECT OF SPATIAL FREQUENCY ON FIXATION DISPARITY

Angeli Mehta, John P. Frisby and Ian M. Strachan

Department of Psychology   Department of Ophthalmology
University of Sheffield   Royal Hallamshire Hospital
Sheffield S10 2TN   Sheffield S10 2JF

The role of spatial frequency (SF) tuned channels in
vergence control was investigated using a fixation
disparity paradigm and objective eye movement recordings.
No significant influence of stimulus SF content on
fixation disparity was observed. The implications of
this result for SF-tuned models of stereopsis are
discussed.

INTRODUCTION

The central feature of Marr and Poggio's theory of stereopsis (5) is that
large and small disparities are processed with low and high SF tuned
channels respectively. The theory limits the disparity range for each
channel to about +/- the width of its receptive field centres. The advantage of doing this is that it drastically reduces the number of ambiguous
matches each channel needs to resolve, although at the price of usually
placing out of range at any given moment a good deal of medium/high SF
information. Hence the model incorporates a vergence control structure
in which the outputs of relatively coarse SF/disparity channels are used
to drive the vergence changes needed to bring the higher SF/smaller
disparity units into matching range. A sequence of such vergence changes
permits the recovery of all the disparity information, which then contributes to the construction of a complete depth map of the scene held in a
buffer called the 2&1/2D sketch.

Both the computational theory underlying the Marr/Poggio model and its
psychophysical support as a model of human stereopsis have since been
called into question (6). For example, Frisby and Mayhew (4), using a
nonius line technique to record vergence position, demonstrated that
relatively high spatial frequencies (7 c/deg) can initiate vergence eye
movements from disparities as large as 28 mins arc. This is well outside
the range predicted by the Marr/Poggio model: triggering a vergence movement to a disparity of that size would require a channel tuned to about 1
c/deg. Corroboration of this result using objective infrared eye movement
recordings and findings from studies involving disparity discrimination
tasks (refs. in 6) suggest that the initiation of vergence eye movements
cannot be limited solely to mechanisms displaying close coupling between
SF and disparity range. Rather, the evidence suggests that if
SF/disparity channels are involved in guiding vergence, at least some of
the medium/high channels must show a much broader disparity tuning than
that specified by the Marr/Poggio model.

The objective of the research described here was to explore further the role of SF tuned channels in vergence control using the fixation disparity experimental paradigm.

FIXATION DISPARITY

In spite of the accuracy to which vergence eye movements can null retinal image disparity, errors of vergence called 'fixation disparity' do occur. This term reflects the fact that the errors are residual post-fixation retinal image disparities. Fixation disparities in normal observers are usually small misalignments which are not associated with diplopia as they lie within Panum's fusional areas.

Ogle's (7) classic fixation disparity experimental paradigm uses a sub- jective nonius line technique to record vergence position while prisms are imposed to stimulate the disparity vergence system. That is, the prisms induce a lateral movement of the image in each eye and a vergence movement is initiated to maintain binocular fixation of the target. A central feature of the paradigm is that although the prisms induce a new 'vergence distance', they leave the 'accommodation distance' unaltered because the true target distance remains unchanged. Consequently, the prisms have the effect of inducing a mismatch between the signals to the vergence control system from the accommodation and disparity mechanisms. Experimentally induced fixation disparity is almost certainly caused by this conflict of cues (7,8), with the vergence position adopted being some kind of 'balance point' between the two conflicting sources of in- formation about the depth plane upon which the eyes should be verged.

Ogle's classic studies (7) showed that the size of fixation disparity is greatly determined by the characteristics of the retinal image. This result has recently been extended by Crone and Hardjowijoto (2) who, using the fixation disparity paradigm with random dot targets, corro- borated Fender and Julesz's (3) finding, using stabilised retinal images, that Panum's fusional areas for random dot stereograms greatly exceed those for simple line targets. The present studies represent a continua- tion of this general line of enquiry, by asking whether the size of fixa- tion disparity that can be tolerated is influenced by the SF content of random dot targets. If SF and disparity range are coupled in the human visual system, it might be expected that low SF textures would allow a much larger fixation disparity. This is because low SF/large disparity channels would presumably be able to support binocular fusion over a much larger range, i.e. given the cue conflict interpretation of fixation disparity, the 'pull' of the accommodation signal to fixate on the true depth of the targets would be permitted greater influence, as the dispar- ity mechanisms for the low SF texture would tolerate a larger 'disparity slop' while still permitting binocular fusion. This expectation depends, of course, on the assumption that the strengths of accommodative signals to vergence from low and high SF textures (in the range used here: 1-7 c/deg) are more or less the same, but this does not seem unreasonable (8, p.107-112).

METHODS

Fixation disparity is usually measured by the subject viewing the stimulus through a pair of prisms while performing a nonius alignment task for vergence position to be recorded. Various differences from that

classical approach were adopted in the present study: (a) forced convergence and divergence changes were imposed not with prisms but by altering
the positions of targets presented on XY displays under on-line computer
control (with the advantage that results were free from any anomalies due
to optical distortion produced by prisms); (b) objective measurements of
fixation disparity were obtained using a binocular infrared eye movement
recording technique (1) (thereby avoiding problems to do with the validity of the subjective nonius line technique, with its unnatural superimpo-
sition of rivalrous lines on the targets); (c) measurements were taken of
the whole vergence response as subjects tracked a target whose disparity
varied continuously (the continuous change simulated the smooth change of
disparity created when Risley prisms are used to induce fixation disparity).

Stimuli were random dot stereograms presented on a pair of Tektronix 604
XY displays that formed the left and right fields of a Wheatstone stereo-
scope. Binocular fusion was obtained by means of a pair of front-silvered
mirrors, with the subject's viewing position firmly controlled with a
bite bar and a head restraint. The stimuli were seen in the midline at
eye level in a fronto-parallel plane whose 'true' (accommodation) distance was c.40cm. The stimuli subtended a visual angle of 5 x 5 deg and
comprised 64 x 64 pixels, with each pixel set to one of 64 brightness
levels (space average luminance 17.17 cd/m^2). Each stereogram had one
central pixel brightened up to serve as a binocular fixation point. SF
filtered stimuli were obtained by 2D computer controlled filtering of a
random dot black/white texture. Filters were bandpass with a centre-
frequency-to-bandwidth ratio of 5 and centre frequencies of 1.7 and 7c/deg
for the low and high frequency textures respectively (see (4) for an
illustration of the appearance of the stimuli). Following filtering,
stimulus contrast was scaled up to equal that of the unfiltered original
(RMS contrast=0.34). This ensured that any differences in the relative
effectiveness of the different stimuli to drive vergence could not be
attributed to a difference in stimulus contrast. Contrast at stimulus
edges was ramped in over about 4 pixels to eliminate high SFs otherwise
formed at their borders with the neutral grey surrounds. Stereograms
displayed simple planar surfaces (no central area within the random dot
texture was set to a different disparity from its surround as in (2),
thereby avoiding problems to do with the subject fixating different depth
planes at different times). The surrounds of the XY displays were masked
using grey card to remove conflicting cues. When fused, the masks
appeared in the same depth plane as the faces of the XY displays.

Each experimental session began with a calibration of the infrared trans-
ducers for sensitivity to horizontal and vertical (x,y) eye movements
while the subject fixated successively dots shown at the four corners of
the area used on the displays to show the experimental stimuli. This procedure was repeated at the end of each session and the mean sensitivities
from both calibrations used in calculating subsequent eye positions. Comparison of data from the two calibrations also allowed detection of unac-
ceptable drift in transducer sensitivities over the session.

Following the initial calibration, the session continued with the presentation of random dot stereogram targets. Each target first appeared for 1
sec at the 40cm accommodation distance, during which time the subject had
to take up fixation on the central bright pixel. (Use of this spot for
fixation seemed unimportant; trials when it was eliminated and the

subject asked to adopt fixation upon a central texture feature produced
similar results). This initial position was termed the 'zero' position as
it was the position in which no accommodation/disparity vergence mismatch
was imposed. The target was then moved laterally in equal but opposing
small steps on each display, such that it appeared to the subject to move
in depth either towards or away from him along the midline (at 53min arc
disparity per sec), disappearing when it reached a position 5.3 deg
disparate from the starting zero position. His task was to fixate the
target throughout its track, trying to maintain a single fused percept by
allowing his eyes to converge or diverge, as appropriate. Each target
movement lasted 6 secs, during which time the apparatus recorded vergence
positions at 500 points 12 msec apart (termed the 'raw data points'). For
the purposes of data reduction and averaging out noise, 10 eye positions
were computed from each complete raw data record (in addition to the
measure taken of the initial zero vergence position at the start of each
presentation). The 10 eye positions were obtained by averaging samples
of 10 adjacent raw data points, with the centres of each sample evenly
distributed at 600 msec intervals along the vergence track.

The sequence of convergent/divergent tracking directions required of the
subject occurred in random order, with four measures of vergence
responses of each type obtained for each stimulus. The subject initiated
each presentation with a hand-held button switch when he was ready.

The 4 subjects were practiced psychophysical observers with normal
stereopsis (Frisby Stereotest). They all had acuities of 6/6 or better
for the viewing distance of 40cm, JPF achieving this by wearing his usual small myopic correction.

PRELIMINARY EXPERIMENTS

A) Before carrying out fixation disparity experiments using the method
just described, errors of vergence for each of the subjects were measured
for a set of presentations that involved no accommodation/disparity conflict. This was done to obtain baseline data for comparison with data
collected later using the experimental fixation disparity paradigm. Thus
vergence was measured while the subjects were asked to fixate the
unfiltered and SF filtered random dot stereograms presented at 5 true
distances around the zero 40cm position, with values selected to mimic
the apparent distances created in the fixation disparity experiments
(range 26-79cm). This was achieved by physically adjusting the positions
of the displays, so that there could be no artificial conflict between
disparity and accommodation. It was found that the mean recorded vergence
error rarely exceeded 10 mins arc and averaged 5 mins arc across subjects.
This measured error could either reflect a naturally occurring fixation
disparity (7) or the resolution of the eye movement recording system.
Either way, this preliminary study indicated that in subsequent experiments any recorded vergence error greater than c.10 mins arc could safely
be taken to represent a fixation disparity in response to the experimentally imposed disparity vergence demand.

B) A second preliminary experiment investigated the cue-conflict account
of experimentally induced fixation disparity. It did so by measuring fixation disparity curves (for the unfiltered stimulus set at 40cm) with and
without alteration of the accommodation requirement by the imposition of
+0.5 D lenses to both eyes. Technical difficulties experienced in making

satisfactory eye movement recordings from all subjects while the +0.5 D
lenses were in place regrettably limited this experiment to a single
subject (SPS). However, incomplete observations from other subjects make
us believe that his results are not untypical and hence worth reporting.

The experiment was carried out over 2 sessions held on different days,
one with the lenses and one without. In each session the subject's task
was the same - to fixate the central bright spot and to track the target
as it moved towards or away from him in depth. Four convergent and 4
divergent target movements were shown in each session.

Fixation disparity curves with and without the +0.5 D lenses (figure 1)
were significantly different (F=6.8, df=10, p<0.001). Errors of divergence
were somewhat smaller and errors of convergence appreciably larger
with the lens than without it. This result supports the idea that an
important factor underlying experimentally induced fixation disparity is
the imposed accommodation/disparity cue conflict. This is because plus
lenses converge light, with the result that the lens of the eye adopts an
accommodative state appropriate for an object apparently further away.
The imposition of plus lenses should therefore increase accommodation/
disparity vergence conflict in the convergent zone, and decrease it in the
divergent zone, with the size of fixation disparity following suit. This
is what was observed.

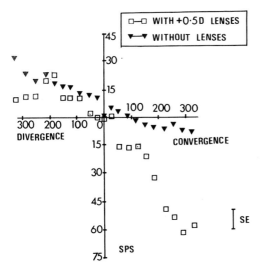

Figure 1. Fixation disparity curves for an unfiltered random dot target viewed with and without +0.5D lenses by subject SPS (see text). Abscissa: divergent/convergent disparity of target (mins arc). Ordinate: esodisparity/exodisparity vergence errors (mins arc). The s.e. bar shows the mean of standard errors calculated separately for the individual points.

The results reported above are consistent with the findings of Ogle (7)
who noted that the addition of spherical lenses changed the position of
the fixation disparity curve along the ordinate. The results are also in
keeping with the findings of Semmlow and Hung (9) who found that if accommodative influences were excluded (by means of an optical pinhole) a
considerable reduction in fixation disparity is observed.

EXPERIMENT: THE EFFECTS OF SF ON FIXATION DISPARITY

The methods described above were used to measure in a single experimental
session fixation disparity curves for the unfiltered and SF filtered targets. A different random order of presentations was used for each subject, with four repeated measures of each fixation disparity curve
recorded per subject.

Individual fixation disparity curves for the 4 subjects are shown in figure 2 for the 3 stimulus conditions. Each point is a mean of positions
sampled from 4 vergence tracks.

The first result worthy of mention is that the fixation disparity curves
obtained here using objective eye movement recordings fall within the
range of different types discovered using subjective nonius line techniques (2,7). Objectively measured fixation disparity curves are rarely
recorded, particularly in response to on-line stimulus presentations, no
doubt because of the relative difficulty in obtaining them. Hence it is
of some interest that the present experiment found curves in keeping with
those familiar from the customary nonius line procedures. Of course, a
proper comparison between the two types of measure was not an objective
of the experiment. Nor have we space here to discuss fully the detailed
shapes of the curves, except to say that those for SBP, AM and SPS would
all seem to fall within Ogle's (7) 'Type 1' normal category, whereas the
rather noisy curves of JPF might reflect the fact that he has spent a
good deal of his time over the last 20 years crossing his eyes to observe
complex random dot stereograms, with possible disruption of his
accommodation/disparity vergence synkinesis!

A second and more important result was the large retinal image disparity
which the subjects proved able to tolerate while maintaining binocular
fusion. For example, the disparity tolerated by SPB was c.60 mins arc,
convergent or divergent, although this did not represent his limit as the
apparatus did not allow large enough disparities to exceed his diplopia
(breakdown of fusion) point in the convergent zone (this was also true
for other subjects). This disparity is considerably larger than the range
typically reported for simple line targets and thus the experiment confirms the conclusion (2,3) that random dot targets considerably extend
the size of Panum's fusional areas.

Finally, the main result from the experiment was that fixation disparity
curves for the different random dot textures were in general similar. In
particular, no subject showed significant differences between fixation
disparities measured for low and high SFs. This could reflect the absence
of close coupling between SF and disparity range, of the kind required by
the Marr/Poggio model of stereopsis. If so, the present data represent a
confirmation of that conclusion, previously arrived at using large
disparity changes (4,6), from a task requiring the tracking of small
shifts in disparity.

ACKNOWLEDGEMENTS. We would like to thank Stephen Pollard and Philip
Stenton for their patient service as subjects, and John Mayhew for his
advice. Angeli Mehta was supported by a Medical Research Council
postgraduate studentship.

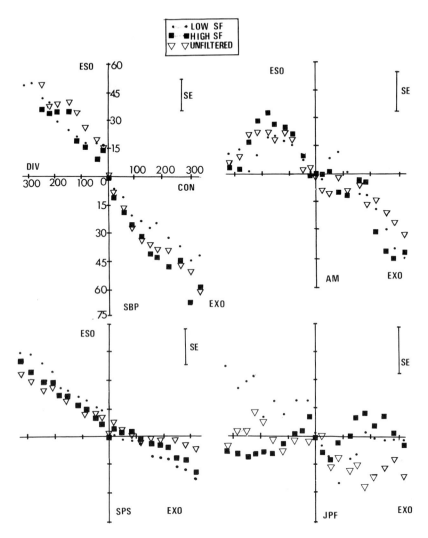

Figure 2. Fixation disparity curves for unfiltered, low SF filtered and
high SF filtered random dot targets for 4 subjects. Labels as described
for figure 1. AM's divergent curves return to zero vergence error at
extreme values due to averaging of large fluctuations in vergence error,
possibly indicating the onset of her diplopia point. SBP experienced
diplopia at the extremes of his divergent tracks for high SF and
unfiltered targets - hence the missing data points.

272   *A. Mehta et al.*

REFERENCES

(1) Brown, C.R. and Mowforth, P., An improved photoelectric system for two-dimensional eye movement recording, Beh. Res. Meth. & Instrn. 12 (1980) 596-600.

(2) Crone, R.A. and Hardjowijoto, S., What is normal binocular vision?, Doc. Ophthal. 47 (1979) 163-199.

(3) Fender, D. and Julesz, B., Extension of Panum's fusional area in binocular stabilised vision, Jrnl. Opt. Soc. Amer. 57 (1967) 819-830.

(4) Frisby, J.P. and Mayhew, J.E.W., The role of spatial frequency tuned channels in vergence control, Vis. Res. 20 (1980) 727-732.

(5) Marr, D. and Poggio, T., A theory of human stereopsis, Proc. R. Soc. Lond. Ser. B 204 (1979) 301-328.

(6) Mayhew, J.E.W. and Frisby, J.P., Psychophysical and computational studies towards a theory of human stereopsis, Art. Intell. 17 (1981) 349-386.

(7) Ogle, K.N., Martens, T.G. and Dyer, J.A., Oculomotor Imbalance in Binocular Vision and Fixation Disparity (Lea and Febinger, 1967).

(8) Schor, C.M. and Ciuffreda, K.J., Vergence Eye Movements: Basic and Clinical Aspects (Butterworths, 1983).

(9) Semmlow, J. and Hung, G., Accommodative and fusional components of fixation disparity, Invest. Ophth. & Vis. Sci. 18 (1979) 1082-

Theoretical and Applied Aspects of Eye Movement Research
A.G. Gale and F. Johnson (Editors)
© Elsevier Science Publishers B.V. (North-Holland), 1984

SACCADES IN HUMAN INFANTS

Louise Hainline

Department of Psychology
Brooklyn College, City University of New York
Brooklyn, New York 11210
U.S.A.

Fast eye movements in infants and adults were examined
with a corneal reflection recording system. When viewing stimuli with high attentional value, infants make
fast movements whose main sequences are comparable to
those of adults. Both infant saccades and the fast
phases of optokinetic nystagmus have adult-like characteristics, suggesting that the saccadic system is functional early in life. For less arousing stimuli, infants show slower saccades, and a high proportion of
saccadic oscillations.

The oculomotor system supports a number of functions to cope with change in
the visual world, produced either by the motion of objects with respect to
the organism, or by motion of the organism in the world. Of the various
oculomotor subsystems (vergence, pursuit, nystagmus, saccades), for humans,
the saccadic system is most important for appreciation of detailed visual
information. Saccades point the high-resolution region of the retina, the
fovea, sequentially at selected features in the environment. Sequences of
saccades and fixations have been used to study how information about the
world is processed. Piaget's (1952) widely accepted theory of development
views cognitive and perceptual development as dependent on the infant's motoric exploration of his world, including visual exploration. Since even
newborns appear to scan their surroundings actively (e.g., Haith, 1980),
analyses of how the eye movements of young humans are related to visual
stimuli may provide a means of understanding the development of higher mental processes. Moreover, this approach bypasses the problem of how to study
a non-verbal organism with limited motoric responses. Changes with age in
visual inspection patterns for simple geometric forms (Salapatek, 1975; Pipp
and Haith, 1977; Hainline and Lemerise, 1982) and for more complex and significant stimuli such as faces (Maurer and Salapatek, 1976; Haith, Bergman,
and Moore, 1977; Hainline, 1978) have been used to index perceptual development.

Although we too had done studies attempting to relate changes in visual
scanning to cognitive and perceptual development (e.g., Hainline, 1978;
Hainline and Lemerise, 1982), we were becoming somewhat uneasy about one of
the assumptions generally included in such work, namely, that developments
in the saccadic system itself are not so substantial as to confound attempts
to study more "central" phenomena. Perhaps the changes in scanning that had
been observed were simply an indication of the maturation of the saccade-
generating system rather than the "mind" beyond it. Evidence began to mount
that oculomotor systems were far from fully mature in infants. Vergence
(Aslin and Jackson, 1979), optokinetic nystagmus (Atkinson and Braddick,
1981; Naegele and Held, 1982; Turkel, Hainline, and Abramov, 1982), and

smooth pursuit (Kremenitzer, Vaughan, Kurtzberg, and Dowling, 1979; Aslin, 1981) all apparently undergo considerable growth during the first few months of infancy. Even the available studies on the saccadic system (Aslin and Salapatek, 1977; Salapatek, Aslin, Simonson, and Pulos, 1980) reported that when saccades were elicited by small targets, infants typically localized them with a series of short, equal-sized steps, rather than a single large saccade or a large saccade followed by a short corrective one, as seen in adults. In addition, our work on the structural development of the infant retina (Abramov, Gordon, Hendrickson, Hainline, Dobson, and LaBossiere, 1982) suggested that foveal immaturity in early infancy might interfere drastically with processes such as saccades, that could depend on a high level of foveal resolution.

We decided that it was important to study explicitly the characterisitics of infant saccades, both as they occur in free scanning of scenes and in the more artificial elicited-saccade paradigm. Here I will be reporting primarily on free-scanning saccades. Saccades in adults have a number of well studied properties (see e.g., Carpenter, 1977). One of these, the "main sequence", relates the highest velocity attained by the eye during the movement to the amplitude of the movement (Bahill, Clark, and Stark, 1975). In order to study infant saccades, we first compared their peak velocity vs. amplitude main sequences to those of adults studied in the same apparatus.

We recorded eye movements from both infants and adults with an infrared corneal reflection system (Applied Sciences Laboratory; Model 1994) which we have modified to work with infants (see Hainline, 1981). In order to collect data from infants, a "positioner" holds the infant subject against his shoulder with his back to the display being viewed by the infant; he steadies the head and positions the right eye for recording. An infrared beam is directed at the subject's eye by a dichroic beam splitter. An infrared sensitive TV camera provides an image of the eye in which changes in the direction of the eye's optic axis result in displacements of the corneal reflection with respect to the "bright" pupil. The TV image is analyzed by a microprocessor to estimate the X and Y coordinates of point of regard at a rate of 60 Hz. The system is relatively insensitive to head movements within the field of view of the camera. It has a resolution of 1/2 degree, and, for infants, an uncalibrated accuracy of $\pm$ 4 degrees. We have devised a spatial calibration technique for infants which can improve this accuracy (Harris, Hainline, and Abramov, 1981), but since not all infants remain in an alert state for the calibration session, calibration data are not available for all infants. For the purposes of the current discussion, it is important to note that, even uncalibrated, the system produces data which can be directly compared between infant and adult subjects. Between infancy and adulthood, there are changes in the center of rotation of the eyeball and in corneal curvature, both of which could distort estimates of a corneal reflection eye tracker. We have analyzed the effects of these physical differences between infant and adult eyes by measuring the performance of our system with "adult" and "infant" artificial eyes and found that these changes essentially offset each other, making uncalibrated estimates from the eye tracker comparable for infants and adults (see Hainline, Turkel, Abramov, Lemerise, and Harris, 1983).

Quantitative characteristics of fast eye movements are optimally measured by a recording device with a high bandwidth (100 Hz or more). Unfortunately, most systems of this sort, because of the constraints necessary for their use, are inappropriate for recording eye movements from human infants. TV-based systems, because of their frame rate, are of necessity of lower

bandwidth; in order to use such a system to study infant saccades, we also
performed a calibration for distortions in velocities due to bandwidth
limitations and machine noise. By moving an artificial eye through simulated saccades of known amplitudes and velocities and analyzing the instrument's output, we constructed nomograms to correct the measured velocities
of saccades (see Harris, Abramov, and Hainline, 1983); use of these nomograms brings the peak velocities measured from adults into ranges reported
from higher bandwidth devices, and so enables us to proceed with some confidence on our analysis of infant saccades, even with a TV-based system.

We recorded the eye movements of infants and adults to two sets of stimuli.
One set consisted of simple black and white geometric forms (circles,
squares, and triangles) ranging in size from 5 to 30 degrees on an unpatterned background (Hainline and Lemerise, 1982). The other stimuli were
Gibsonian texture gradients composed of black and white lines or juxtaposed
checkerboards with different sized elements (Gibson, 1950); these texture
stimuli filled the 45 degree X 45 degree viewing area. Data from 11 adults
and 39 infants between 14 and 151 days of age were analyzed in detail.

Our initial question was whether infants were capable of making saccades
like those of adults while freely scanning the environment. Therefore, an
operator viewing an analog representation of the output of the eye tracker
selected for further analysis only those movements which were fast and relatively free from noise. 75% of the infant and 90% of the adult fast movements were included, about 1200 movements from each age group. As the examples in Fig. 1a, b, and c illustrate, many of the infant fast movements were
indistinguishable from those of adults (Fig. 1f). However, infants' movements had two unusual characteristics. Some infants, some of the time,

Fast Eye Movements

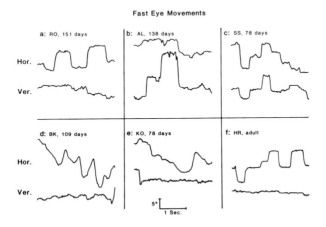

Fig. 1: Fast eye movements of infants and adults. Traces show horizontal and vertical eye position vs. time. (a) Horizontal movements - 151 day old infant. (b) Vertical movements - 138 day old
infant. (c) Oblique movements - 78 day old infant. (d) Fast oscillations - 109 day old infant. (e) "Slow" saccades - 102 day old
infant. (f) Horizontal saccades - adult.

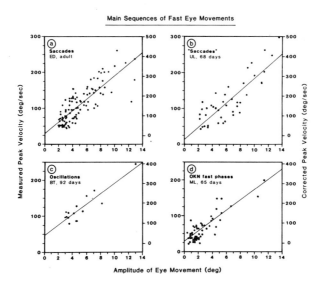

Fig. 2: Typical main sequences for peak velocity vs. amplitude of
eye movements. Left axis: uncorrected peak velocities. Right
axis: peak velocities corrected by our calibration nomogram. Solid
line is a linear regression fit to the data. (a) Main sequence
for an adult; the corrected slope is 22.8, with a correlation of
.83. (b) Main sequence for non-oscillating saccades - 68 day old
infant; corrected slope is 25.9, with a correlation of .82.
(c) Main sequence for fast oscillations - 65 day old infant; corrected slope is 19.7, with a correlation of .93. (d) Main sequence of the fast phase of OKN - 65 day old infant; corrected
slope is 19.4, with a correlation of .79.

showed movements much slower (i.e., much lower in slope) than would be expected given the movement's amplitude; compare, for example, the slopes of
the movements in Fig. 1e with equivalent-sized movements in the infant record in Fig. 1a or in adult record 1f. No adults made saccades this slow.
Also, under most normal conditions, adult saccades are separated by a minimum of 200-250 msec (Carpenter, 1977; Vaughan, 1983). Infants, however,
often made successive fast movements with inter-movement intervals of 50
msec or less (Fig. 1d). These oscillations account for about 40% of the
infant movements. They generally occurred in groups of 2 or 3 movements and
tended to be episodic rather than continuous. Most of the time, if oscillations occurred, infants also made non-oscillatory movements during the session.

Our primary method of analyzing the properties of infants' movements was to
compare their main sequences with those of our adult subjects. Adult peak
velocity vs. amplitude main sequences are linear up to amplitudes of about
15 degrees, after which velocity begins to saturate (see, e.g., Westheimer,
1954). Most of the eye movements measured were within the range of the linear portion of the main sequence. Linear regressions were fit to the peak

velocity and amplitude data. For all the adults, and all but a few infants, these regressions were significant. A typical adult main sequence with a slope of 22.8 is shown in Fig. 2a. In this figure, the left ordinate shows the uncorrected peak velocities while the right ordinate shows the values after calibration; note that this correction also gives intercepts in the vicinity of zero (see Harris, et al., 1983). A main sequence for non-oscillatory movements for an infant is presented in Fig. 2b; the corrected slope of this function is 25.9. Fig. 2c shows a main sequence for another infant's oscillations; the slope is 19.7. There were no significant differences, within a given stimulus type, for infants' main sequence slopes for oscillatory and non-oscillatory movements. Both types of movements were equally well-fit by similar main sequence functions; by this criterion, we judge the oscillations to be back-to-back saccades. For comparison, Fig. 2d illustrates a main sequence derived from the fast phase of optokinetic nystagmus (OKN) in an infant. Again the slope (19.4) is in the same range observed for other infant fast movements, perhaps pointing to a common mechanism of eye movement control.

There was a very marked effect of stimulus on the slope of the infants' main sequence functions. For the textures, the infants' main sequence slopes (for both types of movements) were similar to those of adults (a mean corrected slope of 27.6 for infants versus 24.3 for adults). When infants were looking at geometric forms, however, a different pattern emerged. Form stimuli elicited a higher proportion of oscillations, and the main sequence slopes for forms were significantly lower than those from infants viewing textures or from adults. For the forms, the mean infant slope for non-oscillatory movements was 12.1, for the oscillations it was 14.2, and for both movement types combined, it was 13.3. Again, the slopes for the oscillations and non-oscillatory movements did not differ from each other, although both were significantly different from the other infant values and from adult values, confirming that the oscillations were probably back-to-back saccades, regardless of the stimulus being viewed. Since the eye movements to the forms fit a main sequence just as well just as those for the textures, we felt that we were observing saccades, albeit slow saccades. Regressions of relevant variables over infant age failed to reveal developmental trends on most saccade parameters over the ages studied here (14-151 days).

We were rather surprised at the finding that at least some of the time, infants can make saccades which resemble those of adults. Human infants are renowned for their comparative motoric immaturity. The data suggest that the saccadic system is one of the first finely coordinated motor systems. But perhaps this should have been expected, given that the major mechanisms for saccadic control appear to be located in the brain stem reticular formation (Robinson, 1981). While cortical development is substantial over the course of infancy (Bronson, 1974), the older parts of the brain, including the reticular formation, are fully functional at birth and exert a proportionally greater influence over the infant's behavior than they will in childhood or adulthood. This influence can readily be observed in the infant's sleep cycles, but even in periods of wakefulness, there are cyclic episodes of greater or reduced attentiveness probably related to reticular arousal (Kleitman, 1963). In addition, during waking states, many characteristics of visual stimuli influence infant attention and arousal, both by their physical features and by their novelty (see e.g., McCall and Kagan, 1967). Simply put, infants are more likely to attend to complex than to simple stimuli, and to novel ones in preference to familiar ones. We feel that the simplest explanation of why slow saccades and oscillations occur

more often for our form stimuli is that the geometric forms were less interesting, and thus less likely to maintain reticular arousal, than the textures were; the fact that there was a higher percentage of usable data for
the textures supports this position.

In order to explain how changes in arousal could result in slow saccades and
oscillations, we recruit Robinson's (1981) local feedback model of saccade
generation, which he has explicitly related to oculomotor control mechanisms
in the brain stem. The model posits that saccades involve a feedback system
in which a signal specifying desired position of the eye in the head is compared with the instantaneous eye position to derive an error signal. During
a saccade, burst cells fire as long as there is a non-zero error signal, at
which point pause neurons stop the saccade. Robinson notes that the model
automatically produces saccades with appropriate velocities and durations,
and that minor modifications in a few parameters can explain a number of eye
movement disorders usually associated with brain damage. In particular, he
mentions that slow saccades, seen in disorders of the pontine reticular formation, can be caused by a lowering of the slope and amplitude of the function determining firing of the burst neurons. If the burst neurons are assumed to reflect general reticular activity, in a lowered state of attention, infants' slow saccades may result from similar factors. In support of
this possibility are data showing that drugs and fatigue lead to slow saccades, even in adults (Aschoff, 1968; Becker and Fuchs, 1969; Bahill and
Stark, 1975; Abel, Traccis, Troost, and Del'Osso, 1983). In Robinson's
model, saccadic oscillations can occur if the pause neurons fail to fire
appropriately at the end of the saccade. Again, lowered arousal may reduce
the general "tuning" of the saccadic system, leading to variability in the
timing of burst and pause cell activity. Since infants seem, under some
conditions, to "get the cobwebs out" and make good saccades, their problem
is more likely temporary fluctuations in arousal than consistent immaturity
of the relevant saccade-generating structures.

There are still unresolved issues about how well the infant saccadic system
functions. For example, how can the results of Aslin and Salapatek (1975)
and Salapatek, et al. (1980), described earlier as evidence for saccadic
immaturity, be reconciled with the results presented here? We feel that the
hypometric saccades described by those authors probably result from differences in how infants estimate spatial position rather than from differences
in how they execute saccades. In an elicited saccade experiment, the subject is asked to make a saccade to a target in an eccentric location in an
otherwise unpatterned field. An adult has the benefit of a well-articu-
lated, high-resolution, retinotopic map and many years of feedback about how
large a movement is required to reach the target. The infant may be more
like a swimmer trying to estimate how far it is to a distant island, who
swims and swims some more and still hasn't reached his goal. Estimating
distance over unfilled space requires practice which the infant has not had.
We were struck by the fact that with our highly contoured stimuli, we rarely
(less than 5% of the time) observed the "step" saccades described by the
previous studies. In free scanning studies, we could not, of course, specify what the infant's "target" actually was; however, since mean saccade
amplitude did not differ between infants and adults, infants seemed to be
"aiming" at targets within the same ranges as the adults were. In elicited-
saccade experiments we currently have underway with infants, we have observed a greater frequency of "steps" than in our free scanning studies.
Even in the elicited-saccade paradigm, however, infants seem to be making
more appropriate-sized saccades than had been reported. Comparison across
stimulus situations nevertheless suggests that infants may be more dependent

on landmarks for saccadic accuracy than are adults.

It appears, then, that despite the relative immaturity of the retina, and particularly, of the fovea, that infants can make acceptable saccades. It is our impression, however, that older infants are better than younger ones in the ability to fixate. Fixational control is dependent on receiving an error signal caused by slippage of the retinal image over the retina. If the infant fovea lacks a high degree of resolution, more slippage would occur before being detected and possibly corrected. We observe far more drift and jitter in the fixations of younger infants, consistent with their lower level of foveal development.

To return to the problem raised initially, our analysis of infant saccades supports the hope that analysis of scanning patterns may be a useful tool for assessing developments in perception and information processing in infancy. Further, we feel that studies of eye movement control in the young human may supplement our understanding of how those systems function in adults just as studies of clinical patients with eye movement disorders have.

Acknowledgements

This work was supported in part by grant EY03957 from the Eye Institute of the National Institutes of Health and by grants 13484, 661078, and 662199 from the PSC-CUNY Research Award Program of C.U.N.Y. The nature of the work is necessarily collaborative, and I gratefully acknowledge the assistance of Israel Abramov, Christopher Harris, Elizabeth Lemerise and Joseph Turkel.

REFERENCES

(1) Abel, L.A., Traccis, S., Troost, B.T., and Dell'Osso, L.F., Saccadic variability: contributions from fatigue, inattention, and amplitude, Invest. Ophth. Vis. Sci. Suppl. 24 (1983) 272.
(2) Abramov, I., Gordon, J., Hendrickson, A., Hainline, L., Dobson, V., and LaBossiere, E., The retina of the newborn human infant, Sci. 217 (1982) 265-267.
(3) Aschoff, J.C., Veranderungen rascher Blickbewegungen (saccaden) beim Menschen unter Diazapam (Valium), Arch. Psychiat. Nervenkr. 211 (1968) 325-332.
(4) Aslin, R.N., Development of smooth pursuit in human infants, in: Fisher, D.F., Monty, R.A., and Senders, J.W. (eds.), Eye Movements: Cognition and Visual Perception (Hillsdale, New Jersey, L. Erlbaum, 1981).
(5) Aslin, R. and Jackson, R., Accommodative-convergence in young infants: Development of a synergistic sensory-motor system, Can. J. Psychol. 33 (1979) 222-231.
(6) Aslin, R.N. and Salapatek, P., Saccadic localization of visual targets by the very young human infant, Perc. Psychophys. 17 (1975) 293-302.
(7) Atkinson, J. and Braddick, O., Development of optokinetic nystagmus in infants: an indication of cortical binocularity?, in: Fisher, D.F., Monty, R.A., and Senders, J.W. (eds.), Eye Movements: Cognition and Visual Perception (Hillsdale, New Jersey, L. Erlbaum, 1981).
(8) Bahill, A.T., Clark, M.R., and Stark, L., The main sequence: a tool for studying human eye movements, Math. Biosci. 24 (1975) 191-204.
(9) Bahill, A.T. and Stark, L., Overlapping saccades and glissades are produced by fatigue in the saccadic eye movement system, Exp. Neurol. 48 (1975) 95-106.
(10) Becker, W. and Fuchs, A.F., Further properties of the human saccadic

system: eye movements and correction saccades with and without visual fixation points, Vis. Res. 9 (1969) 1247-1258.

(11) Bronson, G., The postnatal growth of visual capacity, Child Dev. 45 (1974) 873-890.

(12) Carpenter, R.H.S., Movements of the Eyes (London, Pion, 1977).

(13) Gibson, J.J., The Perception of the Visual World (Boston, Massachusetts, Houghton Mifflin, 1950).

(14) Hainline, L., Developmental changes in visual scanning of face and face-like patterns by infants, J. Exp. Child Psychol. 25 (1978) 90-115.

(15) Hainline, L., An automated eye movement system for use with human infants, Behav. Res. Meth. Instr, 13 (1981) 20-24.

(16) Hainline, L. and Lemerise, E., Infants' scanning of geometric forms varying in size, J. Exp. Child Psychology 33 (1982) 235-256.

(17) Hainline, L., Turkel, J., Abramov, I., Lemerise, E., and Harris, C., Characteristics of saccades in human infants, manuscript submitted to Vis. Res. (August, 1983).

(18) Haith, M.M., Rules That Babies Look By (Hillsdale, New Jersey, L. Erlbaum, 1980).

(19) Haith, M.M., Bergman, T., and Moore, M.J., Eye contact and face scanning in early infancy, Sci. 198 (1977) 853-856.

(20) Harris, C.M., Abramov, I., and Hainline, L., The dynamic calibration of eye movement recording systems, manuscript submitted to Vis. Res. (August 1983).

(21) Harris, C.M., Hainline, L., and Abramov, I., A method for calibrating an eye-monitoring system for use with human infants, Behav. Res. Meth. Instr. 13 (1981) 11-20.

(22) Kleitman, N., Sleep and Wakefulness (Chicago, University of Chicago Press, 1963).

(23) Kremenitzer, J.P., Vaughan, H.G., Kurtzberg, D., and Dowling, K., Smooth-pursuit eye movements in the newborn infant, Child Dev. 50 (1979) 442-448.

(24) Maurer, D. and Salapatek, P., Developmental changes in the scanning of faces by young infants, Child Dev. 47 (1976) 523-527.

(25) McCall, R.B. and Kagan, J., Attention in the infant: the effects of complexity, contour, perimeter, and familiarity, Child Dev. 38 (1967) 939-952.

(26) Naegele, J.R. and Held, R., The postnatal development of monocular optokinetic nystagmus in infants, Vis. Res. 22 (1982) 341-346.

(27) Piaget, J., The Origins of Intelligence in the Child (New York, International Universities Press, 1952).

(28) Pipp, S. and Haith, M.M., Infant visual scanning of two and three dimensional forms, Child Dev. 48 (1977) 1640-1644.

(29) Robinson, D.A., The use of control systems analysis in the neurophysiology of eye movements, Ann. Rev. Neurosci, 4 (1981) 464-503.

(30) Salapatek, P., Pattern perception in early infancy, in: Cohen, L.B. and Salapatek, P. (eds), Infant Perception: From Sensation to Cognition, Volume 1 (New York, Academic Press, 1975).

(31) Salapatek, P., Aslin, R.N., Simonson,J., and Pulos, E., Infant saccadic eye movements to visible and previously visible targets, Child Dev. 51 (1980) 1090-1094.

(32) Turkel, J., Hainline, L., and Abramov, I., Orientational asymmetries in reflexive eye movements in infants, Inf. Behav. Dev. 5 (1982) 243.

(33) Vaughan, J., Saccadic reaction time in visual search, in: Rayner, K. (ed.), Eye Movements in Reading (New York, Academic Press, 1983).

(34) Westheimer, G., Eye movement responses to a horizontally moving visual stimulus. Aach. Opthalmol. 52 (1954) 932-941.

Theoretical and Applied Aspects of Eye Movement Research
A.G. Gale and F. Johnson (Editors)
Elsevier Science Publishers B.V. (North-Holland), 1984

EFFECT OF DISPLAY SUBTENSE ON EYE MOVEMENT SEARCH

P J Goillau

Royal Signals and Radar Establishment,
O1 Division, Malvern, UK.

A series of 3 studies is described in which military
observers searched a number of real and artificial scenarios
for a designated target. Search times and acquisition probabilities were measured for unaided, fixed-head, binocular,
eye movement search. Under specified values of background
and target parameters, 50% probabilities of correct target
location and subsequent identification were achieved with
display angular subtenses of 23° and 30° respectively at the
eye. Search performance approached a maximum with subtenses
in excess of 45°. The apparently conflicting recommendations
of previous workers are due to the comparison of experiments
with dissimilar background, target, and other parameters.

The importance of contextual cues was emphasised, as were the
dangers of extrapolating from the results of an abstract search
task to the real-world. Segmenting the search area by superimposing a grid produced a small though non-significant
improvement in performance. A significant correlation was
obtained between scores on one of the search tasks and the
Embedded Figures Test, suggesting a possible predictor of
search ability.

INTRODUCTION

With the publication of the Alvey Report (Dept of Industry, 1982) there has
been a renewed interest in the man-machine interface for information technology and display systems. The combined effect of display size and viewing distance, viz the angle subtended by the display at the eye, is an
important design consideration for all electro-optical systems. Yet in view
of its relevance to e-o system design, there is a surprising paucity of data
regarding display size for optimum observer search efficiency. What data
exists gives rise to conflicting recommendations. The present study arose
from the need to specify the eyespace field of view (FOV) for a particular
imaging system. However, the main aim of this paper is to demonstrate a
practical technique by which various display parameters can be optimised
for human search.

The literature relating to search and display size is reviewed elsewhere
(Goillau, 1982; Carr et al 1983). 2 distinct experimental paradigms
are evident: first where a cutout "window" is held over a display whose
contents remain constant in size (eg Enoch, 1959); and second where the
size of the display together with its contents is varied either directly,
or by adjusting the observer-display distance (eg Farrell & Anderson,
1973). It is clear, in the light of previous work, that the choice of an
optimum display angular subtense is not a simple matter. It will depend,
amongst other factors, on the display type (rastered/non-rastered), the
search task (abstract/real world), the target subtense both in absolute
terms and in relation to the display subtense, the amount of display
clutter, and on the task itself. In particular the interaction between
display subtense and target subtense may account for the conflicting findings of Enoch (1959) and Bell et al (1981) on the one hand, and Farrell
and Anderson (1973) on the other. The use of eye movement measures, search
times or preference ratings by different workers also contributes to the
conflicting recommendations.

Farrell and Booth (1975) attempt a compromise when they recommend a display
subtense of more than 9°, at least 36°, with a preferred value of 45-55°,
but not more than 60°. Clearly, such a global statement is not of practical use to the equipment designer who needs a more precise definition of
optimum display size for a given system. It is to this gap in our knowledge that the present study is addressed. Thus, the basic experiment
measured search times and probabilities for soldiers searching photographs
of realistic military scenarios in which target vehicles had been deployed.
The angular subtense of the pictures was varied over the critical range from
3° up to 60° at the eye.

METHOD

The experimental apparatus comprised a metal display board against which
photographic stimulus material could be held magnetically. A roller blind
of black cloth was mounted above the display board and could be pulled down
to obscure the photograph. Situated 0.67 m from the board were a forehead rest and dental bite-bar, both clamped onto a rigid metal framework.
The display board and metal framework were in turn supported on a heavy,
rigid wooden baseplate. The subject sat in front of the metal framework.
With his brow against the forehead rest he bit onto the bite-bar and looked
at the stimulus material through a pair of ACS EM-130 spectacle lenses,
which were also fixed to the metal framework (Figure 1). The lenses were
positioned 14 mm in front of the observer's eyes and did not interfere
with his vision unduly. The display board was illuminated by a single 60W
tungsten filament lamp placed above the metal framework, and an average picture luminance of 20 cd m^{-2} was achieved. Measurements were made around
the centre of the picture using a Gamma photometer with a 3° acceptance
angle. The observer was supplied with a push button by which he could signal target detection. The button gave an audible warning when pressed, and
stopped an electronic timer previously initiated by the experimenter. In
this way search times could be measured.

12 soldiers acted as observers in this study. All had AFV recognition
experience and none wore spectacles. Their eyesight was tested using the
Keystone Ophthalmic Telebinocular vision screener. In all cases, the
soldier's vision was found to be adequate in terms of monocular and binocular acuity and binocular vergence.

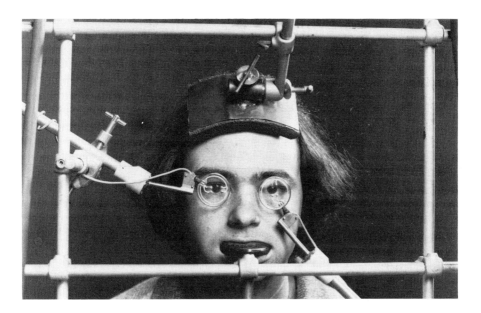

Figure 1   Observer looking through ACS EM-130 eye movement sensors

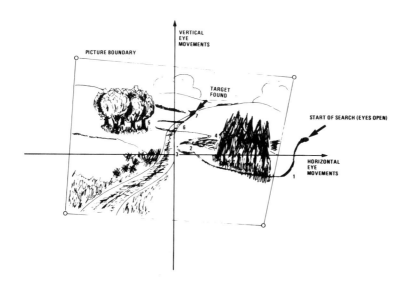

Figure 2   Typical scene with eye movement record superimposed

STIMULUS MATERIAL

3 sets of photographic stimulus material were developed, and are detailed below:

Part (1) Search as a function of picture subtense

9 high definition photographic negatives of single British Army vehicles deployed on different parts of Salisbury Plain were used as stimulus material. These were similar to the scenes used by Smyth (1980) in his earlier study of eye movements. The pictures were judged to be of approximately the same difficulty as regards vehicle location. The maximum target dimension occupied a mean value of 1.23% of the overall picture width, with a standard deviation of 0.24%.

9 semi-matt, black and white, positive contrast square prints were produced from each negative. Their horizontal and vertical dimensions when viewed from 0.67 m subtended 3°, 6°, 9°, 12°, 17°, 23°, 45° and 60° at the eye. From previous work these values were chosen as covering the crucial range. The experimental design was such that an individual observer saw each picture at a different subtense. The presentation order of the picture and subtense combinations was randomised.

Part (2) Search of an artificial scene as a function of grid structure

2 computer generated patterns of randomly positioned, black crosses were printed against a white background. A single vertical cusp positioned amongst the crosses served as the target (see Hockley, 1980). When viewed from 0.67 m, the search area subtended 42° horizontally by 17° vertically. The cusp target subtended 13 minutes of arc vertically by 8 minutes horizontally. The first 6 observers searched the unmodified patterns. For the remaining 6 observers the search area was divided into 8 equal rectangles by a single horizontal and three vertical lines. The object was to determine whether artifically structuring the search process had any effect on detection performance, as suggested by previous workers (Reilly and Teichner, 1962).

Part (3) Colour picture search for an artificial target

Several colour panoramas of Salisbury Plain were also available, their angular subtenses being the same as the artificial patterns of Part (2) above. A single, black, non-serif letter 'E' target was superimposed at one of 4 selected locations on each picture. The letter 'E's subtended 13 minutes of arc vertically by 10 minutes of arc horizontally when viewed from 0.67 m. The chosen locations were 2 woods, one to the left and one to the right of the picture centre (context targets), and 2 foreground locations, again one to the left and one to the right of the centre (non-context targets). In practice, a target vehicle could well have appeared at the context locations, but could not have appeared at that scale at the non-context locations. One each of the context and non-context pictures was divided into 8 equal rectangles as in Part (2). Each observer saw every picture. After the first 6 soldiers had been tested, the target's locations were swapped from left to right, and vice-versa. In this way it was planned to cancel out any effect due to target position and to examine the effect of the grid structure and context/non-context targets on search performance.

PROCEDURE

2 observers were tested per day in separate sessions each of 2½ hours
duration. On arrival the soldier's vision was screened and the nature of
the experiment explained. A bite-bar was prepared for each observer. Setting up and calibration of the eye movement equipment then took place. A
detailed experimental briefing was read to each observer. The experiment
was then conducted. There were 16 experimental runs in all, each lasting a
maximum of 90 seconds. The runs were as follows:

i   a practice run using an easily discernible target vehicle
ii   the 9 different picture subtenses (Part 1)
iii   the 2 randomly-distributed cross patterns with cusp target
       (Part 2)
iv   the 4 letter 'E' search tasks (Part 3)

For each run, the procedural order was:

1   observers' eye movements calibrated using a standard cross pattern
      of central and ± 10° horizontal/vertical marks on the roller blind
2   experimenter exposed picture and started timer
3   observer searched picture for target and pressed button to signal
      target located, which stopped the timer
4   observer's eye movements again calibrated using standard cross
      pattern
5   observer disengaged from bite bar and indicated target location
      using light pen, and attempted vehicle identification
6   experimenter noted responses and gave feedback to observer on
      correct target position and identity.

During each run the observer's eye movements were recorded on an X–Y flat-
bed plotter. An example is given at Figure 2. However, it must be stressed
that from a practical viewpoint the main dependent variables were seen as
acquisition times and probabilities. After each run the observers were encouraged to make comments on their performance and on the experiment in
general.

After the experimental work had been completed, the Embedded Figures Test
or EFT was administered to each observer (Witken et al, 1971). This is
a timed test of ability to extract simple geometric shapes from a pattern
of complex shapes, and can be regarded as measuring aptitude to distinguish
a target from a more complex stimulus of which it forms a part. Thornton,
Barrett and Davis (1968) found that the EFT scores correlated positively
with target detection performance on a complex military search task, as have
a number of other workers. However, Jones and Seale (1971) found that the
Witken's test was not sufficiently sensitive to discriminate between aircrew,
and Lintern (1976) also failed to find any significant correlation between
target acquisition performance and the EFT on a military-type task.

RESULTS

Part (1)   Effect of picture subtense on search of real-world scenes

i   Search probabilities

Figure 3 shows the percentage number of observers able to (a) correctly

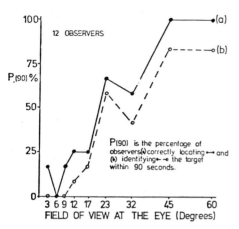

Figure 3   Probability of target
location and identification as a
function of picture subtense at the eye

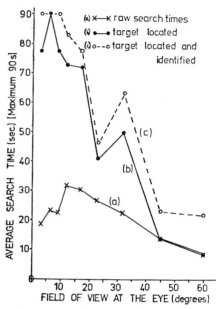

Figure 4   Average search time to
locate and identify target as a
function of picture subtense at the eye

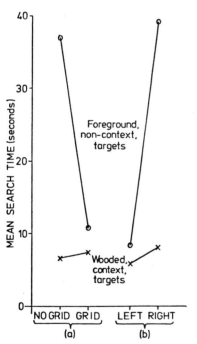

Figure 5   Average search time to
locate letter 'E' target as a function
of target location and grid structure

locate, and (b) correctly locate and subsequently correctly identify, the target vehicle within 90 seconds. It can be seen that there was a general upward trend in the percentage of correct locations, and locations + identifications, as picture subtense increased. Smoothed 'ogive' curves can be fitted through the 2 plots and 50% points estimated. These indicate approximately 23° for 50% probability of correct location, and 30° for 50% correct location + subsequent correct identification. As noted previously target subtenses were held constant at 1.2% of background subtense.

An appropriate technique to test for the statistical significance of any differences between correct/incorrect responses at the various subtenses is Cochran's Q test for dichotomous data (Siegel, 1956). For the location data, an extremely significant difference (χ^2 = 50.6, p $<$ .001) was obtained between the 9 picture subtenses, but no significant difference emerged between the 12 observers (χ^2 = 10.6, p $>$ .05). Similarly, for the location + identification data an extremely significant difference (χ^2 = 52.5, p $<$ .001) was obtained between the picture subtenses, but no significant difference could be found between the observers (χ^2 = 15.84, p $>$ .05). Clearly, the observers were of comparable ability as regards their search performance, and the main differences were due to the picture subtense alone.

ii   Search times

3 separate curves are shown in Figure 4. Each plotted point is the mean search time averaged over 12 observers for each picture subtense. The curves show:

(a)   the raw unmodified search times (not taking into account whether the target was correctly located or identified).

(b)   as (a), but with a penalty time of 90 seconds substituted on each occasion that an observer's reported target location was incorrect.

(c)   as (b), but with a penalty time of 90 seconds substituted on each occasion that an observer correctly located a target, but was subsequently unable correctly to identify the vehicle.

It can be seen from Figure 4(a) that the unmodified search times reached a maximum of 32 seconds at 12°, falling away at picture subtenses of 23° and above and also at 9° and below. The difference between subtenses was highly significant at beyond the 1% level ($\chi^2 r$ = 21.78, p $<$ .01) as shown by Friedman's 2-way ANOVA by ranks test (Siegel, 1956). The reduction in mean search times at the larger picture angular subtenses was to be expected, but the fall-off at the smaller subtenses is surprising. It may well have been due to the subjects giving up after a short while and making a guess. The latter was suggested by the higher number of incorrect responses with the smaller pictures as the eye's resolution limit of 1 minute of arc was approached, and is borne out by the modified search time data of Figure 4(b). Again there was an extremely significant difference between subtenses ($\chi^2{}_r$ = 45.06, p $<$ .001) as indicated by the Friedman's test. In addition a Page's L trend test (Page, 1963) gave a very significant downward trend (L = 3151.5, p $<$ .001) in the mean search times with increasing picture subtense. This trend was only just beginning to flatten at 60° angular subtense, suggesting that even angular subtenses of this magnitude may be beneficial. A similar trend is evident from the correct location + identification data plotted in Figure 4(c).

Part (2)   Effect of grid structure on an artificial search task

The results here indicate that structuring search by segmentation of the
picture area using grid lines gave a small, though statistically non-
significant improvement in search performance. The single cusp target
was correctly detected amongst the field of randomly-distributed crosses
in 33% of presentations without the grid structure, rising to 50% of
presentations when the grid lines were added. Additionally, mean search
times dropped slightly from 75.3 seconds with no grid structure to 67.2
seconds when a grid was superimposed. These figures reduced to 46 seconds
and 44.4 seconds respectively if those trials were excluded where no detection had occurred after the maximum allowed search time of 90 seconds.
Although the sample sizes are really too small to perform any meaningful
statistics, the results do agree with previous workers who reported the
beneficial effects of a moderate degree of structure of the search area
(Reilly and Teichner, 1962). It should be noted that the unmodified mean
search times in this task were somewhat higher than the mean picture search
times of Part (1) for a comparable display area.

Part (3)   Colour picture search for a letter 'E' target

The mean search times from this part of the experiment are plotted in
Figure 5(a). It should be noted that the mean search time for the foreground (non-context) target with grid structure superimposed is somewhat
high due to 2 observers who failed to detect the target after 90 seconds,
and a further 2 who incorrectly located the target within the 90 seconds
allowed. These were awarded penalty times of 90 seconds. It can be hypothesised from considerations of expected versus unexpected target locations,
and from the effect of grid structure previously demonstrated in Part
(2), that the predicted order of search times should be: Context targets
(Grid, No Grid) followed by Non-context targets (Grid, No Grid). In fact
Page's L trend test confirmed that such a trend did underlie the data and
was statistically significant (L = 163, p < .05).

However, the above analysis does not tell the whole story. Eye movement
analysis revealed that many observers followed a top-down, left-to-right
scanning strategy, and so target location within the picture area is of
critical importance. Figure 5(b) demonstrates that targets to the left of
the picture obtained lower search times than did targets to the right, as
did wooded (picture centre) targets compared with foreground (bottom of
picture) targets.

Embedded Figures Test (EFT) Scores

No statistically significant correlation could be found between the EFT
measures and the search times or probabilities described in Parts (1) or (2).
However, analysis of the data from Part (3) proved more fruitful. Spearman's
Rho non-parametric correlation test (Siegel, 1956) yielded a statistically
significant correlation between the EFT mean solution time per item and
the mean letter E location times (r_s = +0.52, p < .0 5, 1-tailed test).

DISCUSSION

The first experiment on search times and acquisition probabilities as a
function of angular subtense suggests that 50% probabilities of correct

target location and subsequent identification are achieved with display subtenses of 23° and 30° respectively at the eye. Optimum performance is achieved with subtenses of 45° or more. These figures assume military ground-based scenarios viewed through black and white, non-rastered displays of luminance in the region of 20 cd.m^{-2}, and with vehicle targets occupying 1-2% of the display subtense. Previous work suggests that there is little to be gained from subtenses greater than 60° (Farrell and Booth, 1975). These recommendations are consistent with those resulting from other visual tasks. Cohen et al (1976) showed that contrast sensitivity increases rapidly with display sizes up to 6.5°, and thereafter rises much more slowly with display diameters up to 60° and beyond. Moreover, most of this increase occurs at low spatial frequencies (5 cycles/degree and below), shifting the maximum sensitivity towards coarser gratings as display size increases. In addition, there are other benefits from the use of large display subtenses. Home (1981) comments that the recovery time to view a display of luminance between 3 and 100 cd.m^{-2} after exposure to bright daylight can be reduced by making the display angle subtended at the eye as large as possible.

Turning now to the second and third parts of the experiment, the importance of contextual cues in real-world as opposed to abstract search tasks has been highlighted. Search times for comparable display areas were lower for both real-world scenes, and real-world scenes with letter targets, than for a purely abstract search task. This is no doubt largely due to the removal of context cues in the latter. In an abstract task the target can occur anywhere within the display area, rather than at one of a limited number of likely locations. Thus a greater search effort is required and in consequence search times are longer. Caution should therefore be exercised in extrapolating from the results of abstract laboratory search tasks to real-world problems.

There is some evidence from the present study, as well as from previous work, that structuring the search area to a moderate extent can improve search performance. Not all observers used the grid structure, but those that did commented that it helped them to keep track of where they had already looked.

Finally, there is limited evidence from this study coupled with some previous research that the Embedded Figures Test may be a useful predictor of target acquisition performance in certain search tasks. The value of the test could only be ascertained by increasing the sample size and employing many more observers than were available for this study.

CONCLUSIONS

1 A series of 3 studies was conducted in which experienced military observers searched a number of real and artificial scenarios for a designated target. The main objective was to demonstrate a practical technique by which various display parameters could be optimised for human search.

2 For non-rastered imagery presented in black and white on a square format at luminances of 20 cd.m^{-2} and containing a Salisbury Plain scenario with target vehicle occupying 1-2% of the display width, 50% probabilities of correctly locating and subsequently identifying the vehicle were obtained at display subtenses of 23° and 30° respectively at the eye.

3 Optimum search performance was achieved with display subtenses in excess
of 45° at the eye.

4 It is suggested that context cues from real-world scenes resulted in
shorter search times than in an abstract search task with comparable search
area.

5 Structuring the search by superimposing a grid structure over the scene
improved search performance slightly.

6 The Witken's Embedded Figures Test may be of use as an aptitude test in
predicting an observer's innate ability at certain specific search tasks.

REFERENCES

1 Bell J B, Holman L K B, and Paul B R (1981). An analysis of search
strategies using eye movement recording techniques.   British Aerospace
Dynamics Group, Bristol Division, Report BT12171.

2 Carr K T, Megaw E D and Goillau P J (1983). Eye movements and visual
search: a bibliography. Unpublished MOD (PE) Report.

3 Cohen R W, Carlson C R and Cody G (1976). Image descriptors for displays.
Princeton: RCA Laboratories.

4 Department of Industry (1982). A programme for advanced information
technology - the Report of the Alvey Committee. London: HMSO.

5 Enoch J M (1959). Effect of the size of a complex display upon visual
search. J.Opt.Soc. America 49, 280-286.

6 Farrell R J and Anderson C D (1973). The effect of display size on
image interpretation performance. Boeing Company, Seattle, Washington,
Document D180-19056-1.

7 Farrell R J and Booth J M (1975). Design handbook for imagery interpretation equipment.   Seattle, Washington: Boeing Aerospace Company.

8 Goillau P J (1982). Effect of display subtense on eye movement search.
Unpublished MOD (PE) Report.

9 Hockley A T (1980). What the observer's brain tells the observer's eye:
a study of differences in eye-movements in two visual search tests.
Unpublished MSc Thesis, Dept of Engineering Production, Univ of Birmingham.

10   Home R (1981), Visual adaptation from high to low photopic luminances
Unpublished MOD(PE) Report.

11   Jones D M and Seale S J (July 1971). An investigation of the relation-
ship between Witkens Embedded Figures test and Target Acquisition
Performance. British Aircraft Corporation, Human Factors Study Note
Series 7 No 10.

12   Lintern G (1976). Field independence, intelligence and target detection
Human Factors 18, 3, 293-298.

13 Page E B (1963). Ordered hypotheses for multiple treatments: a significance test for linear ranks. J. Amer. Statistical Association 58, 216-230.

14 Reilly R E and Teichner W H (1962). Effects of shape and degree of structure of the visual field on target detection and location. J. Opt. Soc. America 52, 2.

15 Siegel S (1956). Non-parametric statistics for the behavioral sciences. NY: McGraw-Hill.

16 Smyth A (1980). Eye fixation distributions of observers searching complex scenes. British Aerospace Dynamics Group, Stevenage/Bristol Division, Report ST23810.

17 Thornton C L, Barrett G V and Davis J A (1968). Field dependence and Target Identification. Human Factors 10, 493-496.

18 Witken H A, Oltman P K, Raskin E and Karp S A (1971). Embedded Figures Test Manual, Palo Alto, California: Consulting Psychologists Press Inc.

Any views expressed are those of the author and do not necessarily represent those of the Department.

Section 5

MEDICAL IMAGE PERCEPTION

Theoretical and Applied Aspects of Eye Movement Research
A.G. Gale and F. Johnson (Editors)
© Elsevier Science Publishers B.V. (North-Holland), 1984

MEDICAL IMAGE PERCEPTION

INTRODUCTION

Alastair G. Gale,
Division or Radiology, Queen's Medical Centre,
Nottingham, England.

Diagnostic radiology utilises various techniques for displaying the internal morphology of the body either in a static or a dynamic manner. The problem of medical imaging is that despite the technological advances medical diagnosis is still subject to human frailties. Nowhere was this more wryly commented upon than by Llewellyn-Thomas (1976) in describing one of the techniques: "It is an incredible machine, a triumph of science and engineering, and yet in the end some jerk can be looking at it and miss something" (p. 352).

Various mass surveys and experimental studies have found error rates as high as 30%. This type of finding has spurned the experimental investigation of the diagnostic process. Most research has concentrated on chest radiography if for no other reason than the vast number of these radiographs which are examined each year. The following papers document the most recent work in this area.

Kundel, Nodine and Toto report a lung nodule search task where the display was directly related to the observer's line of sight in a manner similar to McConkie and colleagues (this volume). A single chest radiograph was presented to observers so that on some occasions a target nodule was present. However this could only be seen via a variable sized 'window' which was yoked to the observer's gaze location. The observers were unaware of this window and the arrangement meant that whilst peripheral vision of the radiograph was unaffected it was of no use in aiding nodule detection. The authors report that as window size increased so the number of nodules scanned increased. Performance also increased but levelled off with window sizes larger than 3.5 degrees. With increasing window size the time to fixate the nodule decreased. This type of study, where the relative contributions of foveal and peripheral vision are teased apart, represents the current state of the art and these early results promise further exciting developments. Carmody also investigates the effect of attending to selected portions of the chest radiograph. Manipulation of the display such as by tachistoscopic presentation or segmented viewing alters the observer's decision making ability. Such studies essentially prevent the comparison of a suspected abnormality site with other display areas. The author reports a study which directly investigates the effect of permitting or preventing such comparative scanning. In a 'constrained' experimental condition observers had to determine whether a nodule was present within a given area of a chest radiograph with the display being terminated if their gaze strayed beyond the defined limits. Thus peripheral vision of the rest of the display was possible but foveal examination of such areas was prevented. In a second 'comparison' condition the observers were allowed to compare this area with other parts of the radiograph. Results demonstrated that the comparison viewing condition produced superior performance. The ability to adequately compare one part of a radiograph with another therefore seems to be an important factor. As Carmody points out - how then can radiologists be encouraged to use this technique? The author goes on to address the complicating issues raised by the findings that training observers to use such a comparative technique does not drastically improve performance and furthermore that eye movement studies of radiologists show little evidence of such comparison viewing in clinical practice.

One of the problems of radiological reporting is that it is carried out at different times of the day. In other tasks involving visual search, performance has been found to be affected by circadian variation. Gale et al. investigate this factor in a pulmonary nodule detection task at three times of the day. Using medical students no variation in performance was initially found but when gaze position upon nodule detection was incorporated into the analysis to give a location response then a drop in sensitivity between morning and lunch session was found. In contrast no variability in the observer's self-assessed ability was reported. Some, mainly minor, variability in eye movement parameters was also reported. The authors argue from a cognitive standpoint that such results may indicate a change in strategy by the observers at different times of the day. Whilst extolling care in extrapolation these findings do indicate the need to investigate circadian variation further.

The final paper in this section is by Papin, Metges and Amalberti and reports three studies. The first is again concerned with examining chest radiographs and complements the others. The authors then extend the recording technique to the dynamic situation of the ultrasound examination. Finally the usefulness of recording visual behaviour as a training aid is considered.

These papers demonstrate the variety of research currently being undertaken to examine the problems of medical image perception. Considering the great financial outlay involved in purchasing any particular patient imaging system and the amount of technical effort exerted in producing good quality images it is always surprising that relatively little attention is paid to the psychological factors involved.

Reference

Llewellyn-Thomas, E. Advice to the searcher, in: Monty, R.A. and Senders, J.W. (Eds.) Eye movements and psychological processes (Erlbaum, Hillsdale 1976).

Theoretical and Applied Aspects of Eye Movement Research
A.G. Gale and F. Johnson (Editors)
© Elsevier Science Publishers B.V. (North-Holland), 1984

EYE MOVEMENTS AND THE DETECTION OF LUNG TUMORS IN CHEST IMAGES

Harold L. Kundel, Calvin F. Nodine, and Lawrence Toto

Departments of Radiology, University of Pennsylvania (HK and LT)
Philadelphia, Pa. 19104
Department of Educational Psychology (CN), Temple University
Philadelphia, Pa. 19122

INTRODUCTION.

There is a large body of data showing that competent radiologists and chest physicians overlook about 30 percent of lung tumors on chest radiographs, tumors that are clearly visible in retrospect. These error rates which were first studied systematically in the late 1940's and early 1950's have not changed with improved technology. A recent report by Muhm et al.(1983) from the Mayo Lung Project, part of a long term prospective study of the value of cancer screening in heavy cigarette smokers over age 45, indicated that 45 of 50 peripheral lung cancers were visible retrospectively in chest radiographs. Clearly, lung cancer would be detected earlier by radiography if these errors of omission were eliminated.

It had been suggested that many of these errors could be attributed to faulty search and we reasoned that if the progress of search could be followed by recording eye position then areas of the chest that were not "adequately" searched could be identified and given a second look. This was called "feedback assisted visual search".

This approach was predicated on the hypotheses that inadequate or incomplete search was responsible for errors and our first experiments were designed to test this hypothesis. This, in turn, required a definition of "adequate" search of a chest radiograph. The scanning behavior of experts looking at chest images follows rough patterns but it is not stereotyped (Gale and Worthington (1983),Kundel and Wright (1969)). The sequence and distribution of fixations is modified by experience(Kundel and LaFollette (1972)), task (Kundel and Wright (1969)), and the diagnostic information in chest image (Kundel (1974)). The scope of the problem was narrowed by limiting the stimuli to chest images with or without a single lung tumor nodule and by limiting the task to searching for nodules. "Adequate" search was then defined as complete coverage of the lungs by some "useful visual field" about the size of the fovea.

In order to test this hypothesis, eye movements of four experienced subjects were recorded while they viewed chest images for 20 seconds(Kundel,Nodine and Carmody (1978)). A set of ten images, four normal and six containing a single lung nodule was used. The starting position was varied six times so that at the end of the study each subject had seen each image six times for a total of 240 trials. The 20 false negative errors (misses) were divided into three classes using two criteria, inclusion in a 2.8 degree "useful visual field" and a fixation dwell time on the nodule of greater than 600 milliseconds (Table 1).

TABLE 1.

CLASSIFICATION OF FALSE NEGATIVE USING USEFUL VISUAL FIELD AND DWELL TIME.

DWELL TIME IN SECONDS WITHIN 2.8 DEGREES OF THE NODULE	NUMBER OF NODULES	PERCENT OF NODULES	TYPE OF ERROR
NOT FIXATED	6	30	SEARCH
LESS THAN 600 MSEC.	5	25	RECOGNITION
MORE THAN 600 MSEC.	9	45	DECISION

Obviously, the number of scans in each error category can be changed by modifying the criteria. However, the interesting result is that most of the missed nodules were fixated by a moderately small useful visual field and many of them for a relatively long time. Search, at least in its mechanical definition, appeared to be less important as a source of error than recognition or decision processes.

We also observed that the eyes frequently moved directly to the nodule even over a relatively long distance as shown in figure 1A. This suggested that the nodule was discovered by the peripheral vision and then fixated by the fovea for purposes of verification. The notion that foveal fixations are used for verification was reinforced by the observation that comparisons scans increased in frequency with decreasing confidence in the positive decision (Carmody,Nodine and Kundel (1981)). An example of a comparison scan is shown in figure 1B.

FIGURE 1.

A   B

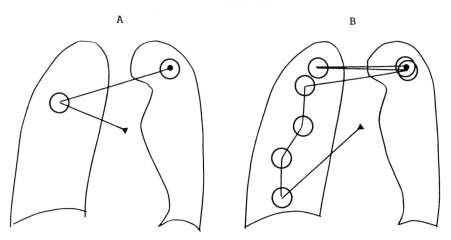

These observations led to the preliminary work that is reported here. We wondered what would happen to nodule detection if the visual system were deprived of either foveal or peripheral vision during search for a lung nodule. Experiments of this sort have been performed in reading (McConkie et al. (1978), Rayner and Bertera (1979)) and using drawings displayed by television (Saida and Ikeda (1979)) but not with radiologic images containing small targets and not with a system that made the window invisible to the subject.

The system described below was developed to uncouple central from peripheral vision during search tasks and a preliminary experiment was done using two experienced subjects who viewed 24 chest images, 12 of them normal and 12 containing single tumor nodules. The nodule was visible to the central vision but not to the periphery. The size of the central window was varied and detection accuracy, time to nodule fixation and mean interfixation distance were measured.

Because of the nature of our problem (lung cancer detection) where reporting a normal image abnormal (a falsely positive response) may be as significant to the patient as reporting an abnormal image as normal (a falsely negative response), we are concerned with overall measures of performance rather than just correct response rates. Therefore, subjects are asked to structure their decisions using a five level confidence rating scale (Swets (1979)) so that receiver operating characteristic (ROC) curves can be computed. The area under the ROC curve is generally considered the most useful overall measure of performance (Hanley and McNeil (1982)).

THE EYE MOVEMENT CONTINGENT DISPLAY SYSTEM.

The apparatus for uncoupling the stimuli presented to the central and peripheral vision is designed about an image array processor[1] which generates a standard 525 line television display. It is controlled by an LSI 11/23 microprocessor[2] and serviced by a 40 MByte Winchester Disk. The eye movements are recorded using a limbus reflection technique[3] and X,Y coordinate pairs are digitized under program control.

The image array processor is programmed to generate a circular window that tracks the axis of the gaze using eye position data. The processor can also display two images simultaneously, one inside and one outside the window. If the images are identical the subject is unaware of the the window whereas if they are different, the inside image is seen by the central vision while the outside image is seen by the periphery. This is illustrated in figure 2.

FIGURE 2.

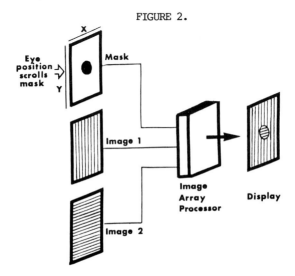

The eye movement recording is divided into 20 millisecond segments. During each segment 18 X,Y coordinates are digitized, averaged, and corrected for nonlinearity. The corrected values are used to update the position of the central window. The correction is done on the basis of a preliminary calibration in which the subject fixates in turn 25 calibration points in a 5 by 5 array. Two polynomials, one for X and one for Y are then computed and used to correct subsequent fixations. After the viewing session which is typically 15 to 20 seconds, the subject fixates the four corner points and the center point. If the allignment is adequate, usually within one degree, the data are considered valid for analysis. The correction procedure and the accuracy that can be achieved using only a chin and head rest to restrain the subject has been described elsewhere (Carmody, Kundel and Nodine (1980)).

EXPERIMENTAL METHOD.

Two experienced subjects who were also familiar with the apparatus and, therefore, aware of the presence of the window although it could not be seen directly, viewed the display at 70 cm. while restrained by a chin and head rest. At this distance the full display subtended 22.5 degrees. Their eye movements were recorded during a 20 second search for a nodule. They were told in advance that half of the images would be normal and that the nodules would be roughly 1 cm. (about 0.8 degree) in size. They were instructed to terminate the display as soon as they had made their decision about the presence or absence of a nodule in the image. They gave their response using a five level confidence scale where 1 was definitely normal, 2 was probably normal, 3 was indeterminate, 4 was probably abnormal, and 5 was definitely abnormal. If the decision was 3, 4, or 5 a location also had to be given.

The rating scale data were reduced using the method of Dorfman and Alf as described by Swets and Pickett (1982). This program uses a maximum likelihood estimator to compute the parameters of the ROC curve. The curves were plotted and the true positive rate at a 15 percent false rate was used as a normalized correct detection rate. The area under the ROC curve and its variance are considered a better measure of overall performance since it is less sensitive to differences in the slope of the curves.

A single chest image was used so that the the only stimulus variable was the nodule. The sequence of presenting the nodule or non nodule containing images and of the window size was randomized. Each nodule was seen with each window size once. The window sizes were 1.75, 3.5, 5.25, and 8.75 degrees. In addition, a display without any window was shown and is called the "open" window.

RESULTS.

The measures of overall performance as a function of the size of the central window are given in table 2.

The effect of the size of the central window can also be appreciated from table 3 which shows the percentage of the nodules reported as positive or negative that were scanned by the window and the percentage of those that were scanned that were then reported. We use the term "reported" rather than the more commonly used "detected" because the nodule may have been detected but then rejected as a valid signal.

TABLE 2

GENERAL PERFORMANCE AS A FUNCTION OF CENTRAL WINDOW SIZE.

WINDOW SIZE IN DEGREES	PERCENT TRUE POSITIVES AT 15 PERCENT FALSE POSITIVES	AREA UNDER ROC CURVE	
		MEAN	VARIANCE
1.75	28	.37	.012
3.5	59	.74	.013
5.25	60	.72	.013
8.75	74	.79	.009
OPEN	56	.75	.008

TABLE 3

NODULE DETECTION AS A FUNCTION OF CENTRAL WINDOW SIZE.

WINDOW SIZE IN DEGREES	PERCENT OF NODULES ACTUALLY SCANNED BY THE WINDOW	PERCENT OF NODULES SCANNED BY WINDOW THAT WERE REPORTED
1.75	38	44
3.5	67	69
5.25	96	61
8.75	100	75
OPEN	100	62

The mean and the modal interfixation distance and the time required to fixate the nodule directly are given as a function of window size in table 4. Direct fixation was defined as bringing the axis of the eye to within one degree of the center of the nodule for a continuous duration of at least 300 milliseconds. Note that direct fixation is not the same as including the nodule within the window. Only nodules that were fixated directly were included in the computation. Data from reported and unreported nodules were pooled.

TABLE 4

INTERFIXATION DISTANCE AND TIME REQUIRED TO FIXATE A NODULE DIRECTLY AS A FUNCTION OF CENTRAL WINDOW SIZE.

WINDOW SIZE IN DEGREES.	INTERFIXATION DISTANCE IN DEGREES.		TIME FROM START OF SCAN UNTIL FIXATION ON THE NODULE.	
	MEAN	MODE	MEAN	(ST.DEV.)
1.75	3.6	(1.4)	8.8	(3.7)
3.5	3.5	(1.2)	8.7	(5.9)
5.25	4.1	(2.1)	4.5	(2.0)
8.75	4.6	(1.2)	4.1	(3.7)
OPEN	5.6	(1.3)	3.5	(3.2)

DISCUSSION.

It is important to make a distinction between perception of the image and the search process. We assume that the image is perceived globally (Kundel and Nodine (1975), Kundel and Nodine(1983)). This means that within the first few fixations the experienced viewer has perceived the overall configuration of the chest image and the major abnormalities. The subsequent movement of the axis of the gaze serves to verify the details and to search for small abnormalities that may be suspected on the basis of a priori information. In medical imaging this information is gleaned from the patients history or from the results of other examinations. The scanning pattern is usually regular but can be disrupted easily by the discovery of unanticipated features.

In this experiment, the subjects were denied the use of peripheral vision to search for small targets. Since they were unaware of window size except perhaps after the target was detected, they could not compensate by modifying their scanning strategy. In addition, they could always see the global features of the image and they could examine the normal areas in detail.

When the central window is 1.75 degrees or about twice the nodule diameter the measures of general performance and the specific detection of nodules are affected adversely. This is clearly due to a failure of coverage of the image as shown in Table 3.

When the area of the central field is increased, two things occur. First, and not surprisingly, the percentage of the nodules actually scanned increases. Complete coverage of the lung area so that all of the nodules are included in the window occurs between 5.25 and 8.75 degrees. Second, there is an improvement in performance that is greatest between 1.75 and 3.5 degrees but then seems to level off. Although more nodules are scanned, the percentage of the scanned nodules reported is relatively stable suggesting that window size has little effect on recognition and decision processes.

The window size does, however, effect the efficiency of search as shown by the time required to fixate a nodule which decreases with increasing window size.

The true positive rate at 8.25 degrees is higher than that for the open field. This may be an experimental variation due to the small number of data points or an artifact caused by a peripheral "pop out" effect. When the window is large the probability of the window edge cutting through the nodule is high and the sudden change of the stimulus makes the nodule peripherally conspicuous. This "pop out" effect was apparent to the subjects and efforts to modify the window edge profile so as to blur the effect are currently underway.

CONCLUSIONS.

These data show that a field of about 5 degrees can effectively scan the entire lung region of the chest for a small target in about 20 seconds. However, scanning even by a relatively small window does not guarantee detection. These data also reinforce the previous observation that most nodules are missed not because of faulty coverage of the image by the useful visual field but because of recognition and decision processes that either fail to convert the sensation into the appropriate perception (Kundel and Nodine (1983)) or having achieved the conversion elect to ignore the nodule.

REFERENCES.

Carmody DP, Kundel HL, Nodine CF. Performance of a computer system for recording eye movements using limbus reflection. Behav Res Meth & Instr. 12 (1980) 63–66.

Carmody DP, Nodine CF, Kundel HL. Finding lung nodules with and without comparative visual scanning. Percept & Psychophys 29 (1981) 594–598.

Gale AS, Worthington BS. The utility of scanning strategies in radiology. in: Groner R. Menz C. Fisher DF. and Monty RA. (eds.), Eye Movements and Psychological Functions: International Views. (Erlbaum, Hillsdale, NJ.,1983).

Hanley JA, McNeil BJ. The meaning and use of the area under a receiver operating characteristic (ROC) curve. Radiology 143 (1982) 29–36.

Kundel HL, Wright DJ. The influence of prior knowledge on visual search strategies during the viewing of chest radiographs. Radiology 93 (1969) 315–320.

Kundel HL, LaFollette PS. Visual search patterns and experience with radiological images. Radiology 103 (1972) 523–528.

Kundel HL. Visual sampling and estimates of the location of information on chest films. Invest Radiol. 9 (1974) 87–93.

Kundel HL, and Nodine CF. Interpreting chest radiographs without visual search. Radilogy 116 (1975) 527–532.

Kundel HL, Nodine CF, Carmody D. Visual scanning, pattern recognition and decision making in pulmonary nodule detection. Invest Radiol 13 (1978) 175–181.

Kundel HL, Nodine CF. A visual concept shapes image perception. Radiology 146 (1983) 363–368.

McConkie GW, Zola D, Wolverton G., Burns D. Eye movement contingent display control in studying reading. Behav Res Met & Instr. 10 (1978) 154–166.

H.L. Kundel et al.

Muhm JR, Miller WE, Fontana RS, Sanderson DR, Uhlenhopp MA. Lung cancer detection during a screening program using four-month chest radiographs. Radiology 148 (1983) 609–615.

Rayner K, Bertera JH. Reading without a fovea. Science 206 (1979) 468–469.

Saida S, Ikeda M. Useful visual field size for pattern perception. Percept & Psychophys 25 (1979) 119–125.

Swets JA. ROC analysis applied to the evaluation of medical imaging techniques. Invest Radiol. 14 (1979) 109–121.

Swets JA, Pickett RM. Evaluation of diagnostic systems. Methods from signal detection theory. (Academic Press, New York, 1982).

FOOTNOTES.

1. Gould DeAnza IP6400.
2. Digital Equipment Corp.
3. Eye Trac Model 200, Gulf and Western Co.

ACKNOWLEDGEMENT.

This work was supported by Grant #CA 32870 from the National Cancer Institute, USPHS.

Theoretical and Applied Aspects of Eye Movement Research
A.G. Gale and F. Johnson (Editors)
© Elsevier Science Publishers B.V. (North-Holland), 1984

LUNG TUMOUR IDENTIFICATION: DECISION-MAKING
AND COMPARISON SCANNING

Dennis P. Carmody

Saint Peter's College
Department of Psychology
Jersey City, New Jersey 07306
U.S.A.

Eye movement studies have shown that during search for
lung tumours radiologists tend to survey chest images
by using a circumferential scanning pattern. At times
this pattern includes local comparison scans between
suspected tumours and the normal features of the image.
This study prevented such comparisons to determine
their necessity. The results support the view that
comparisons are used to resolve discrepancies, to identify abnormalities, and to modify the criteria used in
the interpretation of image patterns.

INTRODUCTION

Physicians often use radiological examinations to make decisions about the
health and subsequent care of their patients. The resulting decisions are
not always valid; as many as 30% of the small tumours recorded on chest
images are not reported, and there is considerable variation in interpretation between image readers[1]. Reader error has been attributed to variations in perceptual processing, such as an incomplete scan of the image, or
variations in the criteria used to classify an image as normal or abnormal
(2-5). Several studies of the eye movements of radiologists have described
the relationships between scanning patterns and decision-making (4,6-8).

At a typical viewing distance of 70 cm, a standard chest image (35 by 43 cm)
subtends a 30° by 37° visual angle. Readers move their eyes over images in
a series of rapid jumps, or saccades, which permit the fovea, the central
area of the retina having the highest acuity, to resolve details. It is
then possible to record the position of the eyes during radiographic search
as a method of determining which image areas receive foveal attention.
Typical findings from those records are that readers often fail to fixate
abnormalities and do not report them, or they fail to report an abnormality
that is fixated[4]. Similar errors of search and recognition were reported
in studies of viewers searching for word targets in line drawings(9,10).

When given a search task, such as lung tumour identification, radiologists
tend to scan a chest image with a circumferential pattern (illustrated in
figure 1A) in which several general regions of the chest receive visual
attention. Large portions of the lung fields of normal images are not
scanned by the circumferential pattern although readers reported an adequate review(6). When images are abnormal, radiologists tend either to
fixate quickly the abnormality and exclude other image areas (as in figure
1B), or do not scan the abnormality. At times the resulting interpretation
does not include all the abnormalities evidenced on the image. These errors

of omission are termed <u>Scanning Errors</u>; an estimated 10-30% of the tumours
overlooked in practice are attributed to these errors(4).

Some experimental attempts to reduce scanning errors have forced readers to
fixate specified image areas. In one study, the axis of gaze was directed
to particular locations and viewing was limited to a single tachistoscopic
fixation(11). Many of the directly viewed tumours were not reported
although all were identified when the locations were later specified in
free search. Thus, decision-making was impaired when viewing was limited
to a single directed fixation. In the second study, radiologists inspected
chest images for tumours under two conditions which allowed multiple fixations: segmented search, in which images were divided into six segments
and viewed piecemeal, and standard viewing(12). Search by segments was
thought to reduce scanning errors, leading to improved decision-making
performance. Although tumour identification rates were similar in both
conditions, there were 37% more false positives in the segmented viewing.

The results of the two studies suggest that a valuable aspect of decision-
making was prevented by the mechanical manipulations. Reviews of the eye
movement patterns during search for tumours have shown that, at times,
certain image areas are compared with other areas, either in the same lung
or in the opposing one(12). When such a comparison involves a set of
saccades starting with a suspected area, directed then to another image
area, followed by a return to the suspected area, the sequence is termed a
<u>Comparison Scan</u>. Such scanning strategies were prevented in the two

Figure 1. Simulated chest images with scanning patterns shown as a series
of fixation points connected by lines (top) and a 2.8 degree visual field
centered about fixations (bottom): circumferential pattern on a normal
image(A); an immediate fixation of a tumour in the right lung without
searching the entire image(B); a tumour detected after comparisons between
the tumour in the base of the left lung and the normal right lung(C).
(Reprinted from (12) with the permission of Investigative Radiology and
the J.B. Lippincott Co.)

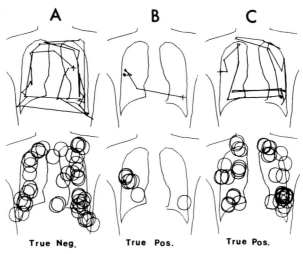

studies involving mechanical manipulations. The tumours which were not reported in the tachistoscopic study were detected by radiologists during free search which involved comparison scans(13). Segmented viewing also prevented comparisons of suspected tumours with the lung fields in the rest of the image. Comparison scans are assumed to be used to establish a normal reference against which suspected tumours are tested. Such shifts in the axis of gaze between target and background features are not unique to radiology and were observed in an embedded figures task using line drawings(14) and in a perceptual differentiation task(15).

Based on the results of free search studies and the effects on performance by mechanical manipulations, three components are identified during search for tumours: selection of areas for foveal viewing; prolonged fixation duration for feature extraction; comparison of local ambiguities with other normal features. The tachistoscopic study directed observers to specific areas for brief fixations without comparisons, thereby disrupting all three components. The piecemeal study allowed selection of restricted areas for an unlimited time but prevented comparisons. What is the effect of directing attention, by mechanical means, to tumour areas, and allowing prolonged viewing with comparisons? Will more accurate decisions be made by eliminating faulty search, and are the comparisons necessary for decision-making?

METHODS AND MATERIALS

Subjects. Three male volunteers from the Pendergrass Imaging Laboratory of the University of Pennsylvania served as subjects; one was a radiologist and two were experienced film readers. Subjects were selected on the basis of having normal vision, tolerance for the recording glasses, and familiarity with the task of tumour detection.

Materials. The test series was a subset of images used in a previous study (12). It consisted of 60 35mm slide copies of chest images which were photographed from ten original, normal, posterior-anterior chest films, three of men and seven of women. Six copies were made of each original film: one copy was normal and five copies contained simulated tumours of 1 cm diameter, of homogeneous texture, with a mean contrast of .10. There were five discrete levels of tumour edge gradient averaging 2.6, 4.4, 5.0, 7.2, and 9.2 mm(16). Simulated tumours with small edge gradients appeared sharp, and with larger edge gradients, the tumour border appeared fuzzier and the tumour appeared less distinct relative to the chest image. Within the series, there were 20 tumour locations, with five different edge gradients at each location, yielding 100 unique tumours.

Chest images were preceded and followed by a clear mask containing four dots (0.5 cm dia.) arranged in a square (5 cm on edge) with a central dot. A set of 20 masks were spatially matched to tumour locations such that the tumour was located randomly within the square window.

Subjects were to make decisions about the presence or absence of a tumour within the defined window. A set of 100 normal images was generated; 50 normal images were copies of the ten original normal chest images and contained no tumours; 50 additional images were abnormal images matched to the preexposure masks such that the defined frame superimposed a normal area. This allowed analysis of the effects of peripheral tumours on the decisions made about the defined normal areas.

Apparatus. Subjects sat 70 cm from a projection screen while their vertical and horizontal eye position was monitored by a corneal reflection technique using a set of glasses (Gulf and Western, Model 200-1) interfaced to
a PDP-11/40 microcomputer. Eye position was sampled 50 times/second and
compared with the position of the window. Subjects were restrained by a
combination chin and forehead rest to reduce head movements.

A set of shutters was attached to the lenses of two slide projectors. One
shutter was normally open and attached to the projector displaying masks.
A second shutter, normally closed, allowed display of the chest images.
Under computer control, the shutters were pulsed simultaneously such that
subjects would see either the mask or the chest image. Chest images were
rear-projected to normal size (35 by 43 cm). The simulated tumours subtended about 0.8 visual degrees when viewed from 70 cm.

Procedure. There were two experimental conditions which had common basic
procedures: Constrained Viewing and Comparison Viewing. While their eye
position was monitored, subjects looked directly at each of the four corner
dots of a mask, depressing a switch that marked eye position at each dot.
This, in effect, defined a 4° square window. Subjects were to inspect that
area for tumours. When the subject fixated the central dot and depressed
the switch, the shutters were pulsed allowing a view of the chest image
without the dots. Eye position was monitored, and the chest image remained
available for inspection until either eye position exceeded the window
frame or the subject depressed a switch when he had a sufficient view; then
the shutters were again pulsed, replacing the chest image with the dots.
Subjects gave a confidence rating which indicated their decision regarding
presence or absence of a tumour in the defined area: 5= definite tumour;
4= possible tumour; 3= ambiguous decision; 2= possible normal; 1= definite
normal. Thus, peripheral information was available but suspected areas
could not be compared by the axis of gaze with image areas outside the
window. This condition was named Constrained Viewing and 2400 decisions
were collected. Confidence ratings and time to decision were recorded on
a permanent file.

In the second condition, Comparison Viewing, the same subjects saw the same
images preceded and followed by the masks, but were permitted to leave and
return to the windowed area thus allowing comparison scans. Viewing time
was limited to five seconds or when the subject terminated the trial. Confidence ratings, time to decision, and the number of times the eye position
crossed the window border were recorded on permanent file for the 1400
trials of this condition.

Design. Each subject viewed in one session a random arrangement of the 200
stimuli in the test set. The radiologist (R) viewed the test set three
times in the Constrained Condition and once in the Comparison Condition.
The two film readers (F1, F2) viewed the test set under both conditions:
Constrained, F1-6 times, F2-3 times; Comparison, F1-3 times, F2-3 times.
The number of sessions per subject was determined by availability, with
each session requiring about 90 minutes.

RESULTS

Two measures of performance were available in both conditions for analysis:
confidence ratings and decision time. In the Comparison Condition, the
number of border crossings was examined.

Table 1.   Decision-Making Performance

	Viewing Condition		
	Constrained	Comparison	Standard*
Decision Rates			
Lung Tumours			
True Positive	62.9%	70.9%	61.4%
Ambiguous	12.6	6.9	-
False Negative	24.5	23.0	38.6
Normal Areas			
True Negative	85.3%	94.8%	92.7%
Ambiguous	10.7	3.7	-
False Positive	4.0	1.4	7.3
ROC Performance**			
d'_e	1.683	2.028**	1.508
Area	.874	.907**	.852
ΔM	2.333	3.514**	1.870

* From separate study (ref. 12).
**All measures between Constrained and Comparison Conditions are signifi-
cant at the p<.01 level as tested by signal detection methods (ref. 17).

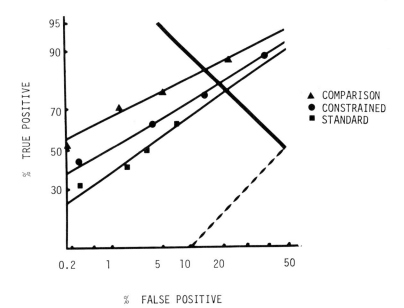

% FALSE POSITIVE

Figure 2.   Receiver operating characteristic (ROC) curves for the
Crossing and Constrained Conditions in this study. The curve for
Standard Viewing was based on performance in a separate study
(ref. 12). Performance measures are given in Table 1.

Confidence Ratings. The five-point rating scale was sectioned into three
categories: Positive decisions= ratings 4 & 5; Ambiguous decisions =
rating 3; Negative decisions= ratings 1 & 2. The two viewing conditions
were compared and the results are given in Table 1. Comparison scans
allowed observers to identify 8% more tumours by resolving ambiguous decisions. An improvement was made in decisions about normal areas: false
positives were reduced, ambiguous decisions decreased by 7%, and true
negative decisions improved by 9.5%. Thus, comparison scans improved
accuracy and were used to resolve ambiguity. These accuracy measures can
vary with the criteria used by observers; therefore signal detection
analyses were performed(17). Results are given in Table 1 and Figure 2
illustrates the ROC curves. All measures show superior performance for the
Comparison condition over the Constrained condition. The images in this
study were a subset of those used in a previous free search condition(12);
the ROC performance of that study is included in Figure 1. The differences shown between Free Search and the two directed conditions of this
study are speculative because the same observers were not used. With this
cautionary note, the effect of directing search could be interpreted as
reducing the component of error attributed to a failure to fixate tumour
containing areas.

Normal Areas. The two subsets of normal image areas were compared: normal
areas on normal images and normal areas on images with tumours peripheral
to the area under study. Confidence ratings were similar for normal images
($\bar{X}$ = 1.57) and for normal images with peripheral tumours ($\bar{X}$ = 1.62, $\underline{t}$
(1198) = 1.02, p > .05). There were no differences in the time to decision
between images with peripheral tumours ($\bar{X}$ = 3.60 seconds) and normal images
($\bar{X}$ = 3.49 seconds, $\underline{t}$ (1198) = .661, p > .05). Apparently the decisions
about normal areas were not influenced by the presence of peripheral
tumours. Confidence ratings given to normal images in the Constrained Condition were analyzed for differences among the ten original chest films.
The resulting 1200 decisions were evaluated by a one-way Analysis of Variance. Significant differences were found between films ($\underline{F}$ (9,1190) =
9.871, p < .01) suggesting that some chest images elicit more ambiguous
decisions or false positives than other images.

Edge Gradient. The effect of tumour edge gradient on performance was
examined by the number of crossings in the Comparison Condition. A one-way
Analysis of Variance showed a significant effect of edge gradient ($\underline{F}$ (4,695)
= 14.560, p < .01). For sharper edge gradients, there were fewer border
crossings. All subjects evidenced more crossings as edge gradient
increased, although there were individual differences in crossing activity.

DISCUSSION

Identification of lung tumours in radiographic images requires an adequate
search strategy for the extraction of relevant image features and the
application of decision-making criteria developed from knowledge of how
normal anatomical features are imaged and disrupted by the presence of a
tumour. Failure to report a tumour has been explained by two models: the
application of stringent reporting criteria, or faulty search. On the
basis of this study, faulty search is confirmed as a contributor to error:
readers performed more accurately when directed to specific areas than they
did in free search. Comparison scans improved decision accuracy, reduced
the proportion of ambiguous decisions, and were used more frequently for
less distinct tumours. If comparison scans are a necessary component of
the interactive process of search and decision-making, how can radiologists

be encouraged to use them more frequently?

Recently one experimental technique trained film readers to adhere to a search strategy in which similar areas in each lung field were systematically compared(18). There were no measurable differences in the decision-making performance for the trained readers compared to a control group which adopted a free search technique. Although many radiological textbooks advise students to use directed search, which is a deliberate sequence of viewing and includes comparisons, most readers adopt a free search technique (see 19 for a review). Experimental evidence suggests that more than 80% of the scanning patterns do not follow a directed search plan, and that radiology residents and instructors use comparison scans less than 4% of the time whether the image is normal or shows pathology.

If the forced use of comparisons does not improve performance, and if comparisons are not used frequently, then why are they important? The evidence in this study suggests that decision accuracy is reduced when such scans were experimentally prevented. A possible explanation is that viewers adopt with experience the ability to recognize the range of the patterns of normal features and how abnormalities, such as tumours, vary the patterns. The free search strategy used by experienced viewers, which appears random, is organized by a cognitive plan responsible for the circumferential survey and the local comparisons. Film readers bring to the search task knowledge of how anatomical structures are represented on the image and of the expected range of normal variation. Images which evidence features within the normal range would receive survey examinations with few comparisons. When the features are beyond the range of normal, prolonged visual attention and comparisons between suspected areas and other anatomical areas would be used to classify the area as normal or abnormal. The comparisons are the result of the inability to decide. Comparison scans are used to resolve discrepancies, to identify abnormal variations, and to modify the range of normal features which the reader then uses in subsequent interpretations of radiographic images. The resulting search strategy is modified by experience and the knowledge of how normal structures can appear rather than by a forced manipulation of viewing sequences.

Perhaps a systematic strategy of viewing images is useful when training students to attend to the complexity of features shown in a radiographic image. Such a strategy, although stressed by textbooks and taught by radiology instructors(19), is not evidenced in the scanning patterns of either residents or their instructors during visual search for lung tumours. The issue which requires continued investigation is how does the knowledgeable observer adopt an efficient search strategy and how can students be trained to use one.

This research was supported by Grant CA-3870-03, the National Cancer Institute, NIH.

REFERENCES

[1] Smith, M.J., Error and variation in diagnostic radiology (Thomas, Springfield, 1967).
[2] Kundel, H.L., Peripheral vision, structured noise and film reader error, Radiology 114 (1975) 269-273.
[3] Tuddenham, W.J., Visual search image organization and reader error in roentgen diagnosis: studies of the psychophysiology of roentgen image perception, Radiology 78 (1962) 694-704.
[4] Kundel, H.L., Nodine, C.F., and Carmody, D., Visual scanning, pattern recognition and decision-making in pulmonary nodule detection, Investigative Radiology 13 (1978) 175-181.
[5] Swennson, R.G., Hessel, S.J., and Herman, P.G., Omissions in radiology: faulty search or stringent reporting criteria, Radiology 123 (1977) 563-567.
[6] Tuddenham, W.J., and Calvert, W.P., Visual search patterns in roentgen diagnosis, Radiology 76 (1961) 255-256.
[7] Kundel, H.L., and Wright, D.J., The influence of prior knowledge on visual search strategies during the viewing of chest radiographs, Radiology 93 (1969) 315-320.
[8] Kundel, H.L., and LaFollette, P.S., Visual search patterns and experience with radiological images, Radiology 103 (1972) 523-528.
[9] Nodine, C.F., Carmody, D.P., and Kundel, H.L., Searching for NINA, in: Monty, R.A., Fisher, D.F., and Senders, J.W. (eds.), Eye movements and the higher psychological functions (Erlbaum, Hillsdale, 1978).
[10] Nodine, C.F., Carmody, D.P., and Herman, E., Eye movements during visual search for artistically embedded targets, Bulletin of the Psychonomic Society 13 (1979) 371-374.
[11] Carmody, D.P., Nodine, C.F., and Kundel, H.L., An analysis of perceptual and cognitive factors in radiographic interpretation, Perception 9 (1980) 339-344.
[12] Carmody, D.P., Nodine, C.F., and Kundel, H.L., Global and segmented search for lung nodules of different edge gradients, Investigative Radiology 15 (1980) 224-233.
[13] Carmody, D.P., Nodine, C.F., and Kundel, H.L., Finding lung nodules with and without comparative visual scanning, Perception & Psychophysics 29 (1981) 594-598.
[14] Boersma, F.J., Muir, W., Wilton, K., and Barham, R., Eye movements during embedded figures tasks, Perceptual and Motor Skills 28 (1969) 271-274.
[15] Vurpillot, E., The development of scanning strategies and their relation to visual differentiation, Jrnl. of Experimental Child Psychology 6 (1968) 632-650.
[16] Kundel, H.L., Revesz, G., and Toto, L., Contrast gradient and the detection of lung nodules, Investigative Radiology 14 (1979) 18-22.
[17] Swets, J.A., and Pickett, R.M., Evaluation of diagnostic systems: Methods from signal detection theory (Academic Press, New York, 1982).
[18] Gale, A.G., and Worthington, B.S., The utility of scanning strategies in radiology, in: Groner, R., Menz, C., Fisher, D.F., and Monty, R.A. (eds.), Eye movements and psychological functions: international views (Erlbaum, Hillsdale, 1983).
[19] Carmody, D.P., Kundel, H.L., and Toto, L., Comparison scans while reading chest images: taught, but not practiced, Investigative Radiology (in press).

Theoretical and Applied Aspects of Eye Movement Research
A.G. Gale and F. Johnson (Editors)
© Elsevier Science Publishers B.V. (North-Holland), 1984

CIRCADIAN VARIATION IN RADIOLOGY

Alastair G. Gale, David Murray, Keith Millar & Brian S. Worthington
Academic Department of Radiology, Queen's Medical Centre,
Nottingham and Department of Behavioural Sciences, Queen's
Medical Centre, Nottingham (K.M.).

Circadian variation in radiological performance was examined by
recording observers' saccadic eye movements as they searched
for pulmonary nodules on chest radiographs at 3 times of the
day. Conventional R.O.C. analysis yielded no evidence of
circadian variation. However an alternative analysis incorporated eye position information upon nodule detection and
demonstrated a drop in sensitivity after lunch. Eye movement
parameters were also examined for time of day variation. The
results are discussed with regard to current theories of
circadian variation. It is argued that the findings
demonstrate a strategy change over the day and implications
for radiology are considered.

INTRODUCTION

It is well documented that errors of omission as high as 30% occur in diagnostic radiology. These errors are caused by various factors such as faulty
visual search, inadequate recognition of abnormalities or inappropriate decision making (Kundel, Nodine and Carmody, 1978). A recent review of
research in this area is given by Gale, Johnson and Worthington(1979). An
inherent problem of radiological reporting is that it must of necessity
occur at various times of the day and so circadian variation may be an additional source of error in diagnostic ability.

Performance on many cognitive tasks varies through the day (Colquhoun, 1971).
Of particular relevance is the finding that the speed of serial visual
search tends to increase over the day, the increase apparently paralleling
the circadian variation in body temperature (see Monk, 1979). However,
while in the past it has been common to relate performance changes directly
to changes in physiological arousal, more recent considerations have shown
that circadian changes in performance efficiency have a far more complex
basis (Folkard and Monk, 1983). It now seems evident that changes in efficiency may be better explained by a multi-oscillator model of performance
rhythms rather than a uni-dimension arousal theory. The nature of the
strategies that subjects bring to bear, and perhaps the nature of the task
itself, will affect performance according to the times of day: for instance
Monk (1979) has shown that free visual search (similar to the search carried
out by radiologists) displays marked circadian differences when compared to
serial visual scanning. Such contrasting results cannot easily be accomo-
dated within a simple arousal theory.

This paper details an experimental investigation of a pulmonary nodule

detection task using chest radiographs. Observers were presented with a
series of radiographs and asked to detect the presence of a small opacity.
A rating response procedure was used and also the observer's eye movements
were recorded. As well as conventional ROC analysis a novel technique of
incorporating eye movement data into this analysis was used. Performance
at three times of the day was examined. On the basis of the serial visual
search tasks it was hypothesised that lesion detection performance would
vary over the day with a dip in performance in the post lunch period
(Craig, Baer and Diekmann,   1981).

METHOD

The microprocessor controlled experimental arrangement has been fully
described elsewhere (Gale, Johnson and Worthington, 1982). 35 mm slides
were prepared of normal A-P male chest radiographs. Using a 3 channel projection tachistoscope these were rear-projected onto a screen which was
monitored by a television camera. A second camera viewed a Tektronix 604
oscilloscope. The composite video from the two cameras was fed to a monitor
which the subject   viewed. A spot of light of known size, brightness and
location could be generated on the oscilloscope and superimposed over the
slide of a normal radiograph so creating an artificial coin lesion. The
displayed radiographs on the monitor subtended 20 degrees x 20 degrees at
the subject's eye and the lesions were 1 degree in diameter.

From a series of 100 normal A-P male chest radiographs half were randomly
selected to be presented as containing a single coin lesion. For each
selected 'abnormal' film a lesion location was randomly chosen within the
lung fields with the constraint that over all such films there was about
equal distribution of lesions in each stimulus quadrant. Using artificial
lesions in this way removes any bias to subjective anatomically important
areas which can occur when searching for real abnormalities (Kundel, 1974).

The detectability of lesions is related to their conspicuity (Kundel and
Revesz, 1976) and so a pilot study was first carried out to equate this
factor across the abnormal stimuli. A group of additional subjects were
presented with the series of abnormal stimuli such that the lesion was
initially not visible. Each subject then increased the brightness of the
lesion until he detected it. This procedure was repeated a number of times
(see Gale and Worthington, 1983). The mean brightness of each lesion
across all subjects was then taken as an empirical measure of equivalent
conspicuity. The similarity to clinical presentation of the lesions was
finally checked by a consultant radiologist.

PROCEDURE

11 male and 10 female third year medical students served as subjects. For
each one the experiment was run over a consecutive 3 day period in 3
sessions as follows: morning (M), post lunch (L) and evening (E). These
3 times were approximately 8.30 a.m., 1 p.m. and 5 p.m. To minimise practice effects the subjects were divided into 3 groups, each beginning at a
different time of day. The order of presentation of the stimuli was randomised for each subject.

Subjects were instructed that they were to be shown a series of A-P male
chest radiographs, some of which would contain a single coin lesion located
anywhere in the lung fields. If they detected a lesion they were to press

a response button whilst looking directly at the lesion. If they considered the film normal then they were to press the same button as soon as
they had reached this decision. Pressing this button terminated the stimulus which was presented for a maximum of 10 seconds. Subjects then used
one of 5 response buttons to indicate their degree of confidence in whether
a lesion was present or not. These buttons were labelled: definitely
lesion; probably lesion; probably normal and definitely normal.

Before beginning the experiment subjects were first familiarised with the
experimental procedure and the coin lesion appearance. The stimulus sequence was as follows. A dark I.S.I. field was presented for 5 seconds of
thesame mean brightness as the stimuli. During the last second a central
white fixation point appeared which the subject was instructed to fixate.
This was then replaced by the stimulus radiograph. As soon as they responded or after a maximum of 10 seconds the blank field was again presented
until 5 seconds after they had rated the stimulus when the next slide was
presented. Each session was completed within 30 minutes.

Head movements of the subject were minimised by the use of a chin rest.
Saccadic eye movements (∓1deg.) were recorded on videotape using a NAC
Eyemark and silicon diode camera which also recorded head movements
(Johnson, Gale and Worthington, 1978). Calibration trials were run for each
subject at the end of each session. The videotape was later replayed
through the microprocessor which analysed the subjects' line of sight in
real time. The data were then transferred to a minicomputer for subsequent
analysis.   .

After each session oral temperature was recorded sublingually and subjects
rated their performance on a visual analogue scale.

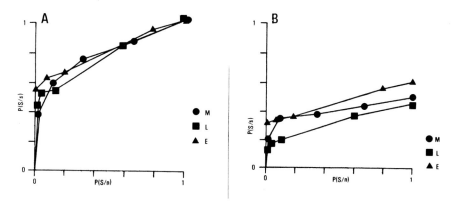

Figure 1. Illustrative data for one subject. A. The ROC curves obtained
at the three test times. B. The curves obtained after correction using
eye position information.

RESULTS

Receiver operating characteristic (ROC) curves were generated from each set
of subjects' responses (Fig. 1A). The area under the ROC curve (Swets and
Pickett, 1982) was used to generate a sensitivity measure employing the
2 arc sine $P(A)^{\frac{1}{2}}$ transformation and an overall measure (B) of the subject's
bias in each test session was calculated (McNicol, 1972). Variations in
these two variables across the three times of day (M-L,L-E,M-E) were examined across all subjects. Using the Wilcoxon test no significant change in
bias or sensitivity was found.

A problem with conventional ROC analysis with this type of stimulus is that
a true positive (TP) score awarded to a rating response which correctly
indicates lesion presence may not necessarily mean that the subject detected the lesion. He may have incorrectly identified some other area of radiographic mottle as a lesion. To overcome this problem Starr, Metz, Lusted &
Goodenough (1975) introduced the localised ROC curve where the observer had
to give the location (e.g. stimulus quadrant) in addition to the usual
rating response. In this analysis a TP response must also be correctly
localised.

In the present experiment the subjects' eye movements were recorded and
they were instructed that on detecting the lesion they had to fixate it and
respond at the same time. Thus a measure of lesion location can be attained by considering the fixation position when the subjects terminated the
stimulus which were then followed by an appropriate TP response - either
'definitely' or 'almost definitely' a lesion. This obviates the need for
the subject to make an additional localisation response. In adopting this
approach some critical area around the measured fixation position has to be
assumed such that if the lesion falls outside the area it is scored as a
'TP incorrect location'. This area is functionally equivalent to the useful field of view (Mackworth, 1965). Altering the criteria for this cut-
off value generates a family of curves. As the acceptance area increases
so the curve (EMROC) approaches the normal ROC curve. As the area is decreased then it reaches a point where it approaches the limits of the eye
movement registration technique as well as the size of the abnormality. The
cut-off value employed was 2.8 deg. as Kundel, Nodine and Carmody (1978)
have previously demonstrated that 90% of lesions are detected within this
distance of the fixation point. A detailed exposition of EMROC analysis is
given elsewhere (Gale, in preparation).

Using these corrected ROC curves (Fig. 1B) the bias and sensitivity values
were again analysed. A significant decrease in sensitivity (Wilcoxon test,
$p < .05$) was found between the morning and post lunch test session. No significant variation in bias was shown. The effect of considering the localisation factor was to reduce the overall mean number of TP responses by
20% (Fig. 3A). The effect on sensitivity and bias measurement was to show
the conventional ROC approach led to a 20% underestimate of bias and a 40%
over-estimation of sensitivity.

The eye movement data itself was examined for circadian variation. Using
the TP and the TN stimuli the following factors were analysed: saccadic
length, fixation time, number of saccadic movements and duration of viewing.
The variation in these are shown in Fig. 4. Only the drop in fixation time
for the TN cases between the M and L times was significant (Wilcoxon test,
$p < .05$).

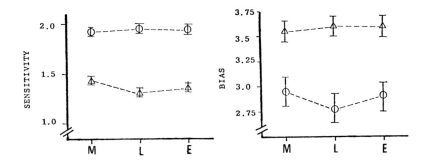

Fig. 2. Mean variation in Sensitivity and Bias at the 3 test times from the original ROC analysis (○) and from the eye movement corrrected analysis (△).

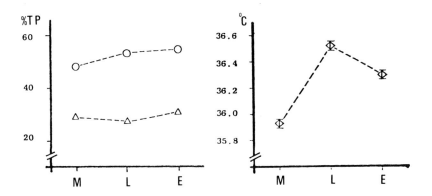

Fig. 3. A. Mean variation in the percentage of True Positives at the 3 times of day for the uncorrected data (○) and eye movement corrected data (△). B. Mean circadian variation in temperature.

Oral temperature increased from M to L sessions and then dropped slightly in the evening (Figure 3B). The morning temperature was significantly lower than the other two test times (Wilcoxon test, $p < .01$). No significant variation in subject confidence over the day was found.

DISCUSSION

The initial ROC analysis of the performance data at the 3 times of day yielded no significant variation in bias or sensitivity. This suggests that circadian variation can be discounted as a factor causing error in diagnostic radiology. However, using an analysis which corrected for lesion localisation revealed a significant drop in sensitivity between the post lunch and the morning sessions. This implies that abnormalities would be missed

A.G. Gale et al.

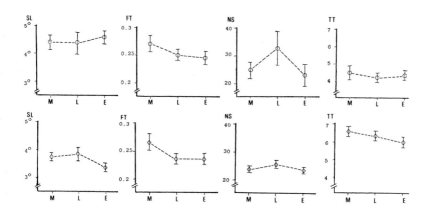

Fig. 4. Mean circadian variation for true positive (○) and true negative
(◇) responses in the following eye movement parameters: SL-saccade length
FT-fixation time; NS-No. of saccades; TT-total viewing time.

at this time which would have been detected in the morning. Oral temperature significantly increased from the morning to both of the other test
times with a slight evening fall. The observers sensitivity was therefore
lowest when their temperature was at a maximum. This result is in contrast
to the usual finding from serial visual search that performance increases
with temperature over the day. The drop in sensitivity may reflect the
post lunch dip reported in other tasks (e.g. Colquhoun, 1971; Craig, Baer
and Diekmann, 1981). In Monk's (1981) free search study although no evidence of an increase in performance over the day was found search times
did vary depending upon target positioning. At midday inner targets were
found faster whereas outer targets took longer to detect. This finding was
not predictable from the serial search work. This together with our present results implies that circadian effects may well be task related making
it difficult to extrapolate findings across different visual search studies.
No variation was found in subjects' self assessed confidence indicating
that although they significantly detected fewer abnormalities they perceived their performance as unchanged. This finding is in contrast to
Craig (1979) who reported a correlation between this variable and oral
temperature variation.

The eyemovement parameters themselves also varied with the time of day
although only the mean fixation time on the TN stimuli showed a significant
decrease between the M and L test sessions. TP fixation time showed a
similar although non significant trend. In contrast the mean saccade
length increased with the TP and decreased with the TN stimuli. The number
of saccades for both stimuli somewhat mirrored the variation in oral
temperature. With time of day the task of detecting TP and TN stimuli
was performed slightly faster. At the same time the mean fixation time
decreased. This suggests that the subjects altered fixation time as a means
of performing the task. Upon target detection fixation location with regard to lesion position did not significantly alter with time of day as
shown by the effect of correcting the TP responses which produced a similar
decrement in such responses across the 3 test times. Thus the change in

misclassification, i.e. drop in sensitivity, is not a result of an alteration in the functional visual field with time of day.

As noted above, several explanations of circadian variation in performance
have been proposed. A simple 'arousal' explanation is unlikely (Eysenck
and Folkard, 1980; Folkard and Monk, 1983) but the more recent and complex
theories of Hamilton, Hockey and Rejman (1977) and Folkard and Monk (1983)
provide more satisfactory explanations for the diverse and often discrepant
experimental findings.

In considering diurnal changes in performance speed Monk and Leng (1982)
have proposed a strategy based explanation which posits that the amount of
processing performed at each decision point in the task will affect the
performance speed. Such a resource re-allocation mechanism is related to
attentional selectivity. Saccadic eye movements serve to alter selective
visual attention about a display although fixation location is not necessarily precisely related to attention location. In the present study variations in saccadic eye movement parameters were found over time of day
although generally these were minor. Do such variations demonstrate an
alteration of strategy? If visual search is an antecedent to making a
diagnostic decision in some linear theoretical system (e.g. Swensson, 1980)
then variations in the parameters affecting search must alter the final
decision itself as different stimulus evidence will be gathered. We propose that the present results may well be evidence of a strategy change with
circadian variation.

CONCLUSIONS

There is a need to reduce errors in diagnostic radiology and although care
should be taken in extrapolating from this experiment the results imply that
circadian variation should be taken into account. These results were
obtained from matched groups of medical students and it is important to
extend this to practising radiologists to see whether they show similar
variation. It is highly unlikely that this study will result in the re-
scheduling of reporting sessions as other hospital factors are also pertinent . The best that can be hoped for is that clinicians be aware of such
variability, particularly as the present findings indicate they will not
realise any alteration. The task of detecting a single pulmonary nodule
is but a part of normal diagnostic practice where the radiologist may be
searching for various possible target classes. Thus circadian effects in
typical clinical reporting may well be far greater than found here.

REFERENCES

(1) Colquhoun, W.P. Circadian variations in mental efficiency, in:
     Colquhoun, W.P. (ed.) Biological Rhythms and Human Performance
     (Academic Press, London, 1971).

(2) Craig, A. Discrimination, temperature and time of day, Human Factors
     21 (1979) 61-68.

(3) Craig, A., Baer, K. and Diekmann, A., The effects of lunch on sensory-
     perceptual functioning in man, Int. Arch. Occup. Environ. Health 107
     (1981) 1-10.

(4) Eysenck, M.W. and Folkard, S.   Personality, time of day and caffeine:
     some theoretical and conceptual problems in Revelle et al. J. Exp.
     Psychol. 109 (1980) 32-41.

(5) Folkard, S. and Monk, T.H. Chronopsychology: circadian rhythms and
     human performance, in Gale, A. , and Edwards, J. (Eds.).
     Physiological correlates of human behaviour (Academic Press, London
     1983).

(6) Gale, A.G., Johnson, F. and Worthington, B.S. Psychology and
     radiology in: Oborne, D.J.. Gruneberg, M.M. and Eiser, J.R.S.
     (Eds.), Research in Psychology and Medicine: Vol. 1. Physical
     Aspects (Academic Press, London, 1979).

(7) Gale, A.G., Johnson, F. and Worthington, B.S. Research in medical
     image perception - a microprocessor application, in: Paul, J.P.
     Jordan, M.M., Ferguson-Pell, M.W. and Andrews, B.J. (Eds.)
     Computing in Medicine (Macmillans, London, 1982).

(8) Gale, A.G. and Worthington, B.S. Scanning strategies in radiology
     in Groner, R., Menz, C., Fisher, D.F. and Monty, R.A. (Eds.)
     Eye Movements and psychological factors: International views.
     (Erlbaum, Hillsdale, 1983).

(9) Hamilton, P., Hockey, G.R.J., and Rejman, M. The place of the
     concept of activation in human information processing in: Dornic
     S. (Ed.) Attention and Performance Vol. 6. (Halstead, New York,
     1977).

(10) Johnson, F., Gale, A.G. and Worthington, B.S.: Microprocessor
     equipment to analyse eye movements during radiograph scanning in
     Proceedings of I.E.R.E. conference on Microprocessors in Automation
     and control. 41. (1978) 93-96.

(11) Kundel, H.L. Visual sampling and estimates of the location of informa-
     tion on chest films, Investigative Radiology. 9 (1974) 97-93.

(12) Kundel, H.L., Nodine, C.F. and Carmody, D. Visual scanning, pattern
     recognition and decision making in pulmonary nodule detection,
     Investigative Radiology 13 (1978) 175-181.

(13) Kundel, H.L. and Revesz, G. Lesion conspicuity, structured noise and
     film reader error, Am. J. Roentgenol. 126 (1976) 1233-1238.

(14) McNicol, D. A primer of signal detection theory (Allen and Unwin,
     London, 1972).

(15) Mackworth, N.H. Visual noise causes tunnel vision, Psychon. Science
     3 (1965) 67-68.

(16) Monk, T.H. Temporal effects in visual search in: Clare, J.N. and
     Sinclair, M.A. (Eds.) Search and the human observer (Taylor and
     Francis, London, 1979).

(17). Monk, T.H. The interaction between the edge effect and target conspicuity in visual search, Human Factors 23 (1981) 615-625.

(18) Monk, T.H. and Leng, V.C. Time of day effects in simple repetitive tasks: some possible mechanisms. Acta Psychologica, 51 (1982) 207-221.

(19) Starr, S.J., Metz, C.E., Lusted, L.B. and Goodenough, D.J. Visual detection and localisation of radiographic images. Radiology 116 (1975) 533-538.

(20) Swensson, R.G. A two-stage detection model applied to skilled visual search by radioloigst, Perception and Psychophysics 27 (1980) 11-16.

(21) Swets, J.A. and Picket, R.M. Evaluation of diagnostic systems: methods from signal detection theory. (Academic Press, New York, 1982).

Theoretical and Applied Aspects of Eye Movement Research
A.G. Gale and F. Johnson (Editors)
© Elsevier Science Publishers B.V. (North-Holland), 1984

USE OF NAC EYE MARK BY RADIOLOGISTS

Jean Paul PAPIN* - Pierre (J) METGES** - René (R) AMALBERTI*

* Centre d'Etudes et de Recherches de Médecine Aéronautique
  5 bis Avenue de la Porte de Sèvres 75731 PARIS CEDEX 15
  ** Hôpital d'Instruction des Armées BEGIN. Service de Radiologie
  69 Avenue de PARIS 94160 ST MANDE

SUMMARY

Three experiments jointly conducted by a department of research on visual
perception and a hospital radiology department are described and discussed.
A gaze direction recording device, the NAC EYE Mark recorder was used to
characterize typical visual behaviors of radiologists and non radiologist
physicians during lung radiography (Exp. 1) and abdomen echography (Exp. 2).
Only very general behaviors were evidenced. The significance of doctor/
patient relationships was also analysed in Exp. 2. These same common visual
behaviors were evidenced and commented to young physicians (Exp. 3) viewing
lung radiographs. While emphasizing the attractive nature of the films the
discussion underlines the difficulties inherent to the interpretation of
such documents : the NAC records gaze shifts and pauses, but is unable to
explain their causes. ·

INTRODUCTION

Often, when you look at someone examining a picture, you can guess the
course, the difficulties or the easiness of analysis only by observance of
eye movements.
In fact, eyes are often preferential means to pick up information. Their
mobility seems to witness, on a first analysis, a controlled and logical
search, tied to the way to perform the current task.
If such intuitive analysis was good, visual behavior would be highly identical for one operator repeating many times the same task.
It would even be possible to find identical features in visual behavior
between two operators performing the same task.
Of course, if you want to argue this fact, you need an objective means to
record eye movements.
To verify this hypothesis in radiology, we used an oculometer display called NAC Eye Mark Recorder. This display allows continuous recording ,on
a movie or video support, the scene taking place in front of the operator.
An indicator (such as an arrow) superimposes gaze direction on the general
scene.
In a first step, we studied a static model of radiography : the lung radiograph. It was in 1981 - (1).

I - USE OF NAC EYE MARK IN CHEST RADIOGRAPHY

In the first experiment, we used a relatively well known model for the description of radiologists' eye movements : chest radiography.
Some important landmarks of this work can be briefly reviewed.
In 1963 Lewellyn Thomas (9) showed the highly individual features of radiologists' visual behavior. In 1969 Kundel & Wright (7), and in 1972 Kundel
& Lafolette (8) emphasized the importance of the context (knowledge of the
patient's clinical history).

Carmody et al. in 1981 (5) showed the great importance of compasture scanning strategy to differentiate modules from anatomical structures and Gale & Worthington (1983 (6)) showed that there is no tie between the visual behavior recommended in handbooks of radiology and the reality of visual search. Considering such results, a comparative study between two groups of subjects (ten non specialist physicians and eleven confirmed radiologists) was conducted.

1-1. *Methods*

Each subject had to successively interpret three lung radiographs (one normal and two pathological pictures) without the help of a medical document. During these interpretations, gaze direction was recorded on videotape.
To make an objective analysis of gaze direction, we selected a number of marks :
  - quality (zone) of location,
  - duration of location,
  - chronology of the change of location,
  - general organisation of visual search.

1-2. *Results*

The results analysis shows that each group can be described by typical features which are :

I-2.1. For the non radiologists, mean duration of lung radiograph interpretation is quite constant (2' ± 20") for normal and pathological pictures.
Mean duration of location is also quite constant ($53^{100} \pm 10^{100}$). Physicians explored many times the same area and never saw the soft parts. Exploration paths often run from top to bottom to periphery and up through the mediastinum. Passage from one site to another is generally done by proximity criterion. This group made many interpretation errors (5).

I-2.2. For the radiologists, duration of lung radiograph exploration is shorter (1'5" ± 10") and duration of location is longer ($96^{100} \pm 15^{100}$). This group explored all the areas, generally only once and spent more time observing the central part of the lung which is used as a visual pivot. Visual pivot is transfered on conflictual location of the picture each time there is a problem, and as long as this problem is not solved. This group did not make interpretation errors.

1-3. *Discussion*

We found the same results as those reported in the bibliography :
Strictly speaking, there is no common visual behavior between the members of each group and, of course, between the two groups.
Nevertheless, there is a way to explore a picture which is used by specialists (and not by non specialists). This way can be characterized by :
  - a complete checking of areas,
  - a long pause on each location ,
  - a use of precise location (generally the upper part of mediastinum) as a visual pivot,
  - a global analysis at the beginning and at the end of the visual search.
The role of global analysis has already been evidenced in other visual tasks (helicopter flying in 1980, Papin et al (10).
Indeed, a long gaze location on the center part of a picture can evidence an unusual mark even if it is in the peripheral visual field. Today, there is

no valuable explanation of information processing in this case.
Finally, we can argue the fact that there is a common visual behavior, for
the specialists and another one for the non specialists. They seem to be
"rules of reading", independant of the nature of the negative. We shall
call such characteristics "higher features of visual behavior". The final
interest would be to search in future experiments how these general behaviors become mentally operational as specialised knowledge is acquired.
In other words are they consequences or inversely conditions for the acquisition of skilled interpretation ? Depending on the answer, they could be
taught at a more or less advanced stage of studies to be efficient. It
would also be possible to have a better understanding of the intellectual
mechanisms used by the interpreting specialist.

II - USE OF NAC EYE MARK IN ABDOMINAL ECHOGRAPHY

One of the NAC eye Mark properties is the great mobility of the apparatus ; another is to provide a continuous movie of gaze direction.
Consequently, we use the display, combined with gestural and manual recording, to study the echograph work station.
The interest of the method,in addition to permit qualifying the visual
behavior, is to know the consequences of two new facts :
- picture analysis in dynamic situation (dynamic search of the good
picture),
- direct presence of the patient (problem of the doctor/patient relationship).

II-1. Methods

Three physicians examined two patients (hospitalized in the same hospital).
Physicians had respectively 4,4 and 9 months experience in liver echographic techniques. Each one had to examine both patients. A few medical documents were available to them.
The first patient had hepatitis without echographic signs. The second
patient had lithiasis with echographic signs.
The morphology of the two patients was similar.
The echographic examination was conducted in three steps :
1 - Greeting of the patient, medical examination, complementary questions, and short explanation of echographic techniques.
2 - Examination with use of a mechanic, handheld transducer. This
transducer permits rapid anatomic exploration with detection of
indicator points (as vena cava). Pictures are continuously monitored using a TV monitor.
3 - An analysis of selected pictures with the help of frequency and
location probes, used in real time (B Mode) follows the second
part. It is the real interpretation part of the examination.
Analysis of gaze direction recordings was made with the same criteria as
in lung radiography. Simply, we added the duration of each part of the
examination.
Voice recordings and gestural recordings allowed completion of the understanding of visual behavior, essential for the doctor/patient relationship.

II-2. Results

Result analysis first shows that visual behavior greatly varies between
physicians and for the same physicians, between patients : e.g., time of
the examination varies from simple to double.
Nevertheless, the three echographs find the relevant echographic information about the patient's illness and it is possible to isolate a few
common visual features :

1 - The patient's face is observed from 3 to 5 % of the overall examination time. TV monitor and transducer are watched from 65 to 80 %.
2 - During the construction of images (phase 2) gaze follows the different changes on the TV monitor. From time to time, gaze stops on
a possible relevant information and does not explore the other
parts of the picture ; the next image is immediatly watched to confirm or to invalidate current relevant information. Such technique
recalls the pivot visual behavior observed in lung radiograph explorations.
3 - When good images are selected, gaze fixes the center part of the
TV screen and stays on it.
4 - In case of a pathological case, examination is longer and more
echographic images are made. The mean checking time of image is
also longer.

II-3. Discussion

For the psychologist, common features we find are maybe higher features of
behavior as in lung radiography. Nethertheless, their banality (...fol-
lowing the image construction, fixing the TV image, etc...) raise great
interest in interpretation of such phenomenon. The fact that gaze direction
is fixed on the center part of TV during all the interpretation can reduce
the interest tied to NAC Eye Mark display. This is a great problem in using
the NAC Eye Mark and evidences the fact that visual behavior is only part
of the general behavior which is also a part of knowledge. Consequently,
information processing can occur without eye movement.
For the clinician, the study of voice communication and gestural behavior,
combined to visual behavior shows that doctor/patient relationships are
very poor. Patient anxiety increases with the poor knowledge of the technique. A great effort must be quickly made to improve the technique and to
humanize relationships. Finally, reproach can be made to this display
system from an ergonomic standpoint (buttons, distance of the TV monitor).

III - USE OF NAC EYE MARK IN PEDAGOGY

Because of the nature of the NAC Eye Mark recordings (videotape), it seems
tempting to use these as a pedagogical means. Such approach has already
been used in other branches of instruction :
    In sport, in 1981, Bayless (4) and Rippoll and coll (11).
    In flying, in 1981 also, Spady and coll (12).
In radiology, as in sport or flying, training is a big problem. On one
hand, much time is needed to train a good specialist. On the other hand,
it is often necessary to give sufficient training in a short time to
allow daily practise for the generalist (case of lung radiography). The
numbers of interpretation errors show that such an approach is far from
being mastered.

III-1. Methods

To simplifiy interpretation we used again a relatively well known model of
radiography :
the 10 x 10 cm radiophotograph (a systematic control size of lung radiograph).
It is a complex sorting task that young physicians never study in their normal education course. Consequently, special training is given in a one year
military applied session ("Brevet de Médecine Aéronautique") - (Aviation
Medicine license) for young military physicians.
To conduct this experimental approach, we select 40 radiophotographs in
two bobins of 20 exposures, with the help of the radiology department of

the BEGIN Military Hospital (PARIS) and of the radiology department of
CPEMPN headed by Doctor PUECH (Centre Principal d'Expertise Médicale du
Personnel Navigant - PARIS).
Each bobin includes:
- 5 obvious pathological lungs,
- 5 pathological lungs more difficult to diagnose,

- 10 normal lungs.
The viewing order of pictures is controlled by a random table.
Special films and one normal film (100 radiophotographs) are examined by
four radiologists. Direction of gaze of three of them is recorded on
videotape. Then, these recordings are analyzed with the respective readers
(radiologists) and a psychologist to assess visual behavior features.
The fourth radiologist is only interviewed and not recorded. He allows us
to verify the hypothesis built on the three other specialists'visual behavior.
Subsequently, two didactive movies are made. The same process is used for
both :
     1 - Two radiologist's gaze recordings appear successively for one or
         many negatives
     2 - comments are made shortly to emphasize each idea
     3 - drawings give at the end of the sequence a summarizing idea of be-
         havior used by experts
     4 - the third radiologist's gaze recording of this point appears a
         last time to illustrate the drawing.
Durations of movies are respectively 7'30" and 10'.
In the same time, at the beginning of the school year, a first test is proposed to 24 young military medical students.
Notation judges first the good or the bad sorting. In addition, it pena-
lizes neglects, always serious in a sorting task, and minimizes excess
control requests. Such notation is kept for the two tests (at the beginning
and at the end of the school year).
At the end of the first test, subjects are divided into two groups with
identical means and standard deviation. The first group (control group) has
a normal teaching course. The second group (experimental group) has the
same teaching course reinforced by presentation of the movies.
A second interpretation test is performed at the end of the school year to
judge progress, if any.

III-2. *Results*

Means of the two groups stay quite the same for the two tests (Student
test N.S. at 05). There is no better performance for the experimental group.
In addition, there is a great instability in the individual scores ; the
best subjects lower their mean and the worst subjects improve their mean.
The end result is that everybody has a comparable   mean score at the end
of the school year
Only the standard deviation differs between the two groups with a tendency
to reduction for the experimental group (but no statistical significance).
Duration of interpretation is very long : 20' for the control group, 18'
for the experimental group. These times are double that of specialists
(9' ± 30")

III-3. *Discussion*

The construction of this experiment permits study   of two different problems :
        - the validity and the transmissibility of higher features found in

the first lung radiograph experiment
- the interest of introduction of visual behavior movies in teaching
courses.
III-3.1. The validity and the transmissibility of higher features
We found again the higher feature showed in the first experiment. No more
but no less. This fact attests the effectiveness of such visual behaviors.
The knowledge of these higher features does not appear as a sufficient
condition for good interpretation.
Probably, we should have greater retrospective to have a proper view of the
matter. Surely, we were too ambitious believing that many years of practise
would be learnt in 6 months.

III-3.2. The interest of video learning

It must be emphasized that students welcomed the method with enthusiasm.
This method uses a very attractive media : the movie, and constitutes a
priviliged means of communication and self teaching (cassettes).
Up to a point, weight of verbal comments must decrease progressively to let
the drawings alternating with large sequences of NAC. So, accustomed by
an Image Society, where TV monitors appear everywhere, it would be possible
to make one's own judgement on the quality of visual strategies shown.
In this case, the NAC would be used for its teaching properties (film of
visual behavior without subsequent digit processing ; the difficulties
inherent to this last phase would then be avoided (processing technique,
etc...).

CONCLUSION

  For the last three years, our team which pursues the same objectives,
focused its interest on three complementary aspects :
- the description of visual behaviors which differentiate experts and
  beginners (1981-(1)) : how visual behaviors become organized as a
  function of experience ?
- the teaching of these behaviors (1983-(3)) : is it possible to speed
  this learning and reduce the number of errors by teaching precedently
  isolated visual behaviors ?
- finally, the study of visual behaviors in new radiology techniques-
  especially echography - to assess the influence of the image dynamics
  (succession of images on TV screen) and the role played by the patient's
  presence (1982-(2)).
These studies have various degrees of interest for several specialties :
- clinicans, radiologists, physicians hope to draw out rules of
  "efficient reading" and thus diminish the number of interpretation
  errors (nearly 30% accordint to YERUSHALMI (13)),
- researchers and psychologists hope to generalize the results obtained
  with complex images and to have a better understanding of relationships which exist between visual behavior and intellectual mechanisms
  involved in image analysis.
Consequently, results observed in these three experiments must be
analysed as a function of the specialty of the clinician or psychologist :
For clinicians the NAC seems to offer dramatic new ways to evidence visual
behaviors. It permits isolating a few specific traits of a population
having the same level of experience and can therefore suggest some advice
or remarks that contribute to the effort to diminish the number of interpretation errors. These visual behaviors could be taught to student radiologists more capable of incorporating the observed behavior into their
knowledge. This special course would complement the regular teaching

program. Another way radiologists can use the NAC EYE MARK recorder is to visualize their own visual behavior in order to become aware and analyse their mistakes, particularly as far as the doctor/patient relationship is concerned, since the number of times when the radiologist looks at his patient dons an affective and human meaning.

For psychologists, the radiologists' experience is a consequence of learning and training. General behavior (visual, gestural, verbal,...) is only a direct observable consequence of the use of knowledge. Likewise, visual behavior is one expression of the general behavior and no more. Nac Eye Mark today represents one of the better tools to objectively study visual behavior. In most radiology tasks, numbers of eye movements and their amplitudes are sufficient to emphasize the way to pick up information and to isolate a large part of the needed information (directed search). Limits of the Nac appear for the understanding of the pick up of information when gaze stays many times on the same point. Then, it is impossible to know what is exactly shown : a detail, a general (time to think) sight to the negative... or nothing !

Limits of the Nac also appear if you want to qualify the information processing and the final decision made by radiologists. In this case, Nac Recordings can only by used as objective and common documents which constitute the base of other psychological techniques of cognitive exploration.

REFERENCES

1 - PAPIN J.P., METGES P., HERNANDEZ C. : L'exploration visuelle de radiographies du thorax. Rapport CERMA 81-12 (LCBA), juillet 1981.

2 - PAPIN J.P., FAVRE Y., METGES P. : Analyse de la tâche du médecin "échographiste". Rapport CERMA 83-04 (LCBA), avril 1983.

3 - PAPIN J.P., METGES P., AMALBERTI R., DELAFONTAINE S. : Experience de pédagogie audiovisuelle à partir de l'enregistrement de la direction du regard en radiologie. Rapport CERMA, 1983 - In press.

4 - BAYLESS M.A. : Effect of exposure to prototypie skill and experience in identification of performance error. Perceptual and motor skills, 52, (2), 1981, 667-70.

5 - CARMODY D.P., NODINE C.F., KUNDEL H.L. : Finding lung nodules with and without comparative visual scanning. Perception and psychophysics, 29, (6), 1981, 594-98.

6 - GALE A.G., WORTHINGTON B.S. : The utility of scanning strategies in radiology in Eye movements and psychological functions. International wiews Laurence Erlbaum, 1983.

7 - KUNDEL H.L., WRIGHT D.J. : The influence of prior knowledge on visual search strategies during the viewing of chest radiographs, Radiology, 93, 1969, 315-20.

8 - KUNDEL H.L., LAFOLLETTE P.S. : Visual search patterns and experience with radiological images, Radiology, 103, 1972, 523-528.

9 - LLEWELLYN, THOMAS E., LANSDOWNE E.L. : Visual search patterns of radiologists in training, Radiology, 81, 1963, 288-92.

10 - PAPIN J.P., NAUREILS P., WIATZ A. : Analyse de la direction du regard d'un pilote d'alouette III effectuant une série de vols tactiques en vision monoculaire et avec champ visuel réduit. Rapport CERMA, 80-18, (LCBA), septembre 1980.

11 - RIPOLL H., BARD C., PAILLARD J., GROSGEORGE B. : Caractéristiques de la centration de l'oeil et de la tête sur la cible et son rôle dans l'exécution du tir en basket-ball. New paths to sport learning and excellence. ISSP 5th World Sport Psychology Congress, 1982, 32-36.

12 - SPADY A., JONES D., COATES G., KIRBY R. : The effectiveness of using real time eye scanning information for pilot training. 26th Annual meeting - Proceedings of the human factors society - 1982.

13 - YERUSHALMY J. : The statistical assessment of the variability in obser-ver perception and description of roentgenographie pulmonary shadows. Radiologic clinics of North America, I, 381-90.

Section 6

VISUAL SEARCH AND VISUAL INSPECTION

Theoretical and Applied Aspects of Eye Movement Research
A.G. Gale and F. Johnson (Editors)
© Elsevier Science Publishers B.V. (North-Holland), 1984

VISUAL SEARCH AND VISUAL INSPECTION

INTRODUCTION

Murray A. Sinclair
Department of Human Sciences, Loughborough
University of Technology, Loughborough,
England.

In the section are gathered together a number of papers addressing visual search. They cover the range of problems, from investigations of information-processing within the observer (Mockel & Heemsoth, Nattkemper & Prinz) through a discussion of methods of measuring visual lobes (Bellamy) to applied work where search is actively involved as a vital component of operator performance within systems (Hillen, Papin and Sinclair).

In the more theoretical papers, Mockel and Heemsoth discuss the importance of prior information in controlling the direction of gaze; in their case it is knowledge of what to look for that controls where the observer looks in the acquisition of knowledge to assess performance. They chose a moving information display (an athlete), thereby ensuring that the locus of maximum information in the visual field was continually changing. This is a clever idea, worthy of further use. Their results indicate that the gaze is indeed directed more efficiently the more one knows about the task, as might be expected.

Nattkemper and Prinz discuss the effects of redundancy in a letter-matrix search task. Their initial hypothesis is that the more predictable is the sequence of non-target letters, the faster will search occur for the target letter. Revising this hypothesis rapidly in the light of their results, they conclude that redundancy does indeed help, once the rules governing the redundancy have been mastered. They pose the question of how long it takes to learn such syntactical rules; one suspects that the answer has been provided by Shiffrin and Schneider (1977) and Schneider and Shiffrin (1977) with the learning plateau occurring after some thousand trials. The paper indicates rather well that the complexities of search go beyond the usual two- or three-variable paradigms so often studied in the laboratory.

Bellamy's paper is mainly a review paper, with a brief report of some experiments she has carried out. The problem that has beset her, and many others concerned with real-life tasks, is how to predict performance in a search task. Frequently, the aim is to establish a performance norm, usually embodying notions of the 'Ideal Operator', which can act as a reference point against which human (or machine) performance can be compared. The common approach is to build a model in which the concept of the 'visual lobe' plays a prominent part. The problem then becomes one of assessing the size of this visual lobe, and then including it in a search strategy to predict performance. The main part of her paper is concerned with the assessment of size and her discussion of methods to do this is linked with overall performance measures. It is a very useful short review. The rest of the papers are far more applied. Hillen discusses the problem of measuring search performance in the inspection of dessert apples. This task is fairly typical of conveyor-paced inspection tasks, differing from manufacturing tasks in that the apples are rotating and are less uniform. He provides a succinct description of the difficulties that may be encountered in practice.

Papin presents a digest of his work over a number of years. He has been
concerned with information-processing by pilots and the search strategies
they use. This is a practical version of the paradigm investigated by
Mockel and Heemsoth and his results converge to the same general conclusion;
experts are more skilled in their search for relevant information than
novices. He then goes on to discuss his current work, in which he is
trying to build his more specific conclusions into training programs for
pilots. With simulators, or without, pilot training is very resource-
intensive and his attempt to optimise training effectiveness is to be comm-
ended.

Finally, Sinclair discusses a quality inspection task in the aerospace industry where the management were contemplating the introduction of an
electro-optical device to replace some human inspectors. The paper reports
the result of a comparison study, and comes to the conclusion that a symbi-
osis between man and machine probably represents the best approach; the
electronic device, albeit a prototype, does not match human performance and
cannot attain the required quality performance either.

Taken as a group, the papers illustrate the problems of research in this
field, and the lacuna between pure and applied research. Given the effort
currently being expended in both human and robot vision, it is to be hoped
that the main gap, that of quantitative models of the observer, will soon
close.

References

Schneider, W. and Shiffrin, R.M. 1977. Controlled and automatic human
    information processing: I Detection, search and attention. Psychol.
    Rev. 84(1), 1-66.

Shiffrin, R.M. and Schneider, W., 1977. Controlled and automatic human
    information processing: II Perceptual learning, automatic attending
    and a general theory. Psychol. Rev. 84(2). 127-190.

Theoretical and Applied Aspects of Eye Movement Research
A.G. Gale and F. Johnson (Editors)
© Elsevier Science Publishers B.V. (North-Holland), 1984

MAXIMIZING INFORMATION AS A STRATEGY IN VISUAL
SEARCH: THE ROLE OF KNOWLEDGE ABOUT THE STIMULUS
STRUCTURE

W. Möckel   C. Heemsoth
Fach Psychologie   Fach Sport
            Universität Oldenburg
            West-Germany

We tested the hypothesis that the degree of knowledge
about a biological motion pattern determines the locations of eye fixations: With increasing knowledge more
fixations should occur at points where a maximum of information can be received about the performance of the
motion if no special task is given to the subjects. Three
groups of subjects with different degrees of knowledge
about the motion pattern in shot putting watched a shot
put in a film. Eye movements were registered with the
NAC-IV-Camera. The mean number of fixations at points
with maximal information about the performance of the
motion as defined by experts increased with increasing
knowledge about the motion pattern.

INTRODUCTION

The selectivity of the visual information processing system gives rise to
the question for criteria of the selection of special parts of the stimulus structure as points of visual fixation. The amount of information one
can get from special parts of the stimulus structure seems to be an important property for becoming a target of eye fixation (Mackworth and
Morandi (1967) Antes (1974)). The amount of information one can extract
from parts of the stimulus structure does not only depend on the stimulus
but also on the cognitive conditions of the perceiving person: The degree
of knowledge influences the amount of information one can extract from a
stimulus pattern. The different degrees of knowledge should result in different eye fixation patterns. Welland(1969) suggests that the perceiving
person tends to develop an economic way of information uptake during a
process of learning that should modify the fixation pattern. We tried to
show this for persons inspecting a pattern of human body motion: Subjects
with different degrees of knowledge about the motion pattern in shot putting
watched a shot put. We tested the hypothesis that with increasing knowledge
more fixations occur at points of the athlete's body where a maximum of information about the quality of the motion can be obtained.

METHOD
Subjects

The subjects were 48 students of sports or psychology of the University of
Oldenburg, participants of a course for coaches of athletics as well as
athletes and coaches for athletics. The subjects were divided into three

groups according to their knowledge about the motion pattern in shot putting.
After taking part in the experiment they had to write down a record of the
motion pattern. This record was compared with a check list containing 11
components of the motion pattern in shot putting:

1. Preliminary position, back towards direction of the throw
2. Crouch
3. Shift, streching of power leg
4. Power position
5. Streching of the right leg
6. Streching the body from the bottom to the top
7. Body is opened from twisted position
8. Release, straight foreward
9. Fixing of left body side
10. Putting hand releases
11. Recovery at the bends

We then counted the number of components that had been mentioned in the
record and thus received the following groups of subjects:

o - 3   components: minimal knowledge
4 - 7   components: medium knowledge
8 - 11 components: experts

with the sample sizes:

minimal knowledge : 22 subjects
medium knowledge : 14 subjects
experts   : 12 subjects

The subjects saw part of an instruction film about shot putting made by the
"Institut für Film und Bild in Wissenschaft und Unterricht", Göttingen,
W. Germany, called "Leichtathletik-Kugelstoßen-Rückenstoßtechnik" (Athletics-
Shot putting-O'Brian Technique). The picture was projected onto a screen so
that it was 122 cm wide and 77 cm high. This means a visual angle of $33,9^o$
or $21,8^o$, respectively. The upright body of the shot putter was to be seen
in a visual angle of $18,6^o$. Eye movements were registered with the NAC IV-
eye-mark-recorder. With this camera a systematic registration error occurs
Böcker and Schwerdt (1981) Nickel, Ruoff, Schlottke,(1972) Möckel and Wendt
(1972). With increasing eye movements relative to the subject's head, the
difference between the marker and the real fixation point increases, too
(up to about 4 degrees of visual angle with some older type III models). We
allowed our subjects to move the head to keep the movements of the eye relative to the head small. We could reduce the registration error to a mean
of about one degree (estimated from samples of fixations of points on a test
picture), which according to Young and Sheena (1975), may mainly be due to
variations in cornea shape, tear fluid or astigmatism.

Ten points of the motion pattern were selected each after a significant
change in position and we tried to be sure that for these ten points positions containing a lot of information about the quality of the motion performance were considered. All ten points have been presented as fixed frames
to three experts (coaches) who where asked to mark the one or two most informative points of the body. (An eye fixation of such a point will be called
"hit" later.)
Table 1 shows the points of the body mentioned by the experts:

Table 1. Shot put: O' Brian-Technique

		1. Expert	2. Expert	3. Expert
	1. Preliminary position	Shoulder, right arm	Bowl	Bowl
	2. Balance position	Hip	Hip	Hip
	3. Crouch	Hip	Hip	Knee, free leg
	4. Shift free leg out-streched	Knee, free leg	Knee, power leg	Knee, power leg
	5. Power position - right foot on ground	Axis:Shoulder -Hip-Foot	Shoulder	Hip
	6. - left foot on ground	left foot	left foot	left foot
	7. Arm strike pos. power foot heel turned out	Hip	Hip	Hip
	8. Arm strike pos. Arm 90°	Shoulder Hip	Shoulder Hip	Shoulder Hip
	9. Release	Elbow Hand	Hand	Hand
	10. Recovery	Bends	Bends	Bends

(Phase sequence from Tidow (1981))

Procedure

Before taking part in the experiment the subjects were asked if they were normal sighted. Those with refractive errors only took part if they wore contact lenses.
After calibration of the NAC-eye-mark-recorder the experimenter said:

" You are taking part in an experiment for the analysis
of eye movement behavior, the only task you have is
looking at the following pictures."

After registrating the eye movements the subjects were asked to write down
a record of the motion pattern.

Figure 1 shows the experimental set up.

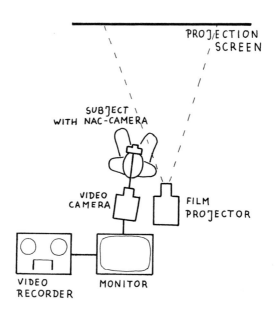

PROJ/ECTION
SCREEN

SUBJECT
WITH NAC-CAMERA

VIDEO
CAMERA

FILM
PROJECTOR

VIDEO
RECORDER

MONITOR

Figure 1
Experimental Set up

Design

We had three levels of the factor "knowledge about the motion pattern "
and two dependent variables: The absolute frequency of fixations and the
frequency of congruence of fixations and point of maximal information as
indicated by the experts (hits). For both variables a comparison of means
of the three factor levels should be done.

RESULTS

We found normal distributions and homogenity of variance in both sets of
data. We took the absolute frequencies of fixations and the frequencies of
hits as representations of an underlying theoretically continous variable
as it is accepted by Winer (1971), at least for the analysis of "numbers

of hits". We did an analysis of variance of the absolute frequencies as shown in Tables 2 and 3.

Knowledge about the motion pattern

	minimal	medium	experts
mean	23,27	21,43	20,75
standard deviation	5,77	7,71	5,12
sample size	22	14	12

Table 2
Absolute frequencies of
fixations

	SS	df	MS	F	sign.
factor "knowledge"	60,27	2	30,14	0,77	-
error	1760,04	45	39,11		
total	1820,31	47			

Table 3
Analysis of variance

There was no difference between the three groups, so we did an analysis of variance of the absolute frequencies of hits as shown in figure 2 and tables 4 and 5.

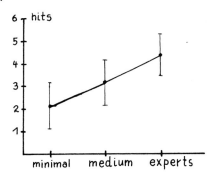

Figure 2
Means of hits for the three groups of subjects

Knowledge about the motion pattern

	minimal	medium	experts
mean	2,14	3,14	4,58
standard deviation	1,13	1,03	0,90
sample size	22	14	12

Table 4
Frequencies of hits

	SS	df	MS	F	sign.
factor "knowledge"	46,70	2	23,35	21,42	++
error	49,22	45	1,09		
total	95,92	47			

F-values for contrasts: 8,56 between minimal and medium ++
                           13,38 between medium and experts ++

Table 5
Analysis of variance and contrasts for hit-frequencies

There is a significant difference between the three factor levels. We could affirm a significant increase in hits from the first to the second and the second to the third factor level by analysis of contrast: the higher level of knowledge goes together with a higher number of hits.

DISCUSSION

There is a long discussion if strategies in visual search are developed or modified with changing cognitive states (Just and Carpenter (1970), Groner (1979), Cohen (1981)) and if possible variations in eye movement behavior - necessary for developing strategies - are limited by physiological mechanisms(Levy-Schoen (1981)).We think with many others that there exists a limited number of degrees of freedom for changing eye movement behavior with changing cognitive states. A principle for a strategy of eye movements which can be very meaningful for an organism seems to be the maximation of information uptake (Mackworth and Morandi (1967), Yarbus(1967)) The amount of information that can be extracted from a visual stimulus structure does not only depend on the stimulus but also on what a subject knows about the stimulus. In our case the motion pattern of an athlete's body in shot putting is the stimulus. The amount of information about the performance that can be extracted depends on the knowledge about the motion pattern: somebody with more knowledge about the motion pattern who tries to maximize information will have a different eye movement pattern from somebody with less knowledge. He should know more points in the motion pattern of the athlete where he can get information about the performance of motion just before the fixation as well as information about the performance of

motion in the direct future (biomechanical dependencies). In consequence
these points should be targets of fixations more often than with people who
don't have that knowledge about the motion pattern. We could show that this
is the case for three groups of subjects with different degrees of knowledge:
with increasing knowledge points of maximal information are more often a
target of fixation. This supports the hypothesis that - having no special
task - the subjects develop a strategy of maximizing information about the
performance of the athlete. This is a contradiction to the point of view
that more knowledge - which usually means amore complex view of the problem-
leads to a selection of a special part of e.g. a motion pattern as formulated as a hypothesis by some experts. This might occur as an effect of a
special task given to the subjects by an instruction: looking for mistakes
in the motion pattern or taking the pattern as a model for a shot put that
has to be done by the subjects themselves. We are testing the interaction
of knowledge and tasks in a current experiment.

REFERENCES

ANTES,J.R. 1974. The time course of picture viewing.
J.exp.Psychol. 103,62-70

BÖCKER,F.,SCHWERDT,A. 1981. Die Zuverlässigkeit von Messungen mit dem Blick-
aufzeichnungsgerät NAC Eye Mark Recorder 4.
Zeitschr.f.exp.u.angew.Psych.,3,353-373

COHEN,K.M. 1981. The development of strategies of visual search. In:
FISHER,MONTY,SENDERS, Eye movements: Cognition and Visual Perception, Hillsdale, L.Erlbaum

GRONER,R. 1979. Die Analyse von Denkstrategien auf der Grundlage kognitiver
Elementaroperationen. In: UECKERT, RHENIUS, Komplexe menschliche
Informationsverarbeitung. Huber, Bern

JUST,M.A.,CARPENTER,P.A. 1976. Eye fixations and cognitive processes
Cognitive Psychology, 8, 441-480

LEVY-SCHOEN,A. 1981. Flexible and/or rigid control of oculomotor scanning
behavior. In: FISHER,MONTY,SENDERS, Eye movements: Cognition and
Visual Perception, L.Erlbaum, Hillsdale

MACKWORTH,N.,MORANDI,A. 1967. The gaze selects informative details within
pictures.
Perception and Psychophysics, 2, 547-552

MÖCKEL,W.,WENDT,D. 1972. Bericht über den Eye Mark Recorder. Manuscript
Universität Hamburg

NICKEL,P.,RUOFF,B.A.,SCHLOTTKE,P.F. 1972. Technisch-theoretische Funktions-
analyse eines Gerätes zur Registrierung von Blickbewegungen
(NAC III). Manuscript, Universität Tübingen

TIDOW,G.1981. Modell zur Technikschulung und Bewegungsbeurteilung in der
Leichtathletik.
Leistungssport 4, 264-277

WELLAND,E.J. 1969. The effect of two density ratios and two background ratios
on the visual search performance of two achievement groups
Diss., Edmonton

WINER,B.J.,1971. Statistical Principles in Experimental Design
          McGraw-Hill, New York

YOUNG,L.R.,SHEENA,D.,1975. Survey of eye movement recording methods
          Behavior Res.Meth.&Inst., 7, 397-429

YARBUS,A.L.,1967. Eye movements and vision
          Plenum Press, New York

Theoretical and Applied Aspects of Eye Movement Research
A.G. Gale and F. Johnson (Editors)
© Elsevier Science Publishers B.V. (North-Holland), 1984

COSTS AND BENEFITS
OF REDUNDANCY IN VISUAL SEARCH

Dieter Nattkemper
Wolfgang Prinz

Department of Psychology
University of Bielefeld
West Germany

We report some results concerning fixation duration during search for a target in lists of different levels of redundancy. Contrary to our expectations we found that redundancy does not facilitate processing per se. Some implications of these
results concerning the mechanisms of making use
of redundancy are discussed.

In continuous visual search tasks Subjects scan through a rectangular matrix of letters or digits in search for some predefined target letter. Figure 1 shows a search list of the type
we used in our experiment.

```
LXTWHVFXMLWTHMHVXKLFWKLNVMFHXLWMFKVL
HKLWHVKFNVHFKNMHXKFTVWFKXWTMFHXWKHNV
NHFKVNMTHXKNTVFHNWKTLVXFWTXLWTMNXLMW
HLFVWTFNHKMVHLMWVXMWTXVNHMKXLWTFNLMH
MXTNMHLKTHVKFXLVHWTLNVKXNLKVWXHMTHHM
HVHFMHWNTHMFKHLXHKNMHLXVWFKTNWKMVNXW
LKVWNTVKHFXLWNTXMWVHXMWHTMHHXFLNMKVW
VFHXLWTHXLWHKFVXTKFNTXKLWVXKLVMHFNKT
MKVLFNWLKFXLWTHMXLKVMTWNMTWNMTVKMHFT
WXTNKMXNLVWMFNLHVXTWVHTFNMXLTFWLXVHN
TMVLXWVLXHTNMHVKFMHXNKHVTKWVLFXWTFHH
LWXHFHKXLWKNTFHKXHVLKMXLNTHHFKTVFLWK
VNFWVKTXLHFTKHVNKFHMWVKFTVMKXWHMTKVW
WTNKHTXLVWXLFHKXLWTKVHFXLKWTHHWXTVFH
TVNLHXVKHWLTHHXVLMHWVXLMWVKFXWHMHVWT
NVXWTHMXFVMXFLVNWKVTLKFHHWKXVFHKXVWF
HHMKVFXKLVNWMTLHMHKVFNKMTHVXLKMHVKFT
XLNWFHXLNKTLNXNFLNXHLMXWLNMWVXMFVHXF
MNHFTHVHTFNVWXLTVMHXWLTVWNTXLKVFXKHL
LWVFTHHKVXLWFXHTWVHFXNVHTFMXHLWTHLNH
```

Figure 1
Example of a search list

Basically it consists of a sequence of letters which are randomly drawn from a predefined set of nontarget letters (e.g.
W T N M H K V F X L). At one randomly chosen location in the
list the nontarget letter is replaced by a target letter (D or
Z) which is thus hidden among the nontargets. The Subject is
instructed to go through the list as in reading, i.e. row by
row, and to indicate the detection of a target letter by pressing a button.

Since NEISSER's early experiments (1963) it is well known that
search speed is critically dependent upon the complexity of the
context. One of the means of controlling context complexity is
by varying the size of the context category. Search lists that
are generated by random sampling from a number of nontarget
letters vary in their complexity dependent upon the number of
context letters.
Another means of controlling complexity is by introducing rules
for the selection of the context letters. Complexity at a
given location of the search list (local complexity) is reduced
if one restricts the random draw at each position to a local
subset of only n out of the total set of m alternatives. By
this means one can generate lists of different levels of (symbol)redundancy.
The degree of local complexity - the redundancy level - is controlled by two factors. One of them is subset size: the number
of nontarget letters out of which one is randomly drawn for a
position in the search list. The other is subset extent: the
number of successive positions supplied from the same subset.
Varying the latter factor one can manipulate the degree of
context redundancy in the following way:
If subset extent is defined as 1 for instance, one position of
the list is supplied from a given subset of nontarget letters,
then the content of the actual subset is changed by a displacement of the copy string by one unit relative to the total
sequence. The following position of the list is supplied from
the new subset of nontargets.
If subset extent is defined as 4 for example, four successive
positions of the search list are supplied from a given subset
of nontargets. Only after the contents of four adjacent positions has been defined the subset of nontarget letters is
changed by a displacement of one unit, thus generating a new
subset differing in one symbol from the last which now
supplies the following four adjacent positions of the list. It
is obvious that in the second case the level of redundancy of
the search list is much higher than in the first example.
(See figures 2 and 3 for an example of those type of lists.)

We believe that using the above described principles for the
construction of search lists in systematic way will help us to
gain knowledge about the mechanisms Subjects develop during
processing of redundant strings of symbols. Under this perspective the work I am going to describe should be interpreted
as a study concerning experimental simulation of Subjects'
operating on a primitive syntax. We assume that the mechanisms
of making use of redundancy can be investigated better under
these reduced conditions than under those conditions which we

find using material of natural language.

```
NHKMXLFTXLHMVFKLWTNMWKNFHKXWNMHKVFXL
XWNMHKVFXKWTNMHKVFXLWFTNHKVFXLWFTMHK
MFKXVFNWHNKFXLWTNMHKVMXLWTXMHHVKXLFT
XMWNVFXLWTNMHKVMXLWTNMHKVFXLWFTMHKVF
VLWTNMHKNFXVWTLMHNVMFLXTNWHTMFHLWFTL
HMVFXLWFNMTKNFXVWLNMHTKFVLWTNMHKVMXL
FWTMHNVMKLWXHTHKNFXLWTNMHKVFXKLTXWHH
NKVLXTNMHKVFXLWTNMWTHMXKWTNMHKVFXLVF
LMNKVFXKWFXTHMVKXLWTNLHMVKFXVTNMHKVF
XFWTNMWKHFXKLWTMHKVFXLWFNTWHKMVLWTNM
WNHFVLWTNMWKNFVLWTXLMNVKFLVXNTHKNFXK
XTWMHNVKXLFWHTHKVFXKWLNHTKVMHKLXNWMK
NVXKWTXNHKVFXLDTNMHKNFVXLWNMHKVFXKLT
TMHKVFXLWTXMHNVMXLWTNMHKVFXLWFNMHKVF
XVWTXNHMVFHKLFNWHMKFXLWTNLWTKFHLWFNT
MKVFHXVFTMHKVFXKWTNMHKNFXLWTNMHTVFXL
VXWMTKVFXLWTNLHMVFXLWTXMWKHMVLXTNMHT
NFKLWTNMHKVFXLWTNMHKVFXKLWHMHTVKHXFT
LTHNVFXKVTLMHKNVHLXTNMWHVFXKWLNMHKNF
XVLTWMHKVFXLWTNMHKVFHLWTXMHKVFHKLXNM
```

```
WTHNMKTHNKMHFVMHXFKHXFKLWFVXWTFXNWLX
NWMTHNWMHTKNHVMNKVHFXKHFLVKXWVLXTFLX
NLXWNLTXMWTNMHTKVNMHFKMVHXFVKLFXVWLF
XTFLHWXLNMWTHMWTKHMTVNHMVFKMHVFKLXVF
XVLWFXTLNWXTLNWTMHWTNHKTVHKNMFVHXFKV
LXKVWXFVTLFWXTLWHNTLWHHMTKNHMVKNFVKM
KHFXLKFVLWXFTWLFTNWLTNWLMHWNTHKNVHMN
VHKMFXHVKXLVWXLVWXLFTWLXNMLTHMWNTKMN
HVZMFVHMKFXVKLFXWVLFWXTFLXNWTMLNHWMT
MNTKMNVKFMVHXFVKXLFVWXLVFTWXNLTXNLWT
MNTHKNMTVNKHMVKHXVFHKXLVFWXVTWFXTWLX
LMTNWHMNTKMHNVKMFVHKFXHKLXVKWLXVFWTL
LNWXTLMWHTMWKTHNKMHVFKHMFVKHLFXVXLFV
FLWTXNLWMNTLWNMHKTMNVHKMVFKMVXHFKLXV
XLFVWTFXLTNWHLNWTMHWTNHHVKNMVFKHXVFH
KXVFLWVXTLFXWTNXLTNWMTNWKTHMNVKHFMVH
HFXKVLFKVWXFLTXFNWXTLNMWHTMWNTKMVHNK
FMHKFXVHLFVKWLFXTWFXLNTWMNLWHNTWHKTN
NVNKMVHFKVXHLKXVWLXVWTXLWTNLMWTLMHWN
THKMHHVKMFVKHXFKLVFKLXWVFTXWLNXWTNLM
```

Figures 2 and 3
Examples of search lists of the lowest
and highest level of redundancy

In earlier experiments (PRINZ, 1979; PRINZ, 1983) we measured search time/row and detection distance. (Detection distance indicates the distance between the actually scanned row out of which a Subject detects a target letter and the row where the

target letter is located.) We could confirm our expectations
that search time/row and detection distance are dependent upon
the statistical structure of the search list:
The higher the redundancy the faster Subjects scan the list
and the larger the area over which the occurence of a target
letter can be controlled (control area).
We interpreted these results as a confirmation of our conception that Subjects make use of redundancy by developing
mechanisms which are constructed in a way which allows to facilitate the processing of individual items of a string both by
knowing the temporally preceding events and by knowing the
rules for the construction of the sequence of events. We proposed two mechanisms to be active during fixation in order to
integrate information during processing of redundant strings
of letters:
Besides a mechanism which makes use of repetitions of identical
letters in the actually focused part of the search list
(spatial integration), we assumed a second mechanism which
makes use of correlations between letters (temporal integration). The process of temporal integration can be described in
the following way:
During fixation Subjects successively scan the available elements in order to test if they are nontarget letters. The test
consists in comparing the representation of stimuli with a
memory representation which has been established during the
actual or preceding fixation pause/s: each identified element
has got a prime the strength of which monotonically decreases
over time. The time needed to test the status of each element
is dependent upon the strength of the prime of the actual
memory representation. Strength of the prime depends upon two
factors. It is the stronger the more frequent and the more
recent an identical or correlated symbol has been found in the
actual or preceding sample.
In the case of redundant lists Subjects profit from the predictability of single elements of the list by reducing the
time needed to test the elements of a sample. According to our
conception the differing performance in processing redundant
vs random strings of symbols results from differing strength
of the prime of the memory representations. We assume that
during each fixation Subjects not only process the found symbols but also prepare for the following fixation by construc-
ting a memory representation of the future string of elements
which in return can facilitate the processing of future elements if it is adequate.
From the saving of time to test the status of single elements
of a sample results that per time unit more letters can be
fully processed. This should lead to a compensatory reduction
of fixation duration dependent upon the statistical structure
of the search list.

METHOD

Apparatus and procedure - The experimental sessions were in all
parts controlled by Real Time Host-Satellite Computer System
Modular Computers Classic CPU 7840/7810.

Subjects were seated in front of a fast display (Hewlett & Packard 1311B display with P 31 phosphor controlled by 1351A Graphic Generator). Their heads were fixed at a viewing distance of 56 cm. During the scan the recordings of the horizontal EOG component were taken in order to identify saccades and periods of fixation. - The details of the data acquisition procedure and saccade detection algorithms are described elsewhere (HALPAAP & NATTKEMPER, in preparation).

In this experiment we used four types of search lists. In all search lists presented the nontargets were selected from the ensemble W T N M H K V F X L. There were two targets: D and Z, but only one of them could occur in a given list. As the subject was not informed in advance which one would occur, he/she had to search for both of them simultaneously. The relation of the occurence of D: Z was 6 : 3; one of ten lists did not contain any target at all. These search trials served to validate the saccade detection program because they offered an opportunity to check the observed against the expected number of return sweeps (i.e. counted by the program vs. expected on the basis of the number of rows in the list). The position of the target was randomly chosen with the exception that it never occured in the first row or the first and the last column of the search lists.

Each type of search lists was represented by ten lists in every experimental session. The first type consisted in search lists the letters of which were randomly selected from the predefined ensemble of nontarget letters (condition 1). As in search lists of the other types the random selection was restricted in so far as repetitions of identical symbols at a region of four adjacent letters were not allowed. For the generation of the further three types of search lists, we used the above mentioned principles: Subset size - the number of nontarget letters from which one is randomly drawn for a position of the search list - was held constant (subset size = 5) for all types of lists. In order to manipulate the degree of context redundancy we varied subset extent - the number of successive positions supplied from the same subset. Out of the subset of five nontarget letters areas of 1, 2, and 4 positions (conditions 2, 3, 4) were supplied before the content of the actual subset was changed by a displacement of the copy string by one unit (see figures 1, 2, and 3 for examples of lists of different levels of redundancy).

Four Subjects were engaged in seven experimental sessions. Each of the four types of lists was represented in every session by a block of ten lists of the same type. Within each Subject the sequence of conditions was varied over sessions according to a latin square. Between Subjects we varied the sequence of conditions in experimental sessions too according to a latin square. The first three sessions served as training sessions, sessions 4 to 7, which were arranged on consecutive days were considered for the analysis of data.

An experimental session consisted of five sections. As we use velocity information for the detection of eye movements in an initial calibration step the individual mean velocities of saccades of one degree and their mean amplitudes have to be found in order to define a velocity criterion for the saccade detection program. After having defined the criterion value the Subject pressed a button and a search list was presented on a display. The search lists consisted of 20 (rows) × 36 (colums) = 720 symbols. They covered about 18.5 × 17.2 cm on the screen, corresponding to an angular size of 18.5 × 17.2 degrees at the viewing distance of 56 cm. Two horizontally adjacent symbols corresponded to an angular size of one degree.

After the Subjects had processed all lists of two of the four types (20 search lists) we repeated the above mentioned calibration routine using the initially defined criterion value in order to check of the criterion had been stable over time. This was the case for all Subjects. We therefore assume the eye movement data to be reliable. After having checked the reliability of saccade detection Subjects continued to search through the lists of the remaining two types. An experimental session ended in the repetition of the initial calibration task. For all Subjects the criterion value turned out to be stable over the time of an experimental session.

RESULTS

First, we have to state that we were not able to confirm our expectations in all details. The results concerning fixation duration are shown in figure 4.

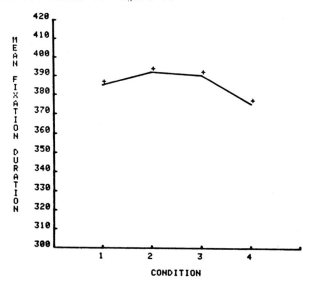

Figure 4

The effects of the statistical structure of the search lists
on fixation duration after saccades to the right are shown in
figure 4. Certainly fixation duration is monotonically decreasing over the three types of redundant lists (significant
linear trend $F = 7.61$, $df = 1/\infty$, $p < 0.01$) but against our
expectations we find that fixation duration is considerably
reduced in the control condition compared to the first two
redundance conditions.

Taken together we have to conclude that fixation duration
reflects the different statistical structure of the four types
of search lists only within the redundancy conditions in the
expected manner. The astonishing aspect of the results consists in the fact that we do not find any significant difference if we compare the control condition with the three
redundancy conditions.

DISCUSSION

At first glance the results of the described experiment were
confusing. Till now we assumed that after a period of training
the introduction of redundancy facilitates the processing of a
given letter at a given position in the search list in any
case. We now find that the possibility of predicting future
elements of a search list does not only not facilitate the
processing of strings of letters - performance was as well in
the control condition as in lists with different levels of redundancy - but even complicates it - lists of the first level
of redundancy seem to be harder to process than those of the
control condition (although the difference between lists of
the control condition and those of the first redundancy condition does not reach significance).

These results become plausible, however, if we assume that
Subjects being confronted with a redundant search list first
have to acquire the exact statistical structure of the actual
list. This should exert an extra load in the beginning of
scanning redundant lists compared to lists which are generated
by random selection out of a given string of nontarget letters.
Search lists of the latter type can be identified at first or
second glance as random lists because there occur combinations
of letters (for example the sequence FWHM) which are not
allowed in redundant lists and because sequences of two letters (for example TV) frequently appear in control lists while
such combinations rarely occur in redundant lists. We therefore assume that Subjects being confronted with a random list
do not even try to find out the statistical structure of those
lists. During fixation duration they are exclusively engaged
in testing if each of the given letters is a nontarget letter.
Another situation results if Subjects are confronted with redundant lists. As easily as they can decide that the actually
presented list is a random list they can state that it is one
of the redundant lists. Now they are not only engaged in
identifying the status of the fixated letters but in addition
try to prepare which letters they will find during the

following fixation. During scanning redundant lists the
initial costs of redundancy should decrease. At the end of
those lists the benefits of redundancy should prevail.

These assumptions can easily be tested by analysing the data
of the first three rows and the last three completely scanned
rows of the search lists. If our revised conception of redundancy holds we should expect (1) fixation duration to be
longer in the first rows than in the last rows of redundant
search lists. We should (2) expect shorter fixation durations
in the last rows of redundant lists than in the last rows of
random lists.

RESULTS

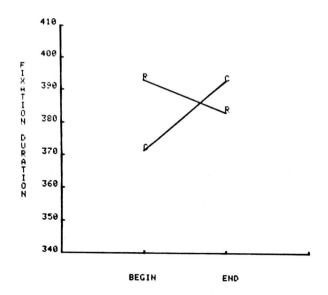

Figure 5

The results (see figure 5) are clearcut with respect to our
expectations. Fixation duration decreases in the course of
scanning redundant list. We also find the expected difference
between the control condition and the redundancy conditions.
The interaction Condition (C vs. R) x Row (begin vs. end)
reaches significance (F = 4.62, df = 1/∞, p < 0.05).

CONCLUSIONS

The results of the present experiment called for a more precise formulation of our conception about the effects of redundancy:
We now must assume that redundancy does not facilitate the processing of strings of symbols per se but even can complicate it. In the beginning of scanning a redundant search list the costs of redundancy prevail because Subjects are engaged in finding out the statistical structure of the actually presented list. Only in the course of scanning Subjects learn to construct adequate memory representations of future strings of letters which in turn facilitate the processing of the actually found symbols.

REFERENCES

(1) HALPAAP, C. & NATTKEMPER, D.: Real time identification of saccades. (In preparation)
(2) NEISSER, U., Decision time without reaction time: Experiments in visual scanning, in: Quarterly Journal of Experimental Psychology, 1963, 31, 376 - 385
(3) PRINZ, W., Integration of information in visual search, in: Quarterly Journal of Experimental Psychology, 1979, 31, 287 - 304
(4) PRINZ, W., Redundanzausnutzung bei kontinuierlicher Suchtätigkeit, in: Psychologische Beiträge, 1983, 25 (in press)

Theoretical and Applied Aspects of Eye Movement Research
A.G. Gale and F. Johnson (Editors)
© Elsevier Science Publishers B.V. (North-Holland), 1984

353

THE APPLICATION OF VISUAL LOBE MEASUREMENT TO VISUAL INSPECTION

Linda J. Bellamy

Ergonomics Development Unit, Department of Applied Psychology
University of Aston
Birmingham
U.K.

Following an introduction to the concept of visual lobe, a general description and evaluation of methods of measuring lobe size is given with particular reference to their application to visual inspection. Problems of relating lobe size or target conspicuity measures to inspection performance are discussed. The overall aim is to emphasise the best method of using visual lobe or target conspicuity data to increase the efficiency of search in inspection tasks.

1.   INTRODUCTION

The definition of visual lobe or target conspicuity area depends upon the way it is measured. Whatever the measurement method, the lobe value obtained is intended to reflect, as accurately as possible, one parameter which determines search time, scanning patterns and target detection rates in visual search tasks. The usual definition is to describe the lobe as the area around the point of fixation within which a particular target can be detected in a single glimpse with a certain specified probability. In general a 0.5 (threshold) probability of detection is used to delineate the boundary to the lobe area, often called the hard shell lobe because it is treated as though nothing will be detected outside this boundary. There is, however, for any particular target/background combination a gradient of detection probability as a function of retinal eccentricity; this is the soft shell lobe.

It has been repeatedly demonstrated that lobe area is inversely proportional to search times and proportional to the probability of detecting the same target in a single eye fixation during search (e.g. Bellamy and Courtney (1981), Bloomfield and Howarth (1969), Engel (1976), Erickson (1964), Johnston (1965)). It is apparent that the more coverage that can be given in a single fixation, the less fixations that are needed to scan a particular area.

The deterioration in peripheral visual acuity with eccentricity is usually attributed to the neurophysiological organisation of the retina and visual pathways. However, the complex interaction between a target and its background, as well as the influence of cognitive factors such as experience or instructions, make the lobe highly elastic in terms of shape and size. For this reason, in applying the lobe concept to real life tasks, a description either at a neurophysiological level or in terms of a mathematical modelling approach is insufficient to explain search behaviour (Megaw and Bellamy (1979)). What is important in an applied

setting is to be able to compare and quantify factors which affect lobe
size and which thereby also affect search behaviour.

Studies of the visual lobe involving real life tasks have generally been
confined to three areas - military tasks, industrial inspection for
quality control, and X-ray examination. Figure 1 illustrates the
potential difficulty in such tasks of trying to relate lobe parameters to
performance. The arrows indicate the influence of one variable on another
derived from case studies or implied by laboratory studies. Although the
example is for industrial inspection tasks the parameters and their
relationships are applicable to the other tasks mentioned above.

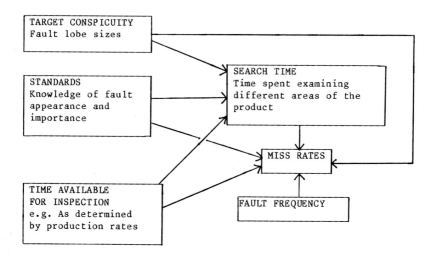

Figure 1. Lobe size and performance: Direct and indirect effects of
lobe size on target miss rates and the confounding effects of other
factors.

There have been few attempts to measure lobe sizes of industrial
inspectors for product faults. Although the major problem in inspection
studies would be to control for the effects of knowledge of standards on
search time and miss rates (cf. The taxonomy of inspection errors and
causes, in Megaw, Powell and Bellamy (1983)), the most likely reason for
the paucity of studies in this area is the difficulty in applying a
suitable measurement technique. In order to elucidate these measurement
difficulties, the various methods which have been used are described and
evaluated.

2   MEASUREMENT OF LOBE SIZE

2.1   Threshold methods

Lobe size is most commonly measured by limiting target/background exposure
to the duration of a single fixation (about 200-300 ms) using a
tachistoscope. The target is randomly presented across a range of

eccentricities, usually in just the horizontal and/or vertical meridian,
using a number of trials for each eccentricity. Detection probability
values can then be calculated for each eccentricity.

An alternative to adjusting target eccentricity is to manipulate target
conspicuity by reducing its contrast to threshold using a visibility meter
(Blackwell (1970)). The reduction in constrast required to achieve
threshold viewing can be used to represent task visibility level for
off-axis viewing. This method has been employed by the Commission
Internationale de l'Eclairage (1980) to quantify the effects of changes in
lighting conditions on visibility level for tasks involving scanning and
search although it does not measure lobe size as defined in the
introduction.

2.2   Eye movement methods

2.2.1 Deviaton from direction of scanning

Mackworth (1976) devised an unusual method of evaluating lobe size by
presenting subjects with varying widths of test strips of target/backgound
material.   He theorised that the smallest width of test strip which
caused the fixation pattern to significantly deviate from a horizontal
scanpath was equivalent to lobe width. As the width of the test strip
increased the proportion of side steps to saccades was also found to
increase.

2.2.2 Interfixation distance

In theory one would expect a direct relationship between interfixation
distance and lobe size because larger lobes mean more coverage can be
given in a single fixation. A number of authors have found a close
relationsip between the two (e.g. Engel (1969), Enoch (1959), Mackworth
(1976)). For example, Enoch obtained a decrease in interfixation distance
when the quality of aerial maps was degraded. However, it should be noted
that interfixation distance does not remain constant during search.

2.2.3 Rating methods

Bloomfield et al (1974) have shown that rated target discriminability on
a linear scale is proportional to lobe size and search time. Carmody,
Nodine and Kundel (1981) found that detection accuracy of lung nodules in
scanning chest X-rays was significantly different when compared according
to three associated levels of visibility rating (low, medium and high),
with the highest miss rates being obtained for the low visibility rated
nodules. Evidence from laboratory experiments at least seem to suggest
that subjective ratings could be used to give some indication of relative
search time and miss rate differences between targets of varying
conspicuity although it is doubtful that subject differences could be
examined using this method.

2.4   Search times

It should be possible to indicate lobe size differences between either
individuals or targets, at least on an ordinal scale, simply by comparing

search times. This has been successfully achieved using a card sorting
task (Bellamy and Courtney (1981)) which requires search; fast sorters are
fast searchers. The speed of search was inversely related to lobe size
(r = -0.92, p<0.01). Test-retest reliability was 0.81 (p<0.001).

3.   EVALUATION OF MEASUREMENT METHODS

3.1   Threshold methods

The brief exposure method is laborious as it requires many measurements
for a single target and individual. It is therefore time consuming and
from the point of view of inspection tasks, the large numbers of
inspectors, fault types and range of fault magnitudes would make
measurement very impractical. However, its value may lie in being able to
determine conditions which maximise target conspicuity, such as changes in
lighting parameters, and quantifying such differences using lobe size
(e.g. Boyce (1981)). Contrast detection thresholds may work just as well
for this purpose. This author (Bellamy (1980)) has found that foveal
contrast detection thresholds, measured with unlimited viewing time using
a visibility meter, significantly correlated with search times for the
same targets (r = -0.55, df = 18, p<0.01). Foveal detection and
unlimited viewing was used because this would be the easiest to apply in
industrial situations, enabling target conspicuity to be measured directly
at any particular workstation. This is important because of the variation
in visual conditions that is often encountered for different inspectors
viewing the same product.

One factor that has not been controlled for in most of the experimental
work is the effect of response bias. It is usual simply to instruct
subjects to report whether or not they can see the target without
considering the confidence with which they are making the response. In an
experiment carried out by Bellamy (1979a) subjects were required to
indicate the certainty with which a target was seen during 250 ms
exposures using the confidence rating method of signal detection theory
(McNicol (1972)). Non-parametric measures of sensitivity (P(A) for single
data points, and P(A) when several data points are available) were used
(Pollack and Norman (1964), McNicol (1972)). ROC curves for one subject
are shown in Figure 2. Hit probability is not only dependent upon
eccentricity but also response bias. For all subjects sensitivity showed
a decrease with eccentricity as would be expected but the lobe sizes thus
measured were much larger than were predicted using the familiar yes-no
technique. Also, the latter technique did not highlight subject
differences so well. One implication of this experiment is that the
apparent increase in lobe size with experience that is found in most
experiments may be partly an artifact of changes in response bias; as
experimentation progresses subjects may become more risky.

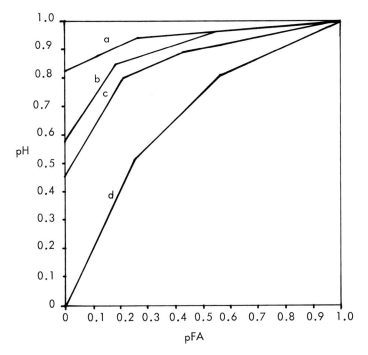

<u>Figure 2.</u> ROC curves for one subject viewing a V target on a background of Xs at 250ms exposures. Eccentricity of target for a = 130', b = 180', c = 230', d = 280'. Sensitivity (P(A)) for a = 0.96, b = 0.92, c = 0.88, d = 0.68.

A final point concerns attempting to equate lobe size with the amount of coverage that may be given in a single fixation during search. If aids are to be designed to optimise search strategy, for example in the use of grids (Eriksen (1955)), this equation must be fairly accurate. Engel's experiments (Engel (1976), Engel and Bos (1974)) suggest that lobe size may decrease by 20-30 % during search. Whether this indicates a change in response bias or a poorer search strategy than the lobe measurement method would imply remains to be investigated. However, if single fixation sensitivity could be equated with sensitivity values obtained from search data it may be possible to use this relationship to estimate lobe size during search.

3.2   Eye movement methods

Eye movements methods are not only limited in terms of having the appropriate equipment but also in their applicability to measuring lobe size. This author repeated Mackworth's (1976) strip-search method using conspicuous (O) and inconspicuous (V) targets on a background of rows of Xs (Bellamy (1979b)). For both V and O targets deviations occurred much earlier than predicted by threshold measurement of lobe size and the V and

O targets were indistinguishable in terms of scanpaths for any particular strip width. Only number of fixations and the higher hit rate for O targets significantly differentiated between target conspicuities (p<0.025). Conversely, the danger of predicting scanning strategies from threshold measurement of lobe sizes is obvious since scanning strategies are not altogether related to target conspicuity although this may be a consequence of insufficient training.

The fact that subjects make involuntary fixations on non-targets, even when instructed not to do so, and particularly when they are very conspicuous (e.g Engel and Bos (1974)) suggests that interfixation distances and lobe sizes will vary according to the conspicuity of non-targets during search. Interfixation distances also do not rema constant during search due to tendencies to adopt overall orientatic strategies preceding detailed examination of likely target areas (Enoch (1959)). In practice, therefore, interfixation distances are unlikely to be reliable indicators of lobe size, particularly for the inspection of more complex products.

3.3   Rating methods

Ratings are an imperfect means of determining conspicuity because they are subject to bias (Poulton (1977)). Nonetheless, they significantly correlated with search times for a simulated inspection task (Bellamy (1980)) giving a better indication of conspicuity than contrast detection thresholds for the same targets. Table 1 shows the results for three methods of rating target conspicity. The seven point scale was the most valid predictor of search times.

Rating method	Correlation with search times	Significance (p)
10 cm line	0.64	<0.005
7 point scale	0.70	<0.0005
Ranking	0.64	<0.005

Table 1 - Correlations between three measures of rated target conspicuity and search times (df = 18). The lower the rating, the more conspicuous the target.

The seven point rating method was then used in a real life context. Fault discriminability was rated by inspectors of plastic moulded car components. The ratings did not correlate with the frequency of customer returns for the same fault types whereas inspectors' subjective ratings of fault frequency did (r = 0.63, df = 9, p<0.05). However, in a study of battery inspectors miss rates were perfectly correlated with discriminability ratings although the company could only provide miss rate data on the basis of three fault categories which limited the testing of this relationship. Returning to Figure 1, these results are explicable in the combined effects of fault conspicuity and fault frequency on miss rates.

The value of the rating method is that it is quick and easy to use
particularly when the number of fault types is large. Since it does give
some indication of the least conspicuous faults this would enable easy
selection of such targets from the fault set for laboratory evaluation of
methods for improving conspicuity. Also it would provide a means of
determining a suitable fault mix in making up test batches for performance
testing.

3.4 Search times

Search times decrease with experience (e.g. Kundel and la Follette
(1972)). It is unlikely that this learning effect would transfer to
unfamiliar search tasks; Bellamy and Courtney (1981) found that
experienced inspectors had great difficulty with the card sorting task
(cf. 2.4) and took much longer to search than inexperienced students. To
investigate the validity of using the card sorting task as a selection
tool, pilot experiments were carried out at the University of
Aston using a simulated and a real task requiring the inspection of solder
joints. The real task was composed of 20 printed circuit boards, half of
which had soldering defects. The simulated task was the same but
represented on a microprocessor (Research Machines 380Z).

The correlations between the card sorting times and task mean search times
for the same twelve subjects were 0.94 ($p<0.005$) for the simulated task
and 0.45 ($p<0.05$) for the real task. The search times for the real and
simulated tasks correlated 0.58 ($p<0.025$). Difficulties experienced in
the real task were mainly concerned with recognising fault types as was
evident in the large number of misses (41% of faulty items) amd false
alarms (21% of acceptable items). In the simulated task there were no
false alarms and only 6% of faulty items were missed. In the former task
the difference between a dry joint and a good joint had a much hazier
boundary than it did in the simulation. So, while the card sorting task
did discriminate between fast and slow searchers, the standards variable
had an independent effect on search times.

Returning again to Figure 1, knowledge of standards is, for real life
inspection tasks, going to confound the effects of individual lobe size
differences on search times. However, because the pilot experiments
suggest that relative lobe size differences are likely to be consistent
across tasks, the card sorting task could be used to isolate the effects
of such individual differences from those of other factors which affect
search times in real life situations. In addition the use of this task
could have significant value in the selection of fast searchers.

REFERENCES

[1] Bellamy, L. J., Pilot study to determine sensitivity to targets in
peripheral vision using procedures of signal detection theory, Internal
report, Dept. Eng. Prod., Birmingham Univ. (1979a).
[2] Bellamy, L. J., Pilot study to investigate a horizontal scanning
technique for evaluating visual lobe size, Internal report, Dept. Eng.
Birmingham Univ. (1979b).
[3] Bellamy, L. J., The relationship between measures of fault
conspicuity and visual search. Paper presented at AVA meeting, Applied
Vision in Industrial Inspection (May 1980).

[4]   Bellamy, L. J. and Courtney, A. J., Development of a search task for the measurement of peripheral acuity, Ergonomics 24 (1981) 497-509.
[5]   Blackwell, H. R., Developments of procedures and instruments for visual task evaluation, Illuminating Engineering 65 (1970) 267-291.
[6]   Bloomfield, J. R., Beckwith, W. E., Emerick, J., Marmurek, H. H., Tei, B. E. and Taub, B. H., Visual search with embedded targets, Report for the U.S. Army Research Institute for the Behavioural and Social Sciences, Virginia U.S.A. (1974).
[7]   Bloomfield, J. R. and Howarth, C. I., Testing visual search theory, in: Leibowitz, H. W. (ed.), Image Evaluation, Procedures of the Nato Advisory Group on Human Factors, Symposium, Munich (1969) 203-224.
[8]   Boyce, P. R., The visual detection lobe and visual inspection, Proceedings of the Comite Espanol de Iluminacion, IV Lux Europa, Granada (May 1981).
[9]   Carmody, D. P., Nodine, C. F. and Kundel, H. L., Finding lung nodules with and without comparitive visual scanning, Perception and Psychophysics 29 (1981) 594-598.
[10]   Commission Internationale de l'Eclairage, An Analytical Model for Describing the Influence of Lighting Parameters upon Visual Performance, Draft CIE Publication 19/2, TC-3.1 (1980).
[11]   Engel, F. L., In search of conspicuity, IPO Annual Progress Report 4 (1969) 89-95.
[12]   Engel, F. L., Visual Conspicuity as an External Determinant of Eye Movements, PhD Thesis, Technische Hogeschool, Eindhoven (1976).
[13]   Engel, F. L. and Bos, Th. M., Visual conspicuity and eye movements, IPO Annual Progress Report 9 (1974) 94-98.
[14]   Enoch, J. M., Effect of size of display on visual search, Journal of the Optical Society of America 48 (1959) 280-287.
[15]   Erickson, R. A., Relation between visual search time and peripheral visual acuity, Human Factors 6 (1964) 165-178.
[16]   Ericksen, C. W., Partitioning and saturation of visual displays and efficiency of visual search, Journal of Applied Psychology 39 (1955) 73-77.
[17]   Greening, C. P., Mathematical modelling of air-to-ground target acquisition, Human Factors 18 (1976) 111-148.
[18]   Kundel, H. L. and La Follette, P. S.,Visual search patterns and experience with radiological images, Radiology 103 (1972) 523-528.
[19]   Leachtenauer, J. C., Peripheral acuity and photointerpretation performance, Human Factors 20 (1978) 537-551.
[20]   McNicol, D., A primer of Signal Detection Theory (George Allen & Unwin, London, 1972).
[21]   Megaw, E. D. and Bellamy, L. J., Eye Movements and visual search, in: Clare, J. N. and Sinclair, M. A., Search and the Human Observer (Taylor & Francis, London, 1979) 65-73.
[22]   Megaw, E. D., Powell, J. C. and Bellamy, L. J., Improving visual inspection performance, in: Kvalseth, T. O., Ergonomics of Workstation Design (Butterworths, London, 1983) 45-64.
[23]   Pollack, I. and Norman, D. A., A non-parametric analysis of recognition experiments, Psychonomic Science 1 (1964) 125-126.
[24]   Poulton, E. C., Quantitative subjective assessments are almost always biased, sometimes completely misleading, British Journal of Psychology 68 (1977) 409-425.

Theoretical and Applied Aspects of Eye Movement Research
A.G. Gale and F. Johnson (Editors)
© Elsevier Science Publishers B.V. (North-Holland), 1984

SEARCHING FOR BLEMISHES ON APPLES

J.R.C. Hillen

Department of Human Sciences
Loughborough University of Technology
England.

The views expressed in this paper are not necessarily
those of the Ministry of Agriculture Fisheries and Food
or of the Department of Human Sciences, Loughborough
University.

The visual inspection of apples has only recently come
under human factors scrutiny in this country. This
paper describes initial work to investigate the applied
problem using visual lobe measures for specific defects
and a simple 'best case' model of inspection. Limited
eye movement recordings are used to show that the 'best
case' is not apparent in real life and highlight certain
individual differences.

INTRODUCTION

Apples can boast one of the earliest references to fruit, when Eve gave Adam
one to eat, but it is only in the past few years that the human factors of
apple harvesting and marketing have been investigated (e.g. Burkhardt and
O'Brien 1979). The work described here is part of a project carried out at
Loughborough, funded by the Ministry of Agriculture Food and Fisheries,
with the aim of improving the British apple. The research has been split
into three main areas:-

1 - Harvesting

2 - Visual Inspection in the Packhouse

3 - Grading Standards Used in the Packhouse.

It is difficult to separate the last two items but this paper will attempt
to concentrate on purely the visual inspection component.

TYPICAL GRADING PRACTICE

Apples are graded when the market requires the fruit, in some cases this is
directly after harvesting but in the main the fruit is taken from a cold
store. Large wooden bins, containing some 720 lbs. of produce, are used
to transfer the apples onto a grading conveyor belt. In the larger packhouses this is done by floating the fruit out of the bins but for the
majority of cases dry tipping is used. The person in charge of the

tipping operation is often responsible for taking out the rots (bad apples which have gone rotten in store).

The apples (mainly Cox's Orange Pippin) may then be polished which removes much of the dried fungicide dip used to increase longevity in the cold store, and imparts a supermarket shine.

The grading table, where visual inspection takes place, is usually next in the line.   This is a rolling top conveyor designed to rotate the apples to exhibit more of the apple surface to the examiners.   (See Fig. 1). The examiners are traditionally female.   The number standing at a single grading table may vary between one and fourteen, the most common number being one or two placed in series.   Apples roll onto the conveyor in an irregular manner and at irregular flow rates.   The colour of the rollers varies from black to grey although the Ministry recommends mid-grey.

After the grading table a mechanical sizing machine channels the fruit into appropriate packing areas where the apples are boxed and weighed ready for market.

Figure 1: Rolling Top Conveyor

THE EXAMINERS TASK

The task at the grading table consists of detecting flawed apples classify- ing the flaw and placing fruit in the appropriate class lane.   Decisions are required on four attributes of each apple, colour, shape, russet and blemish.   The standards for each attribute vary for each apple type and to some extent from farm to farm.   Standards are sometimes checked by an in- house inspector carrying out a 1 percent sample of all boxes destined for sale.   If the farm is a member of a specific marketing scheme the scheme's own inspectors may visit the packhouse once a fortnight for spot checks. Ministry trained inspectors are ultimately responsible for the upkeep of the EEC standards.   Despite these checks it is estimated that approximately 22 percent of fruit is misclassified.

FIELD WORK

During the '82-83' grading season visits were made to several commercial packhouses throughout England to investigate the problem.   Whilst at the packhouse a measure of the examiners binocular visual lobe was attempted. (For a description of the visual lobe see previous paper by L. Bellamy.) To simplify the mathematics involved a 'hard shell' concept was used. The lobe in this case being a two dimensional area around a central fixation point within which objects can be seen on at least a given percent of occasions on which they appear.

To map comprehensively the lobe boundaries is an exhaustive process.   In this study only the major and minor axes were measured off line in an approximate simulation of the on line task of grading apples.

A series of 32 slides of a 9 x 9 grid of cox apples were presented tachisto-scopically at a viewing distance of 900 mm.   Eighty of the apples were perfect, one was defective.   The position of the defect varied at random. Subjects were asked to indicate where they felt the fault was and to give a rating of confidence on a seven point scale in their judgement.   After a five minute rest a further 32 slides were shown.

Two types of defect were used, blemishes which consisted of cracking over an area of 135 sq mm and a markedly discoloured bruise of 100 sq mm. These defects were chosen to represent a range from easily detectable faults to the more subtle appearance fault, such as a bruise.

In classical target acquisition terms a bruise is difficult to define.   The visibility of the target is affected by its position on the apple and the colour of the background.   The colour of the bruise is dependent on its age.   Recent bruises are only visible as indentations on the skin.   Bruises of four to five hours are markedly discoloured, whereas older bruising pro-duces a lighter corky discolouration just below the surface.   The colour across the bruise varies and is also dependent on the original skin colour. Therefore as a standard bruise could not be used for the study the slides were taken of a range of concave bruises.

The objective data has been accumulated for 47 subjects and an average lobe calculated for the larger defect is shown in Fig. 2.   For a 100 sq mm bruise the lobe reduces to the size of a single apple.

Figure 2: 'Average' Lobe for Cracking Defect

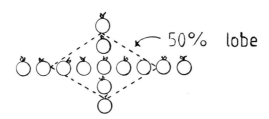

SIMPLE MODEL OF SEARCH

Using the hard shell lobe approximation a simple model based on certain assumptions may be employed to indicate the number of bruises that can be seen by an examiner under specific conditions.   Rotation of the apples has an important bearing on the probability of spotting a blemish.   Assum-ing that an examiner can examine 120 degrees of an apple surface at each glimpse it will require 5 fixations to examine the entire surface of the apple, including some overlap of glimpses.

Assuming the average diameter of an apple is 68 mm, the minimum rotational distance required to exhibit 5 sides of the apple is 285 mm.   Unfortunately the display of the five sides rarely occurs in the desired sequence.   If it

is assumed that any of the three adjacent sides is equally likely to appear
next, then by Monte Carlo methods it will require on average 9 sides to be
displayed for the entire surface to be exposed for examination.   This
requires the apple to rotate over a distance of 569 mm.   In practice due
to the shape of fruit the assumption of each adjacent side is as likely to
be shown is unlikely to hold true.   Again by Monte Carlo methods 60 percent of the apples will have been fully exposed in 569 mm, the remaining
40 percent requiring one or two sides still to be displayed.   On average
12 percent of the apple surface will not be displayed in the 569 mm which
implies the examiner misses the same percentage of blemishes.   The assumption being that the examiner is continuously engaged in the search task.

Assuming the search area for a single examiner, at a specific table, is
450 mm (width of the belt) by 569 mm and is as full as possible with fruit
(an array of 6 by 8 average apples).   Allowing for a viewing angle of 45
degrees the lobe for the major cracking is 350 x 328 mm.   It will require
6 glimpses to optimally cover the search area completely, allowing for
considerable overlap.   Assuming the 'best case' of systematic search with
fixation times of .3 s (Megaw and Richardson 1979) the time required for
complete coverage is 2 s.   For the conveyor in question the time to
traverse the search area is 5.1 s, equivalent to two and a half complete
scans of the belt.   At the centre of the belt where considerable overlap
occurs there will be sufficient glimpses to cover the entire apple surface.
At the sides only 7.5 glimpses occur and only about 83 percent of the side
apples' surface will be seen.   On average, as each row of six apples
passes through, only 94 percent of the available surface will be seen.

Combining the lobe calculation with the one on rotation it can be deduced
that about 20 percent of the total surface will not be exposed for examination by a single examiner.   This calculation requires some very simplistic
assumptions which are not realistic although the model indicates there is a
problem.

The above figure is only applicable to one farm.   Conditions vary markedly
in other packhouses.   Flow rates of apples per examiner may range from 60
to 480 per minute and the physical measurements of the grading table may
differ.

Other factors not taken account of in the model will also affect inspection
performance.   Management style and rates of pay are not uniform throughout
the country.   Pay schemes can be based on a flat rate, flat rate plus bonus
or piece work.   Environmental factors play a large part in examiner
performance,   the range of conditions encountered during the field work is
summarised in the following table.

Ranges of Environmental Conditions Between Packhouses

		Min	Max
Dry Bulb Temp.	^{o}C	7	22
Noise Level	Dba	65	80
Light Vertical	Lux	300	2500
Light Horizontal	Lux	150	1100

EYE MOVEMENT RECORDINGS

In addition to the theoretical 'best case' model it was decided to take eye movement recordings of examiners carrying out the inspection task.   Due to the lack of time available this was not an exhaustive study and few objective conclusions can be reached.   However the recordings can be employed to highlight individual differences between examiners.   The recordings were taken at a Ministry farm, using one male and two female subjects, towards the end of a four hour morning shift.   A NAC eye-mark camera from the University of Birmingham was used for the study.   The examiner being recorded stood third in series at the grading table due to the lack of space for equipment in the grading room.   There were occasional complaints that the half silvered mirrors of the facemask made it very difficult to detect bruising.   No performance data was collected.

From the tape it appears that none of the inspectors used any systematic scan paths to cover the belt.   No obvious pattern could be discerned, due to the irregular flow of items to be inspected.   This observation, although not based on a large sample helps show that the assumption incorporated in the simple model is false.   If the other assumptions hold true the estimate of blemishes missed is evidently conservative.   Two of the subjects had over five years experience of grading whereas the remaining one had just started this season.   The naive grader made more fixations per unit time than the other two, however about 15 percent of fixations were duplications of earlier ones.   In both of the experienced operators such marked duplication was not present.   There is a possibility that the subjects lobe increases with experience although this was not evident from the field work. Rather the decision element may not be as automatic in naive graders as for those with experience.   This supposition strays into the cognitive domain of the project.

The naive examiner picked up a larger proportion of apples to aid inspection and fixated more on the class 2 lane when transferring fruit.   A general philosophy behind grading apples is not to handle the produce more than necessary in order to reduce the chance of bruising.

The naive grader also looked straight ahead at the conveyor whereas the others tended to take a more oblique view of the belt ensuring that any rogue apple stays in sight for a longer period.   This may be due to experience or personal preference.

Summary of the Three Examiners Results

	Mean Fixation Time (s.)	S.D. Fixation Time (s.)	Flow Rate Apples/Min.	Percentage of New Apples Fixated
Sub A	0.45	0.23	177	70
Sub B	0.25	0.20	190	80
Sub C	0.24	0.18	180	55

SUMMARY

Rudimentary lobe measurements, simplistic modelling, and eye movement recordings can be a useful combination when tackling an applied inspection problem.   Eye movements helped to indicate obvious differences between search strategies in a 'best case' model and real life.   The recordings also contained information on individual differences in examiner behaviour. Further work is being carried out at Loughborough to investigate different methods of displaying apples to the examiners.

ACKNOWLEDGEMENTS

I would like to thank Dr E. Megaw for the loan of his time and the eye movement recorder.

REFERENCES

(1) Bellamy, L. 1981, Development of a Search Task for the Measurement of Peripheral Acuity, Ergonomics 24 (7) 497-509.

(2) Burkhardt, T.H. and O'Brien, M. 1979, Human Considerations in Mechanizing Fruit and Vegetable Grading,Transactions of A.S.A.E. 507-509.

(3) Megaw, E. and Richardson, J. 1979. Eye Movements and Industrial Inspection, Applied Ergonomics 10 (3) 145-154.

Theoretical and Applied Aspects of Eye Movement Research
A.G. Gale and F. Johnson (Editors)
© Elsevier Science Publishers B.V. (North-Holland), 1984

USE OF THE NAC EYE MARK RECORDER TO STUDY
VISUAL STRATEGIES OF MILITARY AIRCRAFT PILOTS

Jean-Paul PAPIN

Centre d'Etudes et de Recherches de Médecine Aérospatiale
5 bis avenue de la Porte de Sèvres
75731 PARIS CEDEX 15

SUMMARY
Recording of gaze direction of experienced aircraft pilots with the NAC EYE
MARK RECORDER permits describing an optimal visual behavior in instrument
and visual flight. Results lead to suggestions for the design and correction
ergonomics of new visual information display modes. An analysis of visual
behaviors of pilot students shows that the acquisition of the optimal behavior is progressive and unconscious. To improve the training of these students, it is suggested to use gaze direction recordings in two ways : one
is to show students the optimal behavior of expert pilots, the other is to
show students their own behavior.

INTRODUCTION
The first recordings of visual behavior of aircraft pilots were made by Tiffin and Bromer as early as 1942, and many studies followed, both on aircraft
(Milton, 1952 ; Llewellyn Thomas, 1963 ; Spady, 1977) and on helicopter
(Barnes, 1972 ; Frezell et Hofmann, 1975). However, this type of study
remains of current interest. Indeed, the use of color TV screens or electronic head-up or down displays to provide modern aircraft pilots with the
information he needs to fly raised a double question :

. an ergonomic question : how to provide the crew with only that information needed at a given moment of the flight and in the best
  possible form.

. a pedagogical question : how to teach users efficient information
  pick-up and processing.

To reach these goals it is necessary to know :

. the nature of visual information really used during the various
  flight phases ;

. Whether there are prefential visual strategies to pick-up and process
  this information ;

. finally, whether it is possible to describe, explain and teach these
  strategies if they exist.

To answer these questions it is necessary to know, at every instant, what a
pilot looks at and perceives in his surrounding environment. This is made
possible by the continuous recording of the projection of the gaze axis on
the image of the visual environment of a subject. In the field of aeronautics, one of the preferred techniques is the use of a photooculograph derived
from Mackworth and Mackworth's studies (1958) : the NAC EYE MARK RECORDER.
This device, now well known, was used by the *Centre d'Etudes et de Recherches*

de Médecine Aérospatiale (Aerospace Medicine research Center) in Paris to
conduct several experiments. Some of these experiments are reported in this
paper.

I - ANALYSIS OF VISUAL STRATEGIES USED BY CONFIRMED PILOTS

I-1. - Instrument flying

A study conducted in 1980 by Papin et al, in a fighter aircraft
simulator consisted in recording the visual behavior of twelve pilots representative of a reconnaissance squadron. All pilots had flown 500 to
2500 hours. They had to repeat five times a very stereotyped task : a
ground control approach.
Results show that pilots who are very familiar with this task have good performance scores throughout all tests.
The analysis of exploratory visual behaviors shows that there are no significant inter - or intra - individual differences. There is a common visual
behavior and the strategy of visual exploration may be considered as optimal, even if a better one may exist , since the performance level remains
satisfactory.
This behavior is characterized by the following facts. The instrument panel
is explored in a star-like pattern from a prefered instrument : position indicator which provides information about the aircraft position in space and
heading. The pilot's gaze stops on the position indicator, shifts to another
instrument, then comes back to the position indicator shifting again to
another indicator. Sometimes, gaze stops on an instrument close to the instrument which was just checked before shifting back to the position indicator. Whatever the case may be, gaze never leaves the position indicator more
than three seconds, and in 70 % of cases more than one second. However, gaze
may shift from motor indicators for several seconds. The total amount of
time spent checking the position indicator accounts for 64 % of the total
test time. It is also possible to classify intruments as a function of the
amount of time spent checking them or of the number of times they were checked. We will call these "look-fixations" for two reasons. Firstly, because
when data is analysed, what is considered the sum of all small shifts occuring when gaze shifts back to a dial rather than each individual
"fixation". Second, because in most cases, these look-fixations are likely
to be associated with the pick-up of information, or at least, information
pick-up was possible.
In addition, as there is a strong correlation between the duration and the
number of fixations, it becomes possible to express results as look-fixation
rates for various instruments. In the present case, it is possible to establish a hierarchy for the various instruments : position indicator, airspeed,
altimeter, vertical speed, motor indicators and stand-by horizon.
Another important feature of these results is the average duration of these
look-fixations. An analysis by interviews evidenced that they depend on the
nature of the information to be picked up. A look-fixation of less than
100-150 milliseconds is not long enough to read a dial. In a 100-150 millisecond look-fixation the pilot can check that a hand has not moved. He needs
150-200 milliseconds to perceive a move and 200-250 milliseconds to assess
the angular value of this move. Finally, a pilot needs approximately 400
milliseconds to read a digit value.
These results confirm that to be correct, the pick-up of information on an
instrument panel must be associated with systematic and continuous exploration (information on dials which are not checked is not perceived). An
efficient method is to explore an area in a star-like pattern from a prefered site to which gaze comes back frequently. Finally, information is
more rapidly picked up on symbol displays than on alphanumeric displays.

I-2. Visual flying

The visual strategy of confirmed pilots in visual flying was analysed during a research study (Papin *et al.*, 1981) designed to assess the possibility of flying a tactical mission on a helicopter with reduced visual field and monocular vision. Such flying conditions are encountered when visual aids are used in night flight. The image of the landscape recorded by an infrared camera placed outside of the cockpit appears on a mini TV screen attached to the pilot's eyes. In order to simulate the reduced visual field, a series of masks is used which provides either binocular or monocular vision with visual fields of 60°, 40° or 20°. This system plus the NAC were used to monitor the gaze direction of seven pilots with a good experience of tactical flight during a real flight during daytime. The 20-minute mission consisted in searching a potential ennemy in a hilly terrain while keeping away from the enemy's sight and blows.
Recorded gaze shifts were analysed either as a function of objects looked at outside the cockpit or as a function of the position of gaze projection on the helicopter cockpit.
Results show that all pilots have a nearly identical visual behavior for a given condition of vision and for a flight over a particular type of terrain. One of the most remarkable facts is that whatever the conditions may be, visual exploration is organized in space relative to the canopy. Gaze shifts from one area of the cockpit to another, in a starshaped pattern from the area which corresponds to gaze projection on the canopy when the pilot is at rest and looks straight in front of him. The only case in which gaze shifts depend on the type of object looked at is when the eye encounters a tree or the border of a forest. In this case, it travels back and forth several times from the tree top to the bottom. Another important item is that most look-fixations tend to concentrate in a relatively small area around the area which is again and again looked at under normal visual conditions. The area where 80 % of look-fixations are concentrated on the canopy corresponds to the base of a cone whose point is the pilot's head and which has 30° aperture angle. When the visual field shrinks to concentrate 80 % of look-fixations, the angle of this cone increases. It reaches 60° when the flight test is conducted in monocular vision with a 40° visual field. Concurrently, an increase in the number of look-fixations on the lower half of the cockpit and on the instruments is observed. This change in behavior suggests that pilots search in foveal vision information which they usually capture in peripheral vision and which give them information on their orientation, their altitude, and airspeed. Actually, to compensate this lack of information, they observe more frequently the instruments which provide information on these parameters.
Such observations, i.e. pick-up of information outside the cockpit clues and increased checking of certain parameters in monocular vision, had immediate ergonomic application.

I-3. - Ergonomic applications

During the first flights flown with the prototype system where the image of the landscape recorded by a camera placed outside the cockpit appears on a TV screen without cockpit clues, the pilot seemed to be confronted with spatial orientation, speed, altitude and distance evaluation difficulties. It was therefore recommended to show on the image seen by the pilot bars corresponding to the vertical and horizontal frame of the canopy and also to superimpose on the image heading, airspeed, altitude and changing clim and descent speeds. These changes added during experiments seemed totally satisfactory.

II - ANALYSIS OF VISUAL STRATEGIES USED BY TRAINING PILOTS

The reported example concerns the follow-up of the evolution of the visual behavior of six military air carrier student pilots. Gaze direction of these pilots was recorded at regular intervals during their 18-month training program, during real or simulated flights, but always during the same "Rodeotacan" task. This task of approximately 17 minutes encompasses all possible flight phases.
Results were compared among each other and also with those of a control group of flight instructors who performed the same task. Results show that all students display a common charcteristic behavior at each training phase. At the beginning of the training programm, student pilots explored the instrument panel in a star-like pattern starting from the pitch indicator which corresponds to the position indicator in combat aircraft. However they often looked back and forth between two instruments, stopping for a practically identically short time on each one. Some omissions were also recorded. As the students progressed, the number of back and forth gaze shifts decreased while the duration of look-fixations increased. Omissions gradually disappeared and look-fixation on the most important indicators for a given flight phase became longer. This final behavior is specific to flight instructors. Evidently, the acquisition of this behavior is a slow, totally unconscious process. The question is to know whether it may be possible to actively influence this acquisition process.

III - USE OF GAZE DIRECTION RECORDINGS IN PILOT TRAINING PROGRAMS

To influence actively the acquisition of optimal visual exploration strategies, two axes of research are simultaneously flollowed.
The first consists in creating teaching material which shows and explains these behaviors. In order to do this, once the visual behavior of expert pilots has been recorded during real or simulated flight, a document is prepared which alternates dynamic presentation of these behaviors, explanatory sketches and consequences of most frequent inadequate behaviors, exhibited bu student pilots.
Such documents have been made, in color, for the various phases of helicopter flight instruction, both over countryside and mountainous terrain. However, they have not been distributed to student pilots since the results obtained recently in radiology (Amalberti *et al*, 1983) and aviation studies (Spady, 1982) show that alone, viewing the optimal behavior is not efficient. What is efficient is either to show the students their own behavior, or to let them participate actively when the film is shown. To obtain such active participation during film viewing, special scenes must be prepared and we are currently preparing them for helicopter flight instruction.
The second axis followed in this teaching research is to let students objectivate their behavior. Such an approach is promising but raises numerous problems. Indeed, in its present configuration, the NAC is not adapted for students who are stating their flight instruction or for single-seater combat aircraft students.
However, the method can be used in a simulator to help students acquire emergengy procedures as we demonstrated it in a study on helicopter simulator or as Spady (1982) showed it on a Boeing 737 simulator. This method cans also be used in real flight when a pilot is transferred to a new helicopter or a new aircraft. We could thus help tremendously an air carrier pilot. Instructors were confronted with the alternative to either dismiss the student or make him repeat his courses. An inflight recording helped the

pilot identify his errors and trust his instructors again. He could correct his errors quickly and catch up on what he had missed. This clinical observation is very interesting but it is difficult to make a generalization since actions must be punctual and require great availability. Now, what should be done is to go from research to practical application. To do this specialists must be trained and special equipment must be acquired. It would certainly be a long and expensive process.

CONCLUSION

The NAC EYE MARK Recorder is a valuable tool for the study of the visual behavior of confirmed pilots in visual flight on helicopter or air carrier and all aircraft types in simulators. Such studies help solving ergonomic correction and design problems.
The NAC can be used efficiently to prepare training material for pilot students, and to help certain categories of failing students. However, it should not be used to analyse pilot's work in a single seater fighter aircraft in real flight nor to help beginner students. In order to go fur - ther inthis type of approach in the field of aviation, it will be necessary to wait for the development of recording equipment better adapted to this task.

REFERENCES

- BARNES J.A. : "Analysis of pilots eye movements during helicopter flight" Tech. memo. 11-72, Human engineering laboratory, april 1972.

- FREZELL T.L., HOFMANN M.A. : Comparison of visual performance of monocular and binocular vision during VHR helicopter flight. Aerospace Medical panel, Agard conference proceeding G.B. val CP 1982, n°4 10/1975.

- MACKWORTH Y.F., MACKWORTH N.M. : Eye fixations recorded on changing visual scenes by the television eye-marker. J. opt. Soc. Amer., 1958.

- MILTON J.L. : Analysis of pilots eye movements in flight. Aviation Medecine, February 1952.

- PAPIN J.P., NAUREILS P., SANTUCCI G. :Pickup of visual information by the pilot during a ground control approach in a fighter aircraft simulator. Aviation space, and environnemental medicine, may 1980.

- PAPIN J.P., WIATZ A., GANGLOFF B. : La direction du regard et des alarmes visuelles sur simulateur PUMA 5A 330, Rapport CERMA 80-12, Juin 80.

- PAPIN J.P., MENU J.P., SANTUCCI G. : Vision monoculaire en vol tactique sur hélicoptère. Agard CP 312, mai 1981.

- SPADY A., AMOS . : Airline pilot scanning behavior during approaches and landing in Boeing 737 simulator, Agard CP 240, April 1978.

- SPADY A., JONIS D., COATES G., KIRBY R. : Effectiveness of using real time eye scanning information for pilots training. Proceding HF Society 1982.

- THOMAS E., LLEWELLYN, The eye movements of a pilot during aircraft Landing. Aerospace Medicine, May 1963.

- TIFFIN J., BROMES : Analysis of eye fixations and patterns of eye movements in landing a Piper cub J 3 air plane. Report n° 10 division of research civil aeronautics adminstration, Washington DC, February 1943.

Theoretical and Applied Aspects of Eye Movement Research
A.G. Gale and F. Johnson (Editors)
© Elsevier Science Publishers B.V. (North-Holland), 1984

INSPECTION OF ROLLS IN ROLLER BEARINGS

M. A. Sinclair

Department of Human Sciences
Loughborough University of Technology
Loughborough, Leics. LE11 3TU

The problem with eye movement studies is that though
the eye movements are relatively easy to measure and
the data are prolix, eye movements do not provide much
insight into real tasks without the use of major
assumptions, usually on the nature of information
processing associated with each fixation. These
assumptions are usually justified by curve-fitting,
which does not necessarily validate the model. An
example (Drury and Sinclair, 1983) to illustrate how
eye movements do not necessarily help is discussed.
In addition, the relative failure of a simple automatic
inspection device intended to replace the human
inspector is discussed.

Over the past 15 years, as in many other fields there has been an explosive
growth in the study of vision in human visual inspection and detection tasks.
There have been several studies involving eye movements, but interest in
these has wained fairly rapidly. This has not been occasioned by any
failures in the methods for measuring eye movements - indeed, they are
almost too successful in producing data - but because there has been no
good theoretical answer to the question, 'Now we have the data - what do we
do with it?' Certainly, one may pick up one's statistical club and beat
the data to death to produce probability distributions and other descriptive
statistics, but this is merely surface analysis.

There are some basic assumptions made about which there is general agreement. These are:-

1) The purpose of vision in industrial inspection tasks is to gather
   information relevant to decisions about quality.

2) The purpose of eye movements is firstly to ensure that the visual
   field is scanned, and secondly to bring foveal vision to bear on
   parts of the visual field that are information-rich.

3) The purpose of information-gathering is to make decisions about
   quality based upon a set of internalised standards.

Unfortunately, while these assumptions are clearly inter-related, they do
not form a closed set. This would require a fourth assumption, about which
there is general disagreement:-

4) All other variables are immaterial; in other words, the human being
   acts as a robot.

From introspection, if not from observation, this is clearly untenable. This then implies that eye movements cannot be regarded as indicators of mental activity since the exclusive link between the two has been broken. The only remaining way whereby such a link may be established is by postulating a model which states such a link and then demonstrating the link by some experimental means. An example of this is the scanpath theory, first put forward by Noton and Stark (1971) and considered most recently by Groner et al (1983) at this conference. However the theory is not widely accepted, and in the opinion of this author requires more support from experiments. If one accepts the ideas of Schneider and Shiffrin (1977) and vice versa (1977), including the extension into vigilance (Fisk and Schneider, 1981), then it seems to this author that the only conditions under which such an approach could be tenable would be in situations where the whole process was 'automatic', in their sense. This of course requires that the criticisms discussed by Broadbent (1982) should be disregarded, or otherwise answered. Such a long-term investigation has yet to be carried out.

Another example where an explicit link is built in is the visual sampling model (Senders, 1966) though the model was not envisaged for use in target detection. Other models in this general field of target detection and recognition do not make this explicit link, though as stated before this is due in many cases to the exclusion of psychological and physiological variables entirely from the models. A review article by Greening (1975) gives some excellent examples of this, in early work on models of military target acquisition. The same trend can be seen in more recent models published in the literature (e.g. Silbernagel, 1982, Akerman and Kinzly, 1979, and Kraiss and Kuaeuper, 1982). Of course there are reasons for this in that the relationships between search variables and psychological variables are ill-understood, and are not well quantified at all.

What has not assisted in correcting this situation in the opinion of the author is the tendency to validate models by means of curve-fitting; for example, by examination of the cumulative % of detections (with respect to passage of time). By conflating temporally-dependant data into a global variable such as this most of the fine detail is eliminated and the data then appears to be very well-behaved, and allows of exemplary mathematical treatment. An example of this is the excellent match obtained by fitting exponential curves to cumulative probability of detection data. This is usually performed on the assumption that random search has occurred, and the fit obtained is usually very good even in those situations where it is patently obvious that search is non-random.

An example that perhaps illustrates the lack of direct association (in the statistical sense) between search and decisions is given in what follows, which is a brief report of work to be published elsewhere (Drury and Sinclair, 1983). The work also illustrates particular advantages to be obtained from human visual inspection as opposed to a fairly simple microprocessor-based optical device intended to accomplish the same function.

The project took place in an industrial setting, and was an investigation of the efficiency of inspectors of rolls for roller-bearings to be used in the manufacture of jet engines. All rolls passing through the inspection process were inspected twice; once for dimensional tolerance and once for surface defects. It is the latter operation which was examined. The rolls vary in size; in this exercise only those from 7 to 11mm in length, and 7 to 11mm in diameter were considered. Faults could occur in any of 3 regions - the barrel, the ends, or the curved edges between the ends and the barrel. The latter areas are called the corners. Four classes of fault could occur; pits (pock marks, typically caused by rusting), scratches, nicks and dents, and tool marks (usually a circumferential gouge on a corner, or parallel cuts on an end). Only the last 3 classes are considered hereafter. Reject thresholds for these faults varied depending upon the nature of the fault; for the purposes of this discussion any fault of a size greater than 25/10000 inch in diameter was rejectable, (It should be remembered that even the hardest of metals appear plastic if the unit of measurement is small enough). To measure inspection performance, a test batch of rolls was created. This comprised a set of 110 faulty rolls, with 182 faults upon them, plus a further 52 rolls deemed acceptable. All the rolls were numbered and mapped photographically. During the experiment the rolls were kept in oil-filled tubes to minimise handling damage; even so, the rolls had to be resubmitted to the finishing (SWECO) process to restore the surface finish. The rolls were checked to make sure the maps were still relevant.

Twelve inspectors were used in the experiment. All had had at least 2 years experience, and had undergone a training programme. Their ages ranged from 20 to 50 and, by company policy, had had their eyesight checked and corrected by an optician within the past 3 years. One inspector was tested each day at the beginning of his or her shift (7.00am or 3.00pm). Typically, the test lasted for 4 hours, and took place at a normal workstation in an adjacent room. The workstation consisted of a desk-mounted light fixture with two fluorescent tubes positioned above the inspectors' hands when handling the rolls. At roll level, the illumination averaged 887 lux, with a background illumination of 95 lux. Given the material of the cylinders, there was a glare problem from the surface. Viewing distance averaged 25 cms, which for a fault threshold value of 25/10,000 inch gives an angular subtense at the eye of 50 seconds. To assist in the task, inspectors were given a rectangular glass x4 magnifier. Most inspectors affixed this to their light fixture with rubber bands; one inspector refused to make any use of it. The test batch of rolls was presented on trays, as was normal. Each inspector would then inspect the rolls singly, rotating the rolls in the hands, and using either gloves, a cloth, or barrier-cream-protected fingers, to rub off dust particles. This was a necessary process, as it was frequently difficult to distinguish between dust and a flaw. As each flaw was found the inspector would use a stylus for factual confirmation, and then the type, position, and fault classification was recorded. Normally, as soon as any fault is found, that roll is rejected. In this experiment, because there were occurences of multiple faults on the roll, inspectors were asked to keep on inspecting until they thought they had found all faults (if any). No time limit was placed on this.

Watching the inspectors at work revealed immediately that there was a fairly standard method of search involved. All inspectors organised their search into 3 phases - barrel, corner and end - though the order varied between inspectors. The roll was always held at about 45° to the vertical in the saggital plane for inspection of the barrel. It was then rotated slowly about this axis while the eyes performed vertical saccades up and down the barrel. This produced a zig-zag pattern of fixations around the barrel; typically 3 fixations were used on each zig (start, middle, and

finish). The corner was inspected by holding the roll pointing towards the
body in the saggital plane, and then performing a small, 2-axis rotation
of the roll to allow a reflected bar of light to play all round the corner.
A similar, more oscillatory movement was performed to inspect the end. In
the latter case the fixation pattern was roughly circular, whereas for the
edge the eyes followed the bar of light, since changes in this revealed
the presence of a flaw. These comments are based upon observation of the
eye movements using a small mirror, and are therefore qualitative.

The data obtained was compared with the maps, and was sorted into the fault
classifications. Within each classification it was further sorted into
Hits, Misses, Correct Acceptances, and False Alarms. This was performed
for the defective rolls and the good rolls separately (though in the latter
case, there were no hits, by definition). The hits and the false alarm
categories for each inspector were then plotted graphically on Receiver
Operating Characteristics diagrams (Figs. 1 - 3).

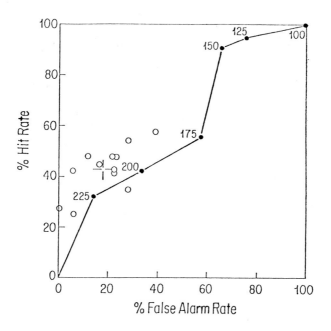

Fig. 1.   R.O.C. for Nicks and Dents

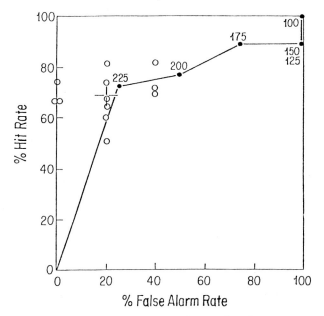

Fig. 2. R.O.C. for toolmarks

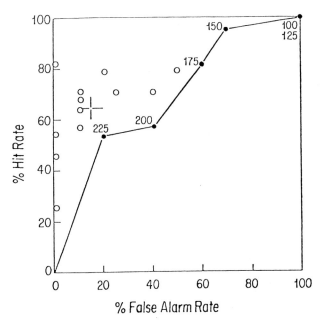

Fig. 3. R.O.C. for scratches

In parallel with this experiment using human inspectors, an evaluation of an optical, microprocessor-based devise was undertaken, which was in a breadboard stage of development. It was intended that this device should replace the inspectors. The sensing element was an array of photodiodes, behind a focussing system. The detection principle was to compare the output of each diode to its 8 nearest neighbours, and to signal any differences that were greater than a set threshold. In machine units, this threshold could be set to any value between 0 and 255; in practice, only the range between 100 and 225 was of any importance. 100 would reject almost every roll; 225 rejected very little. Because a range of thresholds could be used, it was possible to obtain an R.O.C.; this is shown in Figs. 1 - 3.

These three figures indicate that firstly the scatter of the points for the inspectors reveals very different performance among the inspectors despite the very similar search behaviour used, thereby illustrating the points made earlier. The wide scatter is probably due to poor initial training and no subsequent recalibration of the inspectors, allowing them to slip to an inappropriate set of perceived thresholds for rejection. Secondly, there is the comparison between the inspectors and the microprocessor-based device. If one makes use of the usual interpretation of R.O.C. diagrams on the basis of signal detection theory, then those inspectors plotted in the area above and to the left of the machine's R.O.C. have performed better. Since this loosely accords with desirable practice in production, this criterion is used here. It will be seen that in general the human inspectors perform better than the machine. On scratches and nicks and dents, humans are significantly better ($p<0.05$).

A further analysis was undertaken to illustrate the difference between search performance and decision performance by human inspectors. Successful search corresponds to the detection of any fault or flaw in the mapped position on a roll. Successful decision-making corresponds to making the correct decision once a verifiable flaw has been detected. The data allowed this analysis to be performed, and the results are shown in Table I (it should be noted that by multiplying the search performance by the decision making performance for each inspector one obtains the points plotted in Figs. 1 - 3). This table reveals the considerable difficulty in finding flaws below the threshold value, and indicates that this task is near the limits of human performance under normal conditions. Secondly, it also shows the importance of psychological variables in that the decision performance shows considerable aberration.

Type of Faults	Search Function		Decision-Making Function	
	Success on faults	Success on flaws	HR	FAR
Nicks and Dents	56.8%	29.6%	75.2%	60.9%
Tool Marks	75.1%	43.3%	92.4%	46.1%
Scratches	80.9%	37.5%	79.6%	38.2%

Table I. Mean search and decision-making performance of inspectors.

This study is reported much more fully in Drury and Sinclair (1983). Even with the brief description herein, it is hoped that the dangers of inferring from eye-movement data alone, without considering other more global aspects of the system, have been indicated. Furthermore, it is hoped that the power of the human visual system has in some small way been demonstrated compared to a purpose-built device, and that automatic devices are not invariably the best answer to industry's problems (though they frequently are).

REFERENCES

(1) Akerman, A. III, and Kinzly, R.E., Predicting aircraft detectability, Human Factors 21 (3), (1979), 277-291.

(2) Broadbent, D.E., Task combination and selective intake of information, Acta Psychologica 50 (1982) 253-290.

(3) Drury, C.G. and Sinclair, M.A., Human and machine performance in an inspection task, Human Factors (1983) In press.

(4) Fisk, A.D. and Schneider, W., Control and automatic processing during tasks requiring sustained attention : a new approach to Vigilance, Human Factors 23 (6), (1981), 737-750.

(5) Greening, C.P., Mathematical modelling of air-to-ground target acquisition, Human Factors 18 (2), (1976), 111-148.

(6) Groner, R., Menz, C. and Walder, F., Testing the scanpath hypothesis on a local and global level, in : Gale, A.G. and Johnson, C.W., (eds):2nd European Conference on Eye Movements, Nottingham, England.

(7) Kraiss, K.-F. and Knaeuper, A.-M., Using visual lobe area measurements to predict visual search performance, Human Factors 24 (6), (1982), 673-682.

(8) Noton, D. and Stark, L., Scanpaths in eye movements during pattern perception, Science, 171, (1971), 308-311.

(9) Schneider, W. and Shiffrin, R.M., Controlled and automatic human information processing : I, detection, search and attention, Psychol. Rev. 84 (1), (1977), 1-66.

(10) Senders, J.W., A re-analysis of the pilot eye-movement data, I.E.E.E. Trans. Human Factors 7, (1966), 103-106.

(11) Shiffrin, R.M. and Schneider, W., Controlled and automatic human information processing : II perceptual learning, automatic attending and a general theory, Psychol. Rev. 84 (1), (1977), 127-190.

(12) Silbernagel, B., Using realistic target, sensor, and scene characteristics to develop a target acquisition model, Human Factors 24 (3), (1982), 321-328.

Section 7

THE NEUROPHYSIOLOGY OF EYE MOVEMENTS

Theoretical and Applied Aspects of Eye Movement Research
A.G. Gale and F. Johnson (Editors)
© Elsevier Science Publishers B.V. (North-Holland), 1984

THE NEUROPHYSIOLOGY OF EYE MOVEMENTS

INTRODUCTION

Laurence R. Harris
Department of Physiology, University
College, Cardiff.

In an attempt to represent all approaches to eye movements and to bring
together scientists from many disciplines, this conference has included a
section on 'the neurophysiology of eye movements'. Our understanding of
the control of eye movements by the nervous system has received an enormous
surge of progress in the last decade with the introduction to the field of
engineering concepts such as integrators and pulse generators: an introduction due largely to the work of one man, Dr. D.A. Robinson. The vocabulary
and concepts of control engineering have made it possible to define the
problems involved in the generation of eye movements and to construct test-
able hypotheses about how the system might deal with them. Many individual
elements of a hypothetical 'circuit' have indeed now been identified by
recording single cell activity in conscious animals engaged in oculomotor
behaviour. The problem has been moved back one stage - now we have to show
that these elements are actually functioning in the way and context proposed.
One delight of the oculomotor system (and it has many) is that for many
different classes of eye movements, from voluntary saccades to vestibular
compensatory reflexes, several of the control elements seem to be shared.
But we must remember that most models require duplication to be comprehensive enough to cover vertical eye movements.

Pyykko et al. (An oculomotor model of the interpretation of nystagmus) have
described clinical observations to support their model of eye movement control. Clinically there is a wealth of information concerning the role of
the cortex which is the area most often damaged in neurological patients.
The contribution of cortical visual pathways has tended to be overshadowed
in recent years as the brain stem has started to yield its secrets. An
unstated belief has arisen that the brain stem and the cerebellum can control eye movements all by themselves. And indeed it seems that the brain
stem has a large degree of autonomy in the initiation and execution of eye
movements. In the pioneering days of brain research the cortex was proposed
to be the master organ of the brain, with everything else being very lowly
indeed and having almost no initiating capacity of its own. Electrical
stimulation studies (fraught with interpretation difficulties) suggested
that almost the whole cortex was involved in the control of eye movements.
Whilst this view was undoubtedly too simplistic, we are now discovering that
as in many things, the ancients had a point and the cortex cannot, by any
means, be dismissed. Page et als.'s paper also stresses the role of the
cortex, this time with respect to eye-head co-ordination. The role of head
movements in gaze-changing strategies is controversial. Pyykko et al. and
Page et al., take opposite stances as to whether the vestibulo-ocular reflex
is active during head movements.

Models can be tested clinically, neurophysiologically or psychophysically.
Becker and Jurgens (1979) have suggested from psychophysical work that the
timing and amplitude of saccades may be processed separately. It has always
been a puzzle as to why saccades have such a long latency. What is happen-

ing during these 150 msecs? Fischer and Boch's express saccades (which
have relatively short latencies) pose a problem for those who have managed
to find some vital activity to fill the gap (see: Properties of the saccadic eye movement system by J.M. Findlay, this volume).

The parameters of eye movements are modified throughout life to keep up with
the demands of a changing,'ageing nervous system. A remarkable degree of
plasticity in the eye movement system has recently been revealed by inter-
fering with the feedback available. Pyykko et al. (Convergence of different sensory modalities in a single cell of flocculus in alert cat) support
the view that the flocculus (of the cerebellum) is involved in this plasticity. They show that many cells there gather relevant information from
several sensory modalities any of which could provide useful feedback.

Sea-sickness is interesting from two complementary points of view. An
understanding of the interactions between the visual and vestibular systems
will hopefully help us control sea-sickness. Furthermore an understanding
of the mode of action of drugs that reduce sea-sickness should help to
clarify visual-vestibular interaction and its role in the control of eye
movements. Pyykko et al.'s paper (Reduction of nystagmus as a prediction
of efficiency of motion sickness remedies) makes some interesting contributions in both directions, but also illustrates the difficulty in making
conclusions about the site of action of sea-sickness reducing drugs.

Vergence is a class of eye movements that appears to be largely out on its
own. Its underlying control elements are different from all other types.
At least part of the processing needed for their control may be cortical.
Cumming and Judge have approached the problem with an admirable combination
of psychophysics and neurophysiology. They conclude that the cortex
probably has an important role.

So the papers in this chapter look at bioengineering models, the vestibulo-
ocular reflex, eye-movement plasticity, saccades and vergence movements.
They call on clinical, psychophysical and neurophysiological data. However,
I can make no excuse for their essentially disparate nature or for the
predominance of a few select workers. Scientists working in eye movements,
who come to the field with psychophysical, anatomical, engineering, physiological or clinical backgrounds, are keen to exchange ideas. But with the
sheer volume of literature scattered over many journals and the large number
of excellent workers in the field of eye movements, it is difficult to see
how such mutual assistance can in practice be achieved.

Reference

Becker, W. and Jurgens, R. An analysis of the saccadic system by means of
    double step stimuli, Vision Research, 9. (1979) 967-983.

Theoretical and Applied Aspects of Eye Movement Research
A.G. Gale and F. Johnson (Editors)
© Elsevier Science Publishers B.V. (North-Holland), 1984

AN OCULOMOTOR MODEL FOR THE INTERPRETATION OF NYSTAGMUS

Pyykkö, I.[1], Schalen, L[2], Magnusson, M.[2]
and Matsuoka, I.[3]

[1] Department of Physiology, Institute of Occupational Health
Helsinki,Finland
[2] Department of Otolaryngology, University Hospital of Lund
Lund, Sweden
[3] Department of Otolaryngology, University of Kyoto,

Kyoto, Japan

A model for the interpretation of nystagmus is
worked out. A lesion in the vestibular system may
appear as alteration in velocity of slow viz. fast
phase of nystagmus, disconjugation of the nystagmus or as dysrhythmic nystagmus. The visual
system, viz. the cortical and subcortical optokinetic nystagmus seems to provide the eye-head
coordinates and retical error velocity to
calibrate or cancel the head velocity signal
during fixation.

INTRODUCTION

Several models have been made to correlate the oculomotor
disturbances in patients with experimental lesions in animals.
Such models may assist in exploring the function of the system
and in understanding of the disease processes involving
different oculomotor pathways. It must be notified, however,
that the present knowledge of the connections within the
central vestibular system is so vast, that any model may
enclose only a fraction of that information. To exemplify, a
part of the neurons even in the striate cortex change their
orientation by the otolith input (Tomko et al., 1981). In the
following a model of the organization of the central pathways
in the horizontal vestibulo-ocular reflex (VOR)-arch is
discussed with its clinical implications.

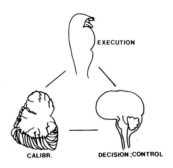

Figure 1
Central processing of nystagmus

The vestibular system is thought to operate at different levels
of hierarchy (Fig. 1). Evolutionarly, the highest level is
governed by the supratentorial brain structures which command
and supervise different orientation reflexes. The cerebellum
has a function analogic to a calibrator adjusting the gain in
each of the reflexes. The brain stem process and execute the
reflexes. Consequently, depending on different operational
levels the central nervous system is capable to compensate
lesions inpinging the peripheral or the central vestibular
system.

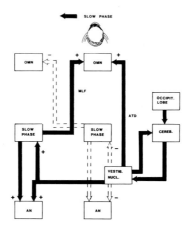

Figure 2
Pathways involved in generation of the slow phase of nystagmus.
Abreviations AN = abducens nucleus, ATD = ascending tract of
Deiter, OMN = oculomotor nucleus,MLF = medial longitudinal
fasciculus.

1. BRAIN STEM MECHANISMS OF NYSTAGMUS

1.1 Neurophysiological aspects

The horizontal VOR-arch stems from the horizontal semicircular
canals and projects to the lateral (LVN) and medial (MVN)
vestibular nucleus. A part of the projections from the LVN
travers through the ascending tract of Deiter (ATD) to the
ipsilateral oculomotor nucleus (OMN) (Fig. 2) (Ito 1975,
Highstein and Reisine 1979) whereas a part of the afferents
originating from the MVN terminates in the contralateral
abducens nuclei (AN). These projections mediate crude information of the head velocity to the medial and lateral rectus
muscle (Highstein and Reisine 1979). Inhibitory pathways
project from MVN to the ipsilateral (AN) and to contralateral
OMN (Yamamoto 1979). In natural conditions, however, the
facilitation of the contralateral AN is mainly caused by
contralateral vestibular nuclei via type II cells (Highstein
and Reisine 1979, Precht 1979).

The horizontal semicircular canal afferents projects to the
oculo-motor nuclei also through polysynaptic pathways (c.f.
Fig. 2) A part of these projections terminates in the contralateral pontine reticular formation (PRF) in the vicinity of
the AN. After synapsing in the PRF the canal afferents travers
via medial longitudinal fasciculus (MLF) to the contralateral
OMN (Furuya and Markham 1982). In addition, a subgroup of
AN-neurons projects through MLF to the contralateral OMN
(Highstein & Baker 1978). These inter nuclear neurons fire in
reversed order for AN motoneurons showing phasic activity.
During saccades and fast phases they pause and during slow
phases they are activated (Precht 1979). The inter nuclear
neurons are mediating the conjugation of the gaze. A lesion in
MLF, and hence in this pathway, produces an internuclear
opthalmoplegia (INO). As a difference for INO, a lesion in AN
involving intra- and inter-nuclear neurons produces ipsilateral
gaze paresis (Henn et al. 1982a).

The slow phase system can be activated also by smooth pursuit
pathway (Fig. 2). Allegedly, the velocity determination of
smooth pursuit and the integration between vestibular and
visual impulses is governed in the cerebellar flocculus.
Consistantly, after a floccular lesion the smooth pursuit
pathway cannot be activated and the tonic output of the
vestibular system cancelled.

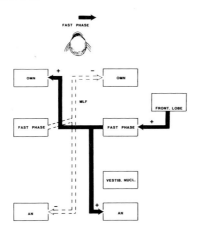

Figure 3
Pathways involved in generation of the fast phase of nystagmus.
Abreviations as in fig. 2.

The mechanism triggering fast phase of nystagmus is not known
in detail. Single unit activity (Keller 1974) and lesion
studies (Cohen et al. 1968) of the paramedian part of PRF,
emphasize the importance of this site for the control of fast
phase of nystagmus (Fig. 3). In addition, the internuclear

388   *I. Pyykkö et al.*

pathway traversing through MLF integrate the activity of
motoneurons between AN and OMN and mediate the conjugation of
the gaze during fast phases of nystagmus (Igusa et al. 1980,
Nakao et al. 1982). Voluntary saccades are also mediated
through the fast phase pathway in PRF Henn et al. 1982b).

1.2 Abnormalities in nystagmic slow and fast phase velocities

Since the velocity signal is derived from semicircular canal a
retardation of slow phase velocity occur in a peripheral and
central lesion. In fact, we have not been able to point out any
significant differences in gain of nystagmus induced by
Barany's rotatory test between patients with a peripheral r a
central lesion (Schalèn et al. 1982, Pyykkö et al. 1983)
whereas time constant of the nystagmus is more reduced in
patients with a brain stem disorder than in patients with end
organ disorder (Pyykkö et al. 1983).

Since reduced peak velocity of fast phase of nystagmus
indicates without exception a damage in brain stem, the
diagnostic value of fast phase peak velocity in setting topo-
graphical diagnosis is great (Pyykkö et al. 1982). The fast
phase abnormalities can be detected by studying the peak-veloc-
ity-amplitude relationship. Especially in lesions affecting the
PRF the fast phase velocity will be primarily reduced. In
extensive lesions of PRF the eyes may deviate in the direction
of the slow phase after vestibular stimulation as frequently
observed in patients with advanced progressive supranuclear
palsy (Steele et al. 1968).

1.3 Abnormalities in the rhythm of nystagmus

Information on eye velocity and position are coded in neurons
of PRF and MLF (Pola and Robinson 1975). The head velocity
information is provided the neurons traversing through ATD
(Highstein and Reisine 1979). Thus, the oculomotor nuclei
receive through various pathways the data needed to process
nystagmus (Fig. 4). Errors in processing are generally
interpreted as dysrhythmia of nystagmus.

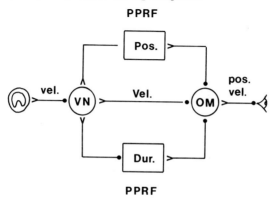

Figure 4
Control circuits determining the end point of nystagmus.

Normally, the end point of nystagmus is largely determined by the position of the eye in the orbit. Thus, in linear regression analysis about 35 percent of end points could be related to the position of the eye (Pyykkö and Dahlen 1983a). The end point of nystagmus can be also determined by time mode. Normally about 26 per cent of the nystagmus beats resulted in Barany's rotatory test is governed by duration of the nystagmus beat (Pyykkö and Dahlen 1983 a). Since the shortest period to visualize an object during fixation is about 100 msec, the time mode is utilized during high nystagmic velocities. Thus the maximum number of nystagmus beats will be limited to about 3 to 4 in one second (Donaghy 1980).

The positional data is presumably derived by the integration of the velocity signal. After a lesion affecting the vestibular system the ability of the amplitude circuit to govern the end point of nystagmus will be weakened (Fig. 5) (Pyykkö and Dahlen 1983 b). For the CNS time mode is more reliable than e.g. amplitude since time is an inherited property of the nerve cell. Patients with a brain stem lesion show poor ability to derive data on eye position since the lesion may encroach mechanisms integrating postional signal from head velocity. Among these patients about 46 per cent of the end points of the nystagmus beats were determined in time mode (Pyykkö and Dahlen 1983 b). Clinically this finding has been denoted as "petit ecriture" of nystagmus.

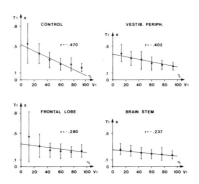

Figure 5
Intrabeat relationship of postrotatory nystagmus in linear regression analysis in normal subjects, in patients with a peripheral vestibular lesion, in patients with a frontal lobe lesion and in patients with a brain stem lesion. Mean, standard deviation, regression line and correlation coefficient are set out. Abreviations T_1 = duration of slow phase, V_1 = velocity of slow phase.

Noteworthy is, that we have not observed any other type of
dysrhythmia of nystagmus e.g. pauses, prominent amplitude
variations or different group formations among patients with a
brain stem lesion which would differentiate them from the
normal subjects (Pyykkö et al. 1983).

1.4 Abnormalities in conjugation of the nystagmus

In the oculomotor model of nystagmus the neuronal pool of PRF
can be divided into two neural subpopulations, one for the
generation of the slow phases (tonic units) and one for the
fast phases of nystagmus (phasic units) (C.f. Keller 1974).
During one type of eye movement, only one type of center n be
active vs. either slow or fast. Furthermore, since eye movements are conjugated, one phase of nystagmus is propulsed by
the activity in one center by one time. Hence, if the commando
over a slow phase is executed ipsilaterally the slow phase
center in the contralateral side mediates inhibition (Shimazu
1979). Furthermore, the fast phase centras are inhibited by
activation of pausing units via PRF. (Hikosaka et al. 1977,
Hepp and Henn 1979).

Disconjugated nystagmus might be expected in brain stem
lesions. However, disconjugated nystagmus with respect to
either slow or fast phase anomalies are rare since the neural
control between the slow and fast phases excerted by commando
units (bursting and pausing neurons) are scattered with tonic
neurons in the PRF. Consistantly, instead of disconjugation of
the eyes the patients with a brain stem lesion show velocity
retardation. The missmatching of the activity between pausing
and bursting units has been verified in single unit recordings
in monkey during drowsiness (Hepp and Henn 1979) and canbe
related to the reduction of fast phase velocities met with in
patients with Huntington's chorea (Starr 1967).

Disconjugated nystagmus can be observed in patients with a
MLF-lesion. In INO the facilitatory information from the contralateral PRF is missing and the activation of the ipsilateral
eye during fast phases is inadequate. Presumably, in effort to
compensate the double image and since inadequate inhibition of
the contralateral OMN, the contralateral eye displays coarse
nystagmus. Nevertheless in bilateral EOG recording patients
with MLF lesions commonly show also reduced velocity in rapid
eye movements (c.f. Barber & Stockwell 1978) confirming the
importance of a detailed velocity analysis of fast phase of
nystagmus.

2. CEREBELLAR CONTROL OF NYSTAGMUS

According to the present model the cerebellum has a function as
a calibrator. Consistently velocity maladaptation, i.e.
hyper- and hypoactivity, may occur in nystagmus after a cerebellar lesions (Baloh and Honrubia 1979). Cogan (1964) and
Honrubia et al. (1981) have observed significant changes in the
continuity of the postrotatory nystagmus after cerebellar

lesions. These findings are in agreement with mismatching of
saccadic amplitude and velocity reported in occurring monkeys
with vermal lesions (Ritchie 1976). Allegedly, the amplitude
information of nystagmus is partly processed in the cerebellum
and therefore lesions in the frequency of nystagmus can be
expected in cerebellar lesions (Pyykkö and Dahlen 1983).

Allegedly, the visual pathways in man is composed of cortical
and subcortical projections (Fig. 6). The cortical projection
is analogic to smooth pursuit pathway and carries information
of velocity error on the retina. The subcortical pathway
carries information on retinal displacement signal i.e. coordinates of the visual field with the reference to the fovea.
Since the cerebellum uses visual influx in order to calibrate
and cancel the vestibular reflexes an impairement in the
integration leads to an inaccuracy in the vestibular gain.

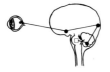 

CORTICAL OKN   SUBCORTICAL OKN

– Fovea   – Periph. retina

– Occip. cortex   – Upper brainstem

– Cerebellum   – Vestibular nuclei

– Brainstem

Figure 6
Optokinetic pathways involved in detection of eye head coordinates (subcortical OKN) and vestibulo-visual interaction
(cortical OKN).

During movement of the head in visual conditions the integration between vestibular and visual influx seem not to be
impaired. The optovestibular nystagmus has in general the same
gain as the vestibular nystagmus (Wennmo et al. 1982). However,
the visual influx can not cancel the vestibular nystagmus e.g.
during smooth pursuit or during fixation with a moving head. As
a result and vestibular responses overrides the fixation reflex
and smooth pursuit becomes saccadic .

A deteriorated ability to produce smooth pursuit activates the
subcortical OKN afferents, which as far as we know, will detect
the displacement of the target from the fovea and actives the
saccadic system to refoveate the target. The subcortical OKN
elicited by stimulation of the periphera retina has poor gain
which may reach 30°/sec in man . Furthermore the rise time is
prolonged resebling the slow bild-up of OKN met with
non-foveate animals (Pyykkö et al 1982).

3. SUPRANUCLEAR MECHANISMS IN THE CONTROL OF NYSTAGMUS

In large temporo-parietal lesions visual suppression of caloric
nystagmus may become defective and light may even enhance the
vestibular responses (Takemori 1981). These findings may be
linked to a decreased attention span and thus reflect a change
in alertness during visual conditions. Consistently, eye movements related neurons have been observed to be activated during
visual interent in posterior parietal association cortex (Lynch
et al. 1977).

In single unit recordings the neurons in the frontal eye fields
(FEF) respond during nystagmus and head-neck coordination
(Bizzi and Schiller 1971, Mohler et al. 1973). A lesion of FEF
may therefore cause a deterioration in the supratenforial control function on nystagmus.

In patients with a frontal lobe lesion the gain of angular
acceleration induced nystagmus in darkness is generally
increased (Dahlen et al. 1980) and the nystagmic response show
more pauses than in present in normal subjects (Pyykkö et al.
1983). In intrabeat analysis of nystagmus the correlation
between velocity and amplitude will become tighter. Thus the
nystagmus will be governed on time mode more than normally
(Fig. 5). The observations indicate a deteriorated amplitude
control of nystagmus. Thus, FEF presumably controls the
excursion of the eye during nystagmus and anticipate the
movement of the surrounding by tiggering the fast phase of
nystagmus (cf. fig. 3).

REFERENCES

Baloh, R.W., Honrubia, V. & Sills, A. 1977. Eye-tracking and optokinetic nystagmus. Results of quantitative testing in patients with well-defined nervous system lesions. Ann Otol Rhinol Laryngol 86, 108-114.

Barber, H.O. and Stockwell, C.W. 1980. Manual of Electronystagmography. 2nd ed. The C.V. Mosby Company, St. Louis/Toronto/London

Bizzi, E. & Schiller, P.H. 1970. Single unit activity in the frontal eye fields of unanesthetized monkeys during eye and head movement. Exp Brain Res 10, 151-158.

Cogan, D.G. 1964. Brain lesion and eye movements in man. In The Oculomotor System (ed. M.B. Bender), pp. 417-423. Harper & Row Publishers, New York.

Cohen, B., Komatsuzaki, A. & Bender, M.B. 1968. Electrooculographic syndrome in monkeys after pontine reticular formation lesions. Arch Neurol 18, 78-92.

Dahlen, A.I., Fex, S., Henriksson, N.G., Pyykkö, I. and Wennmo, C. 1980. Dyspraxia of speech and of eye motility. Acta Otolaryngol (Stockh) 89, 141-144.

Donaghy, M. 1980. The cat's vestibulo-ocular reflex. J Physiol (Lond) 300, 337-351.

Furuya, N. and Markiham, C.H. 1982. Direct inhibitory synaptic linkage of pause neurons with burst inhibitory neurons. Brain res. 245, 139-143.

Henn, V., Büttner-Ennever, J.A. and Hepp, K. 1982a. The primate oculomotor system I. Motoneurons. Human neurobiol. 1. 77-85.

Henn, V., Büttner-Ennever, J.A. and Hepp, K. 1982b. The primate oculomotor system II. Premotor system. Human neurobiol. 1. 87-95.

Hepp, K. and Henn, V. 1979. Neuronal activity preceding rapid eye movements in the brain stem of the alert monkey. In Reflex Control of Posture and Movement. Progress in Brain Research Vol. 50 (ed. R. Granit & O. Pompeiano), pp 645-652. Elsvier/North Holland. Amsterdam.

Highstein, S.M. & Reisine, H. 1979. Synaptic and functional organization of vestibulo-ocular reflex pathways. In Reflex Control of Posture and Movement. Progress in Brain Research Vol. 50 (ed. R. Granit & O. Pompeiano), pp. 431-442. Elsevier/North Holland, Amsterdam.

394   *I. Pyykkö et al.*

Highstein, S.M. and Baker R. 1978. Excitatory termination of abducens internuclear neurons on medial rectus motoneurons; relationship to syndrome of internuclear ophthalmoplegia. J. Neurophysiol. 41, 1647-1661.

Hikosaka, O., Maeda, M., Nakao, S., Shimazu, H. & Shinoda, Y. 1977. Presynaptic impulses in the abducens nucleus and their relation to postsynaptic potentials in motoneurons during vestibular nystagmus. Exp Brain Res 27, 355-376.

Honrubia, V., Baloh, R.W., Yee, R.D. & Jenkins, H.A. 1980. Indentification of the location of vestibular lesions on the basis of vestibulo-ocular reflex measurements. Am J Otolaryngol. 1, 291-301.

Igusa, Y., Sasaki, S. and Shimazu, H. 1980. Excitatory permotor burst neurons in the cat pontine reticular formation related to the quick phase of vestibular nystagmus. Brain res. 182, 451-456.

Ito, M. 1975. The vestibulo-cerebellar relationships: vestibulo-ocular reflex arch and flocculus. In The Vestibular System (ed. R.F. Naunton), pp. 129-146. Academic Press, New York.

Keller, E.L. 1974. Participation of medial pontine reticular formation in eye movement generation in monkey. J Neurophysiol 37, 316-332.

Lynch, J.C., Mountcastle, V.B., Talbot, W.H. and Yin, T.C.T. 1977. Parietal lobe mechanisms for directed visual attention. J. neurophysiol. 36, 649-666.

Mohler, C.W., Goldberg, M.e. & Wurtz, R.H. 1973. Visual receptive fields of frontal eye field neurons. Brain Res 61, 385-389.

Nakao, S., Sasaki, S., Schor, R.H. and Shimazu, H. 1982. Functional organization of premotor neurons in the cat medial vestibular nucleus related to slow and fast phases of nystagmus. Exp. brain. res. 45, 371-385.

Pola, J. and Robinson, D.A. 1978. Oculomotor signals in medical longitudinal fasciculur of the monkey. J. Neurophysiol. 41, 245-259.

Precht, W. 1977. Functional synaptology of brainstem oculomotor pathways. In Control of Gaze by Brain Stem Neurons. Developments in Neuroscience. Vol. 1 (ed. R. Baker & A. Berthoz). pp. 131-141. Elsevier/North Holland, Amsterdam.

Pyykkö, I., Magnusson, M., Matsuoka, I., Ito, S. and Hinoki, M. 1982. On the optokinetic mechanism of peripheral retinal type. Acta Otolaryngol (Stockh) (Suppl) 386, 235-239.

Pyykkö, I. and Dahlen, A.I. 1983a. Intrabeat relationship of postrotatory nystagmus in normal subjects. Acta Otolaryngol. (Stockh). Accepted for publication.

Pyykkö, I. and Dahlen, A.I. 1983b. Intrabeat relationship of postrotatory nystagmus in patients with neurological disorders. Acta Otolaryngol. (Stockh). Accepted for publication.

Pyykkö, I., Dahlen, A.I., Henriksson, N.G. and Juhola, M. 1983. Clinical evaluation of dysrhythmia of postrotatory nystagmus. Acta Otolaryngol. (Stockh). Accepted for publication.

Pyykkö, I., Henriksson, N.G., Wennmo, C. and Schalén, L. 1981. Velocity of rapid eye movements and vertigo of central origin. Ann Otol Rhinol Laryngol 90, 164-168.

Ritchie, L. 1976. Effects of cerebellar lesions on saccadic eye movements. J Neurophysiol 39: 1246-1256.

Schalen, L., Pyykkö, I., Henriksson, N.G. & Wennmo, C. 1982b. Slow eye movements in patients with neurological disorders. Acta Otolaryngol. (Stockh), Suppl. 386, 231-234.

Shimazu, H. 1979. Excitatory and inhibitory premotor neurons related to horizontal vestibular nystagmus. In Integration of the Nervous System (ed. H. Asanuma & W.J. Wilson), pp. 123-142. Igaku-Shoin Ltd, Tokyo.

Starr, A. 1967. A disorder of rapid eye movements in Huntigton's chorea. Brain 90, 545-564.

Steele, J.C., Richardson, J.C. and Olszewski, J. 1964. Progressive supranuclear palsy. Arch Neurol 10, 333-359.

Takemori, S., Ishihawa, M. and Yamada, S. 1981. Cerebral control of eye movements. ORL 43, 262-273.

Tomko, D.L., Barbaro, N.M. and Ali, F.N. 1981. Effect of body tilt on receptive field orientation of simple visual cortical neurons in unanesthetized cats. Exp Brain Res 43, 309-314.

Wennmo, C., Hidfelt, B. & Pyykkö, I. 1983 Eye movements in cerebellar and combined cerebello-brainstem diseases. Ann Otol Rhinol Laryngol. In print.

Yamamoto, M. Vestibulo-ocular reflex pathways of rabbits and their representation in the cerebellar flocculus. In Reflex Control of Posture and Movement. Progress in Brain Reseach Vol. 50 (ed. R. Granit & O. Pompeiano), pp 451-458. Elsevier/North Holland, Amsterdam.

Theoretical and Applied Aspects of Eye Movement Research
A.G. Gale and F. Johnson (Editors)
© Elsevier Science Publishers B.V. (North-Holland), 1984

GAZE ABNORMALITIES WITH CHRONIC CEREBRAL LESIONS IN MAN

N.G.R.Page,H.J.Barratt,M.A.Gresty

MRC Neuro-otology Unit
National Hospitals for Nervous Disease
Queen Square
LONDON WC1

Abnormal patterns of head/eye coordination that occurred in patients with chronic cerebral lesions comprised increased saccade latencies and hypometria contralateral to the lesion. Frequent saccadic intrusions, often with a distinct asymmetry, and an excess of predictive saccades were seen. Head movements were consistent with the saccadic abnormalities and also changes in the waveform and timing of head movements were noted.

INTRODUCTION

Abnormalities of gaze produced by disorders of the nervous system have implications for the functional organization and localisation of the control of head and eye movements. Although the neurology of brainstem mechanisms of gaze is well documented there have been few studies of the effects of lesions of the cerebral hemispheres[1,2]. We report a survey of the abnormalities of head-eye coordination encountered in a series of 21 patients with chronic lesions distributed unilaterally and bilaterally in various sites in the cerebral hemispheres.

Gaze movements were evoked by predictable and unpredictable (in amplitude and duty cycle) stepwise movements of a 0.2° laser target moving in the horizontal plane at a distance of 1.5 m from the vertex. Eye movements were recorded using dc-coupled electro-oculography and head movements in the horizontal plane were transduced by a low torque potentiometer coupled to the head by a tightly fitting helmet. Head and eye signals were electronically summed to produce the gaze function.

The patients studied comprised 5 with unilateral frontal lobe lesions, 11 with unilateral parietal lobe lesions, 2 with involvement of a whole hemisphere and 3 with bilateral lesions. Three patients had cerebral infarcts and the remainder had space occupying mass lesions (benign or malignant tumour). The siting of lesions was confirmed by CAT scanning. Six control subjects were studied whose age range was similar to that of the patients and who had peripheral neurological disease.

FINDINGS

The patterns of abnormalities observed in our patients could be conveniently classified as those of saccade latencies, saccade dysmetria (hypometria), inappropriate saccadic movements and abnormalities of latency and waveform of head movements.

The latencies of saccades to random targets were significantly longer to the
contralateral side than ipsilaterally in patients with cortical visual field
defects or inattention, in one parietal lobe patient with normal vision and
one with a frontal lobe tumour (but extensive oedema). Symmetrically
increased latencies occured in one frontal lobe patient and two with
bilateral lesions. One further frontal patient had significantly longer
ipsilateral latencies, but within normal limits, and two with parietal
lesions had subnormal latencies (figure 1).

Frontal eye field lesions in monkeys produce only temporary effects on
saccades [3], thus it may well be for the reason that our patients with
frontal lesions had chronic disease that saccade latencies were normal.

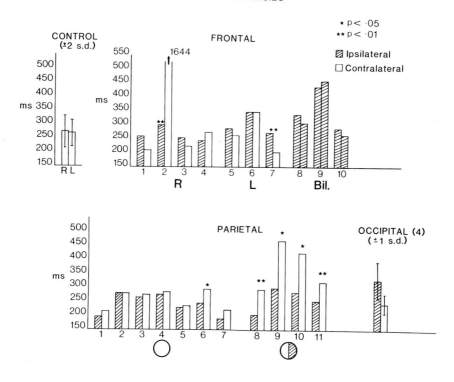

Figure 1

Hypometria of refixation saccades was observed in all patients with
hemianopias or visual inattention, with complex patterns of saccadic
movements being typical in the latter group. Patients who had complex
patterns of saccades during gaze transfer tended not to make accompanying
head movements. A significant degree of asymmetry in hypometria occured in 3
of 7 parietal and 2 of 7 frontal lobe patients with normal vision, in whom
the hypometria occurred away from the side of the lesion. This is reflected
in the higher than normal means and standard deviations of these two groups.

Means of R/L difference of saccades/gaze transference

```
Controls   n=8                    0.11 ± 0.09
Frontals   n=7                    0.29 ± 0.30
Parietals (normal vision)
           n=7                    0.31 ± 0.15
Parietals (innattention or field
     defect)  n=4                 1.73 ± 1.29
```

"Square waves" or saccadic intrusions which took the eyes from and to the
target in an inappropriate manner occurred with greater frequency in patients
but were also present in matched control subjects. In addition, 3 patients
with parietal lobe lesions with normal vision had a marked asymmetry of
square waves in which the square waves were directed contralaterally to the
side of the lesion (example shown in figure 2). These intrusions were
generally of larger amplitude than the square wave jerks occurring in normal
subjects [4] or described in patients with cerebral lesions [5].

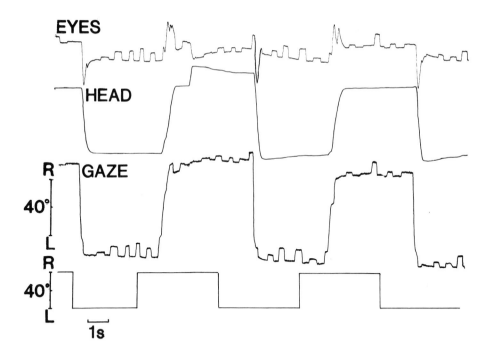

Figure 2
Raw traces of eye and head position with electronically summed representation
of gaze of a patient with a left parietal glioma. Gaze change in response to
predictable 40° target steps is accomplished largely by the head component.
Continuous right-beating square waves occur which are of larger amplitude in
leftwards gaze.

Saccadic intrusions, which may have been predictive in that they were
directed towards and proceeding anticipated target displacements, occurred
more frequently in patients than in controls. For example this pattern of
saccadic intrusions may manifest itself as "gaze instability" (figure 3).
These large intrusions may represent inappropriate or exaggerated attempts at
prediction, failure to suppress inappropriate saccades or be extreme
exaggerations of "square wave" instabilities. Patients with frontal lobe
lesions have impaired ability to suppress unwanted saccades [6].

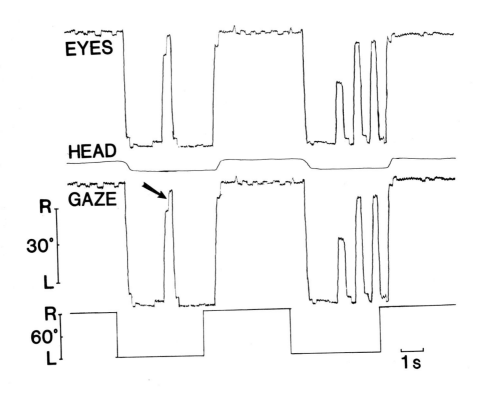

Figure 3.
In a patient with bilateral frontal malignant tumours, large saccadic
intrusions occur on leftwards gaze in response to predictable 60° target
steps. These may consist of several saccades and may approximate the
amplitude of the anticipated saccade(arrowed) but are not accompanied by head
movements. There are also many square waves, visible on rightwards gaze.

Head movements showed abnormalities in that they could occur with much
prolonged latencies (figure 4)(particularly in patients with complex and
hypometric eye movements) or have a multistepped appearance (hypometric head
displacements) (figure 5).

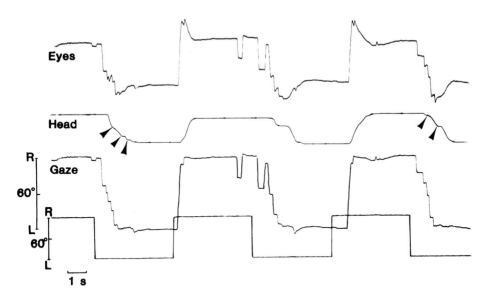

Figure 4

In a hemianopic patient, multiple step head movements (arrowed) occur with hypometric saccades.

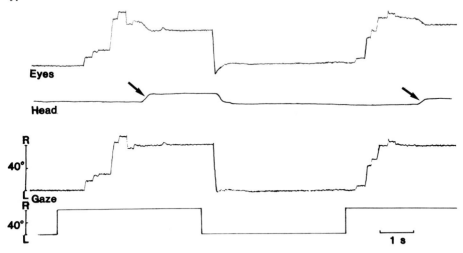

Figure 5

Delayed head movements (arrowed) follow complex multiple saccade eye movements. Coordinated head/eye gaze transfer to the left (ipsilateral to the lesion) is normal.

On occasions, head movements could occur together with the predictive saccadic intrusions demonstrating that the gaze transfer was programmed for a coordinated eye and head movement.

Our general conclusions are that the normal patterns of head and eye movements can be modified by lesions of the cerebral cortex. The major abnormalities are to be found in patients with visual inattention and hemianopia. Patients with even very large frontal lobe lesions may have preserved normal head eye coordination. An excess of saccadic intrusions and predictive saccades may represent a failure of frontal lobes to supress unwanted searching saccades.

REFERENCES

[1]   Herman (Ben-Zur), P. and Atkin, A., A modification of eye-head coordination by CNS disease, J Neurol.Sci. 28 (1976) 301-307.

[2]..Zangemeister, W.H.,Meienberg, O., Stark, L. and Hoyt, W.F., Eye-head coordination in homonymous hemianopia, J Neurol. 226 (1982) 213-254.

[3] Schiller, P.M., True, S.D., and Conway, J.L., Deficits in eye movements following frontal eye-field and superior colliculus ablations, J Neurophysiol. 44 (1980) 1175-1189.

[4]   Herishanu, Y.O. and Sharpe, J.A., Normal square wave jerks, Invest. Ophthalmol. Vis. Sci., 20 (1981) 268-272.

[5]   Sharpe, J.A., Herishanu, Y.O. and White, O.B., Cerebral square wave jerks, Neurology (Ny). 32 (1982) 57-62.

[6]   Guitton, D., Buchtel, H.A. and Douglas, R.M., Disturbances of voluntary eye movement mechanisms following discrete unilateral frontal lobe removals, in: Lennerstrand, G., Zee, D.S. and Keller, E.L., (Eds), Functional basis of oculomotor disorders (Pergamon Press, Oxford, 1982).

Theoretical and Applied Aspects of Eye Movement Research
A.G. Gale and F. Johnson (Editors)
© Elsevier Science Publishers B.V. (North-Holland), 1984

EXPRESS-SACCADES OF THE MONKEY: A NEW TYPE
OF VISUALLY GUIDED RAPID EYE MOVEMENTS
AFTER EXTREMELY SHORT REACTION TIMES

Burkhart Fischer and Rolf Boch

Department of Neurophysiology
University of Freiburg
Germany

Monkeys are trained to fixate a stationary small
central spot and to saccade to a peripheral target.
The saccadic reaction times depend on the temporal
sequence of fixation point off set and target on
set and form a spectrum between 60 and 400 ms
containing at least 4 peaks. If the fixation point
is turned off before the peripheral target appears
(gap duration in the order of 200 ms) a rather
narrow (± 3 ms) distribution of reaction times is
obtained with a peak at 75 ms (express-saccades).
The exact mean value of the reaction times depends
on the amount of daily training (and on the physical
parameters of the target). Reaction times of the
regular saccades may be shortened by as much as
200 ms by daily exercise those of the express-
saccades by about 20 ms.

INTRODUCTION

A voluntary visually guided rapid change of the direction of
gaze is an example of a sensory-motor coordination in which
several processes are involved, which are of major importance
in an understanding of higher brain functions. Among them are
the mechanisms of breaking fixation, selection of a single
target among many others, directing attention to peripheral
stimuli, decision to make one or the other or no movement,
loading and reading of a memory that contains the coordinates
of the location of the target in a retinally or spatially
organized system. The modern methods of recording from single
cells in awake behaving monkeys, trained to fixate a stationary stimulus and to saccade to peripheral targets have
been successfully used to identify neural structures, that
are involved in the initiation of visually guided saccades
(see review by Wurtz et al., 1980).

In an attempt to clarify the contribution of the visual association cortex (prelunate gyrus) in directing attention and/
or gaze (Boch and Fischer, 1983) we have observed, that the
time between the command for a saccade and the time of its
final execution - the saccadic reaction time, SRT - is critically dependent on the temporal sequence of fixation point
off set and target on set and is also subjected to the amount

of daily exercise. These observations led us to the idea,
that one might be able to isolate and identify different
processes that are needed to initiate a saccade by systematically studying the saccadic reaction times under different
visual and behavioural conditions. In fact, we observed a
whole spectrum of saccadic reaction times including a peak at
extremely short times in the order of 70 - 80 ms after target
on set (Fischer and Boch, 1983). Here we describe the conditions under which these so called express-saccades occur in
relation the conditions under which regular reaction times
are observed.

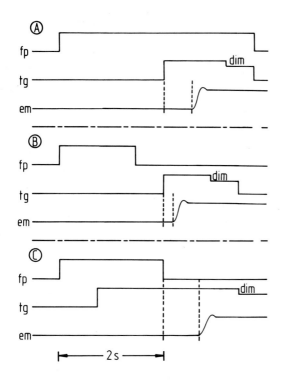

Figure 1

Temporal sequences
of fixation point
(fp) off set,
target (tg) on set
and eye movements
(em). A, B, and C
label the three
paradigms as de-
scribed in the
text.

METHODS

Four rhesus monkeys were trained to fixate a stationary small
light spot using the dimming paradigm. With their head rigid-
ly fixed they were sitting in a primate chair and eye movements were continuously monitored by an infrared sensitive

device with an resolution of 0.1 deg. Saccadic eye movements
were elicited under three different conditions illustrated in
figure 1. In A the fixation point remained visible when the
peripheral target occurred. The animals made saccades to that
target because they had learned that it would be always the
target that received the dimming at some unpredictable time.
In B the fixation point went off before the new target
appeared such that there was a temporal gap of no more than
300 ms were the animals saw none of the two stimuli. In C the
target became visible before the fixation point went off. In
this case, other than in A, the animals kept fixating the
central fixation point, because they had learned that the
fixation point would receive the dimming. Note: animals were
never rewarded for making correct (or any) eye movements,
they rather were rewarded for detecting the dimming.

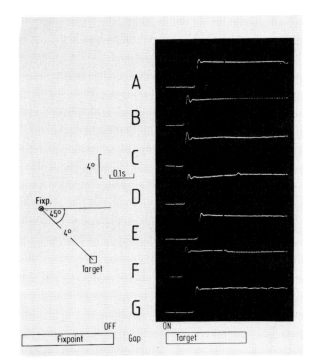

Figure 2

Specimen records of
eye movements (A -
G) using paradigm B
of fig. 1. Spatial
arrangements at the
left and temporal
sequence at the
bottom.

Eye position and the visual display of the animal were both
superimposed on a single TV-monitor on a one-to-one scale. In
addition traces of the eye movement signal were displayed

trial by trial on a storage oscilloscope for visual inspection. Saccades were automatically detected by an electronic
threshold device, which blanked the oscilloscope beam at the
time of the beginning of a saccade (see fig. 2). Other details of the methods are described elsewhere (Fischer and
Boch, 1983).

RESULTS AND DISCUSSION

The basic observation is illustrated by the sample of eye
movement traces shown in fig. 2. The spatial arrangement is
shown at the left, the temporal arrangement below the records
(case B of fig. 1, gap of 220 ms). Under this condition the
monkey produced two different sets of reaction times: In A,
E, and G the reaction times are in the order of 130 ms, whereas in B, C, D, and F values in the order of 75 ms are obtained. The latter saccades - the express-saccades - are real
visually guided eye movements because they occur also when
both the time and the location of appearance of the peripheral target we made unpredictable for the animal. Note also,
that these express-saccades are not followed by corrective
saccades of more than a fraction of a degree (this applies
for most well trained animals).

The distribution of saccadic reaction times in the gap situation (case B of fig. 1) is clearly bimodal as can be seen
from fig. 3 (lower part, peaks labeled B). The left most
narrow peak has a standard deviation in the order of only 3
to 5 ms.

Since the express-saccades occur in a situation were the central fixation point was already extinguished some time before
the peripheral target appeared, the idea came up, that the
animal has already stopped its process of active fixation and
only needs to make a decision and to compute the coordinates
of the new target in order to initiate the next saccade.

In this view the reaction time of 130 ms contains the time
for the decision making and the time for computing the coordinates, whereas the express reaction time of 75 ms contains
only the time for computing the coordinates, the decision
being completed already at the time the target appears.

If this view is correct, one should see express-saccades also
in case A of figure 1, i. e. when the animal had learned,
that the fixation spot is not the relevant stimulus. The corresponding distribution of reaction times is shown in fig. 3,
peaks labeled A. Indeed, express-saccades occur, but at a
lower rate, and the regular saccades have a longer reaction
time (about 160 ms and more) as compared to case B. This can
be explained by assuming that in a considerable number of
trials the animal kept fixating the central target until the
peripheral target occurred, then breaks the process of fixation, makes a decision, computes the coordinates and finally
initiates the movement.

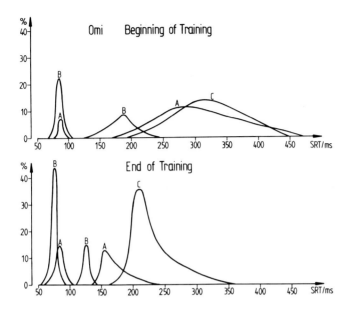

Figure 3

Spectra of saccadic reaction times at the beginning and at the end of training. Letters refer to the paradigms (fig. 1) that produce the different peaks.

If this view holds, one predicts long reaction times but no express-saccades in a paradigm, where the animal always has to fixate the central spot even though the peripheral target becomes visible. The saccade in this case (fig. 1 C) is elicited by the later off set of the fixation point. The corresponding distribution of reaction times is shown in fig. 3 labeled C. Indeed, in this situation no express-saccades were ever seen and the reaction times are rather long, in particular they are even longer as in case A. The explanation for this further increase of the reaction time comes most probably from the fact, that the target became visible long (1.5 s) before the saccade such that its location must be reevaluated(reloaded).We are presently investigating this possibility by having the animals make saccades to targets the location of which must be remembered.

The upper part of fig. 3 shows the results of the same set of experiments in the same animal when it had just learned the corresponding tasks (A, B, and C). Again one sees four peaks but the long reaction times obtained in A and C are only slightly different.

Comparing the spectrum at the beginning and at the end of
training shows that all peaks are shifted to the left but by
different amounts. This indicates that in fact the peaks
represent different processes which are modifiable by daily
exercise. The details of this modifications, in particular
their reversibility, need to be evaluated by further experimentation.

SUMMARY

We have demonstrated the existence of visually guided
saccades after reaction times in the order of 70 - 80 ms
(express-saccades). Express-saccades occur when a monkey is
not fixating anything. Their rate of occurrence increases and
their reaction time decreases with daily exercise. The spectrum of saccadic reaction times contains at least three more
peaks depending on the temporal relation of the off set of
the fixation and on set of the target. The total spectrum of
reaction times is modified by training.

ACKNOWLEDGEMENTS

This work was supported by the Deutsche Forschungsgemeinschaft
(DFG), Sonderforschungsbereich 'Hirnforschung und Sinnesphy-
siologie' (SFB 70, Tp B7).

LITERATURE

Boch, R. and Fischer, B. (1983) Saccadic reaction times and
    activation of the prelunate cortex: parallel observa-
    tions in trained rhesus monkeys. Exp. Brain Res. 50,
    201-210.

Fischer, B. and Boch, R. (1983) Saccadic eye movements after
    extremely short reaction times in the monkey. Brain Res.
    260, 21-26.

Wurtz, R.H., Goldberg, M.E. and Robinson, D.L. (1980) Behav-
    ioural modulation of visual responses in the monkey:
    stimulus selection for attention and movement. Progr.
    Psychobiol.Physiol.Psychol. 9, 43-83.

Theoretical and Applied Aspects of Eye Movement Research
A.G. Gale and F. Johnson (Editors)
© Elsevier Science Publishers B.V. (North-Holland), 1984

CONVERGENCE OF DIFFERENT SENSORY MODALITIES IN A SINGLE CELL OF
FLOCCULUS IN ALERT CAT

Pyykkö, I.[1], Schalen, L[2], and Magnusson, M.[2]

[1] Department of Physiology, Institute of Occupational Health
Helsinki,Finland
[2] Department of Otolaryngology, University Hospital of Lund
Lund, Sweden

The response pattern of 109 floccular cells of
alert, intact cat was examined during different
proprioceptive, vestibular and visual tasks. In
62 neurons covergence of at least two sensory modalities could be confirmed. 14 of these cells
responded to vestibular, visual and proprioceptive
stimuli. Saccadic neurons responded to ipsi- and
contralateral saccades, but with somewhat different latency and intensity indicating that the goal
of the saccades are presumably coded into these
neurons. The muscle spindles and Golgi tendon
organs were the most common site for afferentation
and responses for these receptors were found in 12
neurons. The ipsilateral front limb was the most
usual source for the afferentation. Thus, flocculus governs the visual and vestibular orientation reflexes by gathering different sensory
information from a wide variety of receptor systems to modulate the output of these reflexes.

Several studies have confirmed the close relationship between
eye movements and flocculus. Thus, the activity linked to
saccades (Noda and Suzuki 1979), smooth pursuit (Liesberger and
Fuchs 1977) and vestibular (Ito 1982) and optokinetic
nystagmus (Büttner et al. 1981) has been found in flocculus.
In addition Wilson et al. (1975) showed in paralyzed cat that
flocculus also received information from the neck
proprioceptors. The very same Purkinje cell (PC) responded to
vestibular, visual and proprioceptive stimuli. The influx of
each of these sensory modalities was inhibitory to each other.

Nevertheless, it is evident from the works of Noda and Suzuki
(1979) and Harvey (1980) that a normal function of the cerebellum neccessitates an intact somatosensory influx. Allegedly,
the somatosensory influx through the mossy fiber (MF) pathway
is capable to produce a continuously graded activity of PCs
(Evarts and Thach 1969), which when simultaneously facilitated
through other pathways (e.g. visual and vertibular) may release
a graded inhibition or facilitation of the target cells (Llinas
1975). Evidently a simultaneous afferentation from different
receptor systems is especially important for the calibration

and organisation of the orientation reflexes exerted   by the
cerebellum. It was therefore of interest to study in alert cat
how the different sensory modalities are represented in flocculus and whether the activity linked to these modalities are
converted into the same neuron.

MATERIAL AND METHODS

Totally 21 cat (weight 2,5-3,5 kg) were employed in the present
experiments. Sixteen of the cats were used during elaboration
of the equipments and the recording methods. No data is
included from these cats. The present data is collected from 5
cats.

A special stand was constructed to immobilize the body of the
cat. The legs were hanging freely and the different limb
receptors could be tested manually and screened with a vibrator
(Fig. 1). The head was immobilized with a holder which was
screwed and sealed with dental acrylic cement on the scull. A
teflon cylinder was trephined over the left flocculus. The stand
placed over a turntable fixed to the table.

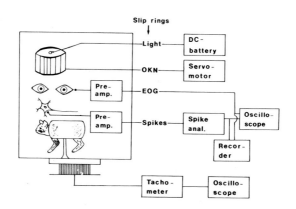

Figure 1
Block diagram of the recording system

To evoke optokinetic nystagmus a planetarium was constructed
which reflected an irregular pattern of small light spots to
the peripheral visual field. Different target velocities of
16, 31, 46, 57 and 72 °/sec were used. Stimulation was
conducted clockwise and counterclockwise.

The vestibular stimulation was caused by turning the table
manually in si .usoidal fashion. If the neuron could be hold
long enough also constant angular acceleration of 2, 4, 8 and
12 °/sec^2 was employed.

Propiroceptive receptors from the neck and limbs were examined
with rotating type of vibrator operating at a constant
amplitude of 0.9 mm (peak-to-peak) and at frequencies of 80 to
125 Hz. The limbs were thereafter examined manually and the
site of the receptor identified. For the neck receptor
identification by vibration was used only.

The visual field was examined for each neuron with a spot of
light and by moving different size of targets in the front of
the cat. For eye tracking small attractive targets were moved
manually in the front of the cat.

Initially the cats were trained during a period of 2 to 4
weeks; first by wrapping the cat into a towel and later by
accustomizing the cat into the stand. After the initial
training period the eye electrodes and head holder were
implanted during bartiturate anesthesia and by using local
anesthetizing agents. Thereafter training was continued
further a period of one to two weeks, after which the cats were
reanaesthetised and a teflon tunnel was stereotactically
trephined over the left flocculus. A stand for hydraulic
microdrive was inserted. One week later the dura was removed
during ketamine analgesia and the recording was commenced.

In recording of the unit activity varnished tungsten electrodes
were used. The size of the tip was 0.1 to $2 \mu m$ with an
impedance of 0.5-2 MΩ. The signal was fed through preamplifier
(amplified 5000 times) into a signal discriminator and
oscilloscope. Selected neuron was recorded on paper (speed 25
mm/s). The spikes were integrated so that four spikes on the
oscilloscope corresponded one spike on the paper.
Simultaneously, the unit activity was monitored through a
loudspeaker. Eye movement signal was amplified 500 times (time
constant 1,5 sec. and uper cut of frequency 30 Hz). Eye
movements were monitored and recorded simultaneously with the
neural activity.

Cells reported in this study fullfilled following criteria 1)
They displaced large negative-positive spikes, either simple or
complex. 2) Cells could be kept isolated during an electrode
travel of $50 \mu m$ or more. 3) Cells showed inegular and usually
high spontaneous activity. Preliminarly we concidered these
cells as PC but later it became evident that a part of these
cells might be a granular layer input elements (tentatively
Golgi cells).

In the end of the recording electrolytic lesions were made in
each cat and a map was reconstructed to track the site of the
recording.

RESULTS

Altogether   150 cells were recorded from which 109 cells did
fullfill the requirements for further analysis. Depending on

the response to different stimuli the cells were devided into
nonresponders, single sensory modality responders and
multiresponders (Table I). Majority of the cells (57 per cent)
responded to more than one sensory modality whereas 19 per cent
of the neurons did not show any activity which could be linked
to the specific stimuli. From the multiple modality cells the
majority responded during vestibular and visual stimulation (65
per cent) (Table 2). The visual neurons encloses both neurons
responding during rapid eye moments (saccades and fast phase of
nystagmus) as well as during slow eye movements (eye tracking
and slow phase of nystagmus) or stimulation of the visual fied.

Table 1.

Response	N
Multiple modality	62
Single modality	26
No responses	21

Table 2.

Multiple modality	N
Vestibular and visual	40
Vestibular and proprioc.	8
Vestibular, visual and proprioc	14

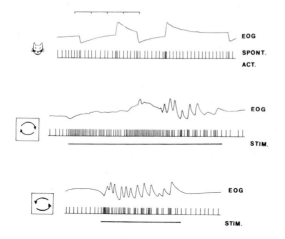

Figure 2
Rapid eye movement neuron. Upper diagram: spontaneous
saccades. Middle diagram: rotation in the light. Lower diagram:
rotation in the dark. Time division 1 sec. Note: one pen
deflexion is integrated activity of four spikes of the neuron.

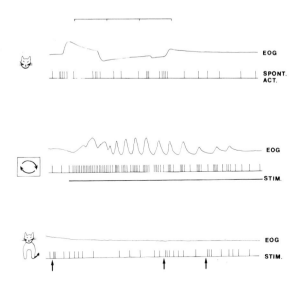

Figure 3.
Rapid eye movement and proprioceptive neuron. Spontaneous
activity, rotation in the darkness, and deflexion of the leg
are exhibited. Abreviations as in Fig. 2.

All neurons responding to saccades responded also to fast phase
of nystagmus. An example of this kind of responsivenes in set
out in Fig. 2. In totally 17 neuron activity linked to rapid
eye movement was observed. Typically saccadic neurons increased
the activity during ipsi- and contalateral saccades. However,
qualitatively differences could be observed in the number of
spikes and in the latency of firing prior to the saccade. These
factors may be attributed to the processing of the goal of the
saccade. It was noteworthy that in light, when the cat fixated
the gage during rotation the nystagmus was suppressed, the
neuron exhibited vigorous activity. When the nystagmus could
not be cancelled in the dark the neuron fired syncronously with
the fast phase of nystagmus. The neural activity usually
preceeded somewhat the release of the fast phase. Fig. 3 shows
an example of a cell responding to saccade, fast phase of
nystagmus and propiroceptive stimulus. Rapid bending of front
limb triggered a short burst of activity which was not linked
to eye movement.

Activity linked to slow eye movements occurred in 30 neurons
(table 3). Majority of the units responded to rotation , some
to eye tracking and or to optokinetic stimulation. In general,
the same responce pattern was observed to clockwise and to
counterclockwise rotation.

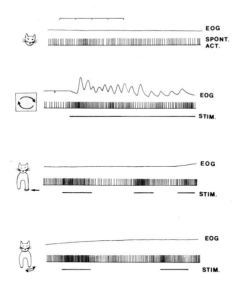

Figure 4

Slow eye movement and proprioceptive neuron. Spontaneous activity, rotation in the dark, distention of the nails bending of the limb is set out. Abreviations as in Fig. 2.

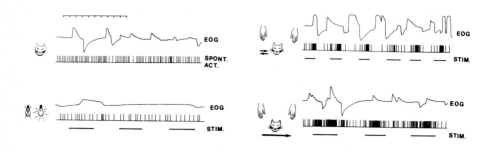

Figure 5.

Visual field neuron. Sopntaneous activity, on-off response to light, on-off response to stimulation of right visual field and unidirectional stimulation of the visual field are set out. Abreviations as in Fig. 2.

Table 3.

Visual	N
Rapid eye movements	17
Slow eye movements	30
Visual field	7
All	54

In these neurons too, a convergence between different sensory modalities was observed (Fig. 4). Usually irregular firing pattern was observed during rotation. Prominent increase in activity occurred when the direction of the rotation was changed especially during phase shift of rotation. During constant angular acceleration the basic frequency was modulated only weakly. During proprioceptive stimulation syncronous activity could be verified in a part of these neurons but not in all.

In 7 neurons stimulation of visual field triggered a response. These neurons had a large visual field and a specific direction of movement triggered the increase of activity (Fig. 5). During to and from movement of the target in the visual field the neurons fired. Usually they paused during stimulation into off-direction and fired vigorously during stimulation into on-direction. These cells did not respond to switching the lights on and off. Furthermore, since a stimulation of the visual field activated thesese neurons, increased activity could be often observed also during saccades.

Neurons responding to proprioceptive stimuli had a receptive field commonly in the ipsilateral front limb (Table 4). Typically, vibration caused a vigorous burst of activity through these receptors. The response was repeatable and the effect persisted as long as vibration continued without showing any signs of adaptation. Some neurons, which usually responded to stimulus with a short burst, had a receptive field in all four limbs and usually a rapid deflexion of the limb released the burst (c.f. Fig. 2). These receptors could be however, found more reliable by manipulation than with vibrator. Activity linked to stimulation of neck receptors, evaluated with vibration only, was detected in 5 neurons.

Table 4.   Table 5.

Proprioceptive	N
Ipsilat. fron leg	13
Multilat (legs)	4
Neck	5

Receptor	N
Muscle spindle	12
Golgi tendon app.	
Joint receptor	3
Skin receptor	3
Unspecified	4

By manipulation a detailed character of the type of afferent receptors could be determined (Table 5). majority of the activity was linked to muscle spindle (MS) or Golgi tendon organ (GTO) and was observed in 12 neurons. Since the MS respond to vibration at much lower threshold levels than GTO we tried to further differentiate between these two receptors. Based on a high sensitivity to vibration of the receptors studied we assume that majority of the receptor classified into MS-GTO-group are in fact, MS-receptors. Typically these receptors were highly specified. A receptor increasing the firing rate by bending of the digits paused when stimulated in the opposite way (Fig. 6).

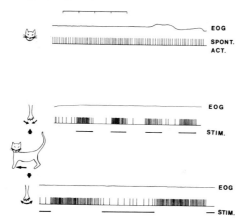

Figure 6.
Proprioceptive neuron. Spontaneous activity, adduction of the digits, and abduction of the digits are set out. During adduction the activity increases 1 whereas during abduction the cell pauses.

Other receptor organs such as joint and skin receptors were few. In four neuron the receptor site remeined undetermined. One neuron responded to the stimulation of the mustaches. Even slight blowing could trigger the response in this neuron. Two neurons responded to sound in addition to other stimuli.

DISCUSSION

In the present study extracellular activity was recorded from
the flocculus of alert cat. In fourteen neurons a convergence
of proproceptive, vestibular and visual stimulus could be
confirmed. Furthermore, in the majority of the cells afferent
influx from at least two sensory receptor system was observed.
Thus, the floccular input constitutes of elementary sensory
convergence which seems to be sufficient for the requirement of
the cerebellar flocculus: to calibrate and organize the
postural and orientation reflexes.

The afferent information can achieve floculus by two ways
either by mossy fiber (MF) or climbing fiber (CF) pathways.
The output can take place only via axons of PCs. The
difference in the organization of the MF and CF is of   ut most
importance (Ito 1975). By extensive convergence and
divergence, the MF-input can modify in graded manner the output
of PC. On average one MF has through parallel fibers a contact
with 25 PCs (Llinas 1975). The tightly coupled CF-input,
mediating mainly visual information is able to modify exactly
right "target" PCs and hence make e.g. during fixation or
tracking, the eyes steady. Our findings confirm those of
Wilson et al. (1975) that the sensory processing can be carried
out in one single PC which may have a very specific function by
coordinating information from different afferent sources.

However, the present data do allow conclusions from the site of
the integration of afferent information. Occasional recording
of MF-activity in flocculus indicated that at least a part of
afferent multisensory influx is carried out in the same MFs.
This woud indicate that integration between different sensory
modalities occur outside of the flocculus. For this part of
integration the vestibular nuclei seems to be the most likely
target organ since the neurons of the vertibular nuclei
integrate semicircular, otolith and proprioceptive influx and
project further to flocculus. A part of the cells in the
vestibular nuclear respond also to visual stimulation.

The MF input is mediated to PC via granular layer elements,
which in turn are in contact with several MFs. It is therefore
possible that a part of the integration would be carried out
also of this stage of information processing (Harvey 1980,
Gilman et al. 1981). Thus, e.g. Golgi cells would furction as a
buffer zone for PCs in grading the activity of PCs. However,
the conclusive evidence for this stage of sensory integration
is not yet evinced.

The third possible site for sensory integration favoured by
observations of Wilson et al. (1975) is in the dendritic tree
of PCs. This would indicate that the information would project
separately trough individual MF and CF pathways to the same

neuron. The net result, i.e. the activation of PC, would hence
depend in mathematical sense, on the number of inhitory and
facilitatory synapses derived from different sensory
modalities. Thus, PCs would need to have a long memory for
spatial summation. In fact, Ito (1982) has pointed out that
after stimulation of CF the change in the resting potential in
PC will be preserved one second or even longer. Thus, a part
of the sensory integration expectedly seems to be carried out
in PCs. The different influx might be especially important for
the learning process of PCs.

In the present work neurons responding to rapid eye movements
were activated during ipsi- and contralateral saccades. This
finding is contrary to the finding of Noda and Suzuki (1979) and
Waespe and Büttner (1981), who showed that the floccular
neurons respond mainly to ipsilateral saccades. However, we
observed differences in the latency and extent of activation in
the neurons responding to rapid eye movements which are in line
with the observations made in pontine reticular formation of
alert monkey (Gisbergen et al. 1981). Consistently, activity
between burst inhibitory (Furya and Markham 1982) and
facilitatory neurons (Igusa et al. 1980) decides the goal of
the saccade. Thus a simultaneous activity of the opposite
floccular PC should be counted and the saccade might be a net
result of the activities between ipsi- and contralateral
flocculus. Since simultaneous recording was not made, this
hypothesis has still to be confirmed. A similar explanation
might also be relevant in nystagmus slow phases and smooth
pursuits.

The orientation and visual exploring of the environment
requires information of the intention of the movement, proper
eye-head cordination and information of the head velocity
during the movement. In the present work all these sensory
influx were observed in the cat flocculus. In addition we also
observed that the very same neurons also responded to
stimulation of the proprioceptive afferents. The results
indicate that the eye movements and the orientation reflexes
are governed by the flocculus neurons which collect data from
several sensory sources and by various neurophysiological
procedures modulates the output of the responses at different
site of the central neurons system.

ACKNOWLEDGEMENT

We wish to thank Dr. Brian F. McCabe, Department of
Otolaryngology and Maxillofacial Surgery University and clinics
of Iowa, Iowa City, Iowa, USA, for providing laboratory
facilities and economic support for conducting the experiments.

REFERENCES

Evarts, E.V. and Thach, W.T. 1969. Motor mechanisms of CNS: Cerebellar interrelations. Ann. rev. physiol. 31. 451-498.

Furuya N. and Markham C.H. 1982. Direct inhibitory synaptic linkage of pause neurons with burst inhibitory neurons. Brain res. 245, 139-143.

Gilman S., Bloedel J.R. and Lechtenberg R. 1981. Disorders of the cerebellum. F.A. Davis Company, Philadelphia, 415 p.

Van Gisbergen J.A.M., Robinson D.A. and Gielen S. 1981. A quantitative analysis of generation of saccadic eye movements by burst neurons. J. neurophysiol. 45, 417-442.

Harvey R.J. 1980. Cerebellar regulation in movement control, TINS, 3, 281-284.

Igusa I., Sasaki S. and Shimazu H. 1980. Excitatory premotor burst neurons in the cat pontine reticular formation related to the quick phase of vestibular nystagmus. Brain res. 182, 451-456.

Ito, M. 1975. Learning control mechanisms by the cerebellum investigated in the flocculo-vestibulo-ocular system. The nervous system. (ed.) Tower D.B. Vol. 1. The basic neuroschiences. Raven Press, New York, p. 245-252.

Ito M. 1982. Eye movements and the cerebellum. Exp. brain res. Suppl. 6. 515-532.

Lisberger S.G. and Fuchs A.F. 1978. Role of primate flocculus during rapid behavioral modification of vestibulo-ocular reflex I. Purkinje cell activity during visually guided horizontal smooth pursuit eye movementsand passive head rotation. J. Neurophysiol., 41: 733-748.

Llinás, R. 1975. The cerebellar cortex. In: The nervous system. Ed. by D.B. Tower, Vol. 1. The basic neurosciences. Raven Press, New York, p. 235-344.

Noda H. and Suzuki D.A. 1979. Processing of eye movement signals in the flocculus of the monkey. J. Physiol., (Lond.) 294: 349-364.

Waespe W. and Büttner U. 1981. Flocculus unit activity in the alert monkey during optokinetic stimulation. Progress in Oculomotor Research. Fuchs A. and Becher W. (eds.). Elsevier North-Holland, Inc. 496-502.

Wilson V.J., Maeda M. and Fuchs J.I. 1975. Inhibitory interaction between labyrinthine, visual neck inputs to the cat flocculus. Brain Res., 96: 357-360.

Theoretical and Applied Aspects of Eye Movement Research
A.G. Gale and F. Johnson (Editors)
© Elsevier Science Publishers B.V. (North-Holland), 1984

REDUCTION OF NYSTAGMUS AS A PREDICTOR OF
EFFICACY OF MOTION SICKNESS REMEDIES

I. Pyykkö, L. Schalen, M. Magnusson, I. Matsuoka

Department of Oto-Rhino-Laryngology, Section of
Neuro-Otology, University Hospital of Lund, Lund Sweden

(1) Present address: Department of Physiology,
Institute of Occupational Health, Helsinki, Finland

(2) Visiting scientist. From Department of Otolaryngology,
University Hospital of Kyoto, Kyoto, Japan

The effects of transdermally administered
scopolamine (TTS-scopolamine, release rate 5 g/h)
and dimenhydrinate (100 μmg) were examined on
caloric, angular acceleration induced and
optokinetic nystagmus in 16 volunteers in a
randomized doubleblind study. All drugs induced a
statistically significant decrease in maximum
velocity of caloric nystagmus. In the rotatory
test, two TTS-scopolamine and dimenhydrinate
reduced the vestibular gain significantly. In the
optokinetic test, all drugs tended to reduce the
responses, but a statistically significant
reduction was found only after two TTS-
scopolamine. The predictability of the efficacy
of the drugs on nausea was best in the caloric
test, followed by rotatory test and lowest in
optokinetic test. Based on the tests about 50 per
cent of the efficacy of the drugs could be
predicted. The results indicate that the drugs
effective against motion sickness reduce the
nystagmic response, which at least partly
explains the mode of action of the drugs.

INTRODUCTION

The pharmacological effect of antimotion drugs such as
scopolamine and dimenhydrinate seems to be linked to their
anticholinergic rather then antihistaminic activity (Chinn &
Smith, 1955; Brand & Perry, 1966). The vestibular nuclei seem
to be the most attractive effector organ for these drugs,
since most of the cells in the vestibular nuclei are
cholinergic and can be selectively inhibited with scopolamine
(Jaju et al., 1970; Matsuoka et al., 1973). Also
dimenhydrinate seems to block the conduction of the impulses
in the vestibular nuclei (Jaju & Wang, 1971). Contradictory
data have been published by Sekitani et al. (1971) and McCabe

(1973), who found no changes in the neural activity of the vestibular nuclei during systemic administration of scopolamine and dimenhydrinate. However, the results are not conclusive since the responses were recorded under barbiturate anesthesia.

In the present work, caloric, rotatory and optokinetic nystagmus were studied in healthy volunteers after transdermal application of scopolamine (Transdermal Therapeutic System of Scopolamine, TTSscopolamine) (Scopoderm, Ciba-Geigy) and dimenhydrinate. The aim of this study was to examine the effects of the two drugs on different types of experimantaly induced nystagmus in man and to examine whether the degree of reduction of nystagmus would correspond to the clinical effect of the drugs.

SUBJECTS AND METHODS

Sixteen paid volunteers with a mean age of 27.5 years (range 2138) who had participated in the previous study (Pyykkö et al. 1983 a) were included.

The following drugs were evaluated: (1) One TTS-scopolamine, programmed to deliver approximately 5μg scopolamine base per hour. (2) Two TTS-scopolamine, each delivering approximately 5 μg scopolamine base per hour. (3) Dimenhydrinate in a capsule containing 100 mg dimenhydrinate and 50 mg caffeine. TTSscopolamine was applied retro-aurally 12 h before each test. Two dimenhydrinate capsules were given, one 12 h and the other one hour before the test. The drugs were given in a double blind, dummy blind design and randomized regarding test order with a Latin Square schedule.

Methods

Eye movements were recorded binocularly in a horizontal plane, using an electro-oculographic technique and analysed with a computer system (Digital Equipment PDP 11-23).

The caloric test was conducted by irrigating the left ear with 350 ml water of 30 degrees C for 90 sec.The subjects was in a recumbent position and the head was supported at 30 degrees. The test was conducted with eyes open in the dark. The maximum velocity for the nystagmus was calculated and were determined for each 10 sec intervals during the response course.

During the rotatory test the subject was seated in a chair accelerating within one second to a speed of 120 degrees/sec. After one minute the chair was brought to a standstill within one second. The test was conducted in the dark, eyes open, both clockwise and counterclockwise. the computer calculated gain (maximum velocity of slow phase of nystagmus/velocity of the chair). The mean value of the responses to right and left was used for each test occasion.

In the optokinetic test a whole-field drum was rotated around
the subject with a constant velocity of 90 degrees/sec for a
period of 60 sec. The subject was instructed to fix the gaze
of the drum. The computer calculated the mean velocity of all
nystagmus beat detected during a period of 60 sec. the mean
value of optokinetic nystagmus to right and left for each
person and occasion was used as a measure of individual
response.

Statistics

The difference between the responses following different
treatments was examined by analysis of variance a.m.
Friedmann. Wilcoxon's signed mid-rank test was used to detect
the level of significance in pairwise comparison. The
relationship between the efficacy of the drugs on nausea and
reduction of nystagmus was sought in linear regression
analysis. Difference between samples were considered
statistically significant when $\underline{p}$ <0.05.

RESULTS

1. Caloric test

The maximum velocity of the slow phase in caloric nystagmus
was 29 deg/sec placebo treatment. During treatment with
dimenhydrinate the maximum velocity was 23 deg/sec, during
one TTS-scopolamine 26 deg/sec and during two TTS-scoplamine
21 deg/sec (table 1). The reduction after active drug-
treatment were statistically differentiated from the
reduction seen after placebo-treatment (F(3)=10.01, p<0.05).
In pairwise comparison, statistically significant differences
were found between placebo vs. dimenhydrinate (p<0.05),
placebo vs. one TTS-scopolamine (p<0.05), and placebo vs. two
TTS-scopolamine (p<0.01). There was no statistically
significant difference in the reduction in caloric nystagmus
between dimenhydrinate and one or two TTSscoplamine.

2. Rotatory test

The mean gain in nystagmus induced by angular acceleration in
the dark was 0.75 during treatment with placebo. The
respective values were 0.67 during treatment with
dimenhydrinate , 0.74 with one TTS-scpoplamine and 0.56 with
two TTS-scopolamine. The changes in gain between different
treatments were statistically significant (F(3)=17.025,
p<0.001). In pairwise comparison the difference between
placebo and dimenhydrinate (p<0.05), and placebo and two
TTSscoplamine (p<0.001) were statistically significant, but
not those between placebo and one TTS-scopolamine. A dose-
response relationship existed between one TTS-scopolamine and
two TTS-scopolamine (p<0.01). Furthermore treatment with two
TTS-scopolamine caused a greater reduction in gain than did
dimenhydrinate (p<0.05).

3. Optokinetic test

The mean velocity of optokinetic nystagmus (OKN) during placebo treatment was 54 deg/sec. The respective values during treatments were 50 deg/sec during dimenhydrinate, 48 deg/sec during one TTSscopolamine and 35 deg /sec during two TTS-scopolamine. No statistically significant difference were observed between the treatments. Nevertheless, in pairwise comparison a statistically significant reduction was observed after treatment with two TTSscopolamine vis-a-vis placebo (p<0.01). Furthermore, a dose response relationship existed between one TTS-scopolamine vs two TTSscopolamine (p<0.05).

4. Predictability of the efficacy of the drugs on nausea based on reduction in nystagmus

In the base line test the individual sensitivity to get motion sickness was related to the intensiveness of the vestibular responses in the rotatory test. The subjects who scored high nausea values in Coriolis manoeuvre (Table 1) had high gain in the rotatory test. The relationship was statistically highly significant (r=0.58, p<0.001).

Fig. 2 exhibits the the correlation between reduction of caloric responses and nausea during treatment with two scopolamine (r=0.56, p<0.01). Based on the result of the caloric test the linear regression analysis could predict about 31 per cent of the respective reduction in nausea.

The correlation between reduction of gain and reduction in nausea in Coriolis manoeuvre during treatment with two TTS-scopolamine was statistically not significant (r=0.36, p=n.s.) The predictability of the efficacy of the drug on nausea was hence not as quite as good in the rotatory test (13 per cent) as in the caloric test.

In prediction of the efficacy of the drugs on nausea in the optokinetic test , OKN turned out to be a poor predictor (fig. 3). No statistically significant correlation between reduction in nausea and reduction in OKN could be ascertained (r=0.02, p=n.s.).

Partial correlation coefficiants were determined to evaluate the predictability of the efficacy of the drugs on nausea from different tests. For the caloric test the partial correlation coefficient was essentially unchanged (r= 0.51, p<0.05), whereas in the rotatory test (r=0.25, p=n.s.) and in the optokinetic test (r=0.04, p=n.s.) it was reduced . Consequently, the best predictability for the efficacy of this achieved by the caloric test whereas the other tests do not significantly contribute to this estimation.

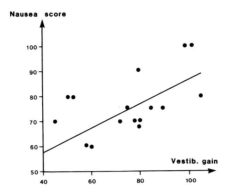

Fig.1. Relationship between vestibular gain and magnitude of nausea in Coriolis test.

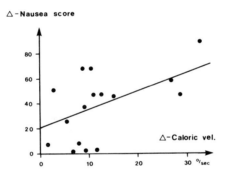

Fig.2. Relationship between changes in caloric nystagmus (max. velocity during placebo - max. velocity during drug) and in nausea (placebo score - drug score) during treatment with two TTS-scopolamine.

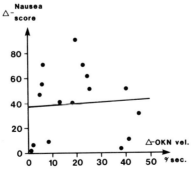

Fig.3. Relationship between changes in optokinetic nystagmus (mean velocity during placebo - mean velocity during drug) and in nausea (placebo score - drug score) during treatment with two TTS-scopolamine.

426   I. Pyykkö et al.

TABLE   I   Changes   in   caloric,   rotatory   and optokinetic
             nystagmus during treatment with different drugs and
             the   effect   of   the   drugs   on   nausea in Coriolis
             manouvre. Mean and   standard   error   of   mean   are
             given.

	Placebo	Dimenhydrinate	One TTS-scopo	Two TTS-scopo
Caloric	30	25	26	21
max vel	3.2	2.9	3.0	2.4
deg/sec				
Rotatory	75	67	74	56
gain	4.5	4.8	4.7	4.3
Optokinetic	54	50	48	35
mean vel	5.1	5.8	4.9	4.0
deg/sec				
Nausea	61	18	40	23
score	7.1	5.9	7.3	5.8

DISCUSSION

The effect of transdermally administered scopolamine and
dimenhydrinate on caloric, angular acceleration induced and
optokinetic nystagmus was studied in a double blind trial.
Treatment with the active drugs caused a systematic reduction
in the different types of nystagmus. Treatment with two TTS-
scopolamine caused the most consistent reduction, whereas
responses after one TTSscopolamine and dimenhydrinate varied
from test to test. The degree to which response diminished
due to a drug, according to subjective evaluation of
experimental nausea and vertigo (Pyykkö et al., 1983 a) was
of a similar magnitude to the reduction in nystagmus and
could be partly predicted from the reduction of the
nystagmus. Hence, presumably efficacy of the antimotion
sickness drugs at least partly depends on their inhibitory
action on vestibular and visual reflexes.

Inhibition of vestibular responses can be caused by reducing
the activity of vestibular end organ, of the vestibular
nuclei or of the higher postural centers. In animal
experiments an inhibition caused by scopolamine has not been
observed to affect the projection of the peripheral nerve or
of the pontine reticular formation (Jaju and Wang, 1971;
Matsuoka et al., 1975). Furthermore, supertentorial brain
structures are not necessary for the genesis of motion
sickness (Bard et al., 1947). Hence the plausable site for
inhibition of the orientation reflexes caused by the motion
sickness remedies is in the vestibular nuclei.

The major effect of motion sickness remedies might not be an
inhibition of the afferent influx as such. Reduction of the
reflexes could also be achieved if the integration between
semicircular canals and otolith afferents and visual and
vestibular afferents would be impaired. A convergence
controlling semicircular canal and otolith reflexes has been
shown to occur in the vestibular nuclei (Fluur and Siegborn,

1973; Graybiel and Lackner, 1980). In fact, in present results the best prediction for the efficacy of the motion sickness remedies was obtained from caloric responses during which mismatching information is achieved from semicircular canal and otolith afferents. Furthermore, inhibition of optokinetic responses can be explained by reduction of the activity of the neurons integrating vestibular and visual influx in the vestibular nuclei (Waespe and Henn, 1979).

In decreasing the neural activity in the vestibular nuclei the motion sickness remedies seem to assist the central nervous system to govern orientation in space during different stressfull situations like those occurring during overexcitation of a receptor system or when conflicting postural information is provided from different sensory channels. In this effort the motion sickness remedies seem to provide alleviation of the symptoms until a full recovery has been achieved by habituation.

REFERENCES

Bard, P., Woolsey, C.N., Snider, R.S., Mountcastle, V.B. & Bromiley R.B. 1947. Delimitation of central nervous mechanisms involved in motion sickness. Abstract. Fed Proc 6, 72.
Brand, J.J. & Perry, W.L.M. 1966. Drugs used in motion sickenss. Pharmacol Rev 18, 895-924,
Chinn, H.I. & Smith, P.K. 1955. Motion sickness. Pharmacol Rev 7, 33-82.
Fluur, E. & Seigborn, J. 1973. Otholith oragns and nystagmus problem Act. Oto-Lar 76 (6) 438-442.
Jaju, B.P., Kirsten, E.B. & Wang, S.C. 1970. Effects of belladona alkaloids on vestibular nucleus of the cat. Am J Physiol 219, 1248-1255.
Jaju, B.P. & Wang, S.C. 1971. Effects of diphenhydramine and dimenhydrinate on vestibular neuronal activity of cat: A search for the locus of their antimotion sickness action. J Pharmacol Exp Ther 176, 718-724.
Matsuoka, I., Domina, E.F. & Morimoto, M. 1973. Adrenergic and cholinergic mechanisms of single vestibular neurons in the cat. Adv Otorhinolaryngol 19, 163-178.
Matsuoka, I., Domino, E.F. & Morimoto, M. 1975. Effects of cholinergic agonists and antagonists on nucleus vestibularis lateralis unit discharge to vestibular nerve stimulation in the cat. Acta Otolaryngol (Stockh) 80, 422-428.
McGabe, B.F. 1973. Central aspects of drugs for motion sicknes and vertigo. Adv Otorhinolaryngol 20, 458-469.
Pyykkö, I., Schalen, L. & Jäntti, V. 1983 a . Transdermally administered scopolamine vs. dimenhydrinate. I. Effect on nausea and vertigo. (In preparation.)
Sekitani, T., McGabe, B.F. & Ryu, J.H. 1971 a. Drug effectsa on the medial vestibular nucleus. Arch Otolaryngol 93, 581-589.
Waespe, W. & Henn, V. 1979. The velocity response of vestibuldar nucleus neurons during vestibular, visual and combined angular acceleration. Exp Brain Res 37, 337-347.

Theoretical and Applied Aspects of Eye Movement Research
A.G. Gale and F. Johnson (Editors)
© Elsevier Science Publishers B.V. (North-Holland), 1984

NEURAL MECHANISMS OF CONVERGENCE AND ACCOMMODATION

B.G. Cumming and S.J. Judge

University Laboratory of Physiology
Parks Road, Oxford
ENGLAND OX1 3PT

Experiments with the awake trained monkey and with human
subjects show that in binocular viewing the dynamic
responses of both vergence and accommodation are determined
predominantly by the disparity stimulus. In the steady
state this is probably not the case, but instead blur cues
provide most of the drive to accommodation and much of that
to vergence. There are neurones in the visual cortex which
could provide the necessary disparity and blur signals but
how and where these sensory signals are processed to produce
the motor signals seen on brainstem neurones is unknown.

INTRODUCTION

To maintain useful binocular vision, man and other primates need both to
converge their eyes and to focus on the object of interest. Moreover,
although these two responses might be thought of as separately governed by
the need to minimize binocular disparity and blur respectively, they are in
fact intimately linked. Accommodation to a monocularly viewed target
induces convergence ('accommodation vergence')(25), and convergence induces
accommodation ('vergence accommodation') even when the target is viewed
binocularly through pin-holes so that blur is unavailable as a cue for
accommodation (10). Functionally, a link between the two responses makes
sense, because there is a nearly linear relationship between the convergence
angle and the dioptric power required to fixate an object.

The classical view of the coordination of accommodation and vergence is
derived from Maddox (19). He was unaware of the existence of vergence
accommodation, and supposed that accommodation was driven entirely by blur.
Vergence was deemed to consist of four additive components: tonic,
accommodative, fusional (i.e disparity-driven) and psychic (i.e due to
knowledge of the target distance). Accommodation vergence was thought to
produce most of the drive to vergence, with the disparity-driven component
providing only a supplement to maintain fusion. Maddox has been enormously
influential, and continues to be so, despite rigorous demonstrations of the
existence of convergence accommodation and of the excellent performance of
vergence in the absence of accommodative drive.

Fincham and Walton (10) proposed a different theory in which accommodation
was entirely the result of vergence, which was itself produced partly by
binocular disparity and partly by unspecified voluntary processes. Semmlow
(37), elaborating an earlier suggestion of Westheimer's (39), has argued
that both disparity and blur-driven components are important in the control

of accommodation and convergence.

Although much is known about accommodation and vergence behaviour from studies of human performance (1, 37, 38), comparatively little is known about the neural pathways and mechanisms that support the behaviour (22). We believe that behavioural and electrophysiological recording experiments with awake trained monkeys, together with associated anatomical studies, are the most promising ways to obtain such knowledge. We will describe some of the preliminary results of such experiments in the course of summarizing what is already known about accommodation and vergence control.

ACCOMMODATION AND VERGENCE PERFORMANCE

Blur-driven and vergence accommodation

Monocularly, at least, accommodation is a sluggish response. The latency in man is about 370 msec and the responses to a step change are roughly exponential with a time constant of about 250 msec (5). It is usually assumed that the poor dynamic response of accommodation is determined by the properties of the lens and the ciliary muscle ('the plant').

Accommodation performance has rarely been studied during binocular viewing. We have discovered (7) that in the monkey we have studied, the dynamics of accommodation are dramatically better binocularly than monocularly (Fig. 1), suggesting that the assumption that it is the plant that limits the dynamic response may be wrong. It should be noted that the gain and phase of the monocular response of our monkey were as good as those in humans (17), making it unlikely that there is anything abnormal about our monkey's monocular performance.

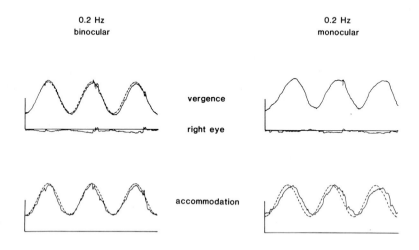

Figure 1. Comparison of dynamic accommodation changes during binocular and monocular viewing of a target moving in depth (2-6 dioptres). Dashed lines stimuli, solid lines responses. Note the substantially smaller phase lag in binocular viewing.

We have also measured vergence accommodation gain and phase in the monkey for frequencies between 0.1 and 1.2 Hz. The phase and gain of vergence accommodation are excellent. In particular the phase properties are much better than those of (monocular) blur-driven accommodation measured in the same monkey with the same apparatus. In other words, in dynamic binocular tracking, disparity cues provide the main drive to accommodation (as they do for vergence).

Vergence

Vergence movements have a shorter latency (160 msec in man according to (30)) and a better frequency response (17) than blur-driven accommodation. Rashbass and Westheimer (30) showed that a step disparity input produces a linear ramp output (in the open loop situation). The slope of the ramp is proportional (with a gain of about 10 deg/sec per degree of disparity) to the size of the disparity step. In other words the system behaves like a simple integrator. We (Cumming and Judge, unpublished observations) have shown that this is also true in the monkey. In both man and monkey, however, the phase of the response to sinusoidal movement in depth is much more accurate than that of a model system with an integrator as the open-loop transfer function. According to Rashbass and Westheimer, the phase improvement is not a consequence of the predictability of the target movement because it does not occur with periodic square wave movements. This implies that the vergence system is sensitive to the velocity of disparity changes, provided the velocity is less than some upper limit. In a step change, therefore, the velocity cue is ineffective but its effect is significant during sinusoidal stimulation at frequencies below 1 Hz.

Accommodation vergence

The vergence associated with accommodation to a monocularly viewed target is linearly related to accommodation. Clinical surveys of large numbers of patients (1) find a population mean gain for accommodation vergence (the actual vergence achieved, divided by the vergence needed to correctly align the eyes during binocular viewing) of about 0.8 in young subjects.

During tracking of monocularly viewed targets moving sinusoidally in depth, accommodation and accommodation vergence are of very similar phase (17). Like accommodation then, accommodation vergence is slower than disparity vergence, and one would therefore expect it to contribute little to dynamic binocular tracking. This interpretation is supported by the observations of Venkiteswaran and Semmlow (37) showing that the dynamics of vergence are similar whether or not accommodation is open-loop.

Prism adaptation

Many studies (e.g. 6, 24, 34) have shown that after wearing a prism in front of one or both eyes so that an additional amount of vergence is required, there is a persistent positive aftereffect: when tested monocularly the subject has more convergence than usual for a given accommodative state. Schor (35) has shown that this adaptation can be described by a slow integrator in parallel with the basic (fast) integrator in the vergence pathway. The slow integrator must be driven from the output of the fast vergence integrator before the accommodation vergence input is added because Schor (35) has shown that vergence produced by accommodation does not induce a prism aftereffect. Judge and Miles (unpublished observations) demonstrated that this is equally true in the monkey.

Schor (36) also argues that the vergence accommodation cross-link must come from the output of the fast vergence integrator, rather than from total vergence output, since prism adaptation does not alter dark focus (26). Miles and Judge's (23) finding that prism adaptation causes a downward displacement of the vergence accommodation relationship has the same implication.

If one incorporates these observations about prism adaptation into an interactive model of the neural control of accommodation and vergence (Fig. 2), then the model predicts several effects other than prism adaptation (36). Consider, for example, a subject who is of the type who adapts nearly completely to prisms. The model predicts in this case that the slow integrator will continue to charge until nearly all disparity-driven vergence is the result of its action. The output of the fast disparity integrator will be very small, and as a consequence vergence accommodation will also be minimal. It is not clear whether or not this is true.

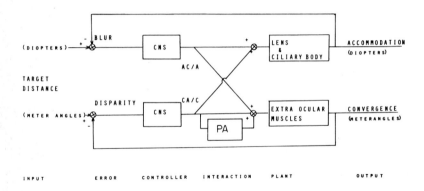

Figure 2. Block diagram illustrating interactions between accommodation and vergence. PA: slow integrator producing prism adaptation. CA/C: vergence accommodation cross-link. AC/A: accommodation vergence cross-link. From Schor (36).

Periscope adaptation

Despite trenchant statements in textbooks and reviews that the gain of accommodation vergence cannot be altered, Judge and Miles (14) have shown that half an hour of experience viewing the world through periscopic spectacles which displace the effective position of each eye about 5 cm laterally, increases the accommodation vergence gain by about 50%. Although these measurements were made by subjective techniques, Miles and Judge (23) obtained the same result when accommodation was monitored with a laser speckle optometer. They also showed that periscopic spectacle viewing decreased the vergence accommodation gain by a comparable degree. Judge and Miles (unpublished observations) have also shown that the accommodation vergence gain of monkeys is raised by short periods of periscopic spectacle experience.

NEUROANATOMY AND NEUROPHYSIOLOGY

Disparity input

Many cells in the striate cortex have been found to respond most strongly to binocular visual stimuli with particular disparities or ranges of disparity. Such cells have been recorded both in the cat (4, 9, 27), and the monkey (28, 29). Other cells in visual cortex have been found which respond to motion in depth (8, 29, 42). Thus cells exist which could be used to supply both information about static disparity and rate of change of disparity to the vergence system.

Sakata et al. (32) reported a category of cells in the posterior parietal cortex of the monkey whose activity is modulated according to the depth of fixation, but they favoured a sensorimotor integration role rather than a motor control function for these cells.

Blur input

Nearly nothing is known about the neural mechanisms that signal blur. Naturally, almost any contrast sensitive visual cell will alter its firing as blur diminishes contrast: this principle is sometimes used by visual neurophysiologists as to establish that an anaesthetised animal is optimally corrected. However, there is no evidence that any particular group of cells have a special importance in the control of accommodation. Bando et al. (3) found cells in the Clare-Bishop area of the anaesthetized cat which fired prior to spontaneous fluctuations of accommodation in darkness, but this observation does not seem to have been followed up yet. Jampel (11) reported that stimulation of a region of the cortex (area 19 and 22) in the anaesthetised monkey produced all aspects of the near response, but this study has not yet been replicated in alert monkeys.

Accommodation output

The ciliary muscle is innervated by fibres from the ciliary ganglion. Because painting the ciliary ganglion with nicotine abolished the pupil response but not the accommodation response to brainstem stimulation, Westheimer and Blair (40) suggested that at least some of the fibres innervating the ciliary muscle pass straight through the ganglion without synapsing there, but this was not rigorously proven.

Bando et al. (2) recorded from 30 cells related to spontaneous fluctuations of accommodation in the brainstem of the anaesthetized cat. The peak discharge rate of the cells preceded the onset of the accommodation response by 270 msec on average. Lovasik and Beauchamp (18) recorded from ten cells in the brainstem of alert cats. These cells responded to accommodation induced by placing lenses in front of the right eye of the animal. Unfortunately, neither of these studies report adequate controls to exclude the possibility that the cells were actually related to vergence.

In the monkey, it has been claimed on the basis of electrical stimulation (12) that the anteromedian nucleus is related to accommodation and the Edinger-Westphal nucleus to the pupil but the demonstration was not so clear as to leave the issue beyond doubt.

Vergence output

It is believed that there is no special set of motoneurones for vergence or
other types of slow eye movement (15, 16). Schiller (33) found seven
neurones dorsolateral to the caudal part of the oculomotor nucleus whose
discharge rate appeared to be proportional to the angle of convergence.
More recently Mays (21, 22) has recorded such cells systematically. They
were not motoneurones. The cells responded both during disparity and
accommodation vergence. Most of the cells increased their firing rate with
convergence but a small number of cells decreased their firing rate with
convergence. The cells were dorsal and dorso-lateral to the oculomotor
nucleus, which suggests that some were not Edinger-Westphal neurones, but
Mays is careful to say that he cannot exclude the possibility that they were
related to accommodation.

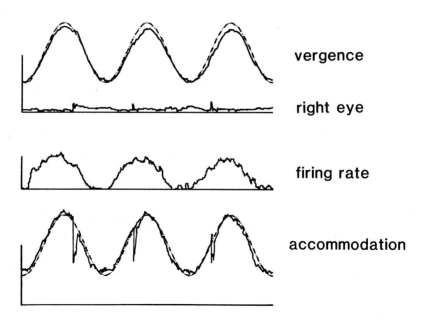

Figure 3. Smoothed firing rate of a brainstem vergence
unit (HER139.10) during target motion in depth (2-6
dioptres, 0.2 Hz, binocular viewing). Vertical
calibration on unit trace is 100 spikes/sec. On
vergence and accommodation traces, solid lines are
responses, dashed lines stimuli. Transient deflections
on accommodation response trace are blink artefacts.

We (13) have recorded from about 80 neurones related to the near response.
See Fig. 3 for an example of the firing pattern of one of these neurones.
Our cells, like Mays', are both dorsal and dorso-lateral to the oculomotor
nucleus. We have been able to characterise some of them as related to
accommodation rather than vergence because we measure the monkey's
accommodation with a dynamic infra-red optometer and give the animal tasks

(prism viewing or periscope viewing) that partially separate the normal tight coupling between accommodation and vergence. We are not yet able to say whether the accommodation related cells occur predominantly near the midline, with the vergence cells more lateral. There does not seem to be a rigid segregation of the two types of cell, because on some recording penetrations we have encountered both types of cell close togther. What is clear is that almost all the cells we have tested fire equally well in normal viewing, accommodation vergence and vergence accommodation. In other words, almost all the cells we see look like output elements, suggesting that the integration of blur-driven and disparity-driven components does not occur in the brain stem.

Prism adaptation

According to (31), Milders and Reinecke (unpublished observations) have found that some patients with cerebellar lesions do not show prism adaptation. Judge and Miles (unpublished observations) tested two monkeys with lesions of the flocculus of the cerebellum, a structure which is known to be critical for some (but not all) types of oculomotor plasticity (41). The monkeys did show some prism adaptation after 30 minutes of prism wearing, but the effects were much smaller than in two control monkeys.

In the brainstem we have not seen cells suitable to form the basis of Schor's slow integrator. This negative observation should be interpreted very cautiously, because we could easily have missed cells with such a long time constant: the difficulty is that during our searching routine of observing a neurone's firing during a brief period of tracking a target moving in depth such cells might well hardly modulate at all.

CONCLUSION

There are neurones in the visual cortex which could provide the information about disparity needed to control the near response, and one has to assume that any contrast sensitive cells could be used to detect blur, but we lack evidence that particular groups of cells or projections from them do constitute the disparity and blur inputs for the near response.

The brainstem contains neurones subserving the final output stages of accommodation and vergence, but our preliminary results suggest that it may be necessary to look elsewhere for the cells which mediate the synkinesis of the near response. We are currently investigating the possibility that such cells may be found in the prestriate cortex of the monkey.

Supported by M.R.C. Programmme Grant 979/49 to Prof. C. Blakemore.

REFERENCES

[1] Alpern, M., Types of eye movement, in: Davson, H. (ed.), The Eye, Vol. 3, 2nd edn., pp 65-174 (Academic Press, New York, 1969).
[2] Bando, T., Tsukuda, K., Yamamoto, N., Maeda, J and Tsukahara, N., Mesencephalic neurons controlling lens accommodation in the cat, Brain Res. 213 (1981) 201-204.
[3] Bando, T., Tsukuda, K., Yamamoto, N., Maeda, J. and Tsukahara, N., Cortical neurones in and around the Clare-Bishop area related with lens accommodation in the cat, Brain Res. 225 (1981) 195-199.

[4]   Barlow, H.B., Blakemore, C. and Pettigrew, J.D., The neural mechanism
      of binocular depth discrimination, J. Physiol. 193 (1967) 327-342.
[5]   Campbell, F.W. and Westheimer, G., Dynamics of accommodation
      responses of the human eye, J. Physiol. 151 (1960) 285-295.
[6]   Carter, D.B., Fixation disparity and heterophoria following prolonged
      wearing of prisms, Am. J. Optom. 42 (1965) 141-151.
[7]   Cumming, B.G. and Judge, S.J., Dynamics of vergence eye movement and
      accommodation in the monkey, J. Physiol. in press 1983.
[8]   Cynader, M. and Regan, D., Neurons in cat parastriate cortex
      sensitive to the direction of motion in three dimensional space,
      J. Physiol. 274 (1978) 549-569.
[9]   Ferster, D., A comparison of binocular depth mechanisms in areas 17
      and 18 of cat visual cortex, J. Physiol. 311 (1981) 623-655.
[10] Fincham, E.F. and Walton, J., The reciprocal actions of accommodation
      and vergence, J. Physiol. 137 (1957) 488-508.
[11] Jampel, R.S., Convergence, divergence, pupillary reactions and
      accommodation of the eyes from faradic stimulation of the macaque
      brain, J. Comp. Neurol. 115 (1960) 371-399.
[12] Jampel, R.S. and Mindel, J., The nucleus for accommodation in the
      midbrain of the macaque, Invest. Ophthal. 6 (1967) 40 - 50.
[13] Judge, S.J. and Cumming, B.G., Brainstem neurones related to
      accommodation and vergence, Soc. Neurosci. Abstr. in press 1983.
[14] Judge, S.J. and Miles, F.A., Gain changes in accommodative vergence
      induced by alteration of the effective interocular separation,
      in: Fuchs, A.F. and Becker, W. (eds), Progress in oculomotor
      research pp 587-594 (Elsevier, Amsterdam, 1981).
[15] Keller, E.L., Accommodation vergence in the alert monkey. Motor unit
      analysis, Vision Res. 13 (1973) 1565-1575.
[16] Keller, E.L. and Robinson, D.A., Abducens unit behaviour in the
      monkey during vergence eye movements, Vision Res. 12 (1972) 369-382.
[17] Krishnan, V.V., Phillips, S. and Stark, L., Frequency analysis of
      accommodation, accommodative vergence and disparity vergence,
      Vision Res. 13 (1973) 1545-1554.
[18] Lovasik, J.V. and Beauchamp, R., Ocular accommodation neurones in
      brainstem of the alert cat, Am. J. Optom. 59 (1982) 785-794.
[19] Maddox, E.E., The clinical use of prisms,
      (John Wright and Sons, Bristol, 1893).
[20] Maunsell, J.H.R. and Van Essen, D.C., Functional properties of
      neurons in middle temporal visual area of the Macaque monkey.
      II. Binocular interactions and sensitivity to binocular disparity,
      J. Neurophysiol. 49 (1983) 1148-1167.
[21] Mays, L.E., Neuronal correlates of vergence eye movements,
      Soc. Neurosci. Abstr. 7 (1981) 133.
[22] Mays, L.E., Neurophysiological correlates of vergence eye movements,
      in: Schor, C.M. and Ciuffreda, K.J. (eds.), Vergence eye movements:
      basic and clinical aspects, pp 647-670 (Butterworth, Boston, 1983).
[23] Miles, F.A. and Judge, S.J., Optically-induced changes in the neural
      coupling between vergence eye movements and accommodation in human
      subjects, in: Lennerstrand, G., Zee, D.S. and Keller, E., (eds.),
      Functional aspects of ocular motility disorders, pp 93-96
      (Pergamon, Oxford, 1982).
[24] Mitchell, A.M and Ellerbrock, V.J., Fixation disparity and the
      maintenance of fusion in the horizontal meridian,
      Am. J. Optom. 32 (1955) 520-534.
[25] Mueller, J., Elements of Physiology, Vol II. Baly, W.
      (Transl.). (Taylor and Walton, London, 1842).

[26] Owens, D.A. and Leibowitz, H.W., Accommodation, convergence and distance perception in low illumination, Am. J. Optom. 57 (1980) 540-550.

[27] Pettigrew, J.D., Nikara, T. and Bishop, P.O., Binocular interaction on single units in cat striate cortex: simultaneous stimulation by single moving slits with receptive fields in correspondence, Exp. Brain Res. 6 (1968) 391-410.

[28] Poggio, G.F. and Fischer, B., Binocular interaction and depth sensitivity of striate and prestriate cortical neurons of behaving rhesus monkeys, J. Neurophysiol. 40 (1977) 1392-1405.

[29] Poggio, G.F. and Talbot, W.H., Mechanisms of static and dynamic stereopsis in foveal cortex of the rhesus monkey, J. Physiol. 315 (1981) 469-492.

[30] Rashbass, C. and Westheimer, G., Disjunctive eye movements, J. Physiol. 159 (1961) 339-360.

[31] Robinson, D.A., Control of eye movements, in: Handbook of Physiology, The Nervous System, Section 1, Vol. 2, pp 1275-1320 (American Physiological Society, Bethesda, 1981).

[32] Sakata, H., Shibutani, H. and Kawano, K., Spatial properties of visual fixation neurons in posterior parietal association cortex of the monkey, J. Neurophysiol. 43 (1980) 1654-1672.

[33] Schiller, P.H., The discharge characteristics of single units in the oculomotor and abducens nuclei of the unanaesthetized monkey, Exp. Brain. Res. 10(1970) 347-362.

[34] Schor, C.M., The influence of rapid prism adaptation upon fixation disparity, Vision Res. 19 (1979) 757-765.

[35] Schor, C.M., The relationship between fusional vergence eye movements and fixation disparity, Vision Res. 19 (1979) 1359-1367.

[36] Schor, C.M., Fixation disparity and vergence adaptation, in: Schor, C.M. and Ciuffreda, K.J. (eds.), Vergence eye movements: basic and clinical aspects, pp 647-670 (Butterworths, Boston, 1983).

[37] Semmlow, J.L., The oculomotor near response, in: Zuber, B. (ed.), Models of oculomotor behaviour (Chemical Rubber, New York, 1981).

[38] Toates, F.M., Accommodation function of the human eye, Physiol. Rev. 52 (1972) 828-863.

[39] Westheimer, G., Amphetamine, barbiturates and accommodative convergence, Arch. Ophthalmol. 70 (1963) 830-836.

[40] Westheimer, G. and Blair, S.M., The parasympathetic pathways to the internal eye muscles, Invest. Ophthalmol. 12 (1973) 193-197.

[41] Zee, D.S., Yamazaki, A., Butler, P.H. and Gucer, G., Effects of ablation of flocculus and paraflocculus on eye movements in primate, J. Neurophysiol. 46 (1981) 878-899.

[42] Zeki, S.M., Cells responding to changing image size and disparity in cortex of the rhesus monkey, J. Physiol. 242 (1974) 827-841.

Footnote

1. Reprinted by permission of the publisher, from Schor/Ciuffreda: Vergence Eye Movements: Basic and Clinical Aspects. Woburn: Butterworth Publishers, 1983.

Section 8

**PSYCHOPATHOLOGICAL AND
CLINICAL ASPECTS**

Theoretical and Applied Aspects of Eye Movement Research
A.G. Gale and F. Johnson (Editors)
© Elsevier Science Publishers B.V. (North-Holland), 1984

PSYCHOPATHOLOGICAL AND CLINICAL ASPECTS

INTRODUCTION

Deborah L. Levy and John M. Davis,
Neurophysiology Laboratory Department of Research,
Illinois State Psychiatric Institute, Illinois, Chicago, U.S.A.

The papers in this section, like those in the rest of this volume, attest
to the robustness of claims on behalf of the ocular motor systems as powerful indices of brain function. The authors represent diverse disciplines,
but each investigator is fundamentally a neurobiologist who capitalises on
the unique window to the brain provided by the study of eye movements.

Neurologists and bioengineers have recognised the value of studying eye
movements for many years. This tradition of linking ocular motor control
to neurological function and of describing the dynamic characteristics of
eye movement systems is reflected in the papers by Johnston, Lass et al.,
Bronstein and Kennard and Buizza et al.

Johnston describes the two forms of inattention that affect the regulation
of visual-motor behaviour. Comparing right hemisphere stroke patients with
normal control, he shows that degree of oculomotor impersistence, or fixation instability, is proportional to severity of visual hemi-inattention and
postulates a similar underlying mechanism.

Starting with the premise that the perceptual and conceptual consequences
of left and right hemisphere brain damage should differ, Lass et al. characterise the information processing of patients with these two conditions.
Using a sequence of changing visual stimuli, they examine the effects of
modality (picture or word) and structure (partial or global changes) on
performance. In addition to using different strategies for processing visual information, aphasic patients are significantly influenced by modality,
whereas for non-aphasics, structure has a dominant effect.

Focusing on a different clinical syndrome, Parkinson's Disease, Bronstein
and Kennard document manifestations of akinesia in the saccadic and smooth
pursuit systems. For both random and predictable target motion saccadic
reaction time is delayed and smooth pursuit phase lag is increased among
parkinsonian patients. Apparently due to compromised anticipatory mechanisms, the predictability of target motion that normally improves ocular
motor precision does not produce significant compensation. These findings
are interpreted as support for separate neural mediation of predictive and
random saccades in the frontal eye fields and superior colliculus, respectively.

Buizza et al. present data on the range of linearity of the smooth pursuit
system. Their findings, that the gain characteristics of ramp responses
are amplitude-dependent, clarify conflicting reports about the saturation
level of the smooth pursuit control system. Sinusoidal target displacements
in contrast, yield gains that remain linear at velocities up to 75 degrees/
second and are not amplitude-dependent for excursions up to 50 degrees.

In psychiatry and psychology the importance of ocular motor dysfunctions in
the pathophysiology of the major psychoses is a relatively new development

stemming from the work of Philip Holzman and coinciding with a heightened appreciation of the relationship between psychopathological conditions and brain function.

Approaching the study of eye movements from the vantage point of cognitive psychology, Done and Frith apply the automatic-strategic paradigm to the study of saccades in schizophrenics. Like other investigators, they find normal saccadic reaction times in schizophrenics, but this result holds true only for "automatic" saccades. In the case of "strategic" saccades, the reaction times of schizophrenics are slowed compared to those of normal controls. The authors also report similar characteristics of the saccadic and psychomotor reaction time responses of schizophrenic patients.

Andersson presents data that further document the presence of impaired pursuit in psychotic patients. Comparisons of psychotic patients and alcoholics yield similarities that strengthen the link between pursuit disruption and an abnormality of brain function, although different loci are implicated in the two disorders.

Levy et al. summarise longitudinal evidence on the effects of psychotropic drugs on smooth pursuit eye movements. They show a clear association between pursuit impairment and treatment with lithium carbonate, in contrast to the independence of pursuit efficiency and treatment with antipsychotic and antidepressant medication. These findings implicate different aetiologies for impaired pursuit in affective disorder patients taking lithium and in schizophrenics.

Theoretical and Applied Aspects of Eye Movement Research
A.G. Gale and F. Johnson (Editors)
© Elsevier Science Publishers B.V. (North-Holland), 1984

VISUAL HEMI-INATTENTION AND OCULOMOTOR IMPERSISTENCE [1]

Cris W. Johnston

Division of Child & Adolescent Psychiatry
University of Minnesota Hospitals
Minneapolis, Minnesota
U.S.A.

The ability to maintain steady fixation for short periods
of time was found to decrease as the severity of visual
hemi-inattention increased in right-hemisphere stroke
patients and matched normal control subjects. Performance
was best in the central area of the visual-spatial field
and worst in peripheral areas. It is suggested that the
paradigm employed reveals the manifestation of a "general"
inattention phenomenon that shares the same underlying
mechanism as hemi-inattention.

INTRODUCTION

The purpose of this study was to discover if the severity of visual hemi-
inattention (VHI) is related to the ability to maintain steady fixation,
and also to determine whether performance in maintaining steady gaze on
specific targets varies from one spatial field to another.

Hemi-inattention is a term used to describe a phenomenon observed in
acutely lesioned unilaterally brain-damaged individuals. It refers to the
failure of such people to report, respond, or orient to stimuli occurring
contralateral to the lesioned hemisphere (Heilman, 1979). This symptom has
been shown to occur in the tactile, auditory, and visual modalities, and has
also been related to asymmetric extremity-motor functioning (Friedland &
Weinstein, 1977). It is generally believed to be most common in right-
hemisphere brain-damage (Mesulam, 1981).

VHI may be demonstrated by asking the patient to perform any of a number of
tasks that require the appreciation of left and right spatial fields.
Letter cancellation tests (e.g. , Diller, Ben-Yishay, Gerstman, Goodkin,
Gordon, & Weinberg, 1974) , and visual matching tests (e.g., Piasetsky, 1981)
are among the more sensitive and better researched methods of demonstrating
VHI. On such exercises the patient exhibits a tendency to miss items on the
left, or to show a bias for selecting or identifying targets the farther to
the right they appear.

The only controlled study of oculomotor behavior in patients with VHI known
to the present author was performed by Chedru, Leblanc, and Lhermitte (1973),
and their measure of interest did not explain the hemi-inattentive behavior
of mildly symptomatic subjects. These investigators discovered asymmetries
in Exploration Time for left and right spatial fields , but only for subjects
with marked hemi-attention. While the most severe cases rarely directed
their gaze into the poorly attended field, the milder cases appeared to show

no such asymmetry. Part of the problem of the limited discriminability for
the parameter Exploration Time may have been due to the lack of a strongly
operationalized definition for hemi-inattention. For example, minor inattention was said to exist when "omissions or displacements in tests of
copying sketches or of writing" (p. 95) were present, yet this criterion was
not elaborated. Moreover, the investigators lamented that their dependent
variable was a rather gross one, suggesting that research of more fundamental aspects of oculomotor behavior was needed.

In attempting to relate oculomotor performance to the hemi-inattentive behavior exhibited by patients with mild VHI, one might postulate that
although these individuals look in the appropriate area, perhaps the manner
in which they visually sample is ineffective. This may be termed "looking
without seeing" (cf., Llewellyn Thomas, 1976), and is probably identifiable
in eye movement studies only by using more precise parameters than Exploration Time.

In 1956 Fisher suggested the term "ocular vacillation" to denote the poor
ability of brain-lesioned patients to maintain steady gaze on specific
targets. This is the oculomotor manifestation of a general phenomenon
commonly known as motor impersistence. Subsequent reports have shown ocular
vacillation, or oculomotor impersistence, to be a symptom of right-
hemisphere or diffuse brain-damage more often than of left-hemisphere
damage (Jenkyn, Walsh, Culver, & Reeves, 1977; Joynt, Benton, & Fogel,
1962). If an individual has difficulty maintaining fixation on an intended
target one possible consequence might be an interference in cognitive processing due to some deficiency or abnormality in the content of a cognitive
unit (see Russo, 1978). The possibility that an asymmetry in oculomotor
impersistence exists that corresponds to the lateralized symptoms evident in
VHI has not previously been studied.

Nevertheless, various explanations have been offered to account for motor
impersistence in general, and oculomotor impersistence in particular. Among
them are notions that it represents a "psychological need to maximize sensory input in an individual who is finding it increasingly more difficult to
maintain control over and contact with the environment" (Jenkyn et al.,
1977, p. 964), that it is a variant of apraxia (Fisher, 1956), that it constitutes the emergence of a primitive "vigilance" mechanism (Berlin, 1955),
or that it represents an impairment of spatial-sensory functioning (Carmon,
1970).

When one considers the last explanation in the context of spatially-related
deficits characteristic of VHI two obvious questions arise: (1) is the
severity of oculomotor impersistence related to the severity of VHI, and (2)
in patients with VHI does severity of oculomotor impersistence improve as the
measure is made farther from the field for which the person is most inattentive? The predicted answer to both questions was yes.

METHOD

Three groups of subjects were studied with 10 subjects in each group. Two
of the groups were composed of right-hemisphere stroke patients. One of
these two groups had marked VHI and the other had mild VHI as determined by
scores on tests described below. The third group consisted of control subjects without a history of neurological disease, and who were equivalent to

the stroke patients in terms of age, visual acuity, and visual field integrity. Fourteen of the 30 subjects were female, with a greater proportion of women found in the control group than in the stroke patient groups. Exclusionary criteria for all subjects included any history of debilitating psychiatric illness, history of drug or alcohol abuse, strabismus, uncorrected visual acuity less then 14/84 at reading distance, or below average scores on Vocabulary and Digit Span subtests from an IQ test (Wechsler, 1981). Control subjects did not have a history of central nervous system neurological disease, and stroke patients did not have a history of central nervous system neurological disease prior to the stroke for which they were presently hospitalized. In the case of the stroke patients unilaterality of lesion was based on evidence from neurological examinations and laboratory test reports such as CT scans. Furthermore, due to symptom instability in the very acute stages following stroke (Kertesz, 1979) only those patients who were at least four weeks post CVA were included in the study.

The previously mentioned cancellation and visual matching tests (Diller et al., 1974; Piasetsky, 1981) were administered in order to yield an index of severity of VHI that could range from 0.0 to 1.0. Calculation of the VHI index is discussed elsewhere (Johnston, 1983). A score of .5 corresponds to no inattention, whereas a score increasingly greater than .5 is proportionately related to the degree of left VHI. The determination of marked and mild groups was based on rank-ordering the stroke patients according to their VHI scores and performing a median split.

The eye movement data were obtained using a photo-electric limbus tracking device mounted on a pair of spectacle frames. Since this apparatus measures vertical eye position from eyelid movement, only data from horizontal measures were used in the oculomotor analyses. In the horizontal mode this instrument has a resolution of .25 degrees and provides a linear response up to 20 degrees to either side of the vertical meridian.

A head-mounted scene holder was used for stimulus presentation (Johnston & Hutton, 1979). This is an apparatus designed specifically for use with subjects who find being restrained for sustained periods of time more objectionable than the slight weight of the headgear. A 3X3 array of tricolor light-emitting diodes (LEDs) mounted in the scene holder provides calibration targets (as well as the possibility of choice reaction time experiments). A drop-cloth may be suspended in front of the subject's eyes which, upon removal, activates a photo-cell marking the onset of scene exposure. Scenes are mounted on light-weight artist's tagboard to be easily slid in and out of the scene holder channels. All signals are recorded on FM tape for subsequent off-line analysis.

The stimuli that were used for the measure of oculomotor impersistence consisted of an array of crosses (embossed on tagboard) placed at positions corresponding to the calibration lights. Crosses were used rather than the LEDs in order to provide very discrete points to fixate (i.e., the intersection of two lines). The distance between crosses in the horizontal plane was 16 degrees; in the vertical plane it was 10 degrees. This 3X3 grid allows for partitioning responses into left, center, and right spatial fields (as well as top, middle, and bottom if desired). Rather than using the drop-cloth to mark the onset of a trial, subjects were instructed to point to a given target and then drop their hand, but to keep staring at the target until told to stop. This procedure yielded greater control over trial onset by verifying when subjects located each target. Ten seconds following the subjects' fixation of the target the same instruction was given to fixate a new target. The sequence of target fixation was counter-balanced according to spatial field independently for the control group and the stroke patients. The same sequence was run through twice for each sub-ject, once at the beginning of the eye movement session, and once approxi-mately 45 minutes later at the end of the eye movement session. The interim involved participation in other oculomotor tasks to be discussed in forth-coming publications.

The data were tape-recorded and played back off-line on strip chart paper for analysis of the total extent of eye movement for each target fixated. Extent of eye movement was obtained by using an analog-to-digital pen to trace over each 10 second epoch of eye position beginning with the moment the target was identified by the subject. The measure thus provided was termed Fixation Stability performance, and represents a numerical index of the extent of oculomotor impersistence a subject manifests. The higher the score, the worse the Fixation Stability performance was considered to be.

RESULTS AND DISCUSSION

Mean VHI scores were .75 (s.d.=.10) for the marked group, .58 (s.d.=.04) for the mild group, and .52 (s.d.=.05) for the normal control group. Although non-directional paired t-tests revealed statistically significant differ-ences in the mean VHI scores between each combination of groups ($p<.01$ for each comparison), this was not the case with more robust analyses for mul-tiple group comparisons. That is, even though the overall one-way analysis of variance (ANOVA) was highly significant (F=28.75, df=2/27, $p<.001$), Student-Newman-Keuls procedure revealed no significant difference between the mild and control groups' mean VHI score. However, significant differ-ences were found between the marked and mild groups, and the marked and control groups ($p<.05$ in each case).

Fixation Stability performance, as evaluated by a two-way ANOVA for group by spatial field (i.e., left, center, and right) revealed a significant effect for group (F=4.83, df=2/27, $p<.05$), and a highly significant effect for spa-tial field (F=10.70, df=2/54, $p<.001$). No interaction between group and spatial field was demonstrated.

The group effect--in other words, the effect for severity of VHI--was as ex-pected, and is further borne out by the finding of a significant positive correlation between Fixation Stability and VHI score for all 30 subjects combined ($r=.65$, $p<.001$ based on a two-tailed t-test).

Figure 1 depicts the consistency of the relationship of Fixation Stability to severity of VHI as it is broken down by spatial field. In each spatial field the marked group performed worse than the mild group, and the mild group performed worse than the control group. Since no interaction was demonstrated between group and spatial field, the data were collapsed across spatial fields and evaluated for differences between the various pairs of groups. This comparison revealed that the use of Fixation Stability perfor-mance is of no greater value in statistically discriminating different levels of severity of VHI than was the Chedru et al. (1973) gross measure, Exploration Time. Mean Fixation Stability scores were 10.26 (s.d.=2.54), 9.13 (s.d.=3.56), and 6.78 (s.d.=.76) for the marked, mild, and normal con-trol groups, respectively. Based on Student-Newman-Keuls procedure, the

only significant difference demonstrated by Fixation Stability for all spatial fields combined was between the marked and control groups ($\underline{p}$<.05),
thus reflecting no better discriminability than Exploration Time.

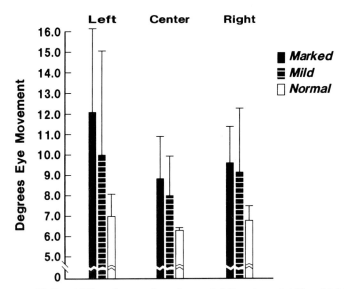

Fig. 1 Fixation stability performance for each group in left, center, and right spatial fields.

To place this result in proper perspective, two important points should be
recalled. First, based on Student-Newman-Keuls test the mild and normal
control groups did not differ significantly in VHI scores. Second, the
simple dichotomization of the 20 brain-damaged subjects diluted the difference in mean VHI scores that might have been obtained by some other cate-
gorization scheme. The fact remains that, as previously noted, a statistically significant positive correlation does exist between severity of
oculomotor impersistence and VHI.

When groups were collapsed and performance in left, center, and right spatial fields was analyzed, an unexpected result was obtained. Performance
was best in the center field ($\bar{x}$=7.78, s.d.=1.32), slightly worse in the
right field ($\bar{x}$=8.65, s.d.=1.50), and worst of all in the left field ($\bar{x}$=9.76,
s.d.=2.56). Significant differences were found for each pairwise comparison based on Student-Newman-Keuls procedure (p<.05 in all cases); however,
the greatest difference (note the mean scores) was between the left and
center fields, not the left and right fields as originally predicted. This
unexpected finding was also consistent across groups (see Figure 1).

The unexpected nature of the spatial field effect prompted the question,
what--if any--are the differences in Fixation Stability performance between
the upper, middle, and lower spatial fields? A two-way ANOVA of group differences (i.e., marked, mild, and normal control) according to this

alternate breakdown of spatial field (i.e., upper, middle, and lower)
revealed a statistically significant effect for spatial field (F=3.89,
df=2/54, p<.05), as well as the aforementioned group effect. There was no
interaction effect.

As before, in each of these spatial fields the marked group performed worse
than the mild group, who in turn performed worse than the control group (see
Figure 2). The nature of the relationship between performance in the upper,
middle, and lower spatial fields was also similar to the relationship
between performance in the left, center, and right spatial fields. For
collapsed groups, Fixation Stability was best in the middle field (x̄=8.34,
s.d.=1.97, slightly worse in the lower field (x̄=8.78, s.d.=2.74), and worst
in the upper field (x̄=9.07, s.d.=2.46). However, based on Student-Newman-
Keuls procedure, a statistically significance difference was obtained only
between the upper and middle spatial fields (p<.05). This may be, in part,
accounted for by the smaller distance separating spatial fields in the vertical plane than in the horizontal plane (10 degrees vs. 16 degrees,
respectively).

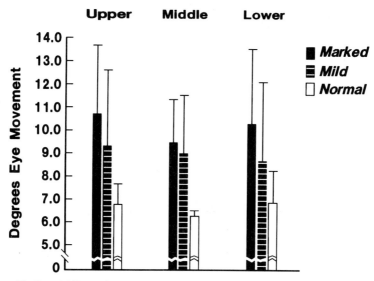

Fig. 2 Fixation stability performance for each group in upper, middle, and lower spatial fields.

In summary, these findings reveal that the degree of Fixation Stability, or
oculomotor impersistence is proportional to severity of VHI, and that Fixation Stability performance is optimal for the central area of a scene,
across all levels of severity of VHI. Although the first finding is what
was predicted, the nature of the spatial field finding is not what was expected. That is, it was thought that in patients with VHI Fixation
Stability performance would improve as the task was moved farther away from
the spatial field for which these subjects were inattentive. In other words,
it was thought that oculomotor impersistence should have been least severe

in the right most spatial field, not the center field, as was found.

One possible explanation is that maintaining steady fixation on a target so simple as a cross is not a cognitively demanding task, and that in such a situation a "general" inattention may be operating in which the subject exhibits a tendency to return gaze to the resting area of central fixation. If the subject is already fixating centrally, the tendency to divert gaze to the resting position is diminished. Moreover, it has been well documented that the likelihood of demonstrating "hemi"-inattentive symptoms is strongly related to the complexity of the task demands (e.g., Piasetsky, 1981).

To test this explanation, an additional analysis was performed to reveal whether, in fact, subjects demonstrated a tendency to divert their line of sight toward the center more often than toward the periphery. For stimuli in the left and right spatial fields a tabulation was made of the number of saccades two degrees or more in length that were made toward the center versus the number that were made in a more lateral direction. A t-test for related measures revealed significantly more eye movements were made toward the center (t=3.57, df=29, $\underline{p}$<.001, one tailed test). This tendency to return gaze to central fixation on cognitively simple tasks may be viewed as poorly self-regulated behavior at the level of general attention.

The problem that remains is to find a parsimonious explanation for the unexpected relationship just described that also encompasses the relationship of Fixation Stability performance to severity of VHI. This is a problem because these relationships are evidently noninteracting. In other words, the behavior that was used to yield an index for VHI has a virtually continuous left to right spatial field gradient, whereas Fixation Stability does not; but at the same time overall Fixation Stability performance is clearly positively correlated with severity of VHI.

A unifying interpretation of these findings resides with the possibility that this paradigm is tapping into a fundamental mechanism that governs self-regulatory visual-motor behavior in general. It is suggested that this mechanism is common to the more specific spatially graded inattentive behavior known as hemi-inattention, as well as to a more general inattention where the tendency is simply to return gaze to the resting area of central fixation. The more disturbed this self-regulatory system is, the stronger will be the hemi-inattention behavior for cognitively demanding tasks, and also the stronger will be the general inattention behavior for cognitively simple tasks.

[1]The author acknowledges Leonard Diller for helpful comments during the formulation and completion of this study, and for providing a stimulating environment at the NYU Institute of Rehabilitation Medicine in which to collect the data. Appreciation is also extended to Susan Erbaugh and Barry Garfinkel who read and commented on an earlier draft of this paper.

REFERENCES

(1)   Berlin, L., Compulsive eye opening and associated phenomena, Arch. Neurol. Psychiat. (Chic.) 73 (1955) 597-601.

(2)   Carmon, A., Impaired utilization of kinesthetic feedback in right hemispheric lesions, Neurology 20 (1970) 1033-1038.

(3)   Chedru, F., Leblanc, M., and Lhermitte, F., Visual searching in normal and brain damaged subjects: Contribution to the study of unilateral inattention, Cortex 9 (1973) 94-111.

(4)   Diller, L., Ben-Yishay, Y., Gerstman, L., Goodkin, R., Gordon, W., and Weinberg, J., Studies on Cognition and Rehabilitation in Hemiplegia: Rehabilitation monograph, No. 50 (New York Univ. Med. Ctr., New York, 1974).

(5)   Fisher, C.M., Hemiplegia and motor impersistence, Journal of Nervous and Mental Diseases 122 (1956) 20-29.

(6)   Friedland, R.P., and Weinstein, E.A., Hemi-inattention and hemisphere specialization: Introduction and historical review, in: E.A. Weinstein and R.P. Friedland (eds.), Hemi-Inattention and Hemisphere Specialization (Raven Press, New York, 1977).

(7)   Heilman, K.M., Neglect and related disorders, in: K.M. Heilman and E. Valenstein (eds.), Clinical Neuropsychology (Oxford U. Press, New York, 1979).

(8)   Jenkyn, L.R., Walsh, D.B., Culver, C.M., and Reeves, A.G., Clinical signs in diffuse cerebral dysfunction, Journal of Neurology, Neurosurgery, and Psychiatry 40 (1977) 956-965.

(9)   Johnston, C.W., Oculomotor Behavior and Visual Hemi-Neglect, Ph.D. Thesis, Dept. of Psychology, City Univ. of New York (September 1983).

(10) Johnston, C.W., and Hutton, J.T., A head mounted scene holder for eye movement research, Behavior Research Methods and Instrumentation 11 (1979) 422-426.

(11) Joynt, R., Benton, A., and Fogel, M., Behavioral and pathological correlates of motor impersistence, Neurology 12 (1962) 876-881.

(12) Kertesz, A., Recovery and treatment, in: K.M. Heilman and E. Valenstein (eds.), Clinical Neuropsychology (Oxford U. Press, New York, 1979).

(13) Llewellyn Thomas, D., Discussion comment, in R.A. Monty, and J.W. Senders (eds.), Eye Movements and Psychological Processes (Lawrence Erlbaum Associates, Inc., Hillsdale, 1976).

(14) Mesulam, M., A cortical network for directed attention and unilateral neglect, Ann. Neurol. 10 (1981) 309-325.

(15) Piasetsky, E., A Study of Pathological Asymmetries in Visual-spatial Attention in Unilaterally Brain-damaged Stroke Patients, Ph.D. Thesis. Dept. of Psychology, City Univ. of New York (June 1981).

(16) Russo, J.E., Adaptation of cognitive processes to the eye movement system, in: J.W. Senders, D.F. Fisher, and R.A. Monty (eds.), Eye Movements and the Higher Psychological Functions (Lawrence Erlbaum Associates, New York, 1978).

(17) Wechsler, D., The Wechsler Adult Intelligence Scale-Revised (Psychological Corporation, New York, 1981).

(18) Weinstein, E.A., and Friedland, R.P. (eds.), Hemi-Inattention and Hemisphere Specialization (Raven Press, New York, 1977).

Theoretical and Applied Aspects of Eye Movement Research
A.G. Gale and F. Johnson (Editors)
© Elsevier Science Publishers B.V. (North-Holland), 1984

EYE MOVEMENTS AND RECOGNITION OF SEQUENCES OF CHANGING OBJECTS IN
THE PICTORIAL VERSUS VERBAL MODALITY

Uta Lass[1], Walter Huber[2], Gerd Lüer[1]

[1]Institute of Psychology, University of Göttingen
[2]Department of Neurology, Technical University of
Aachen, West Germany

Left and right brain-damaged patients as well as normal controls
were asked to find a sequence of changing objects in a multiple
choice set. During scanning, we registered their duration, frequency and alteration of gaze. The test material was systematically varied with respect to modality (picture/word) and structure of change (partial/global). The results show a modality effect for the aphasic group and a structure effect for the non-
aphasic groups. Combination of target objects required significantly more gaze alterations than selection, except for aphasic
patients when they searched for globally changing objects in the
written version. Different combination strategies were found for
words and pictures.

Introduction

The study started from two general assumptions on hemispheric specialization
in human beings that are deducted from behavioral deficits in brain-damaged
patients. A modality effect, in particular a differential impairment of verbal versus visuo-spatial processing, is well documented in many clinical
studies. Given parallel tasks, left hemisphere patients with aphasia have
more trouble with the verbal than with the visual task demands whereas the
reverse is found in right brain-damaged patients (cf. e.g., Warrington and
Taylor 1973, Goodglass et al 1974, Huber and Gleber 1982).In such studies,
the two groups of patients should be comparable with respect to etiology as
well as size and site of lesion.

There are also a few studies that relate assymmetries of hemispheric dysfunction to differences in stimulus structure. Veroff (1978) reported that
aphasic patients find it difficult to arrange pictures representing catego-
rial changes of an object (e.g. a tadpole growing into a frog) as opposed
to pictures representing changes in spatial relations of an object (e.g.,
a golf ball rolling towards the hole), which lead to difficulties in right
hemisphere patients. It remained, however, unresolved whether the differential impairment reflected a difference in perceptual or conceptual processing.

In visual matching tasks, global versus local features of the stimuli seem
to be differentially related to normal right versus left hemisphere functions (Cohen 1973, Patterson and Bradshaw 1975, Martin 1979, Polich 1980).
In these latter studies the contrast of global versus local (or partial)
structure was investigated for states and not for processes as in Veroff's
study. It can be objected that the recognition of processes requires conceptual and not only perceptual analysis. The more such a conceptual task al-

lows for verbal mediation, the more a dissociation between hemispheric functioning will disappear and a general left hemisphere advantage/disadvantage
will emerge.

A shortcoming of Veroff's study was that her results were based on overall
latency and errors taken from small subject samples and a small set of items.
The replication of a similar task by using the registration of eye movements
can avoid the psychometric problems and gives the opportunity of observing
more directly cognitive strategies that lead to a certain solution.

In our study, left and right brain-damaged patients as well as normal controls had to recognize sequences of changing objects from a multiple choice
set. The material was systematically varied with respect to pictorial er-
sus verbal modality and partial versus global structure of change. U   hypothesis in its strongest form predicts an interaction between groups and
both the modality and the structure factor.

Furthermore, the multiple choice paradigm allows for the control of alternative sorting strategies. Finding a correct solution requires both selection
and combination of target objects. The two steps of the analysis should be
reflected by differences in eye movements. Gaze alterations between target
and distractor objects should indicate selection, whereas gaze alterations
between target objects should indicate combination. We hypothezised that decisions on selection should be more intensively controlled with the globally
than the partially changing objects. Decisions on combination should, however, require about the same intensity of control for each type of change.

METHODS

1. Subjects

Included in the study were 15 left-hemisphere (LH) patients with aphasia, 8
patients with right-hemisphere (RH) damage and 15 normal controls. Median
age of the three groups was 51 (range 20 - 61), 48 (32 - 58), 49 (25 - 57),
respectively. Distribution of sex and educational background was comparable
across the three groups. All neurological subjects were outpatients. Their
etiology was mainly vascular, median duration of brain-damage was 6 months
(1 - 24) for the aphasic group and 17 months (3 - 24) for the right-hemisphere group. In both groups, the lesions as assessed by computer tomography
were restricted to the distribution area of the middle cerebral artery. Type
of aphasia was classified by means of the Aachen Aphasia Test (AAT, Huber et
al 1983). The aphasic group included 2 patients with global, 1 with Wer-
nicke's, 6 with Broca's, 4 with amnesic and 2 with non-classifiable aphasia.
In most patients (12), the severity of aphasia was moderate to mild.

2. Material

There was a verbal and a pictorial version of the task. From a multiple
choice set of five, subjects were asked to select those pictures/noun phrases that represent a sequence of changing objects. The number of target objects was always three. The propositional structure (the type of change) was
varied in the pictorial and written version of the task in the same way. The
defining features of the target sequence were either global or partial ones,
as illustrated in the following examples:

Partial changes
(1a) example from written version

target objects	distractor objects
angefangener Brief/'started letter'	weggelegte Zeitung/'put-away newspaper'
halbfertiger Brief/'half-finished letter'	
beendeter Brief/'completed letter'	gelesenes Buch/'read-through book'

(2a) example from pictorial version

leerer Luftballon/'empty balloon'	zerrissener Teddybär/'torn teddy bear'
aufgeblasener Luftballon/'blown-up balloon'	zerbeulter Kreisel/'bumped spinning top'
geplatzter Luftballon/'burst balloon'	

Global changes
(1b) example from written version

dickes Schaf/'fat sheep'	alte Schuhe/'old shoes'
bunte Wolle/'coloured wool'	neuer Hut/'new hat'
warmer Pullover/'warm pullover'	

(2b) example from pictorial version

(großer) Baum/'(big) tree'	Wanne/'bath-tub'
(gestapelte) Bretter/'(stapled) planks'	Eimer/'bucket'
(breiter) Schrank/'(large) cupboard'	

In the written version, partial changes were expressed by variation of adjectives in nounphrases. The same syntactic structure was used for expressing global changes. But recurrance of identical nouns was, of course, only given in the stimuli for partial changes. The pictorial version was parallel in both aspects: There was recurrance of identical objects and there was variation in attributive features of the changing objects.

Besides the three target objects, each multiple choice set contained two distractor objects that belonged to the same semantic category, e.g. reading material, clothing, toys, containers in the examples (1) and (2). With partial changes, each of the three target objects, and with global changes, only the final target object was also an exemplar of this category, although a less prototypical one. Thus, the linguistic structure allowed for the control of conflicting semantic sorting strategies.

There were three items for each of the four parameters. The items were selected from a pool of items that was pretested with students of speech-pathology. In this pilot study, we also found a slight gradation in difficulty among the experimental parameters: partial changes were easier than global ones, pictures easier than words.

3. Procedure

In a pretraining session, subjects became aquainted with the kind of sorting task we used. During the experimental session, the four blocks of three items were given in the same order as in the above examples (1) and

(2). Before each block, the individual noun-phrases/pictures of all three
items were presented in random order. For the aphasic subjects, the noun-
phrases were read aloud by the examiner. By this procedure, we wanted to
make subjects familiar with the stimuli, thus minimizing the influence of
word length, word frequency, reading difficulty and pictorial ambiguity. All
stimuli were presented on slides and the duration of projections was determined by the subjects themselves. The subjects were told to push a button
when they felt sure about the correct solution. The aphasic subjects could
give the solution either verbally or by pointing. The whole testing, including calibration of eye movements of the individual subjects, lasted about
30 - 45 minutes.

The objects (noun-phrases/pictures) were presented in a circular array. The
position of target and distractor objects was systematically varied across
items. The subject was seated 190 cm away from the screen (100 x 75 cm)with
his head fixed. Viewed from this position, the midpoints of the five objects
were separated from one another by a visual angle of at least 7°. Thus, during fixation of one object adjacent objects were out of focus. One gaze
was defined as an uninterrupted sequence of fixations upon the same object
position, defined as an ellipse (30.1 x 19.1 cm) around the object. The stimuli words were written in capital letters with a height of 2 cm. The eye
movements were measured by tracking the corneal reflection center with respect to the pupil center via a video camera (cf. Young & Sheena 1975). The
x and y coordinates were calculated every 40 milliseconds. The coordinates
were recorded and monitored by computer (system DEBIC 80).

RESULTS

1. Decision time and total gaze duration

We compared the actual decision time to the total of gaze duration in order
to detect possible effects of the evaluation method, i.e. the restriction
to eye movements in predefined screen position. Both sets of data were first
submitted to a generalizability study (Cronbach et al 1972), which showed
that the three individual items of each experimental parameter did explain
virtually nothing of the overall variance. (The same result was found for
the other eye movement data we are going to report on.) We therefore felt
justified to sum up across the three items of each block.

The data were then analyzed by a split-split-plot ANOVA (BMDP 4V, Dixon &
Brown 1977) separately for total decision time and for gaze duration. Groups
were taken as plot-factor, structure of change (partial/global) and modality
(word/picture) as the two split-factors. In order to get the most detailed
information, the ANOVA was carried through to the level of simple-simple
main effects.

The results were essentially the same for both sets of data. There was no
significant three-way interaction, but there were two significant two-way
interactions: group x modality, structure x modality. However, none of the
corresponding simple interactions proved to be significant. All significant
simple-simple main effects of the two split factors are marked in table 1,
which gives the mean values for decision time and gaze duration. The modality difference influenced only the aphasic group, whereas the difference in
structure influenced both right-hemisphere patients and the controls. There
were also effects of the group factor, which were further analysed by Mann-
Whitney pairwise comparisons: Throughout, aphasics had significantly higher
values than the controls except for processing pictures with partial changes,

where no differences among groups were found.

means (msec)	APHASICS (n=15)		RH - PTS (n=8)		CONTROLS (n=15)	
	WORD	PICT	WORD	PICT	WORD	PICT
PARTIAL	2770	2160	1720	2150⌝	1100	1250⌝
	1920	1360	1160	⌐1500	690	⌐770
GLOBAL	3180	2120	1680	1180⌟	1000	610⌟
	2240	1300	1150	└ 700	620	└ 330

⌐‾‾⌐ significant simple-simple main effect, ANOVA, p≤.003; α=5/16%

Table 1
Decision time and total gaze duration

The task was rather easy to perform even for the brain-damaged patients as
their rather short decision times and their good accuracy of response shows
(cf. table 2). In a similar metalinguistic sorting task (Huber et al 1983),
which, however, involved syntactic processing, we found for aphasic patients
decision times ranging from 18 - 239 seconds and no correct response.

mean errors (max = 3)	APHASICS (n=15)		RH - PTS (n=8)		CONTROLS (n=15)	
	WORD	PICT	WORD	PICT	WORD	PICT
PARTIAL	0.9	1.3	0.1	1.0	0.1	0.4
GLOBAL	1.8	1.1	0.3	0.3	0.5	0.1

Table 2
Accuracy of response

2. Gaze duration and gaze frequency

Relative gaze durations as well as relative gaze frequencies were summed up
for the three target and the two distractor objects. The data were statistically assessed by four-way ANOVA, with group as the plot-factor and structure, modality, type of object (target/distractor) as the three split-factors. The results were almost identical for the duration and frequency data.
There were no significant four-way and three-way interactions. Only the two-
way interactions group x type of object and modality x type of object were
found to be significant. Contrary to expectation, there were no significant
two-way interactions between group and either structure or modality. The
ANOVA was again carried through to the simplest level of analysis. The corresponding mean values are reported in table 3. All significant simple-simple-simple main effects are marked. The most consistent finding were the significantly lower duration (and frequency) values for distractor than for target objects. An exception is the aphasic group when processing global
changes in the written version. A simple-simple interaction between modality
and type of object under the global processing condition was also found for
the control group. But for them, the difference between target and distractor objects remained significant both for pictures and words.

means in %		GLOBAL		PARTIAL	
		WORD	PICT	WORD	PICT
APHASICS (n=15)	T	63.1*	71.5	72.9*	72.7
	D	36.9°	28.5	27.1°	27.3
RH - PTS (n=8)	T	71.2	75.4	69.3	75.6
	D	28.8	24.6	30.7	24.4
CONTROLS (n=15)	T	68.0	81.2	71.8	80.1
	D	32.0	18.8	28.2	18.9

* * } significant simple-simple-simple main effect, ANOVA, p≥.0011;
o o } α=5/44%
r--¬ close to significant

Table 3
Mean gaze durations on target (T) and distractor (D) objects
(corrected for differences in base rate (3:2))

Next, we considered possible differences between the three target objects.
We distinguished objects with initial, medial and final position in the target sequence. The data were again analysed by a four-way ANOVA, using the
same design as before. For both, gaze duration and gaze frequency, the factor 'type of object' turned out to be in significant interaction with each
of the other factors, namely x groups, x structure, x modality. All significant simple-simple-simple main effects and significant pairwise comparisons
are reported together with the mean values of gaze duration in table 4.

means in %		GLOBAL		PARTIAL	
		WORD	PICT	WORD	PICT
APHASICS (n=15)	I	19.0 < 25.2°		23.4 < 31.9	
	M	24.4* > 23.1		34.5*s> 17.6	
	F	19.7 < 23.3		15.0 < 23.2	
RH - PTS (n=8)	I	23.9 < 31.4		26.3 < 32.6	
	M	29.3 > 18.1		30.8 > 21.7	
	F	18.0 < 26.0		12.3 < 21.3	
CONTROLS (n=15)	I	24.8 <s 42.4°		26.5 <s 42.6	
	M	26.5 s> 15.8		29.0 s> 16.9	
	F	16.8 < 23.1		16.2 < 20.7	

s> <s } significant simple-simple-simple main effect (ANOVA) or pair-
o o } wise comparison (MANN-WHITNEY, WILCOXON Test), p≤.0008;
* * } α=5/57%

Table 4
Mean gaze duration on initial (I), medial (M) and final (F)
objects of the target sequence (corrected as in Table 1)

The most prominent finding is that the medial target object was most intensively looked at with words, but the initial object with pictures. As a consequence, the medial object was more often and longer processed with words
than with pictures, whereas a reversed tendency held for the initial and final object. These differences led to the significant interaction between
type of object and modality.

3. Gaze alterations

For pairs of objects, relative frequencies of gaze alterations were summed
up in both directions. The analysis was done in three steps: First, we considered the relative number of alterations between any two target objects,
between any target and distractor object, and between the two distractor
objects. Second, we determined the number of alterations for each possible
pair of target objects as defined by the position in the target sequence:
initial - medial, initial - final, medial - final. Third, we assessed the
number of alterations between each single target object and the two distractor objects. Statistical analysis was done by three-way MANOVA with
groups as the plot-factor, structure (global/partial change) and modality
(word/picture) as the two split-factors. Types of alteration were taken as
components of the multiple dependant variable.

In the first analysis, the only significant interaction was between the two
split-factors. There was no main effect for the plot factor. Tests for simple main effects (cf. table 5) showed a significant influence of the modali-

MANOVA		TYPES OF GAZE ALTERATION		
simple main effect of	overall	$T \leftrightarrow T$	$T \leftrightarrow D$	$D \leftrightarrow D$
STRUCTURE on word	.0133	.0132	.0216	.0554
on picture	.3049	.5983	.3999	.2843
MODALITY on partial	.1242	.1000	.0882	.2298
on global	.0000	.0000	.0000	.1064

underlinings indicate significant p-values, $p \leq .0125$, $\alpha = 5/4\%$

	GLOBAL CHANGES					
	$T \leftrightarrow T$		$T \leftrightarrow D$		$D \leftrightarrow D$	
means in %	WORD	PICT	WORD	PICT	WORD	PICT
APHASICS	32.7	51.5	46.2	32.6	21.1	15.9
RH - PTS	46.7	57.2	39.7	28.4	13.6	14.4
CONTROLS	39.0	57.3	44.0	32.7	17.0	10.0

Table 5
Influence of modality and structure on gaze alterations
among and between target (T) and distractor (D) objects
(corrected for differences in base rate (3:6:1))

ty factor (word vs. picture) on the processing of global but not of partial
changes. This was found for alterations between two target objects as well
as between a target and a distractor object. The relative frequencies of al-

terations (cf. table 5) make clear that the differences between processing
words and pictures hold in opposite directions. Target-target alterations
were higher for pictures than for words but target-distractor alterations
were lower. These differences were found in each of the three groups.

In the second statistical analysis, no significant interactions were found
but a significant main effect for the split-factor 'modality'. The simple
main effects on each level of the factor 'structure of change' and the corresponding means of relative frequencies are reported in table 6. Alterations between initial and final objects of the target sequences were significantly more frequent in the pictorial than in the verbal version of the
task.

MANOVA	TYPES OF GAZE ALTERATION			
simple main effect of	overall	I ↔ M	M ↔ F	I ↔ F
MODALITY on partial ch.	.2922	.4540	.3854	.0567
on global ch.	.0000	.5325	.0504	.0000

underlinings indicate significant p-values, p≤.0125, α = 5/4%

	GLOBAL CHANGES					
	I ↔ M		M ↔ F		I ↔ F	
means in %	WORD	PICT	WORD	PICT	WORD	PICT
APHASICS	10.4	14.6	8.4	12.0	13.9	25.0
RH - PTS	16.5	16.5	8.7	10.2	21.5	30.6
CONTROLS	13.7	13.4	6.8	11.4	18.5	32.5

Table 6
Influence of modality on gaze alterations between
target objects in initial (I), medial (M) and final (F) position
(corrected as in table 5)

Finally, we considered the frequencies of alterations between single target
objects and the two distractor objects. MANOVA yielded again a significant
two-way interaction for the two split-factors, but no significant main effect for the plot-factor. Simple main effects and corresponding mean frequencies are given in table 7. As expected, alterations between the final
target and distractor objects were significantly higher under the global
than under the partial condition. Contrary to expectation, this was only
true for words and not for pictures as can be seen on vertical comparison
of mean values in table 7. With partial changes, there were significant differences between processing words and pictures for two types of alteration,
but in different direction. Alterations that involved the medial object of
the target sequence were more frequent for words than for pictures. But alterations between the final object and a distractor object were more frequent for pictures. With global changes, the same finding could reliably
be established only for alterations between the medial object and a distractor object.

MANOVA		TYPES OF GAZE ALTERATION			
simple main effect of		overall	I ↔ D	M ↔ D	F ↔ D
STRUCTURE on word		.0063	.4918	.9324	.0005
on picture		.5374	.5483	.2750	.4713
MODALITY on partial ch.		.0005	.0200	.0048	.0053
on global ch.		.0000	.0613	.0001	.3715

underlining indicate significant p-values, p≤.0125, α = 5/4%

	I ↔ D		M ↔ D		F ↔ D	
means in %	WORD	PICT	WORD	PICT	WORD	PICT
		PARTIAL	CHANGES			
APHASICS	11.2	8.7	22.3	10.5	9.3	11.3
RH - PTS	12.9	9.9	17.5	14.6	5.2	11.4
CONTROLS	15.8	9.4	18.7	13.3	6.6	8.8
	└─ ns ─┘		└─ s ─┘		└─ s ─┘	
		GLOBAL	CHANGES			
APHASICS	13.9	8.6	19.0	14.1	13.3	9.9
RH - PTS	15.0	11.1	16.4	9.1	8.3	8.1
CONTROLS	12.3	12.1	20.4	10.9	11.3	9.7
	└─ ns ─┘		└─ s ─┘		└─ ns ─┘	
PARTIAL vs GLOBAL	ns	ns	ns	ns	s	ns

Table 7
Influence of structure and modality
on gaze alterations between target and distractor objects
(corrected as in Table 5; abbreviations see Table 5 and 6)

In order to get information on the relationship between the various types of gaze alterations, rank profiles were calculated. Tests for equality of rank profiles across groups (Lehmacher and Wall 1978) did not yield any significant group differences with respect to any of the processing parameters. By means of the Anderson-Kannemann Test (Kannemann 1976), we furthermore examined for each group whether the frequencies of all possible rank profiles were identical. This was not the case for some of the processing parameters. The analysis was done in the same three steps as before.

In the first analysis, we found significant differences between frequencies of rank profiles for each group under all processing conditions except for the processing of global changes in the verbal version. For the processing of the three other modality/structure parameters, inspection of profile incidence matrices showed that the most common decrease in preferences was: T - T, T - D, D - D.

In the second analysis, a significant difference could be established only for the control group when processing global changes in the pictorial version. The two most frequent profiles had in common that alterations between initial and final target objects had the highest rank order. Such a preference for initial-final alterations was numerically also found for the two other groups and for processing pictures of partially changing objects.

In the third analysis, significant differences were found for the aphasic patients only. They preferred alterations between the medial target object and the distractor object when processing words both with partial and global changes. Contrary to expectation, no preferences were found for alteration between the final target objects and the two distractor objects.

DISCUSSION

Our data only partially support the assumption that the processing of modality and structure of a task gets differentially affected in left- and right-brain-damaged patients. In terms of total gaze duration as well as decision time, aphasic patients processed written material longer than pictures of parallel structure. But, for RH-patients and controls no significant difference in modality was found (cf. table 1).

On the other hand, the difference in structure influenced only the two non-aphasic groups. Contrary to expectation, processing of partial changes took more time than processing of global changes. This difference was found only for pictures and not for words (cf. table 1). Under the assumption that finding the correct solution requires both selection and combination of target objects, it seems that the partial-picture condition was more time-consuming because the two parts of the analysis had to be performed in a serial fashion, whereas when scanning for pictures of globally changing objects decisions on combination are made parallel to selection. Aphasic patients seem to adhere throughout the experiment to serial processing and do not switch to parallel processing routines.

Even the expected modality effect in the aphasic patients was only reliable for objects with global changes (cf. table 1). With partial changes, the recurrance of identical nouns obviously enabled aphasic as well as non-aphasic subjects to use a simple visual matching strategy for the selection of the target objects. Thereby, the difference in total processing time for words versus pictures was reduced. Closer inspection of the accuracy of response data (cf. table 2) directly supports this interpretation. Only under the partial-word conditions, aphasic patients made no selection but only combination errors in contrast to 80%, 37% and 58% selection errors in the other conditions.

Under all conditions except for the partial-picture condition, the aphasic group needed significantly longer decision time (and overall gaze duration) than the two non-aphasic groups. When processing pictures of partially changing objects, the aphasics neither had a modality disadvantage nor had the non-aphasic groups the advantage of parallel decision (and control) with respect to selection and combination of target objects.

Despite these group differences and despite clear differences in overall error rate between the three groups (47%:14%:9%, cf. table 2), target objects were significantly more intensively looked at than distractor objects by each of the three groups (cf. table 3). At this point, the eye movement behavior reveals that the normals as well as the brain-damaged groups were on the way to a correct selection of target objects, i.e. they had understood the concept of a process they had to look for. An exeption is the processing of words with global changes by aphasic patients. Here, the relative values for gaze duration and gaze frequency were not significantly higher on target than on distractor objects.

This finding can hardly result from a cognitive inability in making selections. The performance of the control group exhibited a significant interaction between modality (word/picture) and type of object (target/distractor) in the same directions as in the aphasic group. Obviously, the content of the distractor objects was more prominent under the global word condition than under all other conditions. This interpretation is supported by the analysis of gaze alterations. Aphasics like the two other groups compared significantly more often a target and a distractor object under the global-word than under the global-picture condition (cf. table 5). The reverse was found for alterations between target objects. The difference can not be explained by simply matching for identity: with pictures as well as with words, there were no visual cues for the identification of target objects. Such cues were only present in partially changing objects both in the written and the pictorial version.

The modality effect on processing of globally changing objects seems to be also brought about by different combination strategies. When processing pictures, subjects looked most intensively on the initial target object but when processing words on the medial one (cf. table 4). A parallel contrast was found for gaze alterations. Under the global picture condition, alterations between initial and final target object were outstandingly frequent and the difference to the global word condition was significant (cf. table 6). These findings taken together indicate a "linear and/or periphery-control" strategy for pictures as opposed to a "center-control" strategy for words. The "center-control" strategy can also be seen in the preference for alterations between the medial target and a distractor object (cf. talbe 7). Sorting via exclusion was linked to the "center-control strategy" as the high frequency of target-distractor alteration under the global word condition suggests whereas the "periphery strategy" went together with sorting via inclusion as the preference for target-target comparisons under the two picture conditions shows. Whether preference for inclusion and exclusion are necessary components of the two control strategies cannot be decided.

A final point of discussion is the influence of the linguistic content of the distractor objects. We expected relatively high frequencies of alterations between final target object and a distractor object only under the two global change conditions where the selection of the three target objects and/or the exclusion of the two distractor objects could not be done via visual cueing. However, this expectation was only confirmed for words and not for pictures (cf. table 7). It seems that such a reliance on exclusion is specific to the verbal modality. Consequently, aphasic patients are most likely to have troubles when subjects are required to make selections via semantic exclusion as in scanning for noun-phrases that express globally changing objects.

ACKNOWLEDGEMENT

We would like to thank H.W. Schroiff and D. Sommer for their help in running the experiment, and G. Guillot for his help in the statistical evaluation of the data.

REFERENCES:

(1) Cohen, G., Hemispheric differences in serial versus parallel processing, J. Exp. Psych. 97 (1973), 349 - 356.

(2) Cronbach, L.J., Gleser, G.C., Nanda, H., Rajaratnam, N., The dependabi-
lity of behavioral measurements. Theory of generalizability for scores
and profiles (Wiley, New York, 1972).

(3) Dixon, W.J. & Brown, M.B., Biomedical computer programs, P-Series (Berke-
ley, University of California Press, 1977).

(4) Goodglass, H., Denes, G., Calderon, M., The absence of covert verbal
mediation in aphasia, Cortex 10 (1974), 264 - 269.

(5) Huber, W., Gleber, J., Linguistic and non-linguistic processing of narra-
tives in aphasia, Brain and Language 16 (1982), 1 - 18.

(6) Huber, W., Poeck, K., Weniger, D., Willmes, K., Der Aachener Aphasie
Test (Hogrefe, Göttingen, 1983).

(7) Huber, W., Lüer, G., Lass, U., Processing of sentences in conditions of
aphasia as assessed by recording eye movements, In: R. Groner, C. Menz,
R. Monty (eds), Eye movements: An international perspective (Erlbaum,
Hillsdale N.J., 1983).

(8) Kannemann, K., An incidence test for k related samples, Biom. Z. 18
(1976), 3 - 11.

(9) Lehmacher, W., Wall, K.D., A new nonparametric approach to the comparison of k independent samples of response curves, Biom. J. 20 (1978),
261 - 273.

(10) Martin, M., Hemispheric specialization for local and global processing,
Neuropsychologia 17 (1979), 33 - 40.

(11) Patterson, K., Bradshaw, J.L., Differential hemispheric mediation of
non-verbal visual stimuli, J. Exp. Psych.: Human Percept. Perform 1
(1975), 246 - 252.

(12) Polich, J.M., Left hemisphere superiority for visual search, Cortex 16
(1980), 39 - 50.

(13) Veroff, A.E., A structural determination of hemispheric processing of
pictorial material, Brain and Language 5 (1978), 139 - 148.

(14) Warrington, E.K., Taylor, A.M., The contribution of the right-parietal
lobe to object recognition, Cortex 7 (1973), 152 - 164.

(15) Young, L.R., Sheena, D., Survey of eye movement recording methods,
Behavior Research methods and Instrumentation 7 (1975), 397 - 429.

Theoretical and Applied Aspects of Eye Movement Research
A.G. Gale and F. Johnson (Editors)
© Elsevier Science Publishers B.V. (North-Holland), 1984

PREDICTIVE EYE MOVEMENTS IN NORMAL SUBJECTS AND IN
PARKINSON'S DISEASE

Adolfo M. Bronstein
Christopher Kennard

Department of Neurology,
The London Hospital,
London, E.1
. ENGLAND

Studies of randomly elicited and predictive eye movements in
patients with Parkinson's disease are described.

In young normal subjects the peak velocity of predictive saccades
(PS) is significantly reduced when compared with equal amplitude
randomly elicited saccades (RS). The accuracy of RS, but not of
PS, is determined by target amplitude. Thus RS and PS constitute
two distinct populations of saccades and their possible differing
supranuclear control is discussed.

In performing RS Parkinsonian patients (PP) have a small but
significantly prolonged latency when compared with age-matched
normal controls (NC). During PS these patients show some
ability to predict but unlike NC they were unable to further
reduce saccadic latency with prior knowledge of the target's
predictive pattern. During random (RSP) and predictive (PSP)
smooth pursuit PP show a markedly increased phase lag. The
ocular motor system of PP seems to have a similar "high level"
disturbance to that reported for somato-motor control. Smooth
pursuit "efficiency" (assessed as proportion of time actually
pursuing) in PP was similar to that for NC but both groups
showed a better performance with the head free. The possible
explanation for this is examined.

INTRODUCTION

It has been known for some time that the performance of the ocular
motor system varies according to the nature of the task to be executed.
For instance, when displacement of the visual target becomes predictable
the phase-lag in the smooth pursuit system and latencies in the saccadic
system are drastically reduced (1). Eventually movement of the eye in
advance of the target is seen. In this situation eye movements are no
longer visually guided, in a strict sense, as a result of the retinal
error signal, but are generated by internally pre-programmed motor
commands.

In Parkinson's disease (PD) one of the major clinical signs is akinesia in which there is slowed initiation and execution of movement. An
abnormality of central motor control mechanisms has been proposed to
explain this disturbance, in particular an inability to benefit by
prediction when tracking regularly moving targets (2,3). Flowers and
Downing (4), however, were only able to show this disturbance in
hand but not in eye tracking. This implies that there may be separate
limb and eye "predictor mechanisms" which are differentially affected

in PD, a hypothesis which has not been confirmed. In the studies to be
described we have first investigated in normal subjects the difference
in saccadic metrics when made in response to a randomly and predictable
target. Secondly we have analysed the ability of Parkinsonian patients
to generate predictive eye movements.

PREDICTIVE EYE SACCADES IN NORMAL SUBJECTS

Materials and methods

Eight normal volunteers, aged 18-37 years, with no experience in
ocular motor experiments were the subject of this experiment. Recordings
of eye movements were obtained using the infra-red reflection technique (1)
and displayed on-line by an ink jet recorder. The visual target consisted of a white, light spot (diameter 30' of arc) projected by means
of a mirror galvanometer onto a curved white screen placed 150 cms. from
the subject's head.

In order to elicit random saccades (RS) they were given two sequences
(runs) of unpredictable target jumps between five different positions,
generating target displacements of 7.5, 15, 22.5 and 30 degrees. To
generate predictive saccades (PS) the same target amplitudes used in
the random sequence were each separately delivered at a regular time
interval (0.9 sec) between two fixed points across the mid-line. In
order to avoid fatigue, or diminished level of arousal target sequences
were of short duration: two runs of RS, each of 30 seconds duration,
followed by four runs of PS, one for each amplitude, each of 20 seconds
duration. Subjects were given rest periods of 30 - 40 seconds between
each run. Immediately before each run they were informed whether the
target would move in a regular (PS) or irregular (RS) manner and were
requested to follow it as accurately as possible. Saccadic latency,
amplitude and velocity were measured by hand from chart recordings
produced at 100 mm/sec. Plots of saccadic amplitude/velocity were
then prepared for each subject. After fitting an exponential curve to
these values, the mathematically predicted velocity value for a 25°
saccade was used for statistical comparisons between RS and PS. Saccades
were considered to be predictive if their latency, during a predictive
sequence, was less than 100 msec. RS were defined as those having a
latency of more than 100 msec (typically 150 - 200 msec) during a random
sequence. The term anticipatory saccade will be used for PS the onset of
which occurs before the target movement.

Results

In Figure 1 examples of eye movements from three different subjects
are shown. In Figure 1A two sequential saccades of equal amplitude are
presented; the first was visually triggered with a latency of 200 msec
and has a velocity of 580°/sec, whereas the second one is an anticipatory
saccade, which reached a peak velocity of only 460°/sec. This trend of
PS to be slower than RS was consistently confirmed for all subjects.
The mean and standard deviation for a 25° amplitude saccade is $573^{+}_{-}91\ ^O$/sec
for RS and $479^{+}_{-}79$ for PS (p <0.01, Wilcoxon Signed Ranks Test). Double
saccades as shown in figure 1C sometimes occurred in which the primary
and secondary saccades were of equal amplitude but on the basis of their
latency the former was predictive and the latter visually elicited. In
this situation there is a clear difference in their peak velocities, that
of the primary saccade being slower than the secondary saccade.

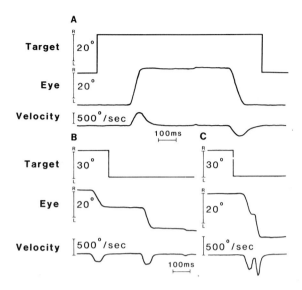

Figure 1.   Examples of predictive saccades in normal subjects.   For
details see text.

In Figure 1B a typical PS is shown in which the primary saccade occurs
in advance of the target displacement and is markedly hypometric (less
than 50 per cent of the target displacement).   This degree of inaccuracy
was never found in our subjects when normal latency saccades were
elicited.   PS also show a very high frequency of hypometric saccades,
i.e. the proportion of normometric saccades for PS is less than 8 per
cent.   PS also show no relationship between accuracy and target amplitude
so that the percentage of normometric saccades remains constant at about
8 per cent for all target amplitudes.   This is in marked contrast to RS
where accuracy depends on target amplitude.   In our subjects the percentage
of normometric saccades for a target amplitude of 7.5° is 51% and falls
to 6% for a 30° target amplitude.

Discussion

The differences found between RS and PS suggest that they may belong
to two distinct populations of saccades (5).   The evidence presented in
our study which shows that saccadic velocity and accuracy differ under
these two conditions strongly suggest that the neural mechanisms and/or
pathways subserving visually triggered (RS) or internally guided (PS)
saccades may differ.   Anatomical, physiological and behavioural observations have suggested that the supranuclear organization of eye saccades
may depend on two different neural structures;   the frontal eye fields
and the superior colliculi depending on whether a "voluntary" internally
generated or visually elicited eye movement is required respectively(6-9).
Within this frame of reference it is possible to suggest that the differences seen between RS and PS could at least be partially due to selective
or preferential activation of these supranuclear structures, the frontal

eye field for PS and superior colliculus for RS.

Similar variations in saccadic velocity and accuracy to those found during PS have been reported when saccades were made in darkness or to an acoustic source (10-12). A common factor in each of these conditions is the absence of a visual target prior to the onset of the saccade. Thus the visual target appears to have a role as a "primer" for optimizing saccadic velocity and accuracy.

PREDICTIVE EYE MOVEMENTS IN PARKINSON'S DISEASE

SACCADE EXPERIMENTS

Materials and methods

Eight Parkinsonian patients (PP) with disease of mild to moderate severity and eight age-matched normal controls (NC) were the subjects of these and the smooth pursuit (SP) experiments. Mean age was 65 in both groups. They were given three different runs of saccadic stimuli in the following order: 1. Random saccades (RS) in time and space. Instruction to the subject was "follow the target as fast and as accurately as possible; it will be jumping randomly on the screen"; 2. Hidden Predictive Saccades (HPS). 20 Predictive Saccades, i.e. totally regular square waves, were unexpectedly and without warning delivered in the middle of an otherwise random sequence. The instruction was the same as in 1; 3. Predictive Saccades (PS). One single run of regular square waves, the instruction given was "instead of randomly moving the target will jump totally regularly between the same two points on the screen". PS in 2. and 3. were delivered at an interstimuli interval of 0.9 seconds, which pilot studies on PP and NC had shown to be optimal to elicit PS. For RS time limits between saccades were 0.6 to 1.2 seconds. Target amplitudes were never greater than $40°$. For head-eye tests the same experimental design was used, inter-saccadic intervals being approximately 60 per cent longer to enable the head to be completely stabilised at the completion of a movement.

Eye movements were obtained with the Infra-red technique (1). Head movements were recorded by having the subjects wear a light helmet connected by a rigid rod to a low torque potentiometer. Gaze (eye in space) was obtained by electronically adding the eye and head signals in a differential amplifier. The subjects were encouraged to vigorously move their head. Analysis of eye, gaze and head movements for RS was by a computer described elsewhere (13). PS latencies were measured by hand. Statistical comparisons were performed using variance and co-variance analysis.

Results

Analysis of RS (Table 1) reveals significantly prolonged latencies in the patient group for eye, gaze and head movements ($p<0.05$). Eye and gaze saccade velocities did not show significant differences between the two groups.

Table 1	Latency and Peak Velocity of Saccades in PP and NC				
	Latency (msec)			Peak Velocity ($°$/sec)	
	Eye	Gaze	Head	Eye	Gaze
Parkinsonian	221 ± 24	221 ± 24	328 ± 41	612 ± 125	563 ± 104
Normals	185 ± 21	193 ± 24	272 ± 57	645 ± 141	543 ± 114

However, the most interesting findings are related to PS. Figure 2
shows the sequential evolution of saccadic latencies for one PP and one
NC in the runs RS, HPS and PS. As can be seen the overall pattern of
behaviour in RS and HPS is not consistently different in the two subjects.
As they pass into the unexpected or hidden predictive sequence both subjects reduce their latency and make two anticipatory saccades (above the
0 msec line). Generation of PS by either subject, therefore, is not
regular or sustained. When the task becomes totally predictive (PS) a
clear difference in behaviour appears; whereas the normal subject now
takes advantage of the new situation and consistently produces short
latency and anticipatory saccades, the Parkinsonian patient continues
anticipating erratically as in the HPS run. One way of quantifying
this phenomenon for the two groups, is by establishing the percentage
of anticipatory saccades found in each task. During the HPS run no
differences are seen (23 and 22% for NC and PP respectively), whereas
during predictive sequence 60% and 29% are anticipatory saccades for
NC and PP respectively.

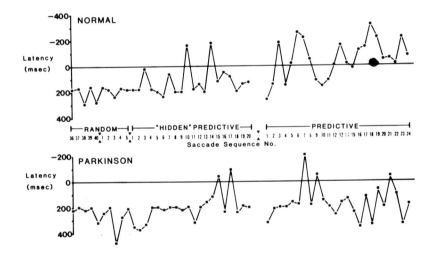

Figure 2. Sequential latency change in a PP and a NC during random and
           predictive experiments. For details see text.

SMOOTH PURSUIT EXPERIMENTS

Materials and Methods

    The same subjects were then studied for their ability to smoothly
pursue random (RSP) and predictive (PSP) sinewaves. The target moved horizontally with a fixed amplitude of 30° at one of five frequencies (0.08,
0.16, 0.32, 0.48, 0.65 Hz). For RSP, after a variable period of 1.3
to 2.6 secs, its frequency randomly changed. For PSP each frequency was
separately delivered. Subjects were instructed as to whether the target
was to move regularly or irregularly. For eye and eye-head tracking
similar stimulus parameters were used. Two different methods of analysis

of smooth pursuit (SP) of eye and gaze recordings were performed. 1.
Phase measurements were made by hand on the paper recording for two
frequencies (0.3 and 0.6 Hz), in random and predictive runs. On
average seven (range 3 to 15) half cycles were measured. 2. Computer
analysis was performed to determine at any particular moment whether the
eye movement was pursuit or saccadic. It then established for each
target velocity the proportion of tracking which was saccadic or SP.
The Turn-Over Velocity (TOV) is the predicted target velocity at which
the proportion of SP (shown at the slowest velocity) drops by 3db. It
gives an objective estimate of the velocity at which smooth pursuit
breaks down into mainly saccadic tracking (14) . 3. In one subject from
the NC group the absolute velocity of the small catch-up saccades (no
bigger than 10° amplitude) during eye alone and head-eye tracking was
measured from the velocity trace obtained by electronic differentiation
of the eye position signal.

RESULTS

The main abnormality found in SP is that PP have a significantly
bigger phase lag than NC (p < 0.05) for eye and gaze tracking (table 2).

Table 2   Phase-lag (degrees) during RSP and PSP at 2 target frequencies

		0.32 Hz.		0.65 Hz.	
		Random	Predictive	Random	Predictive
Eye	Normals	$3.96^{+}_{-}4.02$	$2.04^{+}_{-}3.24$	$20.62^{+}_{-}7.5$	$8.5^{+}_{-}7.75$
	Parkinsonian	$15.72^{+}_{-}4.92$	$9.48^{+}_{-}4.38$	$38.5^{+}_{-}8.0$	$23.25^{+}_{-}8.25$
Gaze	Normals	$6.12^{+}_{-}2.88$	$6.3^{+}_{-}2.04$	$26.5^{+}_{-}7.75$	$14.5^{+}_{-}7.75$
	Parkinsonian	$12.48^{+}_{-}7.2$	$13.08^{+}_{-}7.3$	$32.5^{+}_{-}5.4$	$19.0^{+}_{-}12.75$

This is more evident for the eye alone than for the gaze when the head is
free. During PSP both PP and NC considerably reduce the phase lag of eye
and gaze compared to RSP although PP never reach normal values.

Computer analysis of smooth pursuit efficiency, expressed as TOV shows
that PP are marginally worse than NC, a difference which is not statistically significant (Table 3).

Table 3   Turn-over velocity (TOV) ($^{\circ}$/sec) during RSP and PSP

	Eye		Gaze	
	Random	Predictive	Random	Predictive
Normal	$46.25^{+}_{-}6.01$	$48.75^{+}_{-}7.51$	$47.42^{+}_{-}5.32$	$56.71^{+}_{-}13.18$
Parkinsonian	$41.25^{+}_{-}4.77$	$43.75^{+}_{-}9.19$	$44.37^{+}_{-}6.88$	$55.42^{+}_{-}11.51$

However, TOV values are significantly higher in both groups during PSP
compared to RSP (p < 0.001). TOV for the gaze, in PP and NC, is significantly higher than for the eye, but this is most significant during PSP
(p < 0.05). Figure 3 illustrates these differences in a PP., and further
discussion of this result is in the Appendix.

DISCUSSION

In this study of PP saccadic latencies were found to be increased by
about 15%. This has previously been described for primary and corrective
saccades (15,16), and is considered a manifestation of akinesia. However,
more striking abnormalities were seen with predictive saccades. During

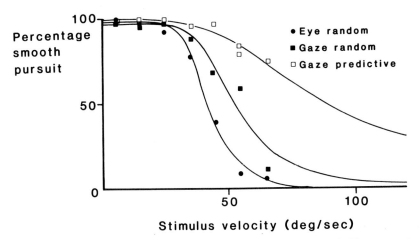

Figure 3. Relationship of percentage smooth pursuit for different stimulus velocities in a PP. The computer calculated TOV was 37.26,
45.85 and 67.80 degrees/sec for eye random, gaze random and
gaze predictive respectively.

the hidden predictive sequence, based on saccadic latency, it appeared that
they recognised the regularity of the target movement, and were able to
make some predictive saccades. In the PS sequence however, when they
were given prior verbal instruction about the regularity of the target
movement, PP did not show any further improvement in their performance.
Thus PP appear to have difficulty in using advance information to initiate
a saccade. This deficit is similar to that found in patients with frontal
lobe lesions (9, 17) and it is interesting that recent studies of cognitive disturbances in PP (18) have also suggested frontal lobe dysfunction. A possible pathophysiological explanation for this finding is the
damage to the dopaminergic mesocorticolimbic pathway, which projects to
the frontal cortex, known to occur in PD (19).

The main defect of the smooth pursuit system in PP shown in this
study is an increase in phase lag probably related to the akinetic phenomenon. During the predictive sequence they were able to improve their
SP by reducing phase lag, but this was insufficient to overcome the
initial deficit. This finding is in agreement with results for hand
tracking experiments (3). However, in a quantitative study of eye tracking (4) during the predictive sequences, the patients reached normal
values of phase lag and in some of the tests showed an even more pronounced
trend than the normal controls to anticipate the target. We believe that
technical differences in the method of analysis of the results could explain these discrepancies.

An examination of the predictive capabilities of the ocular motor
system in PD, therefore, shows a difference between SP and saccadic systems. The former is able to predict to a considerably greater extent than
the latter. This difference may be due to the fact that saccadic anticipation requires the subject to move his eyes towards an unseen target whereas during predictive SP the target is continuously visible. PP may have

470   *A.M. Bronstein & C. Kennard*

difficulty in making movements without direct visual sensory information, a conclusion also reached in hand tracking experiments (20). Thus the ocular motor and somato-motor systems have similar "high level" abnormalities in which the impaired predictive capability reflects a difficulty in generating an "internal model" of the external world.

APPENDIX

The computer techniques used to assess SP showed that gaze performance was better than eye alone. As it had previously been shown that SP gain is the same whether or not head movements take place (21) we decided, as described in Methods, to study the velocity characteristics of the small catch-up saccades during eye alone and head-eye tracking. As shown in Figure 4 the amplitude/velocity relationship of eye saccades in these two conditions was the same. This implies that gaze saccades are not slowed down by

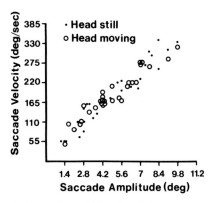

Figure 4. Amplitude/velocity plot of catch-up eye saccades up to 10 degree amplitude during tracking of a sinusoidally moving target in a normal control.

the vestibulo-ocular reflex (VOR) which acts in an opposite direction. Therefore, once the head velocity has been added to eye velocity the resulting gaze catch-up saccades will be faster than that of the eye alone. Since the computer analysis of SP measures the proportion of the time spent in SP as opposed to saccades, the faster and shorter duration saccades of gaze during head free tracking will result in a higher TOV.

Two possible mechanisms to explain why the VOR does not slow down the catch-up saccades can be suggested. Firstly, the VOR may be switched off during combined head eye SP either by an efferent copy of the head movement or by neck proprioceptors (22). Secondly, catch-up saccades may be quick phases of vestibular nystagmus, which are normally not slowed down by the VOR (23).

ACKNOWLEDGEMENT

We are grateful to Dr T. Smith for performing the computer analysis. A.M.B. was a Wellcome Trust Fellow.

REFERENCES

1) Stark, L., Vossius, G. and Young, L.R., Predictive control of eye tracking movements, IRE Trans.Hum.Fact.Electr.HFE-3 (1962) 52-57.

2) Marsden, C.D., The mysterious motor function of the Basal Ganglia, Neurology 32 (1982) 514-539.

3) Flowers, K., Some frequency response characteristics of Parkinsonism on pursuit tracking, Brain 101 (1978) 19-34.

4) Flowers, K.A. and Downing, A.C., Predictive control of eye movements in Parkinson disease, Ann.Neurol. 4 (1978) 63-66.

5) Findlay, J.M., Spatial and temporal factors in the predictive generation of saccadic eye movements, Vision Res. 21 (1981) 347-354.

6) Schiller, P.H., The effect of superior colliculus ablation on saccades elicited by cortical stimulation, Brain Res. 122 (1977) 154-156.

7) Leichnetz, G.R., The prefrontal cortico-oculomotor trajectories in the monkey, J.Neurol.Sci. 49 (1981) 387-396.

8) Collin, N.G., Cowey, A., Latto, R. and Narzi, C., The role of frontal eye-fields and superior colliculi in visual search and non-visual search in Rhesus monkeys, Behav.Brain.Res. 4 (1982) 177-193.

9) Guitton, D., Buchtel, H.A. and Douglas, R.M., Disturbances of voluntary saccadic eye movements mechanisms following discrete unilateral frontal lobe removals, in: Lennerstrand, G., Zee, D.S. and Keller, E.L. (eds.), Functional basis of ocular motility disorders (Pergamon, Oxford, 1982).

10) Becker, W. and Fuchs, A.F., Further properties of the human saccadic system: eye movements and correction saccades with and without visual fixation points, Vision Res. 9 (1969) 1247-1258.

11) Zahn, J.R., Abel, L.A. and Dell'Osso, L.F., Audio-Ocular response characteristics, Sens.Process. 2 (1978) 32-37.

12) Zambarbieri, D., Schmid, R., Magnees, G. and Prablanc, C., Saccadic responses evoked by presentation of visual and auditory targets, Exp. Brain Res. 47 (1982) 417-427.

13) Smith, A.T., Bittencourt, P.R.M., Lloyd, D.S.L. and Richens, A., An efficient technique for determining characteristics of saccadic eye movements using a minicomputer, J.biomed. Eng. 3 (1981) 39-43.

14) Bittencourt, P.R.M., Smith, A.T., Lloyd, D.S.L. and Richens, A., Determination of smooth pursuit eye movements velocity in humans by computer, EEG and Clin.Neurophysiol. 54 (1982) 399-405.

15) Melvill Jones, G. and De Jong, J.D., Dynamic characteristics of saccadic eye movements in Parkinson's disease, Exp.Neurol. 31 (1971) 17-31.

16) Shimizu, N., Maito, M. and Yoshida, M., Eye-head co-ordination in

patients with Parkinsonism and cerebellar ataxia, J.Neurol. Neuro-
surg.Psychiatr. 44 (1981) 509–515.

17) Luria, A.R., Frontal lobe syndromes, in: VinkentP.J. and Bruyn, G.W.
(eds.), Handbook of Clinical Neurology, Vol. 2 (North Holland,
Amsterdam, 1969).

18) Lees, A.J. and Smith, E., Cognitive deficits in the early stages of
Parkinson's disease, Brain 106 (1983) 257–270.

19) Price, K.S.,: Farley, I.J. and Hornykiewicz, O. Neurochemistry of
Parkinson's disease: relation between striatal and limbic dopamine,
Advanc.Biochem.Physchopharmacol. 19 (1978) 293–300.

20) Flowers, K., Lack of prediction in the motor behaviour of Parkinsonism,
Brain 101 (1978) 35–52.

21) Lanman, J., Bizzi, E. and Allum, J., The co-ordination of eye and head
movement during smooth pursuit, Brain Res. 153 (1978) 39–53.

22) Robinson, D. A. and Zee, D.S., Theoretical considerations of the function and circuitry of various rapid eye movements, in: Fuchs, A.F. and
Becker, W. (eds.), Progress in oculomotor research (Elsevier, Amsterdam, 1981).

23) Jurgens, R., Becker, W., Rieger, P. and Widderich, A., Interaction
between goal-directed saccades and the vestibulo-ocular reflex (VOR)
is different from interaction between quick phases and the VOR, in:
Fuchs, A.F. and Becker, W. (eds.), Progress in oculomotor research
(Elsevier, Amsterdam, 1981).

Theoretical and Applied Aspects of Eye Movement Research
A.G. Gale and F. Johnson (Editors)
© Elsevier Science Publishers B.V. (North-Holland), 1984

THE RANGE OF LINEARITY OF THE SMOOTH PURSUIT CONTROL SYSTEM

A. Buizza, R. Schmid, M.R. Gigi

Dipartimento di Informatica e Sistemistica
Università di Pavia
Pavia - Italy

The gain of horizontal smooth pursuit eye movements
was measured during the tracking of ramp and sinusoidal movements of the target. The results indicate
that the range of linear behaviour of the smooth pur
suit control system extends far beyond the limit of
20-30°/s which is often reported in the literature.
Nevertheless, linearity can be masked by the particular choice of the profile and the spatio-temporal parameters of target movement.

INTRODUCTION

Smooth pursuit eye movements (SPEM) can be elicited by asking a subject to
follow by his eyes a small visual target moving regularly in front of him.
In spite of their basic interest and the simplicity of the set-up needed for
their study, SPEM have been so far little explored (at least much less than
saccades) and their characteristics are still poorly understood. The learning capabilities of the smooth pursuit control system (SPCS) (Stark et al.
(1962), Dallos and Jones (1963), Eckmiller and Mackeben (1978), Becker and
Fuchs (1982), Whittaker and Eaholtz (1982)) makes the interpretation of the
experimental results highly intriguing. Actually, most of them need to be
correlated to stimulus predictability, and this correlation requires further
thorough investigation.

It has been stated that SPCS is a time continuous system with a latency of
100-200 ms between the onset of the stimulus and the beginning of the oculomotor response (Rashbass (1961), Dallos and Jones (1963), Robinson (1965)).
The nature and the origin of such a latency, its influence on the dynamic be
haviour of SPCS, and its dependence on learning and prediction are still to
be elucidated.

As for SPCS dynamics, it has been shown that this system behaves as a low-
pass filter with cut-off frequency ranging between 0.4 and 1 Hz, depending
on the predictability of target motion (Stark et al.(1962), Michael and Melvill Jones (1966), Bahill et al.(1980), Melvill Jones and Gonshor (1982)).

Finally, the input-output gain of SPCS is close to one only for low target velocities. A saturation and a cut-off of SPEM velocity have been described in the literature. Nevertheless, the level of saturation is still controversial. A value of 20-30°/s was reported by some authors (Young (1962), Robinson (1965)), whereas some others could record two- and even three-fold higher SPEM velocities (Barmack (1970), Buizza et al.(1981)). It cannot be excluded that different results were obtained since different patterns of target motion were used.

As a first contribution to a better understanding of SPCS characteristics, the dependence of its gain on the amplitude of target displacement was examined in this study for two profiles of target motion, ramp and sinusoids.

METHODS

Five volunteers aged between 21 and 35, without any evidence of oculomotor impairment, served as subjects.

Subjects were seated in the dark, with their head restrained, in front of a cylindrical screen of 2 m in diameter. A LASER beam was front projected on the screen after reflection on a mirror mounted on a galvanometer. The light spot on the screen (visual target) was of about 1 cm in diameter (about 35' of arc, as seen by the subject), and could be moved in the horizontal plane by driving the galvanometer with a function generator. Subjects were asked to track target motion by their eyes.

Two types of target motion were considered.
A) Constant velocity (ramp) displacements from one point on the left of the screen to the symmetrical point on the right. Three amplitudes of target displacement were examined (10°, 30°, and 50°). For each amplitude different target velocities were adopted (10, 15, and 20°/s for the amplitude of 10°; 15, 30, 45, and 60°/s for the amplitude of 30°; 25, 50, 75, and 100°/s for the amplitude of 50°). Each ramp was presented three non-consecutive times within a random sequence. The interval between two successive ramps varied randomly between 5 and 15 s.
B) Sinusoidal oscillations about the central position (subject's medial plane) with peak-to-peak amplitudes of 10°, 30°, and 50°. In order to obtain different peak velocities, three frequencies (0.16, 0.32, and 0.48 Hz) were used for each amplitude. Higher frequencies were excluded since otherwise SPCS could have been made working outside the flat zone of its frequency response. Sinusoidal stimulation was maintained for 10 to 15 cycles and was followed by the next one after 1 min. The order of presentation was random.

Eye position was measured by DC electrooculography and recorded on FM magnetic tape together with target position. Calibration of eye movement was made by presenting the LASER spot at three reference positions right and left. Records were then analysed using a computer programme which removed the sacca-

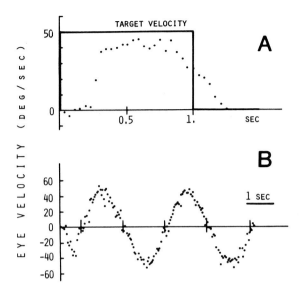

Fig.1 - Typical time course of smooth pursuit eye velocity in ramp (A) and sinusoidal (B) responses

dic components and presented the smooth pursuit eye velocity diagrams on a
graphic display (Buizza and Avanzini (in press)). Response parameters (laten
cy and maximum SPEM velocity in the ramp tests; peak SPEM velocity and phase
shift between eye and target movement in the sinusoidal tests) were evaluated by using the display hair cursors.

RESULTS

Typical examples of the time course of SPEM velocity in the two experimental
conditions considered in the present study are shown in Fig.1. Dots give eye
velocity obtained by averaging the responses of the five subjects. For the
sake of clarity, target velocity is reported by a continuous line only for
the ramp test. For the sinusoidal response only the zero crossing points of
target velocity are indicated by vertical bars.

Two typical features of ramp responses can be noted in Fig.1-A. First, a latency of 150-200 ms separated the onset of the stimulus and the beginning of
the eye movement. Second, the eyes started decelerating significantly before
the end of target excursion, while the target was still moving at constant
velocity.

The sinusoidal response in Fig.1-B started with a latency which was comparable with that of ramp responses, and reached the steady state in less than

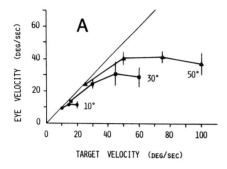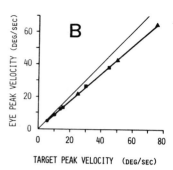

Fig.2 - A. Maximal smooth pursuit eye velocity versus ramp velocity for different target excursions. B. Eye versus target peak
velocity in sinusoidal tests. Same target excursions as in A.

one cycle. Afterwards, no appreciable phase shift between eye and target movement persisted.

Although the amplitude of target excursion was the same for the responses in
Figs.1-A and 1-B (50°), and the peak velocity of the sinusoidal target motion was the same as the constant velocity of the ramp (50°/s), the maximal e-
ye velocity reached in the ramp response was slightly less than the peak velocity of the sinusoidal response. The mean values of the maximal velocity
in ramp responses, and of the peak velocity in sinusoidal responses are plot
ted in Fig.2-A and 2-B, respectively, versus target velocity. The range of
linearity of ramp responses was a function of the amplitude of target displa
cement, and increased with it. For the smallest examined amplitude (10°) the
range of linearity was limited to 10°/s, and eye velocity could not exceed
13°/s. With the largest amplitude (50°), linearity extended up to 40-50°/s,
and eye velocity could exceed 40°/s. A decrease of the maximum velocity could be observed beyond saturation.

In sinusoidal responses, linearity was found in the whole range of target velocities considered for each amplitude of target motion. Moreover, when the
same range of target velocities was explored with different amplitudes of
target motion, no difference related to this latter parameter was observed
in eye velocity. For the same amplitude of target motion, eye velocity in
sinusoidal responses could reach values far beyond the saturation level observed in the ramp responses.

The phase shift between eye and target movement in sinusoidal tests did not
show any progressive evolution through successive cycles. An almost steady
state value was reached within the first cycle. As shown in Fig.3, where the
average phase shift for the five subjects is plotted versus frequency, eye
movement lagged target movement by a few degrees only. No significant variation related to the amplitude of target motion could be observed.

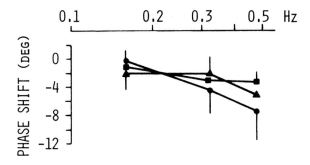

Fig.3 - Phase shift between eye and target velocity
in sinusoidal tests,versus frequency.

DISCUSSION

The most striking result presented in the previous section is the difference
between the diagrams of eye velocity versus target velocity in ramp and sinu
soidal responses. An almost linear gain characteristic was observed in sinusoidal responses, up to the maximum examined target velocity of 75°/s, where
as in ramp responses eye velocity saturated at levels which were dependent
on the amplitude of target motion.

Since repeated sinusoidal oscillations are likely to represent a more easily
predictable target motion than a sequence of randomly spaced ramps of different amplitudes and velocities, one could be tempted to ascribe the apparent
better performance of SPCS during sinusoidal tests to the action of a prediction mechanism which should be ineffective, or poorly effective, during
ramp tests. The following arguments seem to indicate that stimulus predictability is neither the only, nor the main responsible factor for the observed
differences. First, sinusoidal responses reach a steady state condition with
in the first cycle, and no evolution that could be related to learning was
observed in the successive cycles. Second, a lack of prediction cannot justi
fy the amplitude dependence of the ramp response gain characteristics. Third,
a kind of prediction is actually present in ramp responses, as it is proven
by the fact that the eyes start decelerating before the end of target motion.

A more convincing explanation can be found by taking into account the effects produced by SPCS dynamics on ramp responses. As shown in Fig.1-A, the responses to single ramps (pulse responses, in terms of velocity) present an i
nitial latency (τ) of about 150 ms, an exponential increase up to a steady
state condition which is reached in 100-200 ms, and a deceleration which can
preceed the end of the velocity pulse by a time τ' of 100-200 ms. In a first
approximation, eye acceleration can be described by a single exponential
with a time constant T of 40-100 ms. Then the maximum velocity in ramp res-

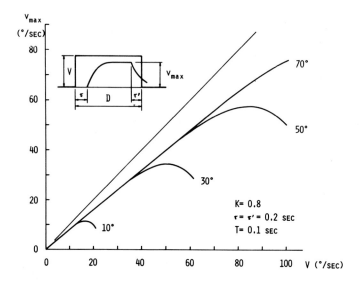

Fig.4 - Values of v-max computed from eq.1 for target excursions of 10° to 70°.

ponses is given by:

$$v\text{-max} = KV(1-\exp(-(D-\tau-\tau')/T))$$ (1)

where (see the inset in Fig.4)
        K= gain factor
        V= pulse amplitude (constant ramp velocity)
        D= pulse duration

Since D= A/V, where A is the amplitude of target displacement, it results
that v-max for a given value of V decreases with A. Then, even under the assumption of system linearity (constant K), the relationship between v-max
and V is nonlinear and amplitude dependent (see the theoretical results in
Fig.4).

As a matter of fact, the ratio v-max/V represents the gain of SPCS only if
the ramp response is given the time to reach the steady state. This is not
the case when small target excursions are covered with high velocities. For
each amplitude A there is an upper bound on target velocity beyond which
the steady state cannot be reached. This is also the upper bound of the range of linearity of v-max versus V characteristics.

The gain computed as the ratio between the peak values of eye and target velocities in the sinusoidal tests always correspond to steady state conditions. Moreover, within the range of frequencies considered in our experimen-

ts, SPCS was always working in the flat zone of its frequency response (see also the invariance of the phase shift given in Fig.3). Thus, the true gain characteristics of SPCS are likely to be represented by the diagrams in Fig. 2-B. For the largest examined amplitude of target oscillation (50° peak-to-peak) the gain is constant up to 75°/s, at least.

This interpretation of our experimental results also provides a simple expla nation for the contrasting results given in the literature about the range of linearity of SPCS. Actually, low levels of saturation were found by authors who used small amplitude ramp target displacements (Young (1962), Robinson (1965)), while higher velocities were recorded by authors who used larger ramps (Barmack (1970), Buizza et al.(1981)).

REFERENCES

1 - Bahill,A.T., Iandolo, M.J., and Troost, B.T., Smooth pursuit eye movements in response to unpredictable target waveforms, Vision Res. 20 (1980) 923-931.
2 - Barmack, N.H., Dynamic visual acuity as an index of eye movement control, Vision Res. 10 (1970) 1377-1391.
3 - Becker, W. and Fuchs, A.F., Predictive mechanisms in human smooth pursuit movement, in: Roucoux, A. and Crommelinck, M. (eds.), Physiological and Pathological Aspects of Eye Movements (W. Junk, The Hague, 1982).
4 - Buizza,A. and Avanzini, P., Computer analysis of smooth pursuit eye movements, in: Groner, R., Menz, C., Fisher, D.F., and Monty, R.A. (eds.), Eye Movements and Psychological Functions (L. Erlbaum Ass., Hillsdale N.J., in press).
5 - Buizza, A., Schmid, R., and Droulez, J., Influence of linear acceleration on oculomotor control, in: Fuchs, A.F. and Becker, W. (eds.), Progress in Oculomotor Research (Elsevier/North Holland, New York, 1981).
6 - Dallos, P.J. and Jones, R.W., Learning behavior of the eye fixation control system, IEEE Trans. AC-8 (1963) 218-227.
7 - Eckmiller, R. and Mackeben, M., Pursuit eye movements and their neural control in the monkey, Pflügers Arch. 377 (1978) 15-23.
8 - Melvill Jones, G. and Gonshor, A., Oculomotor response to rapid head oscillation (0.5-5.0 Hz) after prolonged adaptation to vision reversal, Exp. Brain Res. 45 (1982) 45-48.
9 - Michael, J.A. and Melvill Jones, G., Dependence of visual tracking capability upon stimulus predictability, Vision Res. 6 (1966) 707-716.
10 - Rashbass, C., The relationship between saccadic and smooth tracking eye movements, J. Physiol. 159 (1961) 326-338.
11 - Robinson, D.A., The mechanics of human smooth pursuit eye movement, J. Physiol. 180 (1965) 569-591.
12 - Stark, L., Vossius, G., and Young, L.R., Predictive control of eye tracking movements, IRE Trans. HFE-3 (1962) 52-57.

13 - Whittaker, S.G. and Eaholtz, G., Learning patterns of eye motion for foveal pursuit, Invest. Ophthalmol. Visual Sci. 23 (1982) 393-397.
14 - Young, L.R., A Sampled Data Model for Eye-tracking Movements, Sc.D. Thesis, Instrumentation Lab., M.I.T. (1962).

Work supported by C.N.R., Rome, Italy.

Theoretical and Applied Aspects of Eye Movement Research
A.G. Gale and F. Johnson (Editors)
© Elsevier Science Publishers B.V. (North-Holland), 1984

AUTOMATIC AND STRATEGIC CONTROL OF EYE MOVEMENTS IN SCHIZOPHRENIA

D. John Done and Christopher D. Frith

Division of Psychiatry,
Clinical Research Centre,
Watford Road, Harrow,
Middlesex, HA1 3UJ,
England.

The symptoms of schizophrenia can be explained by
recourse to the automatic - strategic paradigm of
cognitive psychology. Presented here are 2 experiments which show that the familiar pattern of slowed
RT in schizophrenia is only found for strategic
saccades. Tasks requiring an automatic saccade elicit
normal RT's in schizophrenics. Operational criteria
for strategic and automatic processes were validated.

Although there are numerous reports of deficits in the pursuit eye tracking
ability of schizophrenics there is some data (Levin et al., 1981; Iacono,
1981) indicating that saccadic eye movements are quite normal in this
group. The saccadic RT, amplitude, and speed of eye movement all appear to
be quite normal in schizophrenics. However the finding that saccadic RT is
normal in schizophrenics is of great interest since more traditional
methods of assessing RT, primarily manual procedures, invariably find that
schizophrenics are slower than normals (see Nuechterlein, 1977 for
summary). This finding is so universal that retardation is frequently
classed as a symptom of schizophrenia. It has also been a problem in the
interpretation of schizophrenic - normal differences in other reaction time
phenomena e.g. the crossover effect (Nuechterlein, 1977). This retardation
has been attributed to schizophrenic deficits in a number of hypothetical
constructs including inability to establish a segmental set, narrowed
attention, selective filter deficit, response competition and defective
biological motivation. As Cancro reported, "The literature on schizophrenia is a voluminous one, the size of which is in part determined by the
multiplicity of contradictions. There is a marvellous value in this since
one can, on an ad hoc basis, claim virtually anything has been demonstrated
and be correct" (Cancro, 1971, p. 352). There are at least three reasons
for the variability of results that produces this nebulous situation. These
are (a) criteria for patient selection, (b) varied methodology, (c) the
imprecision of constructs (e.g. attention) by which schizophrenic deviance
is measured. With regard to the first problem we believe that patients'
recent drug histories should be more carefully recorded since neuroleptics
have been reported to slow down RT's (Milner and Landauer, 1971) and
anticholinergics have clear effects on certain aspects of cognitive
function (Frith et al., 1983; Tune et al., 1982). We have used the PSE
criterion (Wing et al., 1974) for schizophrenia which avoids interjudge
variability. The problems of methodological variation and imprecision of
theoretical constructs are very much related. Variation of results due to
different methods usually occur because researchers make a priori assumptions about similar processes underlying different behaviours. It is only
when the different behaviours consistently produce different results that

an explanation is invoked which implicates processes. For example the robust crossover effect discovered by Shakow and others to be peculiar to schizophrenics for a finger lifting response appeared in normal adults when the response was a 90° knob twist (Kuehr, 1954). Finally the third problem of theoretical constructs was highlighted by a number of the reviews (e.g. Kopstein and Neal, 1972). In their study a large number of patients were rated on five different tasks (one of which was RT) which had been frequently used as indices of attention. However the highest intercorrelation between any two indices was 0.3. We have made an attempt to obviate these methodological problems by using two versions of the same method so that differences that accrue are not trivial. Now the concepts of "automatic" and "strategic" behaviours, although somewhat nebulous, do allow for some operational definitions. Strategic, complex, intentional processes appear to be carried out in a serial fashion by an "executive". Only one operation can be executed by the limited capacity executive at any one time such that a current ongoing operation will incur "costs" on any other operation that is performed. However automatic processes make no costs on the operation of the limited capacity executive and hence will not interfere with strategic processes. In addition automatic processes operate in parallel to other automatic processes since they do not interfere with each other unless they make demands on the same neural systems or the same effectors. Conversely "contexts" which facilitate or inhibit the speed of processing by the executive will not speed up or slow down the execution of automatic processes. During serial processing each stage has to be complete before the processing at a subsequent stage can commence, and usually a number of stages must be completed before a response can be selected. However during automatic processing all necessary processes are executed independently and in parallel such that the duration of information processing is equivalent to the execution time of the slowest single parallel process. Thus automatic processes tend to be executed with great rapidity and are not subject to a wide variability in RT since each parallel process is a coherent, stable unit; large variability results only when the serial link between units becomes uncertain.

In order to fulfil these demands we have examined the RT of saccades in schizophrenics. Posner (1978) reports, "Saccades seem to be forced upon us reflexively without the need for a conscious decision. On the other hand, we may choose to inhibit saccades or to make them voluntarily even without an external stimulus. There are experimental reasons for believing that changes in eye position to visual stimuli are more automatic (reflexive) than other performances studied in the laboratory such as hand movements" (p. 192). We have designed 2 similar tasks in which S. has to respond by making a saccade. In the first version saccades are elicited at a strategic level and in the second version execution appears to be automatic.

Method

Subjects

Thirteen acute schizophrenics who conformed to the criteria for schizophrenia in the PSE were given the strategic task. Data of one of these subjects failed to be recorded in the automatic version giving a sample size of 12. All of them had been neuroleptic free for at least a fortnight although they were all receiving benzodiazepines. 9/13 were first admissions and hence had never received neuroleptic treatment before. The remaining 4 patients were first admitted, on average, 2 years previously. The mean age of the sample was 27.9 years (range 20 - 43). The control group was made up of nursing staff and vacation students. Data was

obtained from eleven controls for the strategic task, but only ten of these
provided data on the automatic task. Mean age was 29 years (range 19 –
36). None of them had a previous history of any psychiatric illness.

Procedure

The 2 tasks both demanded of the subjects a saccadic eye movement 10° to
the left, or right, of fixation. One version was considered as producing
voluntary or strategic Saccades (SIS task). The other version produced
automatic initiation of saccades (AIS task). In both versions a subject
sat with his eyes approximately 40 cms away from a PET (CBM) VDU. On a
given trial the S. observed a rectangular box in the centre of the screen.
After a time interval (approx. 2.0 secs) a spot appeared in the centre of
the box either alone (50% trials) or together with a warning signal (50%
trials) which consisted of the letter "L" or "R" immediately above the
spot. At the same time two +s appeared 10° to left and right of the
fixation spot. Subjects were instructed that letters above the spot acted
as warning signals so that S. could prepare himself prior to receiving the
imperative stimulus. "L" indicated a leftward movement, "R" a rightward
movement. The time interval (SOA) between the appearance of the spot (the
"Ready" signal) and the imperative stimulus was either 0.25 secs (50%
trials) or 2.0 secs (50% trials). These conditions prevailed in both
versions. However with respect to the SIS task the imperative signal was a
presentation of the letter "L" or "R" <u>beneath</u> the spot after a 0.25 or 2.0
sec SOA. Approximately 0.5 sec after the imperative signal one of the
crosses at the periphery of the screen changed to the letter "O" and the
other cross changed to the digit "∅". S. was instructed that as soon as he
saw the imperative signal he had to move his eyes as quickly as possible in
the appropriate direction and report whether "the letter O had a line
passing through it, or not". S. was also instructed to observe the
fixation spot when it appeared.

In the AIS task the imperative signal consisted of the fixation spot moving
rapidly from the centre of the screen to the peripheral point marked with
the "+". On arrival the "+" transformed into either the letter "O" or the
digit "∅". S's were required again to report whether the character had a
line through it or not. In both versions there were 48 trials. This allowed
for 6 replications of each combination of the 3 conditions (direction of
movement, duration of SOA, presence/absence of warning). All eight combinations (2 x 2 x 2) were presented in a random order before repetition.

Eye movements were recorded with Beckman skin electrodes attached at the
outer canthus of each eye with a reference electrode on the back of the
right hand. Eye movements were amplified and recorded on an Akai 4000 DS
MIL-II tape recorder. The tape record was subsequently analysed using ELSA
(a package designed by Chris Frith for the analysis of psychophysiological
data) on a PDP 11/10. Eye movements were recorded over a 1.08 second epoch
commencing at the onset of the imperative signal. Sampling occurred every
5 msecs. A visual analogue of the eye movement was displayed on VDU along
with an analogue for the stimulus channel. This allowed for visual
examination of eye movements pre- and post- stimulus onset such that
non-central fixations, anticipations, blinks and other erroneous eye
movements could be eliminated before the marking routine analysed the
record. An automatic marking routine was used to standardize the markings
for onset and termination of saccades, thereby avoiding subjective biases.
Output from the routine provided the RT as well as the amplitude and
duration of the saccade.

Results

Only the RT data will be reported here. The average number of SIS trials
excluded from the analysis was 7.46 (15.5%) in the schizophrenics against
2.1 (4.4%) in the control group. These errors were composed mainly of
anticipations (i.e. moving the eyes before the imperative signal) and
misdirections. None were artefacts of the recording technique. However
with regard to AIS trials the error rate was much lower. Out of a total of
576 schizophrenics trials only 10 (1.7%) had to be excluded and out of 480
control trials only 14 (2.9%) had to be excluded.

Repeated measures analysis of variance (ANOVA) were calculated (BMDP2V
1981) for SIS and AIS data separately. Log_{10} transforms were made for
each individual RT and a mean for each of the 4 relevant types of trial
(SOA x warning) was used in the ANOVA. Direction of movement was considered
a necessary condition to increase response uncertainty but is of no great
interest here except for increasing the number of trials for the other
conditions. Therefore we will not consider direction in the subsequent
analysis.

Analysis of strategic saccades (SIS data)

Geometric mean RT's for each group are presented in table 1. Standard
errors obtained from the raw data are also given. The mean RT over all
conditions was slower in the schizophrenic Group, $F_{1,22} = 13.40$, p =
.001. There were also main effects of SOA, $F_{1,22} = 29.30$, p <.001 and
Warning, $F_{1,22} = 60.38$, p <.001. The effect of SOA was caused by a
reduced RT in the 2.0 sec condition (an average reduction of 65 msecs).
The effect of Warning was caused by the markedly reduced RT in the
conditions in which a warning of direction was given (an average reduction
of 89 msecs).

Table 1

Mean RT's (msec) and standard errors for strategic saccades

		+Warning of direction		No Warning of direction	
		0.25 SOA	2.0 SOA	0.25 SOA	2.0 SOA
Schiz (n=13)	Mean	477	418	552	531
	S.E.	31.94	24.36	25.7	21.72
Control (n=11)	Mean	383	279	452	376
	S.E.	34.4	23.26	36.5	32.8

There was a significant 2 way interaction of Group x SOA, $F_{1,22} =$
6.99, p = .01. This appears to be due to a larger group difference when
there was a 2.0 sec SOA than when the SOA was 0.25 secs (mean RT difference
between groups = 147 msec for a 2.0 sec SOA and 97 msec for a 0.25 sec
SOA). There was also a 2-way interaction of Warning x SOA, $F_{1,22} =$
6.03, p = .02. This result was caused by a larger warning/no-warning
difference at the longer SOA. This result is probably best explained by
considering that stimulus encoding and response preparations both take time
and that both procedures can be commenced prior to receiving the imperative
stimulus and hence a faster RT the longer the time allowed for preparation.

Analysis of automatic saccades (AIS data)

Geometric means and standard errors are presented in Table 2.

Table 2

Mean RT's (msec) and standard errors for automatic saccades

		+Warning of direction		No Warning of direction	
		0.25 SOA	2.0 SOA	0.25 SOA	2.0 SOA
Schiz (n=12)	Mean	240	227	241	215
	S.E.	18.84	15.5	15.1	13.2
Control (n=10)	Mean	221	203	236	213
	S.E.	8.56	7.76	8.24	13.61

Unlike the analysis of SIS data there was no overall slowing of the schizophrenic AIS RT's, $F_{1,20} < 1$, $p > .05$. There was only one significant main effect, which was the effect of SOA, $F_{1,20} = 34.92$, $p < .001$. This effect was caused by faster responses with a 2.0 msec SOA (a difference of 20 msecs).

Summary of results

Schizophrenics are much slower in making strategic saccades than age matched normal controls. However both groups produce similar RT's with automatic saccades. The familiar reduction of RT when a warning is given about target location for the impending saccade (see Posner, 1978) was very noticeable for strategic saccades, but not for automatic saccades. The familiar reduction of RT as the SOA increased from 0.25 secs to 2.0 secs was observed for both strategic and automatic saccades. Automatic saccades were executed much faster than strategic saccades.

Discussion

The results of these two experiments justify the separation of strategic and automatic modes of execution. Both tasks make the same demands on the efferent system and the difference between the two tasks in respect of stimulus presentation is only one of manner of information processing. Thus the two tasks differ only with respect to how information is encoded, and how responses are selected and the mediation between encoding and response selection. Now in the SIS task both warning signal and SOA influence RT. The warning signal allows information to build up about the subsequent imperative signal and organization of a specific response. SOA does neither. Posner (1978) found that the SOA effect, or "effect of alertness", was independent of conscious efforts to speed up RT. Thus when S. was given a strong incentive to perform fast in a speed-accuracy trade-off situation the effect of SOA was constant at different values of the trade-off. The SOA is therefore considered to be an automatic subcortical influence and akin to arousal. Posner (1978) also found in a letter match-ing task that the influences of SOA and pre-warning were independent and summative. It is therefore not surprising that the alerting effect of an increased SOA appeared in both strategic and automatic versions. However the effect of reducing uncertainty provided by the warning signal provides a strong indication that the strategic saccades were mediated by a central

executive whereas the automatic saccades were not. In addition the
standard deviations for RT's on the SIS task varied from 77 to 121, with
the schizophrenics having a smaller standard deviation in 3/4 cells. This
compares with standard deviations of 26 - 65 msecs in the AIS task. Indeed
3/4 standard deviations in the control group were 27, 26 and 24 with the
other being 42 msecs. Such small intersubject variability on a RT task
together with the lack of warning effect strongly suggest that the system
mediating between stimulus and response events in the AIS paradigm is
severely restricted.

Frith (1981), Done and Frith (1983) and Callaway and Naghdi (1982) have
used the automatic-strategic description of motor and cognitive function
(cf. Schneider and Shiffrin, 1977 for a theoretical description of the
theory) to make sense of the findings of schizophrenic dysfunctioning. The
Frith model holds that the monitor which selects behaviour routines operates according to a number of functions that cause particular routines to
enter the strategic mode of processing. Normally these functions delegate
the bulk of current routines to operate at the automatic level (e.g.
walking, talking and the subroutines of thinking). However Frith holds
that in schizophrenia the monitor ceases to work satisfactorily such that
these routines that should operate automatically are now executed at the
strategic level. This account neatly describes the disparate schizophrenic
disorders such as hearing one's own thoughts, delusions and disjointed or
incoherent speech. Conversely the inability of the monitor to terminate
routines operating automatically, and enter a strategic mode, could be used
to account for stereotypy, catatonia, lack of volition and withdrawal.

The Calloway and Naghdi hypothesis is more basic than that of Frith's.
They consider the schizophrenics only perform badly on tasks when execution
is at a strategic level. Automatic tasks can be performed normally, or
even better, by schizophrenic subjects. The results from our study on
saccadic eye movements strongly supports the theory of Calloway and Naghdi.
On the other hand the results appear to invalidate the Frith hypothesis.
However Posner (1978) has pointed out that "Saccades seem to be forced upon
us reflexively ... . There are experimental reasons for believing that
changes in eye position to visual stimuli are more automatic (reflexive)
than other performances studied in the laboratory such as hand movements."
(p. 192) Indeed whereas simple incompatible manual RT's are similar to
compatible manual RT's when the response is always made in the same
direction incompatible saccades are always slower than compatible ones
(Posner, 1978). Thus there are varying degrees of flexibility in automatic
tasks. Saccades elicited automatically are not modifiable by cognitive
strategy and hence are probably not accessible to a strategic level even in
schizophrenics, whereas simple manual responses are.

One result that has not been commented on is the SOA x Group effect in the
SIS version. We believe that this effect is to be expected in view of the
"crossover effect" which has been frequently reported (c.f. Neuchterlein,
1977). In brief the crossover effect occurs in RT tasks with predictable
SOA's. Schizophrenics unlike normals fail to improve their RT's in predictable series when the SOA exceeds 2 secs (Huston, Shakow and Riggs, 1937)
although this precise value varies between studies. So the facilitation of
a 2-sec over a 0.25 sec SOA in normals, in a predictable series, would not
occur in schizophrenics and so an interaction should be observed. One
could criticise this argument on the grounds that the series here is
unpredictable in which case the crossover effect should be absent. However
(a) unpredictable series normally use a large number of SOA's, (b) once

0.25 msecs of interval have elapsed without stimulus presentation stimulus onset is predictable. Therefore the use of only 2 very different SOA's allows S. to make predictions about the time of onset. Hence the observed interaction here is not unlike the results obtained in the experiments on the crossover effect.

REFERENCES

Calloway, L., and Naghdi, S., An information processing model for schizophrenia, Arch. Gen. Psychiat. 39 (1982) 339-47.

Cancro, R., Sutton, S., Kerr, J., Sugarman, A.A., Reaction time and prognosis in acute schizophrenia. J. Nerv. Ment. Dis. 153 (171) 351-59.

Done, D.J., and Frith, C.D. The effect of context on word perception in schizophrenic patients (submitted for publication).

Frith, C.D., Schizophrenia : An abnormality of consciousness? In: Underwood, G., and Stevens, R. (eds.), Aspects of Consciousness, Vol. II (Academic Press, London, 1981).

Frith, C.D., Richardson, J.T.E., Samuel, M., Crow, T.J., and McKenna, P.J., The effects of intravenous diazepam and hyoscine upon human memory. Q.J. Exp. Psychol. (in press).

Huston, P.E., Shakow, D. and Riggs, L.A. (1937). Studies of Motor Function in Schizophrenia: II. Reaction Time. Journal of General Psychology, 16.

Iacono, W.G., Tuason, V.B., Johnson, R.A., Dissociation of smooth-pursuit and saccadic eye tracking in remitted schizophrenics. Arch. Gen. Psychiat. 38 (1781) 991-96.

Kuehr, C.A., Schizophrenic reaction time responses to variable preparatory intervals. Am. J. Psychiat. 110 (1954) 585-8.

Kopstein, J.H., and Neale, J.M., A multivariate study of attention dysfunction in schizophrenia. J. Ab. Psychol. 18 (1972) 294-8.

Levin, S., Holzman, P.S., Rottenberg, S.J., and Lipton, R.B., Saccadic eye movements in psychotic patients. Psychiatry Res. 5 (1981) 47-58.

Milner, G., and Landauer, A.A., Alcohol, thioridazine and chlorpromazine effects on skill related to driving behaviour. Br. J. Psychiat. 1971 351-2.

Nuechterlein, K.M., Reaction time and attention in schizophrenia : A critical evaluation of the data and theories. Schiz. Bull. 3 (1977) 373-428.

Posner, M.I., Chronometric explorations of mind (Lawrence Erlbaum Hillsdale 1978)

Schneider, W. and Shiffrin, R.M., Controlled and automatic information processing. Psych. Rev. 81 (1977) 1-66.

Tune, L.E., Strauss, M.E., Lew, M.F., Breitlinger, E., Coyle, F.T., Serum levels of anticholimergic drugs and impaired recent memory in chronic schizophrenic patients. Am. J. Psychiat. 139 (1982) 1460-2.

Wing, J.K., Cooper, J.E., and Sartorious, N., The description and classification of psychiatric symptoms: An instruction manual for the PSE and Catego systems. (Cambridge Univ. Press, Cambridge, 1974).

Theoretical and Applied Aspects of Eye Movement Research
A.G. Gale and F. Johnson (Editors)
© Elsevier Science Publishers B.V. (North-Holland), 1984

EYE TRACKING PERFORMANCE AND ATTENTION IN ALCOHOL ABUSERS AND PSYCHOTIC PATIENTS

SVEN INGMAR ANDERSSON
Department of Community Health Sciences
S-240 10 Dalby
and
Department of Psychology
S-223 50 Lund
SWEDEN

Pendular eye tracking performance of a group (n=20) of alcohol abusers was studied by means of electrooculography, retests being made on 15 of the patients within a three-month period. An electronic pendulum involving a "moving" light of constant speed was employed, there being two different tasks, a less attention demanding task (following a red light only) and a more attention demanding one (besides following the light, having to press a button when the light turned green, at short, irregular intervals). The results were compared with those of psychiatric patients (n = 19) and normals (n=4) whom the author studied earlier. Although it was not possible to distinguish psychotic patients from alcohol abusers on the basis of overall measures of noise/signal ratio and deviation area, microtremor was found to be categorically higher in psychotic patients than in alcohol abusers. Since no differences were observed between psychotic patients and alcohol abusers in patterns of response as cognitive strain increased or over time during a testing session, these aspects of the tracking tasks did not seem to account for the very large differences found in microtremor, leaving the possibility open that disturbance of a central nature was found in the psychotic patients. During eye tracking, macro squarewave jerks were significantly more frequent in alcohol abusers than in psychiatric patients. It is suggested that in alcohol abusers this phenomenon represents a defect in a saccadic pulse generator, triggered by an unwanted supranuclear command signal.

Alcohol in moderate amounts has been reported earlier to impair the two basic types of conjugate eye movements used for tracking a moving target, namely those of the saccadic and the smooth pursuit systems (Drischel, 1968; Franck and Kuhle, 1970; Wilkinson, Kime and Purnell, 1974; Flom, Brown, Adams and Jones, 1976; Guedry, Gilson, Schroeder and Collins, 1975; Umeda and Sakata, 1978; Baloh, Sharma, Moskowitz and Griffith, 1979; Levy, Lipton and Holzman, 1981; Lehtinen, Nyrke, Lang, Pakkanen and Keskinen, 1982). In all of these studies, however, the influences of alcohol on pursuit movements was examined using healthy young people as subjects. In the present study, therefore, the aim was to investigate the effects of alcohol on smooth pursuit in a group of alcohol abusers. Although it is generally believed that even moderate doses of ethanol disrupt smooth pursuit and result in saccadic pursuit, due to compensation for a reduced ability of the eyes to maintain foveation, little is known of the smooth pursuit capacity in excessive users of alcohol.

Levy et al. (1981) studied pendular eye-tracking performance in a group of healthy young non-alcoholics before and after administration of single doses of alcohol. They found that a number-reading task they presented simultaneously, one which Shagass, Amadeo and Overton (1974) and Holzman, Levy and Proctor (1976) reported as reducing eye-tracking impairment in schizophrenics, tended to alleviate the alcohol-induced disruption of eye-tracking performance. In the number-tracking task,   arabic numerals were displayed on the pendulum,   with different numbers appearing as the pendulum oscillated,   the subject being instructed to read the numbers silently while tracking the pendulum.   The authors attributed the effect of the number-reading task to its facilitating the visual fixation mechanism,   which they considered to have been disrupted by alcohol.   At the same time,   Shagass et al.   (op.   cit.) and Holzman et al (op.   cit.) suggested that the number reading task only partially compensates for psychosis-induced disruption of eye-tracking performance,   the magnitude of the "normalizing effect" varying with the extent of the impairment.   This made it appear of interest in the present study to explore the effects of attentional manipulations on the disruption of smooth pursuit performance in chronic alcoholics,   one hypothesis being that the central nervous system dysfunction which alcohol produces is more pronounced in these subjects.

In earlier investigations,   the present author (Andersson,   1983a,   1983b) used electrooculographic methods to study smooth pursuit eye movements and attention in psychotic subjects,   using an electronic pendulum which produced variations in the attentional demands of the task.   Differences were found between schizophrenic and non-schizophrenic psychotics on such measures as noise/signal ratio,   deviation area,   microtremor rate and various of the harmonics.   In the present study,   tasks identical to those used in one of the earlier investigations (Andersson,   1983b) were given to a group of alcohol abusers,   i.e.   to persons suspected of having central nervous system disturbance due to excessive use of ethanol.   Their eye tracking is compared to that of the group of psychiatric patients studied in the earlier investigation just cited,   for all of whom central nervous system involvement,   though conceivable,   is at least not documented.

Method

Subjects

The mean age and the size of the various subject groups is presented in Table 1.   Regarding the psychotic comparison groups,   all the members of which had been remitted to a psychiatric hospital for short-term therapy,   further details have been published elsewhere (Andersson,   1983b).   All of the alcohol abusers were males,   whereas 8 of the 19 psychotic patients (4 schizophrenics and 4 cycloid psychotics) were women.

Table 1.   Characterization of the Subjects

Group	N	Age	
		M	SD
Schizophrenics	9	30.9	9.3
Mano-depressive psychotics	2	38.5	16.3
Cycloid psychotics	8	26.6	5.6
Alcohol abusers	20	41.2	11.3

Among the alcohol abusers (n = 20),   13 subjects (nos.   1-13) were tested twice within the four month period of data collection.   In 9 of the cases tested twice,   retesting occurred within 10 days (subjects 2-10); in the remaining 4 cases (subjects 1,   11-13) retesting came at the end of the data-collection period.   On the first testing occasion,   17 of the 20 subjects had been receiving no medicine for at least a week,   whereas two subjects (nos.   4 and 14) had been administered benzodiazepine derivates,   and one subject (no.   3) had been on antihistamine for at least a week prior to testing.   Seven of the 13 subjects who were tested a second time had at that point received no medicine for at least a week prior to testing, while five subjects were on benzodiazepine derivates (nos. 4-8). In addition, at this second testing subject 6 had been administered klomeliazole, and subject 3 antihistamine.

Eleven of the subjects (nos. 2-9 and 15-17), when tested the first time, had sought out the clinic after heavy ethanol consumption lasting for about a week. Five of the other subjects (nos. 1,10,18-20) had consumed no alcohol within a period of at least two weeks, while three subjects had consumed beer or wine but no heavy alcohol for the week prior to testing. One subject (no. 11) declared he had drunk no alcohol, although a smell of ethyl alcohol was noted. When tested on the second occasion, three subjects (nos. 1,3 and 11) had sought out the clinic due to acute alcohol intoxication. Seven of the subjects (nos. 4, 9, 11, 15, 16 and 19) were characterized as showing mild alcohol abuse.

All but one of the psychotic subjects were receiving phenothiazine derivates or related substances (derivates of thioxanthene or butyrophenone); two of the cycloid psychotics and one of the manic-depressive psychotics were on lithium citrate; the latter patient and one of the cycloid psychotics were on benzodiazepine derivates; five of the cycloid psychotics and two of the schizophrenics were receiving anticholinergic drugs.

Recording Procedure

Exact details of the testing procedures are given in Andersson (1983b). An electronic pendulum, involving a "moving" light of constant speed, provided by special light diodes which could light up either red or green, was employed, the moving spot being red most of the time but becoming green for short, irregular intervals. The task involved three periods of tracking. In the first period (60 sec) the target was red. In the second period, in which there were 180 sec of tracking following a 60 sec pause, the stimulus color changed temporarily from red to green at irregular intervals, subjects being instructed to press a button each time the light changed to green; a green segment, when it occurred, comprised 1/4 of the "swing" of the pendulum in the one direction or the other. The third tracking period, 60 sec in length following a 30-sec pause, involved (as did the first) a red stimulus only.

Eye movements were registered in the AC mode with an electrooculographic technique, using a mingograph (a Beckman R-511A Dynograph Recorder). Silver-silver chloride electrodes were placed at the outer canthi for horizontal recording, and above and below the right eye for recording blinks. A ground electrode was attached to the middle of the forehead. The placement of the electrodes for horizontal recording aimed at minimizing the appearance of blinks on the recordings. Calibration was facilitated by having the subject shift his gaze several times between the two endpoints of the pendulum. The preamplifier of the mingograph drove a Tandberg (TIR) FM tape recorder operating at a 3 3/4 IPS; it was RC-coupled to balance out DC voltage, with a high pass filter (cut-off frequency 0.16 Hz) and a low-pass filter (cut-off frequency 30 Hz) for noise reduction.

Measures Obtained

The data for any given subject were processed in basic accordance with Andersson (1983b) for each of four different slightly over 30-sec segments of the total tracking task separately, each such segment comprising 12 cycles (or slightly more than this) of the pendulum. Periods 1-4 involved successive portions of the pendular task--Period 1 representing the initial segment of the first of the two red-light-only tasks, Period 2 the initial segment of the two red-and-green-light tracking tasks, Period 3 a segment of the latter task commencing 90 seconds after the end of Period 2, and Period 4 the initial segment of the second red-light-only tracking task. Tape-recorded signals were digitized for each of the approximately 30 second segments. Data-processing involved computing eye position P from electrical voltage V at each of the sampling points through solving the differential equation $a \cdot dP/dt - b \cdot V$, where dv/dt represents change in V over time. A correction was employed for drift; this consisted in subtracting from V its mean value within each cycle. Each recording segment was shortened to exactly 12 cycles, a preliminary Fourier analysis being used in identifying cycles here. A more thoroughgoing Fourier analysis was also carried out. The following measures were obtained for each of the four periods: (1) Noise/signal ratio : the ratio between the sum of the squared amplitudes of the 2nd, 3rd, 4th, 5th and 6th harmonics, on the one hand, and the squared amplitude of the major wave (1st harmonic) on the other. For a subject following the movements of the pendulum closely, this measure would be low. (2) Deviation area : the area between the two curves describing the movements of the pendulum and of the subject's eyes, respectively. If the subject followed the pendulum closely, this measure would be low. (3) Microtremor rate : (d(tot)-d(eff))/d(eff), where d(tot) represents the total distance and d(eff) the "effective" distance the eye covered, and where d(tot) and d(eff) represent the sum of all small eye movements as measured through recording eye position at 3.66 msec intervals, and at intervals nine times larger (33 msec), respectively. For each such measure, a summary measure of the subject's overall tracking performance (OTP) during the four periods together was calculated. In addition two types of difference score, Dif-1 and Dif-2, were obtained for each of the measures (1) to (3) above, in accordance with Andersson (1983a). Dif-1, the difference (1+4)-(2+3), where the numbers here refer to the successive periods, indicates the effect on eye tracking performance of changing from the less attention-demanding conditions (red-light-only) to the more demanding (changing light color). Dif-2, the difference (1+2)-(3+4), represents the effect of practice. Using these summary measures, comparisons were made between the alcohol abuser and the psychotic groups, the latter group comprising all three groups of psychotic patients mentioned above. Comparisons were made between occasions A (first testing) and B (retest) for all the above measures.

Results

Inspecting the mingographic recordings for irregularities, the most striking characteristic among the majority of alcohol abusers was the frequent presence of macro squarewave jerks in the same direction as that of the target, a macro squarewave jerk being defined here as a complex comprising an initial saccadic deviation of at least 10^{0} amplitude from actual tracking, followed after normal reaction time (about 200 msec) by a catch-up saccade in the direction opposite to that of the initial saccade. Such deviations as counted from computer graphs of foveal position vis à vis the target (see Figure 1 for an example), were found to a significantly more frequent extent in alcohol abusers than in psychiatric patients, occurring at least four times in total during the 30-sec periods of tracking in 8 of the 20 alcohol abusers but in only 2 of the 19 psychiatric patients (p=0.039, Fisher test), corresponding figures in occasion b being 5 of 13 and 1 of 15, respectively (p=0.023, Fisher test). A typical saccadic performance, on the other hand, was evident in subject 1 when tested approximately 24 after having finished a one week-period of heavy ethanol consumption (Figure 2).

On occasion A, the seven mild abusers were found to have very few macro squarewave jerks (less than 4 in total), whereas eight of the heavier abusers (n=13) demonstrated macro squarewave jerks to a significantly higher

L041

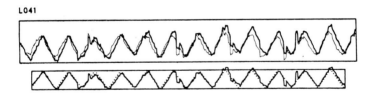

Figure 1. Eye tracking pattern in an alcohol abuser (no.6) during occasion 1 (Period 1) with macro squarewave jerks appearing, all in the same direction as the target. Movement to the right = upwards, to the left = downwards. The lower, more abbreviated record has been corrected for drift and shortened to 12 cycles. Thin curve = initially registered voltage; thick curve = foveal position; dotted, saw tooth curve = stimulus (light) position.

B124

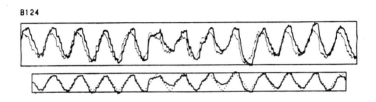

Figure 2. Saccadic eye movements in a alcoholic abuser (no. 1) with heavy alcohol comsumption during the week prior to testing.

extent (range 4 - 15 in total each; p=.010, Fisher test). No such relation was evident when comparing subjects who had consumed alcohol during an extended period prior to testing on occasion A (nos. 2-9 and 15-17) and those who had no such recent history of drinking (in the former group, 5 showing values greater than 4, and 6 smaller than or equal to 4, the corresponding figures for the latter group being 2 and 6, respectively).

In a descriptive sense it could be noted that two of the mild alcohol abusers (subjects 8 and 16) who on occasion A had shown no occurrence of macro squarewave jerks (no. 8) or only one occurrence (no. 16), demonstrated such deviations to a greater extent on occasion B, about one week later (showing in total 9 and 18 such deviations, respectively). At the same time, subject 3 showed 12 macro squarewave jerks in total on occasion A, when not having consumed alcohol for some time, but no such deviation at all (the total pattern being saccadic) when tested on occasion B about one week later, after a period of heavy drinking.

OTP-scores: No significant differences between the psychotic patients and the alcohol abusers were found regarding either noise/signal ratio or deviation area as measured on occasions A and B. Both groups showed relatively high values at first testing as compared with the small sample of normals (n=4) tested once in Andersson (1983a), two of the normals showing the lowest noise-signal ratio of all the subjects. As to occasion A, four of the seven alcohol abusers who had only mild alcohol abuse were found within the range of normals, both as regards noise/signal ratio and deviation area, the four subjects involved being the same in both cases (nos. 19, 14, 16 and 9).

The most striking result was that concerning microtremor rate. Here the difference between the alcohol abusers and psychotic patients was categorical, all the alcohol abusers scoring lower than any of the psychotic patients, whereas the normals all showed values rather similar to those of the alcohol abusers (range being 3.067 - 5.752 for normals, who were tested on occasion A only, its being 2.013 - 5.610 on occasion A and 0.737 - 2.999 on occasion B for the alcohol abusers, and 9.081 - 23.478 on occasion A and 14.749 - 24.284 on occasion B for the psychotic patients).

Dif-1 and Dif-2 scores: For noise/signal ratio performance was found to be better generally under the more than under the less attention-demanding conditions (i.e, Dif-1 scores were predominantly positive), both for occasion A (13+, 7-, n.s., chi^2 here, as in the following) and occasion B (9+, 3-, n.s.). Andersson (1983b) found a similar tendency among psychotic patients, both for occasion A (15+, 4-, p<.02) and occasion B (12+, 3-, p<.05). The results for the alcohol abusers appeared reliable, only 2 of 12 subjects (nos. 5 and 13) showing a different sign (+ or -) on occasion B than A (one subject, in addition, changing from + to no difference), quite in line with what was found in the psychotic group, where corresponding figures were 2 of 15 patients. As to Dif-2 scores on occasion A, alcohol abusers showed a decrease in noise/signal ratio over time (15 decreasing, 5 increasing, p<.05), while in psychotic patients a certain tendency to the contrary was found (7 decreasing, 12 increasing, n.s.). A tendency to perform better on this measure under the more than under the less attention demanding conditions, as well as over time during a testing occasion, was evident also among the 11 patients who, when tested the first time, had sought out the clinic after consuming ethanol heavily for about a week prior to testing. Comparing occasions A and B, changes occurred in 8 of 13 cases, seven subjects changing from + to - (nos. 2, 5, 7, 9, 10, 12 and 18) and one only (no. 13) from - to +.

Concerning deviation area, there was no indication, such as had been observed in psychotic patients (Andersson 1983a, 1983b), that performance was better under the more than under the less attention-demanding conditions (dif-1 scores on occasion A: 10+, 10-, n.s.; on occasion B: 5+, 8-, n.s.). Comparing occasions A and B, only 5 of 13 patients showed the same type of tendency on both occasions (3++, 2--), whereas 13 of the 15 psychotic patients in Andersson (1983b) demonstrated high retest reliability (7++, 6--).

In alcohol abusers, microtremor rate tended to be higher under the more than under the less attention demanding conditions (Dif-1) on occasion A (14+, 6-, n.s.), whereas, on occasion B, the number of increases and decreases was equal (7+, 7-, n.s.). As to Dif-2 scores, there was a tendency for microtremor rate to decrease over time, both on occasion A (12-, 8+, n.s.) and occasion B (8-, 6+, n.s.). Likewise, on occasion A, the above mentioned group of eleven subjects with heavy ethanol consumption just prior to their first testing showed a tendency, although not significant, in the same direction as that of the total abuser group. Andersson (1983b) found microtremor rate in psychotic patients as measured by Dif-1 scores to also be predominantly positive (17+, 2-, p<.001) on occasion A,   whereas on occasion B results were not significant (9+,   6-).   As to Dif-2 scores in the psychotic patients,   a decrease over time was found,   both on occasion A (14-,   5+,   p<.05) and occasion B (12-,   3+,   p<.05).

Discussion

Views differ regarding the phenomenon of macro squarewave jerks.   Couch and Fox (1934),   studying 117 patients with various mental diseases,   reported lapses of attention in 60 of the patients,   who otherwise displayed what was interpreted as normal ocular response.   In pendular pursuit such evidence of inattention included deviations ahead of the stimulus,   followed after a brief period by refixation to the predetermined position of the stimulus.   However, these authors explicitly observed lapses of attention to be extremely rare in records from normal subjects.

Jung and Kornhuber (1964),   investigating tracking movements of the eyes to a horizontal pendulum in patients with various neurological syndromes,   also indicated normal subjects to in some cases demonstrate interruptions of an otherwise sinusoidal curve by quick intermittent fixations of other objects, although no information is given as to number of normals studied.

Examining patients with otoneurological disorders in response to a pendular stimulus,   Benitez (op.   cit.) considered as normal even recordings of patterns which were characterized by a few intermittent non-nystagmic movements superimposed on a sinusoidal curve.   Although it is at least unclear whether a sample of otoneurological patients can be considered as representative of normals, this view seems to have pervaded later investigations (e.g.,   Holzman,   Proctor, Levy,   Yasillo,   Meltzer and Hurt,   1974; Lipton,   Frost and Holzman,   1980). The result reported in the present investigation,   showing psychotic patients to have significantly less macro squarewave jerks than alcohol abusers and also indicating the heavier alcohol abusers to demonstrate these phenomena to a greater extent than mild alcohol abusers,   may instead suggest pathological involvement to lie at the basis for this phenomenon (cf. Dell'Osso,   Troost and Daroff,   1975).   As is evident from inspection of the graphs showing target position vis a vis foveal position (see Figure 1 for an example),   there is no evidence for nystagmic involvement here,   the initial (abnormal) saccades of the macro-square complexes not being correlated with delayed foveal targeting. Rather,   they may perhaps be interpreted in terms of a saccadic pulse generator being triggered by an unwanted supranuclear command signal (cf.   Zee and Robinson,   1978).

Although it was not possible to distinguish psychiatric patients from alcohol abusers on the basis of the overall measures of noise/signal ratio or deviation area,   there is another phenomenon which merits special interest,   namely that of microtremor being categorically higher in psychotic patients than in alcohol abusers.   Although no differences were observed between psychotic patients and alcohol abusers in patterns of response over time during a testing session or as cognitive strain increased,   these aspects of the tracking task did not seem to account for the very large differences found on this measure,   leaving the possibility open that disturbance of a central nature is present in the psychotic patients.

REFERENCES

Andersson, S.I., Smooth pursuit eye movements and attention in schizophrenia and cycloid psychoses, in Groner, R., Menz, C., Fisher, D.F. and Monty, R.A. (eds.), Eye Movements and Psychological Functions: International Views (Lawrence Erlbaum, London, 1983a, in press).

Andersson, S.I., Eye tracking performance and attention in psychotic patients, Documenta Ophthalmologica (1983b) in press.

Baloh, R.W., Sharma, S., Moskowitz, H., Griffith, R., Effect of alcohol and marijuana on eye movements, Aviation, Space, and Environmental Medicine 50 (1979) 18-23.

Benitez, J.T., Eye-tracking and optokinetic tests: diagnostic significance in peripheral and central vestibular disorders, Laryngoscope 80 (1970) 834-848.

Couch, F.H. and Fox. J.C., Photographic study of ocular movements in mental disease, Archives of Neurology and Psychiatry 31 (1934) 556-578.

Dell'Osso, L.F., Troost, B.T. and Daroff, R.B., Macro square wave jerks, Neurology 25 (1975) 975-979.

Drischel, H., The frequency response of horizontal pursuit movements of the human eye and the influence of alcohol, Progress in Brain Research 22 (1968) 161-174.

Flom, M.C., Brown, B, Adams, A.J. and Jones, R.T., Alcohol and marijuana effects on ocular tracking, American Journal of Optometry & Physiological Optics 53 (1976) 764-773.

Franck, M.C. and Kuhle, W.K., Die Wirkung des Alkohols auf die raschen Blickzielbewegungen (Saccaden) beim Menschen, Archiv für Psychiatrie und Nervenkrankheiten 213 (1970) 235-245.

Guedry, F.E., Gilson, R.D. Schroeder, D.J. and Collins, W.E., Some effects of alcohol on various aspects of oculomotor control. Aviation, Space, and Environmental Medicine 45 (1975) 1008-1013.

Holzman, P.S., Levy, D.L. and Proctor, L.R., Smooth pursuit eye movements and functional psychoses; a review. Schizophrenia Bulletin 3(1977) 15-27.

Holzman, P.S., Proctor, L.R., Levy, D.L., Yasillo, N.J., Meltzer, H.Y., and Hurt, S.W., Eye-tracking dysfunction in schizophrenic patients and their relatives, Archives of General Psychiatry 31 (1974) 143-151.

Jung, R. and Kornhuber, H.H., Results of electronystagmography in man: the value of optokinetic, vestibular, and spontaneous nystagmus for neurologic diagnosis and research, in Bender, M.B. (ed.), The Oculomotor System (Harper & Row, New York, 1964).

Lehtinen, I., Nyrke, T, Lang, A.H., Pakkanen, A. and Keskinen, E., Quantitative effects of ethanol infusion on smooth pursuit eye movements in man, Psychopharmacology 77 (1982) 74-80.

Levy, D.L., Lipton, R.B. and Holzman, P.S., Smooth pursuit eye movements: effects of alcohol and chloral hydrate, Journal of Psychiatric Research 16 (1981) 1-11.

Lipton, R.B., Frost, L.A. and Holzman, P.S., Smooth pursuit eye movements, schizophrenia, and distraction, Perceptual and Motor Skills 50(1980) 159-167.

Shagass, C., Amadeo, M. and Overton, D.A., Eye-tracking performance and engagement of attention, Archives of General Psychiatry 33 (1976) 121-125.

Umeda, Y. and Sakata, E., Alcohol and the oculomotor system. Annals of Otology, Rhinology and Laryngology 97 (1978) 392-398.

Wilkinson, I.M.S., Kime, R. and Purnell, M., Alcohol and human eye movements, Brain 97 (1974) 785-792.

Zee, D.S. and Robinson, D.A., A hypothetical explanation of saccadic oscillations, Annals of Neurology 5 (1979) 405-414.

Theoretical and Applied Aspects of Eye Movement Research
A.G. Gale and F. Johnson (Editors)
© Elsevier Science Publishers B.V. (North-Holland), 1984

PSYCHOTROPIC DRUG EFFECTS ON SMOOTH PURSUIT EYE MOVEMENTS: A SUMMARY OF RECENT FINDINGS

Deborah L. Levy[1,4], Richard B. Lipton[5], Nicholas J. Yasillo[2],
James Peterson[4], Ghanshyam Pandey,[3,4] John M. Davis[3,4]

Departments of Psychiatry[1] and Radiology[2], University of Chicago
Department of Psychiatry[3], University of Illinois
Department of Research[4], Illinois State Psychiatric Institute
Chicago, IL, USA
Department of Neurology[5], Albert Einstein College of Medicine
Bronx, New York, USA

INTRODUCTION

In studies conducted prior to the availability of psychotropic drug treatments, impaired smooth pursuit eye movements (SPEM) were a more common finding in schizophrenics than among manic-depressives or other psychiatric patients (Diefendorf and Dodge, 1908; Couch and Fox, 1934). The results of recent studies on never-medicated patients support these early data both in terms of the relative specificity of the eye movement abnormality for schizophrenia and the independence of SPEM impairment from treatment with antipsychotic medications. Karson (1979), for example, reported that pursuit irregularities were rare in never-medicated manic-depressives. In contrast, never-medicated schizophrenics had worse pursuit than their medicated counterparts (Karson, 1979; David, 1980).

Unmedicated patients are more the exception than the rule, however, and most of the evidence bearing on the issue of drug effects is indirect. We know, for example, that short-term drug withdrawal does not change pursuit compared to performance while medicated (Holzman et al., 1974; Spohn, 1981). SPEM abnormalities are also frequently found in schizophrenics who have not received antipsychotic drugs for periods as long as six years (Stevens, 1978). Although a persisting and possibly irreversible drug effect cannot be ruled out when abnormal pursuit occurs in drug withdrawn patients, the studies on never-medicated patients suggest that drug treatment is not a necessary condition for the presence of impaired SPEM. The results of family studies also support this interpretation. As many as half of the first-degree relatives of schizophrenics, none of whom have taken antipsychotic drugs, have abnormal SPEM (Holzman et al., 1974, 1978, 1980; Kuechenmeister et al., 1979). Moreover, there is a significant intrafamilial association between impaired pursuit in a schizophrenic patient and at least one other member of the same family, particularly a parent (Holzman and Levy, 1977).

Antipsychotic drugs are also not a sufficient condition for the presence of impaired SPEM. Just as unmedicated patients may show abnormal pursuit, many patients taking antipsychotic drugs have normal pursuit. High doses of haloperidol (60 mg IM) do not result in abnormal SPEM in patients with normal pursuit prior to drug treatment (Levy et al., 1983a). This finding supports other reports indicating no relationship between measures of SPEM and drug treatment, including no apparent interaction between medication status and diagnosis (Holzman et al., 1974; Shagass et al., 1974; Mialet and Pichot, 1981; Lipton et al., 1983). Moreover, single doses of chlorpromazine and diazepam do not produce qualitative changes in SPEM in normal controls (Holzman et al., 1975), although diazepam has been shown to reduce pursuit gain (Rothenberg and Selkoe, 1981; Bittencourt et al., 1983). Impaired pursuit has not been reported among psychiatric patients receiving diazepam, possibly because eye velocity has not been assessed.

As suggestive as the evidence summarized above may be, psychotropic drug effects on smooth pursuit have not been studied systematically. This issue assumes particular importance in the context of conflicting reports regarding the specificity of SPEM impair-ment for schizophrenia. Despite the consensus that impaired pursuit is a more frequent occurrence in schizophrenics than in nonpsychotic patients or normal controls (Holzman et al., 1973, 1974, 1978; Shagass et al., 1974, 1976; Klein et al., 1976; Kuechenmeister et al., 1977; Pass et al., 1978; Salzman et al., 1978; May, 1979; Pivik, 1979; Karson, 1979; Cegalis and Sweeney, 1979; Lipton et al., 1980; Iacono et al., 1981; Mialet and Pichot, 1981; Tomer et al., 1981), comparisons of schizophrenics and patients with affective disorders have yielded more variable findings. In some studies SPEM impairment was significantly more prevalent in schizophrenics than in manic-depressives (Diefendorf and Dodge, 1908; Couch and Fox, 1934; Holzman et al., 1973, 1974; Karson, 1979). Others, however, have reported no differences between schizophrenics and those with affective disorders (White, 1938; Shagass et al., 1974; Lipton et al., 1980; Klein et al., 1976; Salzman et al., 1978).

Differences in psychopharmacological treatments among diagnostic groups could explain some of these inconsistencies. Particular classes of psychotropic drugs are more commonly used in treating some psychiatric syndromes than others, as in the use of lithium carbonate to treat manic-depressive illness. If a particular class of drugs impairs pursuit and these drugs are given preferentially to patients in certain diagnostic categories, drug effects cannot be distinguished from biological deviations associated with psychiatric syndrome. This problem arose in a recent study by Iacono and colleagues (1982). Patients taking lithium carbonate had higher tracking error scores (root mean square error deviation) than patients who were not taking lithium carbonate. However, the diagnosis of bipolar affective disorder was confounded with treatment with lithium, making it impossible to distinguish between drug and diagnostic group effects. This same methodological problem complicates interpre-tation of studies on patients taking antipsychotic and antidepressant drugs as well. SPEM impairment is most prevalent among chronic schizophrenics (Holzman et al., 1974), who also have the longest histories of treatment with antipsychotic medications. It occurs least frequently among psychiatric patients in nonpsychotics, who are less likely to be exposed to antipsychotic drugs than schizophrenics or manic-depressives.

Although the results of studies on never-medicated patients make it unlikely that pursuit impairment in schizophrenics is drug-induced, the well-documented effects of alcohol and barbiturates on SPEM (Rashbass and Russell, 1961; Norris, 1968; Baloh et al., 1979; Levy et al., 1981) underscore the sensitivity of this ocular motor control system to pharmacological agents. Antipsychotics and antidepressants, as well as lithium, have motor side effects. The antipsychotic drugs cause side effects with some of the neurological features of idiopathic Parkinson's Disease, in which SPEM can be impaired (Shibasaki et al., 1979). If nigro-striatal dopamine blockade by antipsychotics produces oculomotor effects similar to those found in dopamine-depleted Parkinson's patients, the prevalence of pursuit impairment would be inflated in groups on antipsychotics.

We report a summary of our longitudinal data (Levy et al., 1983c,d) on the effects of antipsychotic and antidepressant medications, as well as of lithium carbonate, on smooth pursuit eye movements.

METHODS

SUBJECTS. All subjects were recently admitted to a psychiatric hospital, and most were psychotic at the first testing. Patients who were taking psychotropic medication at admission underwent a two-week drug-free period prior to a baseline testing. Psychiatric diagnoses were assigned based on a Schedule for Affective Disorders and Schizophrenia-Lifetime (SADS-L) interview (Endicott and Spitzer, 1978), using DSM-III criteria (1980). Smooth pursuit eye movements were evaluated before and during treatment with I) an antipsychotic drug, II) an antidepressant drug, or III) lithium carbonate or Lithobid. Group I consisted of 15 patients with the following diagnoses: schizophrenia (9), paranoid disorder (3), personality disorder (2), and schizo-affective illness (1). Three of the schizophrenic patients had never received psychotropic drugs prior to admission. The average age of patients in Group I was 28.3 years (SD, 9.3 years; range, 18-47 years). Baseline testing was conducted a minimum of once and a maximum of four times. Antipsychotic drug effects were assessed after as early as one day of treatment and after a maximum period of two months. The number of medicated tests per patient varied from as few as one to as many as seven. These patients received the following antipsychotic medications; daily dosage range is listed in parentheses: trifluoperazine, 8 (10-60 mg.), fluphenazine decanoate, 2 (.5-2.5 cc; one patient was also treated concurrently with 20 mg. of fluphenazine hydrochloride), thiothixene (15-40 mg.), haloperidol (10 mg.), perphenazine (16-32 mg.), chlorpromazine (400-1000 mg.), loxapine (20-40 mg.), 1 each. The patient who received loxapine was later treated with 300 mg. of thioridazine and was tested on each medication. Group II consisted of nine patients: major depressive disorder (5), personality disorder (2), recurrent depression (1), and minor depressive disorder (1). Two patients in Group II had never before received psychotropic drugs. The average age of patients in this group was 30.4 years (SD, 8.7 years; range, 20-49 years). Baseline testing was conducted a minimum of once and a maximum of seven times. The length of drug treatment at the time of SPEM assessment ranged from one week to 9½ months. The number of medicated tests per patient varied from as few as one to as many as ten. These patients received the following antidepressant medications; daily dosage range is listed in parentheses: amitriptyline, 3 (100-200 mg.), desipramine, 1 (150 mg.), imipramine, 2 (75-225 mg.), fluoxetine, 2 (50-80 mg.), maprotiline, 1 (150 mg.). The seven patients in Group III met DSM-III criteria for bipolar affective disorder. The average age of these patients was 32.7 years (SD, 10.84 years; range, 19-48 years). Four of these seven patients had never before taken lithium carbonate, but all had had previous psychotropic drug treatment. Four patients were tested while receiving lithium alone and three patients were tested while receiving lithium alone, and again during treatment with an antipsychotic (thiothixene, 2; fluphenazine decanoate, 1) and lithium.

PROCEDURE. The procedure consisted of four 50-second trials in which the subject was instructed to follow a yellow cross subtending 1.45 degrees of visual arc displayed on a rear projection screen one meter from the subject. The target was driven sinusoidally at a frequency of 0.4 Hz and an amplitude of 20 degrees of visual arc peak to peak by a servo-controlled galvanometer driven by a waveform generator. Head movement was restrained by means of a chin cup. In one of the four trials, the color of the cross changed at random intervals, and the subject monitored the number of color changes. In previous studies this condition has been shown to optimize peformance (Shagass et al., 1976; Holzman et al., 1976; Levin et al., 1981).

Binocular horizontal eye movements were recorded with non-polarizable silver-silver chloride skin electrodes applied lateral to the outer canthus of each eye and a ground electrode fixed at mid-forehead. The amplified eye movement signal was recorded on FM tape and simultaneously displayed as eye position on a dynograph strip chart. Eye movement tracings were rated qualitatively on the basis of visual inspection as normal or abnormal according to the scheme proposed by Benitez (1970). A rating of abnormal SPEM required no evidence of intact pursuit capability on any of the four test trials, including the color-changing condition. Quantitative assessment yielded a frequency analysis, or natural logarithm of the signal-to-noise ratio (ln S/N), score (Lindsey et al., 1978). This score quantifies power in the frequency band of the stimulus, centered at 0.4 Hz, which is considered "signal", and power in a band between 1.2 and 12 Hz, which is considered "noise". This measurement was originally based on Fourier analysis of digitized eye movement data, but has now been implemented using analog circuitry. Higher ln S/N scores correspond to more efficient SPEM than lower ln S/N scores.

Consensus Brief Psychiatric Rating Scale (BPRS) and Global Assessment Scale (GAS) ratings were made at the time of each smooth pursuit test, including baseline (Overall and Gorham, 1962; Endicott et al., 1976).

Extracellular and RBC lithium ion concentrations were determined by atomic emission spectrophotometry using methods described in detail by Pandey et al (1978).

RESULTS

Age was not significantly correlated with frequency analysis scores in any of the subject groups.

Group I. As Table 1 indicates 6 of 15 (40%) patients had normal SPEM at baseline according to qualitative assessment; pursuit was abnormal in 9 (60%) patients. Seven of these nine patients (78%) were diagnosed as schizophrenic, and two had paranoid disorders. The prevalence of impaired pursuit among the schizophrenic patients was also 78%, a figure that is consistent with previously reported data (Holzman et al., 1973, 1974, 1980). Qualitative ratings did not change in any patient after receiving antipsychotic drug treatment.

Qualitative Rating

	Normal	Abnormal	Ln S/N
Baseline	6	9	$2.401 \pm .410$
Antipsychotic drug	6	9	$2.342 \pm .419^{a}$
Baseline	9	0	$2.743 \pm .343$
Antidepressant drug	9	0	$2.794 \pm .132^{a}$
Pre-Lithium	6	1	$2.520 \pm .311$
Lithium	1	6	$2.208 \pm .348^{b}$

[a]not significant

[b]$p < .05$

Table 1. Qualitative ratings and mean ln S/N scores before and during treatment with antipsychotics, antidepressants, and lithium carbonate.

Two of the never-before-medicated schizophrenics had abnormal pursuit both at baseline and when medicated for up to seven weeks on trifluoperazine (maximum dose: 60 mg/day). The third patient had normal pursuit before ever receiving antipsychotic drug treatment and during the month afterwards, while taking up to 1000 mg/day of chlorpromazine.

Ln S/N scores also showed no change compared to baseline testing. A repeated measures t-test comparing average pre-drug treatment and average post-drug performance revealed no statistically significant effect of antipsychotic drugs on SPEM. Figure 1 illustrates this result. In contrast to the ln S/N scores and qualitative ratings of SPEM, BPRS and GAS ratings did change significantly during antipsychotic drug treatment. Average baseline BPRS scores declined significantly compared to pre-drug evaluations (t(14)=2.86, p < .02), whereas GAS scores rose significantly (t(14) = 3.13, p< .01). These changes in clinical ratings indicate a significant improvement in severity of symptoms during drug treatment.

Group II. Qualitatively normal SPEM were present in all patients before and during treatment with antidepressant drugs. Frequency analysis scores also showed no change. These findings are illustrated in Figure 1 and summarized in Table 1. The two never-before-medicated patients had normal pursuit on all tests.

Group III. Six of the seven bipolar patients had qualitatively normal SPEM prior to taking lithium carbonate. Pursuit was rated as abnormal in five of these six patients during lithium treatment, including the four patients who had never before taken lithium. The onset of the change from normal to abnormal SPEM varied from one day to one year; in most cases it was evident within several weeks. Once it occurred, the

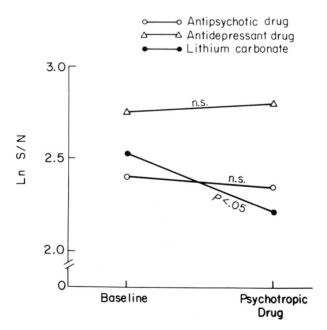

Figure 1: Average ln S/N scores before and during psychotropic drug treatment with antipsychotics, antidepressants, and lithium carbonate.

change persisted throughout all subsequent tests for as long as six weeks. Total BPRS scores decreased significantly compared to baseline (t (6) = 2.704, p< .05). One patient continued to show normal SPEM after taking lithium for one week, and abnormal SPEM persisted in one patient on three tests spanning three weeks of treatment with lithium. This last patient, whose pursuit was abnormal at baseline also, had a history of prior treatment with lithium, phenelzine and carbamazepine and had taken an overdose of diazepam one week prior to admission. Persisting effects from any of these drugs cannot be ruled out as contributing factors in the impaired pursuit at baseline (Levy et al., 1983c).

Paralleling these qualitative changes ln S/N scores also declined significantly compared to baseline performance (t (6) = 2.76, p< .05), indicating reduced pursuit efficiency. Figure 1 illustrates this change. Summary data are listed in Table 1.

We also assessed the relationship between SPEM and lithium ion concentrations in these seven patients and in four additional patients whose pursuit was tested only during lithium treatment. Plasma Li^+ ion concentration correlated -0.53 with ln S/N scores. This correlation did not reach statistical significance. Both RBC Li^+ ion concentration and Li^+ ratio in vivo (RBC Li^+ concentration/extracellular Li^+ concentration) were significantly correlated with frequency analysis scores (RBC Li^+ ion concentration, r = - 0.77, p< .01; Li^+ ratio, r = - 0.73, p <.02). The negative correlation indicates that lower ln S/N scores, reflecting a less precise match between eye velocity and target velocity, were significantly associated with higher RBC Li^+ concentrations and Li^+ ratios. The one patient with normal SPEM in this group of eleven patients also had among the lowest Li^+ concentrations on all three measures.

DISCUSSION

These data provide support for the independence of abnormal SPEM and treatment with antipsychotic or tricyclic, bicyclic, and tetracyclic antidepressant medications while clearly documenting an association between pursuit impairment and treatment with lithium carbonate. Our results are consistent with previous reports on never-medicated (Diefendorf and Dodge, 1908; Couch and Fox, 1934; Karson, 1979; David, 1980) and drug withdrawn (Holzman et al., 1974; Stevens, 1978; Spohn, 1981) patients as well as longitudinal (Levy et al., 1983a) and cross-sectional comparisons (Shagass et al., 1974; Mialet and Pichot, 1981) in demonstrating that impaired pursuit in schizophrenics is not a consequence of drug treatment. We have confirmed the work of other investigators in showing that pursuit tends to be normal in patients with affective disorders who are not taking lithium (Holzman et al., 1974; Karson, 1979; Iacono et al., 1982). Like Iacono et al (1982) we found that treatment with imipramine did not interfere with SPEM proficiency, although our results have the benefit of comparisons with performance preceding drug treatment. The tricyclic antidepressants, amitriptyline and desipramine, the bicyclic, fluoxetine, and the tetracyclic, maprotiline, also did not alter performance compared to baseline despite differences in their neurochemical actions. Similarly, the various types of antipsychotic agents (phenothiazines, thioxanthenes, dibenzazepines, butyrophenones) uniformly produced no effect on pursuit. In view of these findings, comparisons of schizophrenics and patients with affective disorders may yield an increased prevalence of impaired smooth pursuit in both groups. The etiology of the abnormality does not appear to be the same, however. Treatment with lithium carbonate as well as with monoamine oxidase inhibitors (Levy et al., 1983c) is associated with impaired SPEM in some patients whose pursuit is normal prior to drug treatment. Thus, among patients with affective disorders, but not in schizophrenics, there is evidence to suggest that abnormal pursuit is drug-related.

Most of the patients were tested several times before and during drug treatment, including periods when heightened distractability, poor motivation, or other symptoms of acute psychosis could have interfered with task involvement. Clinical ratings showed significant amelioration of psychotic symptoms during treatment with anti psychotics and lithium carbonate. Pursuit deteriorated only in lithium-treated patients, corresponding to periods of significant clinical improvement. This dissociation between measures of pursuit and clinical state indicates that impaired SPEM are not a manifestation of psychotic disorganization.

The findings pertaining to each type of psychotropic drug were remarkably consistent across subjects, whether they had taken comparable medications in the past or had had no previous treatment, and despite diagnostic differences among the patients receiving these drugs. In view of this consistency and the agreement between our findings and those of other investigators, it seems unlikely that chronic drug treatment or a lengthier drug-free interim would alter the interpretation of these results. Additional empirical data are necessary in order to address this question, as well as the related issues of dose dependence and reversibility of drug-related impairment.

Our purpose was to assess the contribution of psychotropic medication to the SPEM abnormalities observed in psychiatric patients. The methods and measures we used are sufficiently sensitive to detect this disturbance in ocular motor control and are comparable to those used in other studies of similar populations. However, it is possible that changes in the gain of the pursuit system would reflect drug-related alterations that are not apparent in our measures. For example, medications may slow pursuit eye movements sufficiently to result in small amplitude saccades that would not be of sufficient magnitude to change qualitative ratings or significantly reduce ln S/N scores. Data that were recently obtained in collaboration with Dr. Robert Baloh and colleagues at UCLA reveal normal pursuit gain (peak eye velocity/peak target velocity), saccadic eye movements, optokinetic responses, and vestibulo-ocular reflex in most patients taking antipsychotic drugs. A variety of abnormalities, including reduced pursuit gain, are present in some patients, but these appear to be unrelated to medication status (Levy et al, 1983b). Nevertheless, the presence or absence of drug effects on SPEM in psychiatric patients receiving these agents for therapeutic purposes may not generalize to nonpatient populations in whom these medications have no pathotrophic neurochemical action.

The mechanism by which normal SPEM deteriorate during treatment with lithium carbonate is unclear. Intracellular concentrations of lithium and perhaps brain levels may be implicated. Reduced pursuit efficiency was significantly correlated with RBC Li^+ concentration, which may reflect brain Li^+ concentration more closely than plasma levels do (Frazer et al., 1973). An interaction between antipsychotic drugs and lithium could also play a role, especially insofar as this combination of medications may affect lithium concentrations. Addressing this issue, as well as the related questions of individual differences in the onset and duration of the effects of lithium on SPEM, require further investigation.

Supported by a Research Scientist Development Award (MH 00336) from the National Institute of Mental Health and by grants from the Benevolent Foundation of the Scottish Rite, Northern Masonic Jurisdiction (Dr. Levy) and the MacArthur Foundation (Dr. Davis).

REFERENCES

(1) Baloh, R.W., Sharma, S., Moskowitz, H., and Griffith, R., Effect of alcohol and marijuana on eye movements, Aviat Space Environ Med. 50 (1979) 18-23.

(2)   Benitez, J.T., Eye tracking and optokinetic tests: Diagnostic significance in peripheral and central vestibular disorders, Laryngoscope. 80 (1970) 834-848.

(3)   Bittencourt, P., Wade, P., Smith, A. and Richens, A., Benzodiazepines impair smooth pursuit eye movements, Br J Clin Pharmac. 15 (1983) 259-262.

(4)   Cegalis, J.A. and Sweeney, J.A., Eye movements in schizophrenics: A quantitative analysis, Biol Psychiat. 14 (1979) 13-26.

(5)   Couch, F.H. and Fox, J.C., Photographic study of ocular movements in mental disease, Arch Neurol Psychiat. 34 (1934) 556-578.

(6)   David, I., Disorders of smooth pursuit eye movements in schizophrenics and the effect of neuroleptics in therapeutic doses, Activitas Nervosa Super. 22 (1980) 155-156.

(7)   Diagnostic and Statistical Manual of Mental Disorders, ed. 3. (American Psychiatric Association, Washington, D.C., 1980).

(8)   Diefendorf, A.R. and Dodge, R., An experimental study of the ocular reactions of the insane from photographic records, Brain. 31 (1908) 451-489.

(9)   Endicott, J. and Spitzer, R.L., A diagnostic interview: The schedule for affective disorders and schizophrenia, Arch Gen Psychiat. 35 (1978) 837-844.

(10)   Endicott, J., Spitzer, R., Fleiss, J. et al, The Global Assessment Scale: A procedure for measuring overall severity of psychiatric disturbance, Arch Gen Psychiat. 33 (1976) 766-771.

(11)   Frazer, A., Mendels, J., Secunda, S. and Bianchi, C., The prediction of brain lithium concentrations from plasma or erythrocyte measures, J Psychiat Res. 10 (1973) 1-7.

(12)   Holzman, P.S. and Levy, D.L., Smooth pursuit eye movements and functional psychoses: A review, Schiz Bull. 3 (1977) 15-27.

(13)   Holzman, P.S., Levy, D.L. and Proctor, L.R., Smooth-pursuit eye movements, attention, and schizophrenia, Arch Gen Psychiat. 33 (1976) 1415-1420.

(14)   Holzman, P.S., Proctor, L.R. and Hughes, D.W., Eye tracking patterns in schizophrenia, Science. 181 (1973) 179-181.

(15)   Holzman, P.S., Kringlen E., Levy, D.L. and Haberman, S., Deviant eye tracking in twins discordant for psychosis: A replication, Arch Gen Psychiat. 37 (1980) 627-631.

(16)   Holzman, P.S., Kringlen, E., Levy, D.L. et al, Smooth-pursuit eye movements in twins discordant for schizophrenia, J Psychiat Res. 14-(1978)111-120.

(17)   Holzman, P.S., Levy, D.L., Uhlenhuth, E.H. et al, Smooth-pursuit eye movements and Diazepam, CPZ, and Secobarbital, Psychopharmacologia. 44 (1975) 111-115.

(18)   Holzman, P.S., Proctor, L.R., Levy, D.L. et al, Eye tracking dysfunctions in schizophrenic patients and their relatives, Arch Gen Psychiat. 31 (1974) 143-151.

(19)   Iacono, W.G., Tuason, V.B. and Johnson, R.A., Dissociation of smooth-pursuit and saccadic eye tracking in remitted schizophrenics, Arch Gen Psychiat. 38 (1981) 991-996.

(20)   Iacono, W.G., Peloguin, L.J., Lumry, A.E. et al, Eye-tracking in patients with unipolar and bipolar affective disorders in remission, J Abnormal Psychol. 9 (1982) 35-44.

(21)   Karson, C.N., Oculomotor signs in a psychiatric population: A preliminary report, Am J Psychiat. 136 (1979) 1057-1060.

(22)   Klein, R.H., Salzman, L.F., Jones, F. and Ritzler, B., Eye-tracking in psychiatric patients and their offspring (Abstract), Psychophysiology. 13 (1976) 186.

(23)   Kuechenmeister, C.A., Linton, P.H., Mueller, T.V. and White, H.B., Eye-tracking in relation to age, sex, and illness, Arch Gen Psychiat 34 (1977), 578-599.

(24)   Levin, S., Lipton, R.B., and Holzman, P.S., Pursuit eye movements in psychopathology: Effects of target characteristics, Biol Psychiat. 16 (1981) 255-267.

(25) Levy, D.L., Lipton, R.B., and Holzman, P.S., Smooth-pursuit eye movements: Effects of alcohol and chloral hydrate, J Psychiat Res. 16 (1981) 1-11.
(26) Levy, D.L., Lipton, R.B., Holzman, P.S. and Davis, J.M., Eye-tracking dysfunction unrelated to clinical state and treatment with haloperidol, Biol Psychiat. 18 (1983a) 813-819.
(27) Levy, D.L., Sakala, S.M., Baloh, R.W., Marder, S., et al, Oculomotor and vestibular function in psychiatric patients, In preparation, 1983b.
(28) Levy, D.L., Yasillo, N.J., Janicak, P.G. and Davis, J.M., Psychotropic drug effects on smooth pursuit eye movements, In preparation, 1983c.
(29) Levy, D.L., Yasillo, N.J., Janicak, P.G., Pandey, G., et al, Lithium carbonate causes saccadic pursuit, Submitted for publication, 1983d.
(30) Lindsey, D.T., Holzman, P.S., Haberman, S. and Yasillo, N.J., Smooth-pursuit eye movements: A comparison of two measurement techniques for studying schizophrenia, J Abnormal Psychol. 87 (1978) 491-496.
(31) Lipton, R.B., Levin, S. and Holzman, P.S., Horizontal and vertical pursuit eye movements, the oculocephalic reflex, and the functional psychoses, Psychiat Res. 3 (1980) 193-203.
(32) Lipton, R.B., Levy, D.L., Holzman, P.S., and Levin, S., Eye movement dysfunctions in psychiatric patients: A review, Schiz Bull. 9 (1983) 13-32.
(33) May, H.J., Oculomotor pursuit in schizophrenia, Arch Gen Psychiat. 36 (1979) 827.
(34) Mialet, J.P., and Pichot, P., Eye-tracking patterns in schizophrenia: An analysis based on incidence of saccades, Arch Gen Psychiat. 38 (1981) 183-186.
(35) Norris, H., The time course of barbiturate action in man investigated by measurement of smooth tracking eye movement, Br J Pharmac Chemother. 33 (1968) 117-128.
(36) Overall, J.E. and Gorham, D.R., The brief psychiatric rating scale, Psychol Rep. 10 (1962) 799-812.
(37) Pandey, G., Baker, J., Chang, S. and Davis, J.M., Prediction of in vivo red cell plasma Li^+ ratios by in vitro methods, Clin Pharmacol Ther. 24 (1978) 343-349.
(38) Pass, H.L., Salzman, L.F., Klorman, R., Kaskey, G. et al, The effects of distraction on acute schizophrenics' visual tracking, Biol Psychiat. 13 (1978) 587-593.
(39) Pivik, R.T., Target velocity and smooth pursuit eye movements in psychiatric patients, Psychiat Res. 1 (1979) 313-323.
(40) Rashbass, C. and Russell, G., Action of a barbiturate drug (Amylobarbitone sodium) on the vestibulo-ocular reflex, Brain. 84 (1961) 329-335.
(41) Rothenberg, S.J. and Selkoe, D., Specific oculomotor deficit after diazepam: II. Smooth-pursuit eye movements, Psychopharmacologia. 74 (1981) 237-240.
(42) Salzman, L.F., Klein, R.H. and Strauss, J.S., Pendulum eye tracking in remitted psychiatric patients, J Psychiat Res. 14 (1978) 121-126.
(43) Shagass, C., Amadeo, M. and Overton, D.A., Eye-tracking performance and engagement of attention, Arch Gen Psychiat. 33 (1976) 121-125.
(44) Shagass, C., Roemer, R.A. and Amadeo, M., Eye-tracking performance in psychiatric patients, Biol Psychiat 9 (1974) 245-260.
(45) Shibasaki, H., Tsuji, S. and Kuroiwa, Y., Oculomotor abnormalities in Parkinson's Disease, Arch Neurol. 36 (1979) 360-364.
(46) Spohn, H., Correlates of eye tracking in schizophrenic patients, Presented at the Eye-Tracking Methodology Conference. (Boston, December, 1981).
(47) Stevens, J.R., Disturbances of ocular movements and blinking in schizophrenia, J Neurol Neurosurg Psychiat. 41 (1978) 1024-1030.
(48) Tomer, R., Mintz, M., Levy, A. and Myslobodsky, M., Smooth-pursuit pattern in schizophrenic patients during cognitive task, Biol Psychiat. 16 (1981) 131-144.
(49) White, H.R., Ocular pursuit in normal and psychopathological subjects, J Exp Psychol. 22 (1938) 17-31.

Section 9

THEORETICAL ASPECTS

Theoretical and Applied Aspects of Eye Movement Research
A.G. Gale and F. Johnson (Editors)
© Elsevier Science Publishers B.V. (North-Holland), 1984

THEORETICAL ASPECTS

INTRODUCTION

Alastair G. Gale,
Division of Radiology, Queen's Medical Centre,
Nottingham, England.

Numerous theoretical aspects of eye movement research are presented in the papers elsewhere in this volume. However the following three papers specifically deal with theoretical aspects foremost. Galley utilises an aspect of saccadic behaviour as an indicator of an underlying state, Groner et al. expand upon the sequencing of fixations whereas Gerrissen proposes a theoretical treatment based on the information capacity of the eye.

Galley presents a task where saccadic eye movements were recorded whilst subjects followed a jumping light spot. Summary data is presented from six studies where this task was implemented using different subject populations and experimental conditions. On the basis of their relationship to the stimulus movement saccades were considered either as anticipatory, reactive or 'too-late'. The velocity of these saccades was found to vary with exercise and circadian variation, to be higher in young children and schizophrenics, slightly slowed by Beta-blocking agents as well as being subject to individual differences. Galley argues that saccadic velocity is a sensitive indicator of activation level following Duffy's theoretical formulation where the term is considered synonymous with arousal. Saccadic velocity is further considered in several of the papers in the section on the saccadic eye movement system.

In viewing pictures and line drawings several researchers have reported some regularity in saccadic eye movement patterns. This patterning of gaze was proposed by Noton and Stark to be related to the internal representation of the scene and developed the concept of the 'scanpath' and its subsequent elaboration into a 'feature ring'. In their paper Groner, Walder and Groner elaborate both the theoretical and empirical problems of the concepts. They propose a useful distinction between local and global scanpaths where the former represents 'bottom-up' information processing and the latter 'top-down' processing. In a face recognition task both types of scanpath were evidenced by most of the subjects. The authors conclude intriguingly that local scanpaths were related to the recognition of a face as being unknown but not to the recognition of a known face. No relationship between global scanpaths and recognition was found.

A different theoretical standpoint is presented by Gerrissen based on a hypothesised retinal array constructed from concentric rings of photoreceptors having small central receptors of high resolution and larger peripheral receptors of low resolution extending up to 15 deg. eccentricity. The main concern of the paper is with the detection of global structure. The author proposes that spatial resolution is dependent upon variations of pupil size and fixation time. Values of these two factors are selected by the visual system to achieve a particular spatial resolution of an object which is linked to the adequate matching to the optimum 'receptor ring'. In this model saccadic movements are seen as part of a system that aligns components of the stimulus image to the retinal array so that the desired resolution or globality is achieved. Consideration of redundancy in the visual image leads onto the concept of a chord   which is then expounded to explain feature extraction.

Taken together these papers demonstrate the breadth of theoretical treatments afforded to saccadic behaviour of the eye.

Theoretical and Applied Aspects of Eye Movement Research
A.G. Gale and F. Johnson (Editors)
© Elsevier Science Publishers B.V. (North-Holland), 1984

AN ACTIVATION PARADIGM USING SACCADIC MOVEMENTS

Niels Galley

University of Cologne,
Psychological Institute,
D-5000 Koln 41

Saccadic eye movements of 98 subjects following a jumping
light spot in continuously shortened intervals were
recorded as EOGs and analysed with computer aid. Response
saccades may be separated in reactive and anticipatory
ones. Mostly there are anticipatory ones, which are more
efficient showing more precision in time and need less
'activation', i.e. have slower saccadic velocities than
reactive ones. In several situations - circadian
variation, physical exercise, Beta-blocking agents - and
in several populations - adults, children, schizophrenic
patients - the jumping spot paradigm has been used. Results
are discussed within the framework of the unidimensional
ativation theory of Duffy (1972).

INTRODUCTION

Saccadic velocity has sporadically been used as an activation index.
Brozek (1949) reported slowing down of the saccades during tiredness. The
same was reported following Diazepam intake (Aschoff, 1967), alcohol intake
(Franck and Kuhlo, 1970) and barbiturate injection by Eckmiller and Mackeben
(1976). These authors reported eye movement recordings together with oculomotor neuron's firing in monkeys and attributed the slowing down to two
factors: the reduction of the tonic impulse rate of the neurons and the
reduction in the number of active motor neurons. On the other hand arousing
stimuli instantaneously elevated the impulse rate of the neurons. So one
can assume, that saccadic velocity may be an excellent "activation" indicator - we use the term in the sense of the unidimensional activation theory
formulated by Duffy (1972) - which does not have a vegetative origin as most
other activation indicators. It is rarely used, however, although there
are several controversies between psychophysiologists concerning under-
versus over arousal states in some populations, which would benefit from
such an activation indicator. Two examples: Shagass, Roemer and Amadeo
(1976) report that in psychotic patients "poor eye-tracking can be improved
by facilitating attentional effort". Therefore one can assume there should
be a low activation state in patients which is increased to an elevated one
by facilitating attentional effort thus enabling better performance
(Brezinova and Kendell, 1977). But there are other reports and discussion
about high activation levels in schizophrenics (Gruzelier and Venables,
1975). What about saccadic velocity when doing an eye movement task? We
found evidence for an elevated saccadic velocity in a small group of
schizophrenic patients. A controversy also exists in children with minimal
brain dysfunction concerning their activation being lowered or increased

(Zentall 1977 , Satterfield, 1972) and no data are available about their saccadic velocities.

Here a task is presented called the jumping spot which people are motivated to do if necessary 20 times and more and which enables the experimenter to collect saccadic velocities and at the same time the performance qualities using several indicators. Selected data of several studies are presented laying the emphasis on methodical aspects.

METHODS

If one wants to collect a higher number of saccades of a subject with nearly the same amplitude and wants the subject to be engaged in doing the task, one obviously has to create such a task that a subject has to match a jump- ing spot in shortened intervals as closely as possible (see Fig. la). Before doing the jumping spot task all our subjects had to follow a slowly accelerated moving light spot. Of course, subjects reacted with following eye movements and the task is therefore not considered here further (Galley, Widwera-Bernsen and Ishak, 1983). But in such a way subjects had experien- ced a feeling of an accelerated responding.

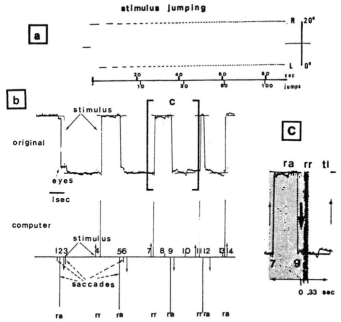

Fig.1 a shows the time course of a jumping light spot, which jumps from the midpoint of an oscilloscope screen to the right (R) stays there for a while, jumps then to the left (L) over a distance of mostly 20°. The in- tervals of the horizontally jumping spot are shortened during the 90 sec duration of one trial.The subject had to match the spot as closely as possible.Fig. 1 b shows how the eyes follow the jumping spot and the computer prog- ram recognise the stimulus jumps and the saccadic eye movements. The saccades 2,5,9,12 and 14 are identified as anticipatory responding saccades (ra), while the saccades 4 ,7 and 11 as reactive responding ones (rr). In c) there is figured the time relation of the saccade 9,in which it would be identified as an anticipatory response (ra = lightly shadowed) respective as a reactive one (rr = densely shadowed).

Stimulus

A small light spot (.2° in diameter and with a luminance of 21 cd/m^3) easily
detectable on the oscilloscope screen but not bright in the luminated room
jumps horizontally in such a manner as figured in 1a. In most studies a big
oscilloscope was used with 40 cm diameter at a distance of 114 cm in front
of the subjects' eyes, but for technical reasons in some studies other
oscilloscopes were used at other distances (see Table 1). The jumping intervals (N = 124 for one trial) seemed to be subjectively linearly shortened
but were objectively as seen in Fig. 2 non-linearly shortened from .2 - 1.4
c/sec. Most adults could match the spot even when it moved very quickly.
but we have learned that for the children aged 9 it moved too fast.

Recording Procedure

Stimulus and dc-EOG (in some studies other biosignals too, which are not
considered here) are recorded on magnetic tape (PCM machine 8 K 60 by Fa.
Johne and Reilhofer) with a total bandwidth of 0 - 50 c/s.

Data processing - saccades

Signals were off-line processed with computer programs devised by the author
on a Nicolet Med 80 computer. The programs used can be outlined only briefly here, but may be requested from the author. The first program identifies
stimulus jumping simply by sign reversals in the signal. Saccades in the
EOG-signal were identified by the program in the following manner: first
in an identification routine the EOG was smoothed, differentiated and points
were identified where velocity became higher than 90°/sec. The correspon-
dent x,y-values of the original signal were stored. Secondly in the evaluating routine these raw values were matched with some conditions which must
be fulfilled to be accepted as a true saccade. When accepted all values
of one trial were stored in a file on floppy disk. Later on these data were
processed by evaluating response saccades - see below -their time relation
stimulus jumping, their mean velocity as well as their maximal velocities.
In this third routine each saccadic velocity was transformed to one of a
15° - saccade because velocity is a function of saccadic amplitude (Bahill,
Hsu and Stark, 1978).

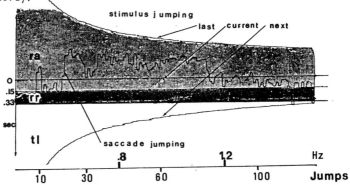

Fig 2 shows the order of all stimulus jumps with the time relations of
their corresponding response saccades as well as of the preceding and following jumps. Here as in every case the first saccades being reactive ones.
Later on reactive responses (rr) and too-late (tl) responses were rare.

Data processing - performance categories

As seen in Fig. 1b the program identifies more saccades than stimulus jumps.
Therefore "response" saccades had to be defined to distinguish them from,
for example, "corrective" ones. They had to be greater than the half of the
stimulus amplitude and had to arise before the actual stimulus jump or not
later than 330 ms afterwards. Such a response saccade could be a reactive
response (rr) saccade if arising from 150 to 330 ms after an actual stimulus jump or an anticipatory response (ra) arising before (see Fig. 1b,c,
2). Saccades arising later than 330 ms or being smaller than half a stimulus amplitude were ignored. If there arose no response saccade until 330
ms after actual stimulus jump a "too-late" response was identified. The
time relations between stimulus jumping and response saccades during a whole
trial (shown in Fig. 2). The first saccades are reactive response saccades.
After the first few reactive saccades anticipatory ones suddenly emerge
and persist with few exceptions until the shortest intervals. At this time
only two too-late responses were identified.

The following categories of eye movement behaviour were defined:

- % responses as anticipatory saccades (% RA)
- % responses as reactive   saccades (%RR)
- % too-late responses (% TL)
- mean and maximal saccadic velocity as an average of all anticipatory
  saccades of one trial (RA - velocity)
- saccadic velocity difference between anticipatory and reactive
  saccades (RR - RA)
- oculomotor reaction time (ocul. R.T.) defined as the median of all
  times between the occurrence of a reactive saccade and the actual
  stimulus jump. Mostly average values of one or more trials are
  processed further on.

Subjects

Saccadic velocities from six studies are reported here. They are listed in
Table 1.

Study	Subjects N	Age Years	Stimulus ampl.	Trials per Ss N	Response sacc. N
1. SCO '80: Schizophr. Pat.	11	32	20	2	2236
Control subjects	11	28	26	2	2373
2. CHI '81: children	43	9	28	1	2772
3. BETA '82: Betablock. Agents	11	26	21	6	6101
4. CIR '82: Circad. Variation	8	28	21	23	22116
5. CIR '83: Circad. Variation	12	29	15	18	20026
6. PHY '83: Physical Exercise	1	32	15	17	1918

Table 1 lists the six studies, which were done from 1980 until 1983 and serve as data basis here.

In study 1, 11 schizophrenic patients (all outpatients, taking neuroleptics)
and 11 control subjects performed the moving and jumping spot task. The
results of the moving spot task will be published in Galley, Widera-Bernsen
and Ishak (1983). In study 2 children (N = 43) of an elementary school
grade 2 - 4 performed the eye movement tasks within a variety of tasks concerning their reading ability and so on. Results are partly published in a
thesis (van Triel 1983). Study 3 was a double blind study with a well
known beta blocking agent (Propanolol) and a newly synthesised one. The

subjects served as their own controls: the trials 1,3,5 were agent free and the first trials on three weekly sessions. On trials 2,4,6 there were administered placebo, Propanolol or the new agent randomly. Study 4 was designed to test the saccadic velocity in an old activation paradigm: circadian variation of several indicators (heart rate, body temperature, tiredness and several performance indicators). Here 8 subjects did the tasks each hour over 23 hours without sleep. Results will be published in detail in Galley and Karl (1984). Study 5 was a replication study of study 4, but allowing the subjects to sleep thus avoiding deprivation. But subjects were instructed whenever they should awake in the night they should request the experimenter to take a trial. Evaluation of the data are still in progress. In study 6 a pilot experiment was undertaken with one subject; after ten baseline trials (served as 'monotomy') four trials were taken after 20 pushups (mild physical exercise) and three after a 400 m sprint (strong physical exercise).

Results

Spontaneous fluctuations of saccadic velocities. As seen in Table 2 most response saccades done in a trial are anticipatory ones. Therefore saccadic velocities are listed as anticipatory ones further on.

Study	Scan values From	RA(^{0}/sec) mean velocity	RA - RR(^{0}/sec) veloc. diff.	%RA	%RR	%TL	Ocul. RT ms.
6. PHY '83	1 subject 1 trial	273 $\pm$ 19	5	83	14	2	195

Table 2. Average values of one trial. Mean saccadic velocities of all anticipatory responses (RA) saccades occurring in the trial drawn in Fig. 2., together with standard deviation, difference of reactive (RR) and anticipatory velocities' mean, percentages of the response categories (TL = too-late response), and the median of oculomotor reaction times (Ocul.RT).

There is a small, but as we shall see, a constant difference between saccadic velocity of reactive to anticipatory saccades indicating that velocities of reactive saccades may be higher than anticipatory ones. As seen in Fig. 4 this difference remains positive over the whole data block of a study (3 CIR 83): every time measured saccadic velocity of anticipatory saccades are somewhat slower than reactive ones. Considerably lower values during the night are measured in maximal saccadic velocities as well as in performance indicators, i.e. increasing too-late responses and oculomotor reaction times. Values climb slowly during the self made interupts in the night and slowly during the morning hours. Saccadic velocities and performance indicators run parallel.

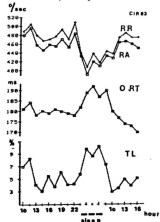

Fig. 4 shows the circadian variation of the maximal velocities of reactive response (RR) and anticipatory response (RA) saccades, of the oculomotor reaction time (ORT) and of the percentage of too-late responses (TL). Activation deterioration runs parallel with performance one.

Stability of saccadic velocity and the reactive-anticipatory velocity difference.

Most values of saccadic velocity reported here are mean values averaged from at least one trial or 124 stimulus jumps. Therefore it is to be proven if the values are stable, regardless how quickly the stimulus jumps. In Table 3 there are listed the values of all trials of the most extensive studies 4 and 5 separated into long jumping intervals (.2 - .8 c/sec)and fast ones (.8 - 1.4 c/sec) respectively.

Study	Scan Values From	RA(°/sec) Scan Velocity	RA - RR(°/sec) veloc. diff.	%RA	%RR	%TL	Ocul. RT ms
4 CIR 82 .2 -.8 c/sec .8 - 1.4 c/sec	8 subjects 170 trials	261 ± 48 270 ± 44*	9 ± 16 11 ± 16	72 ± 17 75 ± 22	20 ± 10 10 ± 9**	8 ± 16 15 ± 17**	187 ± 25 206 ± 31**
5 CIR 83 .2 -.8 c/sec .8 -1.4c/sec	12 subjects 196 trials	276 ± 43 280 ± 42	7 ± 15 5 ± 6 ± 12	79 ± 12 87 ± 11**	18 ± 10 5 ± 6**	3 ± 5 7 ± 8**	178 ± 18 194 ± 26**

Table 3. Average values of the two most extensive studies. Separately there are listed the values of the slow (.2- .8 c/sec) and quick (.8-=.4 c/sec) jumping light spot task. The differences between reactive and anticipatory saccadic velocities show no significant trend, while most other variables are slightly but mostly significantly, influenced by faster jumping of the stimulus.velocity becomes a little faster, percentage of reactive responses reduced, and performance deteriorates: the too-late responses are doubled and oculomotor reaction time worsened.

One can see that the velocity difference is stable, but the other values are slightly influenced by increasing jumping frequency. But if one considers, that the task experienced is more strenuous towards the end these differences are not astonishing: Velocities become higher, reactive responses more seldom, too-late responses more frequent and oculomotor reaction times are lengthened. Note, however, that this is quite the opposite relation as outlined from the same study in Fig. 4. Here in circadian variation the saccadic velocities become slower and while the performance indicators deteriorate.

The standard deviations of the velocities are greater than in Table 2, indicating that there is more variation of indidivual differences than of circadian fluctuations. The standard deviations of the velocity differences indicate that there must be some individuals (or some situations), in whom this difference may be negative.

Study	Scan values From	RA(°/sec) mean velocity	RA - RR(°/sec) veloc. diff.	%RA	%RR	%TL	Ocul. RT ms.
5 CIR 83 .2 -1.4 c/sec circad. var. Individ. diff.	8 subjects 170 trials altogether 1. subject, 2. time 1. time, 2. subject	278 ± 42 ± 14 ± 33	5 ± 14 ± 4 ± 7	86 ± 9 ± 4 ± 5	9 ± 5 ± 2 ± 3	5 ± 6 ± 2 ± 3	180 ± 17 ± 6 ± 12

Table 4. The standard deviations show greater values for individual differences than for circadian variations in the subjects.

In Table 4 therefore of the studies, one is averaged first over all trials altogether, secondly over subjects and then over hours. Thus one can make a comparison between the variation over the individuals and time. These results indicate considerably individual differences, which demand representative populations.

Populations

Saccadic velocities and performance indicators from the jumping spot task

were measured in children and schizophrenic patients. The results are
listed in Table 5 and 6.

Study	Scan values From	RA(°/sec) mean velocity	RA - RR(°/sec) veloc. diff.	%RA	%RR	%TL	Ocul. RT ms.
2. CHI 82 4. grade	20 subjects	216 ± 34	4 ± 12	44 ± 12	15 ± 6	41 ± 13	223 ± 25
3. grade	11 subjects	221 ± 27	-1 ± 10	36 ± 9*	14 ± 4	50 ± 9*	240 ± 28
2. grade	11 subjects	248 ± 32*	-9 ± 43	37 ± 13	10 ± 6*	53 ± 14*	237 ± 41
2. CHI 82 .2-.8 c/sec	43 subjects	228 ± 34	2 ± 13	59 ± 21	16 ± 12	24 ± 19	226 ± 35
.8-1.4c/sec	43 subjects	223 ± 36*	4 ± 17	32 ± 11**	11 ± 6**	57 ± 14**	234 ± 34

Table 5. shows the mean values of the children study. Explanations see text. *= p $<$.05, ** = p $<$.01.

The data of the children is listed as means of grade 4,3 and 2 in the first
rubric and as a function of the increasing responding in the second rubric.
Saccadic velocity becomes higher in young children and performance indicators decreased. Interestingly the difference between reactive and anticipatory saccadic velocities becomes smaller, yet, negative in the young children. The high percentage of too-late responses in the children indicates
the task must have been an overstrain for them. Separation into low and
high rate responding in the second rubric shows its influence on the performances indicators, but not on the saccadic velocities and the difference
between reactive and anticipatory velocity. This indicates that it is more
a factor of individual differences.

Study	Scan values From	RA(°/sec) mean velocity	RA - RR(°/sec) veloc. diff.	%RA	%RR	%TL	Ocul. RT ms.
1. SCO 80 patients	11 subjects 1. trial	304 ± 46	9 ± 14	64 ± 24	18 ± 9	18 ± 14	186 ± 21
control Ss.		267 ± 30*	10 ± 11	62 ± 12	25 ± 10	13 ± 7	206 ± 27
patients	2. trial	294 ± 32	16 ± 16	68 ± 24	16 ± 13	16 ± 12	195 ± 19
control Ss.		251 ± 33**	10 ± 10	65 ± 17	25 ± 14	9 ± 7	194 ± 19
1. SCO 80 patients .2-.8c/sec	11 subjects 2 trials	294 ± 39	14 ± 22	70 ± 23	18 ± 18	12 ± 11	184 ± 21
.8-1.4c/sec		311 ± 41**	1 ± 16*	63 ± 28	16 ± 14	21 ± 18*	201 ± 27**
control Ss. .2-.8c/sec		254 ± 33	7 ± 11	65 ± 18	31 ± 19	4 ± 4	199 ± 22
.8-1.4c/sec		261 ± 31	11 ± 11	63 ± 17	23 ± 11*	14 ± 9**	208 ± 26

Tabel 6. Average values of the Schizophrenics Study.

Schizophrenic patients have higher saccadic velocities than control subjects,
the difference being not very high. They show no deterioration of performance indicators, show no slower oculomotor reaction times, yet a slightly
faster reaction time in this task. Looking at the increasing responding
rate they show deterioration of the reactive-anticipatory difference indicating in the dynamics of their activation pattern. There is a slight but
insignificant trend in patients as in control subjects that the second trial
shows slower saccadic velocities.

Experimental manipulation of activation

In the study with Betablocking agents saccadic velocities are slower after
intake of the agents, as shown in Table 7, but most differences are slight
and non significant.

The effect of the agent on slowing down saccadic velocity is not high. The
difference between Placebo and agents is in the same order as between the
first and the second trial (Baseline vs. Placebo which can be considered as

Study	Scan values From	RA(°/sec) mean velocity	RA - RR(°/sec) veloc. diff.	%RA	%RR	%TL	Ocul. RT ms.
3. BETA '82	11 subjects						
Baselines	1.3.5. trial	257 ± 39	6 ± 8	59 ± 14	24 ± 7	11 ± 12	197 ± 15
Placebo	10 Ss,1trial	243 ± 35	10 ± 13	57 ± 15	26 ± 6	17 ± 13	207 ± 24
Betabl.Prop.	7 Ss, 1trial	231 ± 34**	6 ± 10	52 ± 20	28 ± 12	20 ± 15	205 ± 23
Betabl. xxx	11 Ss,1trial	230 ± 32*	-1 ± 11	57 ± 15	24 ± 6	19 ± 13	203 ± 14

Table 7 shows the mean values of the study with Betablocking agents. Significance notes are related to the baseline values.

an habituation effect. The significant difference in the reactive-anticipatory velocity difference between Placebo and the new agent may indicate
less relief in the agent condition.

The influence of mild and strong physical exercise on saccadic velocity as
well as heart rate and one performance indicator is outlined in Fig. 5.

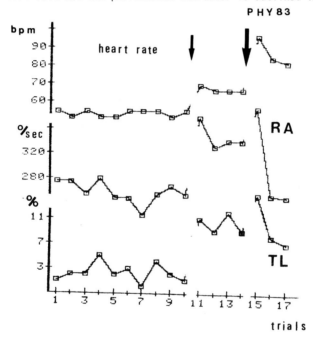

Fig. 5. shows the influence of monotony (baseline), mild (small arrow) and
strong physical exercise (big arrow) on anticipatory saccadic velocity (RA),
performance (TL) and heart rate. While saccadic velocity runs parallel
to heart rate increase, the performance deteriorates.

The results are collected from one subject in a pilot study. Saccadic velocity slows down a little during monotony, is increased after mild physical
exercise, only a little more after strong exercise and then decreased
more rapidly than heart rate. The performance indicator shows deterioration during the higher activation state opposite to the circadian variation
but in line with the demand situation during the higher stimulus jumping
rate.

DISCUSSION

Saccadic velocity is not only sensitive to tiredness (Fig. 4) as was previously reported but to increasing demand when the subject had to respond
more quickly to the accelerated jumping stimulus (Table 3 & 6). It is sensitive to relief when the task is done by following a concept (RR-RA velocity difference in Table 3 - 7, Fig. 4) in contrast to reacting to each
stimulus jump separately. Saccadic velocity is a function of habituation:
it declines in the first repetitions (Fig. 4,5 and Table 6,7). There are
great differences of saccadic velocities between individuals (Table 4 - 7).
This demands representative groups in population studies as well as more
consideration of intraindividual differences.

The small and therefore non representative group of schizophrenic patients
shows two interesting differences to their controls: they had an elevated
saccadic velocity and a waning relief by anticipating when the stimulus
accelerated. Higher velocities as an indicator of elevated activation fits
well to the reported elevated vegetative activation indicators in this population. It contrasts to the note of Schmid-Burgk et al (1982) that patients had no elevated saccadic velocities which is not substantiated in the
report. Out patients showed no significant performance deterioration.
Being a non representative group these data need further confirmation.

Younger children show higher saccadic velocity doing this task, lower
performance and lower relief by anticipating. We prefer a speculative
interpretation, which needs further substantiation, that there is a growing
concentration process in these years, a postulated cortical activation
focussing process, which enables task specific narrowing of the attention
(Galley & Karl, 1984). Saccadic velocities of all children seem to be unrealistically low. Perhaps the regression of their large saccadic amplitude (28^0) to the standard amplitude of a 15^0 saccade is too steep. The
regression may be non linear of different populations have other regressions
for this relation of saccadic amplitude versus saccadic velocity. This has
to be decided in further elaboration of the task. For the moment it is
recommended to use nearly the same stimulus amplitudes, for example 20^0.

The jumping spot task gives additional information about performance of the
subject for example in the percentage categories as well as the oculomotor
reaction times. The too-late category which contains no responses nearly
exclusively, is sensitive for elevating the response rate (Table 3,5,6) as
well as for tired-ness or overactivation (Fig. 4,5). The high percentage of
too-late responses showing children in low jumping frequency indicates the
task being experienced as overstrain.

Considerable interindividual differences must be stated as follows from the
differences in the different populations (Table 3,5,6,7) as well as the
great standard deviations in one populations.

The border between the reactive versus anticipatory response at 150 ms after
actual stimulus jump (Fig. 2) is up to now a definition. The RR-RA velocity
difference enables us to test this definition empirically. Preliminary results show that some subjects may be as quick as 140 ms but most people
are slower. Therefore 150 ms seems to be a good compromise. The reactive
response catergory is a function of an order of jumps: all beginning responses are reactive ones. (Fig. 2). Only after some reactions has one built
up a concept of how the stimulus jumping will continue. Later on the occurence of the reactive response mode seems to be constant. Preliminary corr-

elations show a low correlation between the reactive response value with·
the anticipatory or too-late values, but there are high correlations between
anticipatory and too-late percentages. The oculomotor reaction time mostly
covaries for example with the too-late response.

The question that may be asked is that if the oculomotor reaction time category is a real reaction time because the stimulus is certainly accelerated
but can be anticipated. It is answered by the stable difference in normal
conditions between velocities of reactive and anticipatory response saccades
When reactive saccades permanently show higher velocities than anticipating
ones, it must be a different behaviour catergory. The interesting question
then is whether the corrective saccades may also be separated in reactive
or anticipating ones depending on their velocities has not been proved.

Looking at the covariation of the activation indicator, i.e. saccadic velocity with the performance indicators one had to establish opposite covariation: running parallel in tiredness and habituation, being opposed in increasing demand and overactivation. We have argued for modification of the
activation concept (Galley & Karl, 1984) in that manner, that there must be
at least one stage between the lower brain stem process activation and the
cortical regulation of performance. Performance is not a direct function
of activation level but of concentration.

Reference note

van Triel, M. 1983, Lateralitatsunterschiede bei Augenbewegungen von
    Kindern des 2.-4 Grundschuljahrs, Thesis University of Cologne.

References

Aschoff, J., 1968, Veranderugen rascher Blickbewegungen (Saccaden) beim
    Menschen unter Diazepam (Valium). Arch. Psychiat. Nervenkrkh. 211
    325-332.

Bahill, A.T., Hsu, F.K. and Stark, L. 1978, Glissadic Overshoots are due to
    pulse width errors. Arch. Neurol. 35, 138-42.

Brezinova, V & Kendell, R.E., 1977. Smooth pursuit eye movements of schizophrenics and normal people under stress. Brit. J. Psychiatr. 130, 59-63.

Brozek, J. 1949. Quantitative criteria of oculomotor performance and fatigue.
    J. Appl. Physiol. 2, 247-60.

Duffy, E. Activation, 1972 in: Greenfield, N.S. & Sternbach, R.A. (Eds.)
    Handbook of Psychophysiology; pp 577-622 (Holt, Rinhart & Winston,
    New York.

Eckmiller, R. & Mackeben, M. 1976. Functional changes in the oculomotor
    system of the monkey at various stages of barbiturate anaesthesia
    and alertness. Pflugers Arch. 363, 33-42.

Franck, M.C. & Kuhlo, W. 1970. Die mirkung des alkohols euf die raschen
    blickzielbewegungen (Saccaden) beim Menschen. Arch. Psychiatre.
    Nervenkr, 213, 238-45.

Galley, N., Widera-Bensen, M., Ishak, H.B. 1983. Eye movements in schizophrenics - Revisited in: Groner, R., Menz, C., Fisher, D.F. & Monty, R.A (Eds.). Eye movements and psychological functions: International Views. Hillsdale, N.J. Lawrence Erlbaum Associates pp...

Galley, N. & Karl, G. 1984. Activation and Performance - a Reformulation. Submitted for publication in human Neurobiology.

Gruzeljer, J.H. & Venables, P.H. 1975. Evidence of high and low levels of physiological arousal in schizophrenics. Psychophysiol. 12, 66-73.

Satterfield, J.H. 1972. Physiological studies of the hyperkinetic child. Am. Jour. of Psychiatry 128, 1418-1424.

Schmid-Bursk, M., Becker, W., Diekmann, V., Jurgens, R. & Kornhuber, H.H. 1982. Disturbed smooth pursuit and saccadic eye movements in schizophrenia. Arch. Psychiatr. Nervenkr. 232, 381-389.

Shagass, C., Roener, R.A., Amadeo, M. 1976. Eye-tracking performance and engagement of attention. Arch. Gen. Psychiatr. 33. 121-25.

Zentall, S.S. 1977. Environmental stimuluation model. Exceptional Children. 43, 502-10.

Theoretical and Applied Aspects of Eye Movement Research
A.G. Gale and F. Johnson (Editors)
© Elsevier Science Publishers B.V. (North-Holland), 1984

LOOKING AT FACES:
LOCAL AND GLOBAL ASPECTS OF SCANPATHS

Rudolf Groner, Franziska Walder and Marina Groner

Department of Psychology
University of Berne
Switzerland

The scanpath hypothesis as proposed by Noton and Stark
is briefly reviewed and some of the difficulties in its
theory and empirical investigation are mentioned. A distinction is proposed between "local scanpaths" in the
sense of Stark et al. reflecting consistent patterns of
successive fixations, and "global scanpaths" reflecting
the distribution of fixations over a larger time scale
irrespective of their immediate succession. It is assumed that local scanpaths are regulated by the momentary
fixation and its peripheral information in a moment-to-
moment control mode, and global scanpaths are regulated
by the hypothesis or search plan of the subject and are
assumed to operate in a top-down fashion.
The data of a facial inspection and recognition experiment are analysed and empirical evidence for both
local and global scanpaths is found. Finally, the type
of scanning behavior is set into relation with several
performance measures.

INTRODUCTION

Already early investigators of eye movements during picture viewing
(Buswell, 1935; Yarbus, 1967; Jeannerod at al., 1968) noticed sometimes
very characteristic patterns of fixations which appeared to be relatively stable within the same S(=subject) and identical stimulation, but
showed considerable variance over different Ss(=subjects) and experimental conditions.

Most notably Yarbus contributed several examples of those characteristic
fixation patterns (see e.g. 1967, p. 172 f). Here we prefer to provide
an example from our own research. Figure 1 shows the records of four
subjects each looking at four portraits. Although these plottings do not
represent the temporal order of the fixations, they give a strong impression of an individual regularity in looking patterns.

It was the programmatic work of Noton & Stark (1971a, 1971b, 1971c) who
emphasized these individual regularities and also introduced the term
scanpath: "..... when a subject was freely viewing a pattern during the
learning phase, his eyes usually scanned over it following, intermittently
but repeatedly, a fixed path characteristic of that subject viewing that

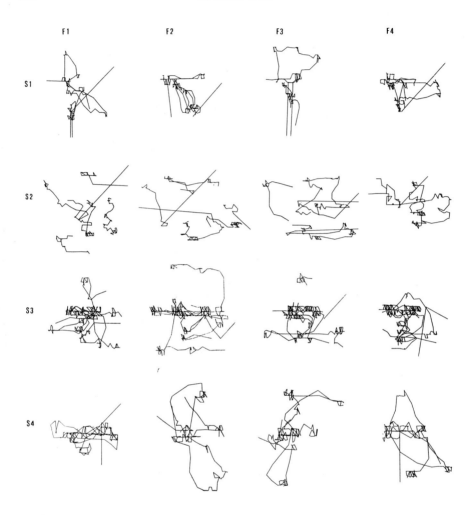

Figure 1

Eye movement records of four Ss (=S1 - S4, rows), each observing four en face portraits (=F1 - F4, columns).

pattern (quite different fixed paths being followed by the subject with
different patterns and by different subjects with the same pattern); and
second, that when the same pattern was presented to the subject during
the recognition phase, his first few eye movements usually followed the
same path he had established for that pattern during the learning phase"
(1971b, 308). In their attempt to establish a theoretical explanation
for the appearance of those scanpaths, Noton & Stark proposed the so called "feature ring" as a format for the internal representation of pictures.
"Feature ring consists of sensory memory traces of subfeatures and motor
memory traces of saccades composing scanpath (Stark & Ellis, 1981,194)".
A picture is coded by a sequence of alternating sensory-motor-sensory-
motor-... elements where the sensory elements are "foveations" representing
'semantic subfeatures of scenes or pictures being observed and motor elements are saccades that represent the syntactical structural or topological
organization of the scene (Stark & Ellis, 1981, 193)".

In the following, we will distinguish between the two aspects of the scanpath hypothesis. First we shall briefly discuss the *theoretical* arguments
against the feature ring and consider alternative models of eye fixation
control. Secondly, there is the *empirical* problem of identifying regular
scanpath patterns which cannot be attributed to statistical artefacts of
random movements.

ARGUMENTS AGAINST THE FEATURE RING AS A CORTICAL PICTURE REPRESENTATION

In the light of neurophysiological research there is hardly any evidence
that pictorial features are stored in combination with their associated
motor signals which would appear as an inefficient and memory space consuming format. On a very early processing stage, it seems plausible, according to reafference theory, that the consequence of the motor activity
is cancelled out from the sensory signal, but already at the primary visual cortex recordings from single cells demonstrate the representation of
pictorial features irrespective of any motor memory traces. With single
cell recordings in macaque monkeys (Perret, Smith, Milner & Jeeves, 1983)
it was even possible to identify highly specialized cells which respond
selectively to such complex patterns as human faces in certain orientations.

It still could be argued that the firing of such a "grandmother cell" was
only initiated through the preceding activation of a feature ring assembly,
possibly over the secondary visual pathway. However, it can easily be
demonstrated that the tachistoscopic presentation of a face, excluding any
eye movements, is sufficient for a correct recognition. Apparently, the
proposed sequence of "foveations" grossly ignores the role of peripheral
information in the construction of an internal pictorial representation.

The following experiment (Groner, Walder & Groner, 1982) might be considered as an "experimental simulation" of feature rings. Our Ss(=subjects) had
attached to their eyes a tube allowing only a foveal view of 2 degrees of
visual angle. They were instructed to look around on the stimulus display
showing a facial portrait with their tubes as much as they liked to. In
subsequent recognition they were unable to recognize those faces, neither
under restricted foveal nor under normal viewing conditions. To another
group of Ss the same portraits were presented parafoveally with 10 deg

excentricity allowing no eye movements. This second group with no foveations
and no scanpaths had a significantly better recognition performance compared with the first group who went through their series of foveations. Therefore it can be argued that experimentally produced feature rings are not
a sufficient basis for subsequent recognition, and therefore do not serve
that purpose they were postulated for. However, it still should be noted
that the cortical representation of moving the tube is not the same as that
from normal eye movements, and therefore this experiment should be repeated using the technology of eye movement contingent stimulus display
(McConkie, this volume).

EMPIRICAL PROBLEMS IN ASSESSING SCANPATHS

If the control of eye movements could be described in deterministic terms,
there would be no difficulty in assessing the regularity of oculomotor
patterns. For a stochastic or probabilistic process, however, several sources of difficulty arise. First of all, it is impossible to formulate a
really adequate null hypothesis, i.e. a prediction of purely random movements. E.g., a random walk model or Brownian motion would certainly not reflect essential properties of saccadic movements like directional preference and a specific spectrum. If we reduce spatial resolution by applying a
gaze grid and temporal resolution by introducing steps of possible transitions, one could not reasonably expect a rectangular distribution of fixations over all grid positions nor equal transition probabilities between
them. The spatial distribution of fixations will rather depend on the arrangement of the display and on stimulus density, where central regions and
highly structured areas will consume more fixations. With regard to transitions, neighbouring areas will be favored compared with widely separated
regions. Therefore, revealing scanpaths by testing serial dependence on
Markovian chains (Stark & Ellis, 1981) suffers from two shortcomings:
(1) Testing independence by the multiplication of marginal probabilities
is based on an inadequate null hypotheses for the random model.
(2) For transitions over more than two steps, there are not enough data
points available for estimating the probability from observed frequencies.

Therefore it seems almost impossible to distinguish a common scanpath, that
is one shared by all subjects, from an adequate model of random movements.
Paradoxically, matters become much easier with *individual* scanpaths, i.e.
those which are specific to single subjects. Here, one can capitalize on
differences between Ss and substitute for the null hypothesis the average
or modal behavior of several different Ss. This group serves as a control
and it can be tested whether the *within*-subject variability is smaller compared with the variance *between* Ss.

THE DISTINCTION BETWEEN LOCAL AND GLOBAL SCANPATHS

In the following, a consistent pattern of consecutive fixations in the sense
of Stark and others will be called a "local scanpath", since the fixations
are in a local way connected to each other. In addition to that the notion of a "global scanpath" will be introduced for the distribution of
fixations over the inspection period on a larger time scale. In the global
scanpath, fixations don't follow each other, but rather reflect a tendency

to concentrate somewhere in the course of the exploration process (e.g. start, middle or end of total viewing time).

Our hypothesis with regard to this distinction is the following: *local scanpaths* reflect the moment-to-moment control of eye movements, regulated by the momentary fixation plus peripheral information in a bottom-up information processing way. The regular succession arises mainly because the process, when repeated, leads essentially always to the same result with respect to the following saccade to be executed.

In contrast, global scanpaths represent the search plan, the expectation or hypothesis of the S(Groner, 1978; Reusser & Groner, 1981; Groner & Groner,1982,1983). The underlying processes are assumed to operate in a top-down manner.

These two organizing principles modify and interact with each other, and that is one of the reasons why the empirical observed scanning patterns appear stochastic.

A distinction somewhat analogous to the one proposed here was made by Kornhuber (1974, p. 403 ff). He calls *tactics* those aspects of voluntary motor responses which are driven by the environment as opposed to *strategies* which are governed by an inner goal. Kornhuber provides a neurophysiological model of these two mechanisms and their interaction.

EXPERIMENTAL SET-UP

Seven Ss participated in a facial inspection and recognition experiment. Black and white portraits of males photographed en face in a standard position and copied in prints of 24 x 30cm were presented to the Ss at a distance of 90cm, thus subtending a visual angle of 15 deg in the horizontal and 19 deg in the vertical axis. Eye movements were recorded by an Eye View Monitor 1994S manufactured by Applied Science Laboratories of Gulf + Western. This apparatus was supplemented by our own computer-interactive calibration and experimental control programs.

In a first session, the initial inspection or learning session, 30 portraits were presented for 5 sec., separated from the next by an ISI of 10 sec. The first five pictures were considered as warming-up trials and excluded from further analysis, leaving 25 portraits to be remembered. One week later in the recognition session, a total of 50 portraits were presented, among them the 25 previously seen pictures randomly interspersed between the new portraits. During the recognition session the inspection time was under control of the S. One of the seven Ss was excluded from further analysis because in the recognition session he mostly spent only one single fixation per picture and was still able to give a high percentage of correct responses. The eye movement records of the remaining six Ss were submitted to a fully computerized data analysis.

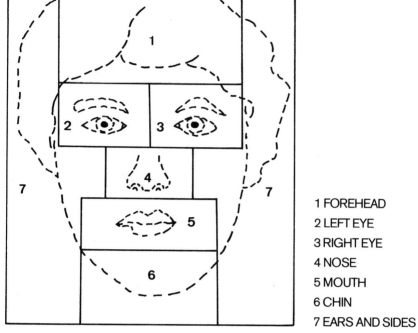

1 FOREHEAD
2 LEFT EYE
3 RIGHT EYE
4 NOSE
5 MOUTH
6 CHIN
7 EARS AND SIDES

Figure 2: Areas of the portrait photos which were analysed as con-
tiguous gazes. Any sequence of fixations falling into the
same gaze area was treated as a single gaze fixation.

In a first step the clusters of the most frequent fixations were identified
Then a gaze grid was constructed accordingly (Fig. 2); our results were very similar to those of Walker-Smith, Gale and Findlay (1977). The entire
display area was divided into seven gaze positions. The choice of such a
grid was based on the results of an earlier pilot study. From each cluster
of fixations in the earlier study, a gaze area was chosen and labeled accordingly.

Initially, the data stream consisted of separate horizontal and vertical
recordings based on a time sampling of 50 Hz. This stream was converted into a sequence of data pairs consisting of a gaze position with its accompanying duration. The computer program also eliminated blinks and recording
artefacts, and it optimized the assignments at the boundary between two
gaze positions. Thus, for each combination of the factors subject x learning vs. recognition x picture we obtained a data string consisting of successive gazes.

ANALYSIS OF LOCAL SCANPATHS

As could be expected, there was hardly a single one out of the 450 data cells
representing individual gaze sequences exactly identical with another one.
This results from the fact that, if there is a stochastic error component,
it will randomly interrupt scanpaths. Therefore a statistical analysis is

necessary, and it should be based on grouped data points in a sufficient number in order to ensure statistical power. In a compromise between theoretical considerations and statistical arguments, we decided to analyse molecular units of scanpaths in the length of three successive gaze fixations, called *triplets*.

In a first step all observed triplets were counted, summing up over Ss, faces, and inspection conditions, and ordered according to their frequency. The resulting rank frequency distribution necessarily starts with its maximum and decreases monotonically thereafter. Moreover, according to Zipf's law, as generalized by Simon (1957) and Mandelbrot (1955), the resulting rank frequency distribution should follow a logarithmic function. With our data, by taking only the highest ranking triplets it still was possible to have included in a sample a considerable part of all observed triplets. Cutting off after the 20 highest ranking triplets from the total of 210 possible combinations left a total of 57% of all triplets observed in the reduced data sample which is the basis of comparisons.

TRIPLETS	1.RU	2.ME	3.BR	4.BK	5.WI	6.TR	TOTAL
2-3-2	7	9	13	36	0	68	133
3-2-3	20	9	10	25	1	46	111
7-2-3	13	2	5	4	0	31	55
2-3-4	8	6	3	19	0	17	53
4-2-3	10	9	3	10	3	17	52
5-2-3	14	13	3	6	3	6	45
2-3-5	21	1	2	8	2	9	43
3-2-4	5	7	0	12	3	13	40
5-4-3	3	2	0	0	15	8	28
3-2-5	5	4	3	7	3	6	28
4-3-2	6	1	0	5	7	8	27
7-5-4	9	2	0	0	8	8	27
4-5-4	1	3	0	4	11	2	21
2-4-3	4	3	1	4	3	4	19

S U B J E C T S

Table 1: The 14 most frequent triplets. Numbers indicate frequencies of observation. Underlinings symbolize significant higher frequency than expected by independence (Chi-square test, $p < .01$)

Table 1 gives the individual observed frequencies of the top 14 triplets. The two triplets with the highest frequencies are left eye - right eye - left eye and right eye - left eye - right eye, followed by ear - left eye - right eye, and so on. On these data, we first tested the null hypothesis of independence between Ss and triplets. This test resulted in some significantly higher entries than expected by chance, represented by unterlinings in Table 1. Due to the low statistical power of this procedure, the results are relatively sparse.

A statistically more sensitive analysis was carried out by computing the correlations between learning and recognition over the relative frequencies in the 20 most frequent triplets.

Subjects	Correlation between learning and recognition within <u>same</u> S	Number of significantly lower cross-correlations with <u>other</u> Ss (p < .05, max = 10)
1. RU	-.269	2
2. ME	.148	3
3. BR	.606*	5
4. BK	.578*	4
5. WI	.704*	8
6. TR	.706*	5
Average over Ss	.184	1

Table 2: Local scanpath analysis. Correlations of relative frequencies over the twenty most frequent triplets.

Table 2 presents in the first column the within-subject correlation between learning and recognition. We see a pronounced difference between the first two Ss who show low and insignificant correlations between the repeated inspection of the same stimuli, compared with the remaining four Ss who exhibit highly significant correlations. In addition to high within-subject correlations, the four Ss with local scanpaths showed a larger number of significantly lower correlations when comparing their triplet frequencies with those of other Ss. In conclusion we propose that subjects 3 to 6 showed elements of local scanpaths, while subjects 1 and 2 did not do so.

GLOBAL SCANPATH ANALYSIS

Again we started with the individual sequences of gazes. Separately for each inspection trial, the gaze sequence was divided into three parts of equal length, representing the start, middle and end of the total inspection period. Then separately for each period the frequency distribution of fixations on the seven gaze positions was counted.

This procedure resulted in a fourdimensional contingency table of subjects x learning vs recognition x gaze position x period. It was analysed by the loglinear model (Upton, 1978) which in a hierarchical way tests for higher interactions between the dimensions of the contingency table. The simplest model being adequate for the data included as one term the three-way interaction subjects x gaze position x period, and as a second term the three-way interaction subjects x gaze position x learning vs recognition. The first interaction confirms the main hypothesis that at least some subjects exhibit different gaze preferences at the three periods.

In order to localize data cells with significantly higher frequencies, again the loglinear model was used 1).

Subject	Gaze Position						
	Forehead	Left eye	Right eye	Nose	Mouth	Chin	Ears
1. RU	-	-	S	-	-	-	-
2. ME	M	-	E	S	S	-	-
3. BR	-	-	-	-	-	-	-
4. BK	-	-	S	-	E	-	-
5. WI	-	-	-	S	-	-	-
6. TR	E	S	-	M	-	E	-
Average of Ss	-	S	-	-	-	-	-

Table 3: Results of the global scanpath analysis. Distribution of fixations on different gaze positions during the three periods of inspection (S.. Start, M...middle, E... end). A letter in a cell denotes a corresponding data frequency significantly higher than expected by chance ($p < .05$; residuals of loglinear model).

Table 3 gives the results of this analysis. The rows represent Ss and the columns gaze positions. The letter S indicates a significantly larger fixation frequency than expected by chance, M indicates a concentration in the middle period and E at the end. Subjects 2 and 6 exhibit a rather strong tendency towards global scanpaths; Ss 4, 1 and 5 show a weaker relationship, and only S 3 fails to show any significant preferences.

COMPARISON BETWEEN TYPES OF SCANPATHS AND RECOGNITION PERFORMANCE

First it should be noted that there is no relation between preferred triplets and the global scanpaths within and across individuals.

In Table 4 a summary is presented of the earlier results together with some performance measures in the recognition task. By comparing the incidence of local and global scanpaths within the same individuals, the two appear practically independent. None of the two scanning types correlate with the solution time (i.e. the average time spent for a face under the recognition condition).

The number of successes and failures in the recognition task has been grouped into hits, i.e. recognizing a portrait correctly as already known;

1) This was done by applying the next more specific model with the three-way interaction subjects x gaze position x learning vs recognition with the two two-way interactions subjects x period and gaze position x period, and testing all residuals (obtained minus predicted) for significance.
A more primitive way of localization could be done by testing separately for each individual and gaze position whether the three frequencies at the start, middle and end are equally distributed by simple chi-square tests. However, this procedure is going to confound the effect under investigation with the other significant three-way interaction which is not predicted by the global scanpath hypothesis.

Subject	Presence of local scanp.	global scanp.	average re-cognition time (sec)	hits	correct reject.	misses	false alarm
1. RU	no	weak	5.86	.46	.38	.04	.12
2. ME	no	strong	2.67	.40	.44	.12	.04
3. BR	yes	no	4.21	.32	.46	.18	.04
4. BK	yes	weak	1.61	.40	.48	.10	.02
5. WI	yes	weak	1.70	.28	.46	.22	.04
6. TR	yes	strong	6.16	.38	.48	.12	.02

Table 4: Relation between scanning behavior and various performance measures.

correct rejection of somebody as unknown; misses, i.e. failing to identify an already presented face; and false alarms, i.e. claiming a new face as already known.

Although the small number of individuals should let us abstain from over-interpretations, there is an interesting relation between local scanpaths and the success measures: there is a positive relation between the incidence of local scanpaths and the probability of correct rejections, and a negative relation between hits and local scanpaths. Stated in simpler words: those Ss who followed a local scanpath were good in identifying somebody as unknown, and relatively bad in recognizing an already seen face. No such relation could be found with global scanpaths.

The empirical work presented in this paper certainly requires continuation and extension to other stimulus domains. Our main point here has been to give some warnings from methodological pitfalls in assessing scanpaths and open the view on a new type of scanning behavior which we have called "global scanpaths".

REFERENCES

(1) Buswell, G.T. How People Look at Pictures. Chicago: University of Chicago Press, 1935.

2) Groner, R. & Groner, M. Towards a hypothetico-deductive model of cognitive activity. In R. Groner & P. Fraisse (Eds.) Cognition and Eye Movements. Amsterdam: North Holland, 1982.

(3) Groner, R., and Groner M. A stochastic hypothesis testing model for multi-term series problems based on eye fixations. In R. Groner, C. Menz, D.F. Fisher, & R.A. Monty (Eds.) Eye Movements and Psychological Functions: International Views. Hillsdale N.J.: Lawrence Erlbaum, 1983.

(4) Groner, R., Walder, F., & Groner, M. Central and peripheral vision in face recognition. Paper presented at the Conference 'Functional Properties of Peripheral Vision', Durham GB, 1982.

(5) Jeannerod, M., Gerin, P., & Pernier J. Déplacements et fixations du regard dans l'exploration d'une scène visuelle. Vision Research, 1968, 8, 81-97.

(6) Kornhuber, H.H. Blickmotorik. In O.H. Gauer, K. Kramer & R. Jung (Hrsg.) Physiologie des Menschen, Band 13: Sehen, Sinnespsychologie III. München: Urban & Schwarzenberg, 1978.

(7) Mandelbrot, B. On recurrent noise limiting coding. In E. Weber (Ed.) Symposium on Information Networks. Brooklyn: Polytechnic Institute,1955.

(8) Noton, D., & Stark, L. Eye movements and visual perception. Scientific American, 1971, 224, 34-43. (a)

(9) Noton, D., & Stark, L. Scanpaths in eye movements during pattern perception. Science, 1971, 171,308-311. (b)

(10)Noton, D., & Stark, L. Scanpaths in saccadic eye movements while viewing and recognizing patterns. Vision Research, 1971, 11, 929-942. (c)

(11)Perrett, D.I., Smith, P.A.J, Milner D., Jeeves, M.A. Visual cells sensitive to face orientation and direction of eye gaze in the macaque monkey. Perception, 1983, 12, A13.

(12)Reusser, M., & Groner R. Informationssuchprozesse bei Globalisationsaufgaben. In K. Foppa & R. Groner (Hrsg.) Kognitive Strukturen und ihre Entwicklung. Bern: Huber, 1981.

(13)Simon, H.A. Models of Man. New York: Wiley, 1957.

(14)Stark, L., & Ellis S.R. Scanpaths revisited: Cognitive models direct active looking. In D.F. Fisher, R.A. Monty and J.W. Senders (Eds.) Eye Movements: Cognition and Visual Perception. Hillsdale N.J.: Lawrence Erlbaum, 1981.

(15)Upton, G.J.G. The Analysis of Crosstabulated Data. Chichester: Wiley, 1978.

(16)Walker-Smith, G.J., Gale, A.G., & Findlay, J.M. Eye movement strategies involved in face perception. Perception, 1977, 6, 313-326.

(17)Yarbus, A. Eye Movements and Vision. New York : Plenum Press, 1967.

(18)Zipf, G.K. The Psycho-Biology of Language. Boston: Houghton Mifflin,1935.

Theoretical and Applied Aspects of Eye Movement Research
A.G. Gale and F. Johnson (Editors)
© Elsevier Science Publishers B.V. (North-Holland), 1984

MODULATION OF GLOBALITY AT THE INPUT
OF THE VISUAL SYSTEM

Jack F. Gerrissen

Department of Industrial Design Engineering
Delft University of Technology
Netherlands

To better understand the transfer of visual information in an ergonomic
context, we investigate the interaction of visual structure and information
intake behaviour. In dealing with the visual world the observer evidences
through his information intake behaviour (head/eye movements, fixation
duration, accomodation, pupil size variations, manipulation of the visual
object, etc. etc.) the intention to concentrate on the components of visual
structure with information contents supporting his knowing process. We
are developing descriptors of both the visual structure and the intake
behaviour in terms that allow systemic modeling of the interaction. This
paper reviews the theory and model and illustrates how eye behaviour may
be interpreted to reveal structure feature extraction at levels of globality.

THE INPUT OF THE VISUAL SYSTEM

We think of the retina as an array of photon counting elements. From considerations on image
intake limitations due to photon noise and 'imperfect' optics, Snyder *et al.* [3] computed the
spatial information capacity for an array of non-overlapping, but touching, photoreceptors. They
found that in order to achieve optimal information capacity within a given integration time of
the eye, the photoreceptors should have their diameter depending on the combination of
average photic intensity and mean (spatial) contrast at the retina. Taking the integration time as
a variable, with the help of their data and computational instruments, one could easily derive a
relation between receptor diameter, integration time and average photic intensity, so as to have
optimal information capacity in the array. The characteristics of this relation are depicted in
fig. 1. We see that for fixed integration time large receptors tend to be optimal for low
intensity and small receptors for high intensity. While, in a situation with only one level of
average photic intensity at the retina, large receptors tend to be optimal for short integration
time and small receptors for relatively long integration time.

It should be noted that the term optimal
information capacity refers to the
highest achievable information capacity
under the given circumstances of average
intensity and contrast at the retina.
The *absolute* value of this maximum
increases for longer integration (resulting
in a better signal to noise ratio) and
smaller optimal receptors (better spatial
resolution). For accurate intake of
spatial contrast we expect the eye to
employ long integration time and small
photo-receptors. But to cope with
situations that do not permit the
employment of either one of these

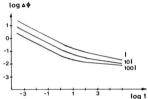

Fig. 1 Optimal information capacity as a
function of average retinal intensity
I, receptordiameter and fixation time,
for given mean contrast.

(e.g. low light conditions and/or time limitations) the retina should also have available a range of larger photoreceptors. In[1], motivated by visual task related and rate-distortion theoretic arguments, we hypothesized a retinal array built from concentric receptor rings with high resolution/low sensitivity (small) receptors packed around the center and low resolution/high sensitivity (large) receptors more towards the periphery (Fig. 2). In recent work on models of the retino-cortical mapping [7-9] based on findings from neurophysiological research [4-6] we find confirmation of the hypothesized receptor arrangement.

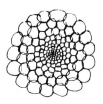

There are indications [4], [10], [11], that this configuration of 'cells' extends up approx. 15 degrees eccentricity and that there is a division into central and peripheral zones for which important aspects of information analysis in the central nervous system are qualitatively different. The theory names these the focal channel and the ambient channel respectively, adopting the Trevarthen terminology [12]. The ambient channel is primarily equipped for target detection while the focal channel concentrates on fast parallel information intake for global structure detection. In the present paper we shall review the functioning of the focal channel.

Fig. 2 Model of retinal array of photoreceptors

With the concentric-cell-layer configuration in the focal zone of our retina model, it will now be shown that variations of pupil size and/or fixation time affect the spatial information capacity of the eye and, in so doing, dynamically determine *the resolution of spatial detail.*

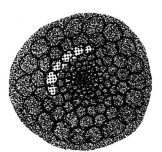 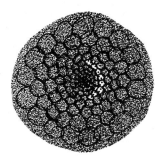

Fig. 3 Optimal receptor for image parts with same intensity and contrast at short integration time (a), and at long integration time (b)

From fig. 1 we see that for given retinal intensity and contrast, at different integration times there are correspondingly different optimal receptordiameters and these are located at different (concentric) rings of the retina. Thus, a choice of integration time implies the allocation of optimal information capacity to one out of the many available receptor rings.

To have the highest achieved spatial information capacity in and around one concentric ring means that: (1) the intake resolution is limited to the maximum spatial resolution obtained in the optimal ring, and (2) the parts of the object field projected upon or near this ring are - in terms of intake rate and distortion - favoured over the rest of the visual world.
In fig. 3 are compared two situations that only differ in integration time, i.e. same intensity and same contrast at the retina. Optimal information capacity is found 'further' from the center at short integration (fig. 3a), than it is at relatively long integration (fig. 3b).
As if the eye probes for image features of a certain size, thereby employing a size-related spatial resolution. In this view of vision, the eye functioning could well be considered an extention of the tactual sense: feeling with the eyes.

When the elements of the image exhibit two or more distinct intensity levels at constant mean contrast, then at the end of the integration period the elements with low intensity are taken in with lower resolution than the elements with the higher intensity (fig. 4). In other words, some elements are taken in more globally than others, depending on their intensity level and contrast. A comprehensive discussion of the contrast-and-intensity related optimal intake and the resulting difference of globality level of processing can be found in [2].

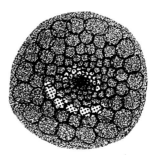

Fig. 4 Optimal receptors for image parts with different intensity and contrast

We summarize our findings in three points :

(1)   the image projected upon the retina from the fixated object field can be broken down into a number of subimages of different average photic intensity and mean (spatial) contrast; these different *retinal visual values* invoke differences of intake resolution and intake moment of information from the corresponding subimages; per level of resolution the optimal intake of subimages with lower retinal visual value is later than the optimal intake of the subimages with higher visual value; the retina preprocesses the information in the different *slices of retinal visual value* at different moments during the eye's fixation period.

(2)   retinal visual value (RVV) is a fundamental interaction variable ; it depends on the objective magnitudes of intensity and contrast of the object subfield in concern *and* the subjective transfer characteristic of the eye, i.e. a function of (i) the subfield's position relative to the focussed target point and (ii) the pupil size ; defocussing an object results in lowering its RVV ; alternatively, given its position in the fixated object field a change of its *object visual value* will change its RVV accordingly.

(3)   with the object of interest in focus the visual system has to decide on pupil size and fixation duration that should be applied ; the pupil modulates the retinal visual value while the fixation duration determines the maximum integration time applied ; the combination of chosen pupil size and choosen fixation time reflects the observers' intention to achieve a certain level of spatial resolution of the visual information that is taken in ; with the product of pupil area and fixation time we associate the term *intended spatial resolution* ISR.

Visual object manipulation, as it is performed by means of head/eye movement and accomodation, is meant to align the various image components with those parts of the retinal array that potentially match the desired globality (resolution) of intake. Pupil and fixation time adjustments are then necessary to achieve the desired resolution of visual value, which alternatively can be described as optimizing different receptor rings for different image components at different moments during the eyefixation.

VISUAL STRUCTURE

For the visual system to function efficiently, it should act upon visual image redundancies with the aim of guessing the message by observing only part of it. We distinguish two types of image redundancy : *local* and *global* redundancy. *Local* redundancy refers to a situation where at several positions in the image, neighbouring image elements may be grouped to form clusters of elements with equal or adjacent perceptual characteristics (color, contrast, etc. etc.). The visual system then does not need to deal with each individual element, but instead evaluates the structure significance of the cluster. Having determined this, there is a high predictability of structure contribution per cluster element. *Global* redundancy relates to predictability of spatiotemporal relations between elements and/or clusters of elements with equal or adjacent perceptual characteristics. Total predictability, i.e., certainty about the relative positions of all elements, occurs when total correlation [13] has been determined. Knowledge about total correlation can be gained from analysis of partial correlations between the elements or clusters. To this end our structure model uses the chord concept [14] as a means to register the spatio-temporal relations between pairs of elements. A chord is taken to be a virtual line between two elements of a pattern. A formal basis of the chordstructure system has been given in informa - tion theoretic terms [1] .

In fig. 5 are depicted three dot patterns and their feature spaces. Since chords are classified by their length and orientation, the chordlength-chordorientation space registers all chordoccurren-ces per chordclass. Chordclass resolution is proportional to chordlength (expressed in unit lengths of interval of the image quantization grid). This size-related resolution articulates the *globality aspect* of structure, while the chordorientation articulates the *configurative aspect.* A chordclass, thus, is an elementary structure feature identifying an elementary configurative relation at a level of globality. The chordlength;chordorientation space is a feature space, where the chordlength axis indexes the globality aspect and the orientation axis indexes the configura-tive aspect of structure. In fig. 5 the black squares in the displayed feature spaces indicate the frequency of chord occurrences in a chordclass ; with this frequency we associate the term *feature intensity* . From fig. 5a we see that over a range of globality levels there are two featu-res with high feature intensity, one oriented at 50 degrees of angle and the other at approx. 140 degrees. Two or more dominant features (due to their feature intensities) at a level of globality form a composite or complex feature. Feature intensity and the consequent potential emergen-ce of complex features stem from the *feature dominance aspect of structure* . From figs. 5b and 5c can be seen that disturbing the detailed regularity while conserving the global shape of the pattern results in a dilation of feature dominance at low levels of globality. The feature space representation of fig. 5c indicates that at a global level there is a high degree of resem-blance with the undistorted pattern (fig. 5a).

CHORDSTRUCTURE AT THE INPUT OF THE VISUAL SYSTEM

Retinal receptor rings are well suited for chord classification, in terms of length and orientation of spatial (perhaps spatiotemporal) relations between pairs of image elements with equal or adja-cent retinal visual value. Chords with lengths approximately equal to the diameter of the ring and with their 'endpoints' projected upon or near to the ring receptors are classified with a size-related resolution. With this our model of the input of the visual system complies with both the globality aspect and the configurative aspect of chordstructure. Chord intake should not be interpreted as a means for the visual system to build an internal representation of something similar to the chordlength-chordorientation space (feature space). We rather hypothesize that by image manipulation (head/eye movements and accomodation) the eye optimizes its infor-mation capacity for features that can be distinguished from other concurrent features at the same level of globality. Feature dominance (defined as a function of feature intensity) influences *feature extraction* as it is applied by the eye at several levels of globality. A consis-tent relation between feature dominance and human feature detection has been confirmed [1] .

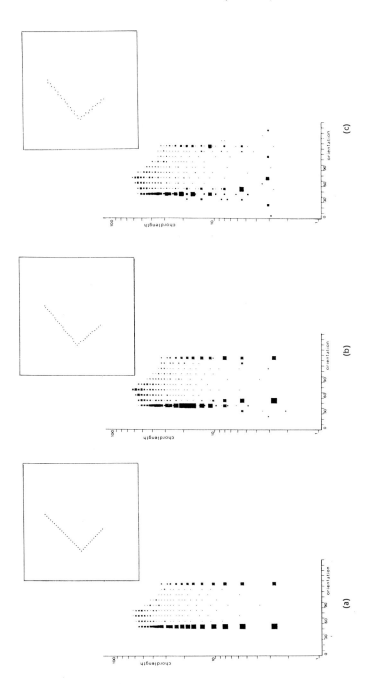

Fig. 5 Dot patterns and their feature space, (a) undistorted pattern, (b) slightly distorted detail, (c) more distortion of detail

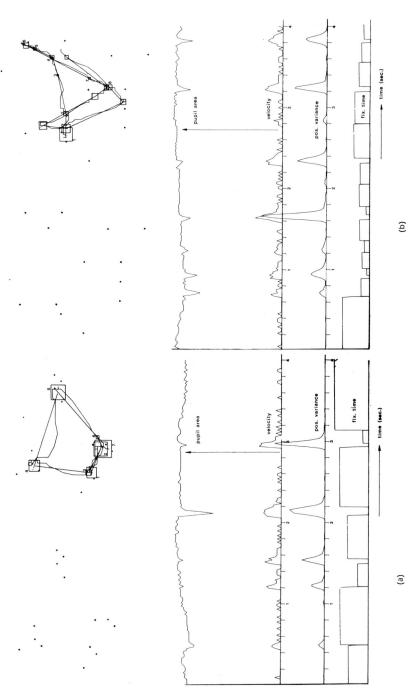

Fig. 6 Eye behaviour registrations from experiments on dominant chord detection

EYE BEHAVIOUR FOR FEATURE EXTRACTION : MODULATION OF GLOBALITY

From experiments that we perform with the objective of testing the validity of our structure and visual input models, two registrations (fig. 6) are used to illustrate how the 'measurement' of eye behaviour may reveal important aspects of feature extraction. The subject in concern had been instructed (after the proper learning phase) to identify the orientation of the chord (feature) he considered dominant in a pattern of white dots which appeared 4 secs. with low contrast on a continuously illuminated white screen. The left upper part of the figures shows the dot pattern in black, the right upper part the same pattern with superimposed eye traject- ories and squares that code fixation position and duration, while in the lower half of the figures we have the eye behaviour parameters plotted against time. Feature dominance was correctly established by the subject (in fig. 6a four occurrences of a short chord extending 0.7 degrees of visual angle and 45 degrees orientation ; in fig. 6b also four occurrences of a verti- cal chord extending 2.8 degrees of visual angle).

We make the following observations from the two registrations :

(1)   pupil size is essentially constant (fast eye movement and blinking cause dips in the registration because of apparent pupil distortion) ;

(2)   the sequences of fixation times suggest that for the pattern with the long chords intake was more global, i.e. the intended resolution (ISR) was lower ;

(3)   there is more accuracy in positioning the line of sight near the center of the chords in the case of the short chords ;

(4)   the eyemovements over the pattern with the long chords seem to reflect a certain degree of uncertainty about the place where a dominant chord could be found.

Observations 1, 2 and 3 indicate that after the initial eye latency scan tactics were set to a level of globality that supposedly contained the dominant feature. The significance of observation 4 relates to difficulties in isolating a feature from a number of equally dominant features at more or less equal level of globality. We illustrate this by means of the pattern and feature space in fig. 7. Both features are now incorporated in one pattern ; the long vertical chord (length - 16) occurs with equal feature intensity as the short slanted chord (length - 5), but isolation of the latter feature is easier. This is in agreement with the differences of 'traffic' that we see in feature space at the two corres- ponding globality levels. In order to extract the short feature, the eye simply adjusts to the proper globality level, but for the long feature the eye has to apply some additional image manipulation to isolate the desired chord from among the concurrent features.

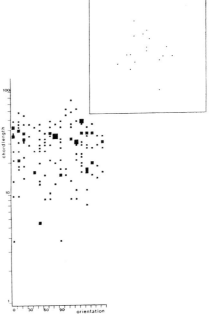

Fig. 7 Dot pattern with dominant features
at different levels of globality

542   *J.F. Gerrissen*

REFERENCES

[1] Gerrissen, J.F., Theory and Model of the Human Global Analysis of Visual Structure, IEEE
Trans. Systems, Man and Cybernetics SMC-12, 6 (1982) 805 - 817
[2] Gerrissen, J.F., Theory and Model of the Human Global Analysis of Visual Structure, part.II:
The Space-Time and Visual Value Segment, manuscript submitted for publication in IEEE
Trans. Systems, Man and Cybernetics (1983)
[3] Snyder, A.W., Laughlin, B.L., and Stavenga, D.G., Information Capacity of Eyes, Vision
Research 17 (1977) 1163 - 1175
[4] Hubel, D.H., Le Vay, S., and Wiesel, T.N., Mode of Termination of Retinofugal Fibers in
Macaque Monkey : An Autoradiographic Study, Brain Research, 96 (1975) 25 - 40
[5] Hubel, D.H., and Wiesel, T.N., Uniformity of Monkey Striate Cortex : A Parallel Relationship between Field Size, Scatter, and Magnification Factor, Jrnl Comparative Neurology,
158 (1974) 295 - 306
[6] Hubel, D.H., and Wiesel, T.N., Brain Mechanisms of Vision, Scientific American, 241 (1979)
130 - 144
[7] Wilson, S.W., On the Retino-cortical Mapping, Int.Jrnl Man-Machine Studies, 18 (1983)
361 - 389
[8] Schwartz, E.L., Columnar Architecture and Computational Anatomy in Primate Visual
Cortex : Segmentation and Feature Extraction via Spatial Frequency coded Difference
Mapping, Biol. Cybernetics 42 (1982) 157 - 168
[9] Braccini, C., Gambardella, G., Sardini, G., and Tagliasco, V., A Model of the Early Stages
of the Human Visual System, Biol. Cybern. 4 (1982) 47 - 58
[10]Poppel , E., and Harvey , L.O., Light-difference Threshold and Subjective Brightness in the
Periphery of the Visual Field, Psychol. Forsch. 36 (1973) 145 - 161
[11]Frost, D., and Poppel, E., Different programming modes of human saccadic eye movements
as a function of stimulus eccentricity : indications of a functional subdivision of the visual
field, Biol. Cybernetics 23 (1976) 39 - 48
[12]Trevarthen, C.B., Two Mechanisms of Vision in Primates, Psychol. Forsch. 31 (1968)
299 - 337
[13]Watanabe, S., Information Theoretical Analysis of Multivariate Correlation, IBM Jrnl
Res. Develop. 4 (1960) 66 - 82
[14]Moore, D.J.H., Seidl, R.A., and Parker , D.J., A Configurational Theory of Visual Perception, Int. Jrnl. Man-Mach. Studies (1975) 449 - 509.

Report

THE RECORDING EYE

Theoretical and Applied Aspects of Eye Movement Research
A.G. Gale and F. Johnson (Editors)
© Elsevier Science Publishers B.V. (North-Holland), 1984

THE RECORDING EYE

The Invited Lecture Given by
Professor E. Llewellyn-Thomas

A Report by

Frank   Johnson

We invited Professor E. Llewellyn-Thomas to attend the Conference and
provide a final lecture which summarised the proceedings and returned us to
the realities of applying eye movement research.

Professor Thomas began his professional career with a degree in Electrical
Engineering, and served during the War as Captain of the Engineers in Radar
and Communications. He then went to McGill Medical School, working with
Penfield as an engineer at the Montreal Neurological Institute. He spent
some time as a General Practitioner in an isolated Nova Scotia island, then
moved to Cornell to work as a Research Assistant in Social Psychology. He
then moved to Toronto to work in Aerospace Medicine developing particular
interest in Man Machine Systems and Pharmacology. In 1961 he joined Norman
Moody at the Institute of Biomedical Electronics and constructed a head-
mounted cinematograph version of Mackworth's eye movement camera. This was
applied to recording eye movements during radiograph scanning and for studies of drug effects on the oculomotor system.

In his review of the Conference Professor Thomas took as his text the thir-
ty-sixth verse of the fifteenth chapter of the first letter of Paul to the
Corinthians: "In a moment, in a twinkling of an eye, at the last trump,
all shall be changed".

The Conference drew together workers from many disparate disciplines who
were reporting on the entire gamut of applications of eye-movement research.
In turn, the new knowledge presented will be applied to many areas of en-
deavour. In pharmacology the papers on eye movements and drug activity are
scattered across many journals, but the instrument design reported by
Frecker et al provides a benchmark standard against which other instruments
will be compared and which is finding use in many new areas of study of
pharmacological agents. The techniques used for this and other areas of
research have much in common and the forum provided by the Conference allowed considerable exchange of ideas.

The eye movement control system is unique. Stimulus and response are both
accessible in time, intensity, direction and magnitude. With output and
input specified the transfer functions may be developed. Feedback occurs
at many levels of sophistication: both negative and positive, feedback,
feedforward of first, second, third and higher order systems. Open loop,
bang-bang servos, proportional and anticipatory control mechanisms are all
present with peripheral data processing and information compression. Peripheral, automatic and central nervous systems are involved at all levels.
All this is constructed without good insulators or conductors. It forms
a general-purpose system which has distant and near sight, is movement
sensitive and has high central acuity. Built for survival in the jungle
we are now accosting the Visual System with many situations for which it
was not designed, such as high speed aircraft piloting, radar scanning

and assembly line inspection.

Professor Thomas's first attempt at eye movement recording was at the
Montreal Neurological Institute helping David Hubel analyse saccades. EOG
was amplified using vacuum tubes to give imperfect recordings, but much
valuable work was done. Investigation of vestibular disorders received
attention at the Conference. In the early 1950's when high-speed flight
was beginning the prevention of motion sickness was crucial. It was rapidly learnt that moving the head was not advised during fast turns, and that
observation led to radical improvements in aircraft instrumentation and
control layout. By 1958 the problems of nausea and vomiting in zero gravity
were of significance and the first physician in space was a vestibular
specialist. Motion sickness in astronauts remains the predominant problem.

Refinements to cockpit layout continued with the use of Mackworth's head-
mounted camera and world-wide standardisation of the design of aircraft
control layouts came out of this work. Such standardisation has not been
successfully applied to automotive control. In instrument reading a fixation is necessary (despite comments at this Conference!) and distractions
are anathema. It is a pity that the agents behind road-side advertising
are not aware of this.

One of the distractions to Professor Thomas as he drove through the
saccade-like English country lanes was the appearance of familiar words out
of their North-American context and meaning. Perhaps one rule of eye
movement research is that you are never wearing a camera at the time of the
most interesting events.

Further work on the detection of lesions was reported by Kundel during
the Conference. In a structured search 30% of the lesions could be missed,
suggesting that thinking about controlling eye movements is incompatible
with effective search. Adquate fixation is not the main problem and an
efficient search may be quite irregular. The radiologist knows that the
experiment is artificial, and the detection thresholds in real life may be
quite different.

Similar difficulties exist when extending such studies to air traffic controllers, perhaps the most important man-machine interface (especially for
the air traveller) and certainly the job carrying the highest rate of pep-
tic ulcers. Studying these tasks in real-life is very difficult, however,
but the successful application of image enhancement and visual pattern
recognition to such expensive machinery demands fuller understanding of the
human processes involved.

The application of eye movement recording to studies of psychiatric disorders received much attention during the conference. Schizophrenic patients
display very different eye movements from normal and quantification of this
is improving, along with the generation of better models. Such models need
to be fertile, but not confused with reality. The studies of aggression
and the application of eye movement recording the recognition of faces were
aspects of the Conference that confirmed the impression that new areas of
work were being developed.

Professor Thomas concluded with a plea for the extension of eye movement
research into the realms of architecture. His visit to the Old World had
perhaps reminded him of the grace and beauty of Renaissance line

and aesthetics. Study of the relationships between perceived beauty and eye fixations may help us to restore that heritage.

With that final fusion of philosophy, engineering and medicine, Professor Thomas concluded the Conference in a note of perspective and challenge.

His lecture will be remembered as a scholarly contribution from one of the Founding Fathers of Eye Movement Studies, and his continued enthusiasm and vigour will be an inspiration for us all.

AUTHOR INDEX

SUBJECT INDEX